D0866742

Chirality in Industry

Chirality in Industry

The Commercial Manufacture and Applications of Optically Active Compounds

Edited by
A. N. COLLINS, G. N. SHELDRAKE and J. CROSBY
ICI Specialties, Manchester, UK

JOHN WILEY & SONS
Chichester · New York · Brisbane · Toronto · Singapore

Other Wiley Editorial Offices

John Wiley & Sons, Inc, 605 Third Avenue,
New York, NY 10158-0012, USA

Jacaranda Wiley Ltd, G.P.O. Box 859, Brisbane,
Queensland 4001, Australia

John Wiley & Sons (Canada) Ltd, 22 Worcester Road,
Rexdale, Ontario M9W 1L1, Canada

John Wiley & Sons (SEA) Pte Ltd, 37 Jalan Pemimpin #05-04,
Block B, Union Industrial Building, Singapore 2057

Library of Congress Cataloging-in-Publication Data

Chirality in industry : the commercial manufacture and applications of optically active compounds / edited by A. N. Collins, G. N. Sheldrake, and J. Crosby.
p. cm.
Includes bibliographical references and index.
ISBN 0471935956
1. Chirality. 2. Enantiomers—Separation. 3. Enantiomers–Biotechnology. I. Collins, A. N. (Andrew N.) II. Sheldrake, G. N. III. Crosby, J.
QP517.C57C47 1992
660'.6—dc20 92–16000
CIP

British Library Cataloguing in Publication Data

A catalogue record for this book is available from the British Library

ISBN 0 471 93595 6

Typeset by Thomson Press (India) Ltd., New Delhi.
Printed and bound in Great Britain by Biddles Ltd., Guildford, Surrey

Contents

List of Contributors

S. Akutagawa
Takasago Research Institute Inc., 5–36–31 Kamata, Ohta-ku, Tokyo 144, Japan

C. R. Bayley
Norse Laboratories Inc., P.O. Box 796, Newbury Park, CA 91319, USA

W. J. Boesten
DSM Research, P.O. Box 18, 6160 MD Geleen, The Netherlands

A. Bommarius
Degussa AG, Organic and Biological Chemistry Research, P.O. Box 1345, D W 6450 Hanau 1, Germany

Q. B. Broxterman
DSM Research, P.O. Box 18, 6160 MD Geleen, The Netherlands

I. Chibata
Tanabe Seiyaku Co. Ltd, Chuo-ku, Osaka 541, Japan

J. Crosby
ICI Fine Chemicals Manufacturing Organisation, Hexagon House, Manchester M9 3DA, UK

J. T. Davis
Genzyme Corporation, 1 Kendall Square, Cambridge, MA 02139, USA

K. Drauz
Degussa AG, Organic and Biological Chemistry Research, P.O. Box 1345, D W-6450 Hanau 1, Germany

E. M. Fox
Genzyme Corporation, 1 Kendall Square, Cambridge, MA 02139, USA

K. Furuhashi
Nippon Mining Co., Ltd, Bioscience Research Laboratories 3–17–35 Niizo-Minami, Toda, Saitama 335, Japan

C. Giordano
Zambon Group SpA, Via Cimabue 26/28, 20032 Cormano (Milan), Italy

U. Groeger
Degussa AG, Organic and Biological Chemistry Research, P.O. Box 1345, D W 6450 Hanau 1, Germany

J. Hasegawa
Kaneka Corporation, 2–4, 3-Chome, Nakanoshima Kitu-ku, Osaka 530, Japan

H. F. M. Hermes
DSM Research, P.O. Box 18, 6160 MD Geleen, The Netherlands

R. A. Holt
ICI Bio Products and Fine Chemicals, P.O. Box 1, Billingham Cleveland TS23 1LB, UK

J. Kamphuis
DSM Research, P.O. Box 18, 6160 MD Geleen, The Netherlands

B. Kaptein
DSM Research, P.O. Box 18, 6160 MD Geleen, The Netherlands

C. G. Kruse
Solvay Duphar Research Laboratories, P.O. Box 900. 1380 Weesp, The Netherlands

J. Martel
Roussel-Uclaf, 102 Route de Noisy, P.O. Box 9, 93230 Romainville, France

M. R. Mischke
Genzyme Corporation, 1 Kendall Square, Cambridge, MA 02139, USA

T. Ohashi
Kaneka Corporation, 2–4, 3-Chome, Nakanoshima Kitu-ku, Osaka 530, Japan

K. Oyama
Tosoh Corporation, 2743-1 Hayakawa, Ayase, Kanagawa 252, Japan

S. Panossian
Zambon Group SpA, Via Lillo del Duca 10, 20091 Bresso (Milan), Italy

D. G. Schena
Genzyme Corporation, 1 Kendall Square, Cambridge, MA 02139, USA

H. E. Schoemaker
DSM Research, P.O. Box 18, 6160 MD Geleen, The Netherlands

G. N. Sheldrake
ICI Specialties, Hexagon House, Manchester M9 3DA, UK

T. Shibatani
Tanabe Seiyaku Co. Ltd, Chuo-ku, Osaka 541, Japan

S. Sifniades
Allied-Signal Inc., 101 Columbia Road, Box 1021, Morristown, NJ 07962, USA

T. Sonke
DSM Andeno, P.O. Box 18, 6160 MD Geleen, The Netherlands

D. I. Stirling
Celgene Corporation, 7 Powder Horn Drive, P.O. Box 4914, Warren, NJ 07060, USA

P. Stutte
Lonza Ltd, Münchensteinerstrasse 38, CH-4002 Basle, Switzerland

T. Tosa
Tanabe Seiyaku Co. Ltd, Chuo-ku, Osaka 541, Japan

N. A. Vaidya
Norse Laboratories Inc., P.O. Box 796, Newbury Park, CA 91319, USA

W. J. J. van den Tweel
DSM Research, P.O. Box 18, 6160 MD Geleen, The Netherlands

M. Villa
Zambon Group SpA, Via Cimabue 26/28, 20032 Cormano (Milan), Italy

A. E. Walts
Genzyme Corporation, 1 Kendall Square, Cambridge, MA 02139, USA

C. Wandrey
Research Centre Jülich, Institute for Biotechnology, P.O. Box 1913, DW 5170 Jülich, Germany

Preface

The industrial production of optically active materials is a topic amenable to orthogonal approaches: one may consider in a systematic way how the various methodologies have been or could be applied or, alternatively, attempt to draw lessons from the study of a series of case histories. We believe both have a place and are necessary if maximum benefit is to be derived. Case histories alone do not necessarily set the solution to a problem in context or give an appreciation as to why other approaches did not afford an industrial and, by definition, an economically viable, answer. They may not indicate the breadth of available techniques.

In this volume, the scene is broadly set by an extended opening chapter which systematically considers the options, relates these to a variety of important targets and illustrates how particular problems or areas of chemistry have been tackled. Following this, the main subdivisions of the book deal with (i) 'classical,' non-biological resolutions, (ii) biological methods (both resolution and asymmetric synthesis), (iii) non-biological asymmetric synthesis and (iv) immobilization and membrane technologies. This more or less follows the historical evolution of practical methods for the large-scale production of optically pure materials.

No attempt has been made to subdivide by product category: pharmaceuticals; animal health products; agrochemicals; electronics chemicals; pheromones; flavours and fragrances. This is the least fruitful approach, and there are many cross-applications, for example fungicides, which may appear under pharmaceutical and agrochemical headings.

The 'chiral pool' is often regarded as a specific subdivision but, as noted in Chapter 1, *all* optically active materials, whether natural or man-made, are, for practical purposes, available for inclusion in the pool. We have not, therefore, used this as a sub-heading, although extensive reference is made in Chapters 13, 14 and 18 to the synthesis of derivatives of several recent, man-made additions to the chiral pool: β-hydroxycarboxylic acids, cyanohydrins and β-lactones, respectively.

Classical resolutions are, numerically, used most often to produce optically active materials, but in terms of product tonnage biological methods currently hold the lead. This is reflected in the proportion of this volume devoted to the latter.

At present, membrane methods are the most useful of the technologies emerging as aids to the production of enantiomerically pure materials, and we believe they merit special consideration. They are used almost exclusively in conjunction with enzymic methods, and are contributing to the development of those methods, for example, by facilitating solutions to the problem of co-factor recycling. Chiral preparative liquid chromatography and the use of non-biological, chirally modified membranes have still to come of age as practical methods.

Within the major subdivisions of biological methods and asymmetric synthesis, there is provided a mixture of case histories, surveys of reaction types and surveys of methods applied to the obtainment of particular structural entities.

All of the contributors are closely associated with, or based in, industry and collectively they have many years experience in applying *practical* methods to the production of optically active materials. Their contributions are illustrated by reference to many different commercially important products or product groups. However, to keep the volume to a reasonable length it was not feasible to include reference to *every* important product. Some, such as HMG–CoA reductase inhibitors, are just starting to be made in significant quantities; in the next few years these will undoubtably enrich the range of examples on which a volume such as this could draw.

An increasing number of 'chiral synthons' are coming to the market place. To some extent these are products waiting to be exploited, and the technologies for their production have in many cases only been operated at a semi-technical scale (1–100 kg). Nevertheless, they comprise one of the springboards from which we shall see the area develop, and examples are included accordingly.

'Industrial'-scale production needs definition. Where optically active materials are concerned, this spans the range from kilograms to 10^5 tonnes p.a.; from highly active materials such as prostaglandins or peptide drugs where only a few kilograms may be needed per annum to, at the other end of the scale, amino acids. In the middle of the spectrum are pharmaceuticals and agrochemicals required in amounts from tens to thousands of tonnes p.a. Hence, in setting our terms of reference to include any optically active materials of commercial interest, we encompass relatively simple petrochemical-scale technologies on the one hand, and complex chemical, physical and engineering methods at the other, high-value, end of the spectrum.

Whilst we hope this volume will prove of particular interest to readers who are professionaly involved in the scale-up of methods for the production of optically active materials, it is hoped that students and researchers involved in a more academic pursuit of optical activity, and who may be less preoccupied with economic optimization, will benefit from some of the facets of 'large-scale' thinking. An economic solution is more likely to be a simple, elegant solution.

Finally, a note on nomenclature. At the time of writing an almost theological debate is taking place in the correspondence columns of *Chemical and Engineering News* concerning the descriptors to be used for chiral materials. We favour the simple, descriptively accurate title of 'optically active' and prefer 'enantiomerically pure' as an alternative to the currently fashionable 'homochiral.'

Both the DL and (*R*, *S*), Cahn–Ingold–Prelog, descriptors are used as appropriate. The older DL convention is still in widespread use for amino acids and carbohydrates.

March, 1992

A. N. C
G. N. S
J. C.

Acknowledgements

The Editors would like to express their gratitude to ICI Specialties for assistance with the production of this volume. The advice of Trevor Laird of Scientific Update at the planning stage is gratefully acknowledged, as is the help of numerous colleagues, in particular Robert Holt of ICI Bio Products and Fine Chemicals. We also thank Sheila Collins for help with the preparation of the manuscript, and all our families for their support and patience.

1 Chirality in Industry—
An Overview*

J. CROSBY
ICI Fine Chemicals Manufacturing Organisation, Manchester, UK

---

* Based on a paper first-published in *Tetrahedron*, **47**, 4789 (1991) and reproduced here by permission of Pergamon Press plc.

Chirality in Industry. Edited by A. N. Collins, G. N. Sheldrake and J. Crosby

1.1 INTRODUCTION

The importance of obtaining optically pure materials hardly requires restatement. Manufacture of chemical products applied either for the promotion of human health or to combat pests which otherwise adversely impact on the human food supply is now increasingly concerned with enantiomeric purity. A large proportion of such products contain at least one chiral centre.

The desirable reasons for producing optically pure materials include the following: (i) biological activity often associated with only one enantiomer; (ii) enantiomers may exhibit very different types of activity, both of which may be beneficial or one may be beneficial and the other undesirable; production of only one enantiomer allows the separation of the effects; (iii) the unwanted isomer is at best 'isomeric ballast'[1] gratuitously applied to the environment; (iv) the optically pure compound may be more than twice as active as the racemate because of antagonism, for example the pheromone of the Japanese beetle (**1**) where as little as 1% of the (*S*, *Z*)-isomer inhibits the (*R*, *Z*)-isomer;[2] (v) registration considerations;[3] production of material as the required enantiomer is now a question of law in certain countries, the unwanted enantiomer being considered as an impurity; (vi) where the switch from racemate to enantiomer is feasible, there is the opportunity effectively to double the capacity of an industrial process; alternatively, where the optically active component of the synthesis is not the most costly, it may allow significant savings to be made in some other achiral but very expensive process intermediate; (vii) improved cost efficacy; (viii) the physical characteristics of enantiomers versus racemates may confer processing or formulation advantages.

That the shape of a molecule has considerable influence on its physiological action has been recognized for a long time.[4] For example, in the early 1900s Cushny[5] demonstrated that one member of a pair of optical isomers could exhibit greater pharmacological activity than the racemate: (−)-hyoscyamine (**2**) was approximately twice as potent as the racemate (atropine) in its effect

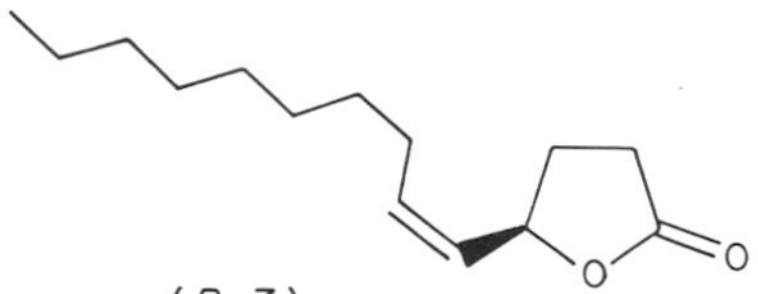

(*R*, *Z*)

(**1**)

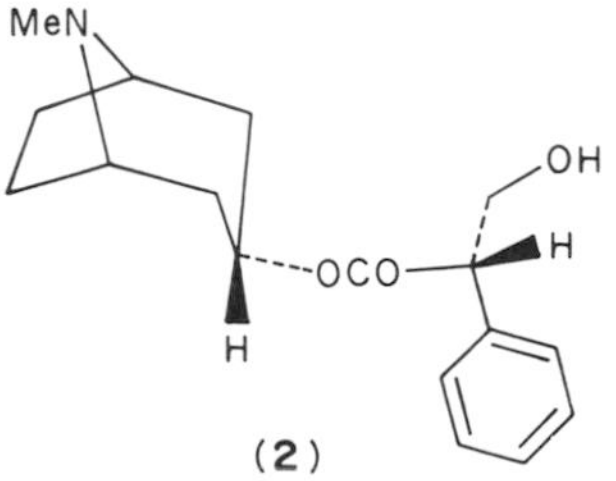

MeN OH H OCO H

(**2**)

on pupil nerve endings. Examples of property differentiation within enantiomer pairs are numerous and often dramatic. A selection is given in Table 1 which emphasizes the reasons for commercial interest and the incentive for producing

Table 1. Differences in the properties of enantiomers

(*S*) bitter taste	Asparagine	(*R*) sweet taste
(*S*) caraway flavour	Carvone	(*R*) spearmint flavour
(*R*,*R*) antibacterial	Chloramphenicol	(*S*,*S*) inactive
(*S*) β-blocking agent *ca* 100 x activity of (*R*)	Propranolol	(*R*)
(*S*,*S*) tuberculostatic	Ethambutol	(*R*,*R*) causes blindness

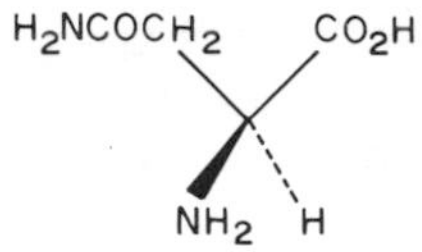

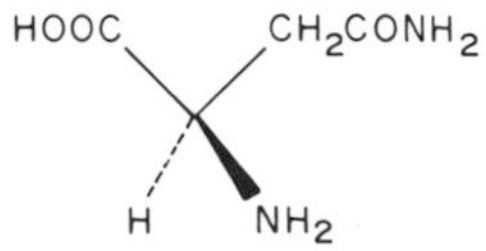

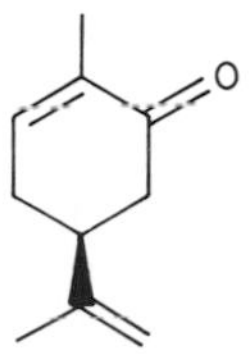

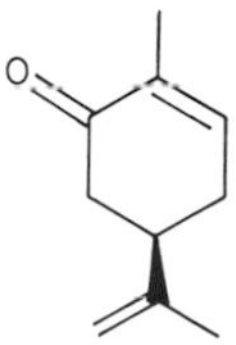

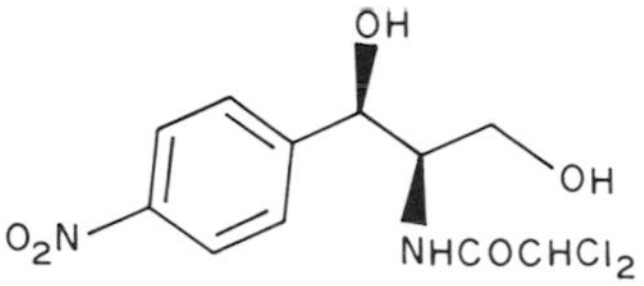

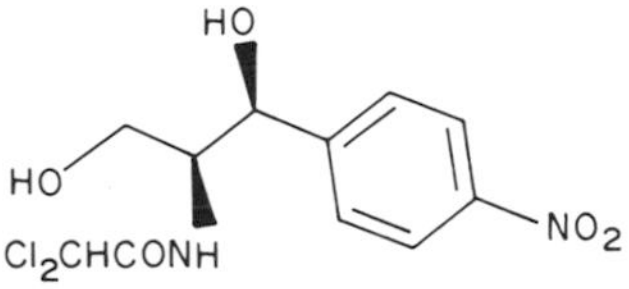

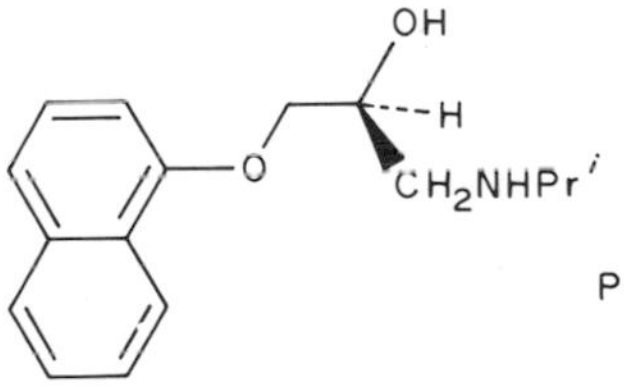

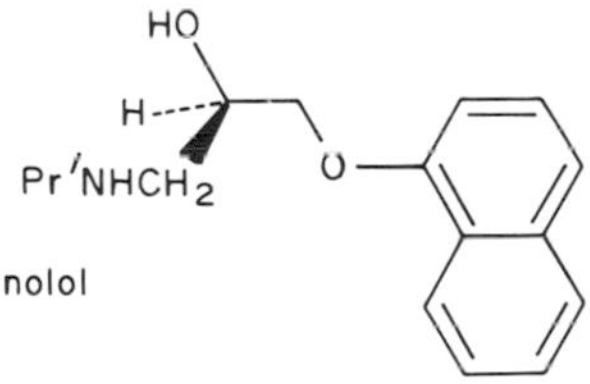

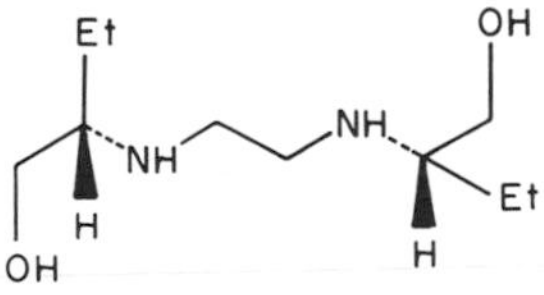

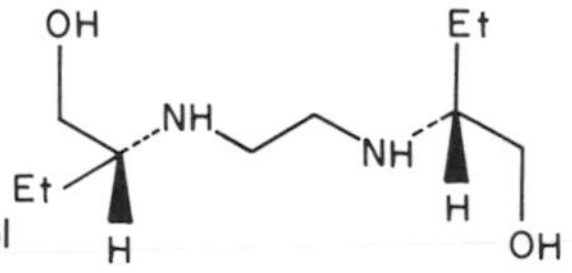

(*continued*)

Table 1. *(continued)*

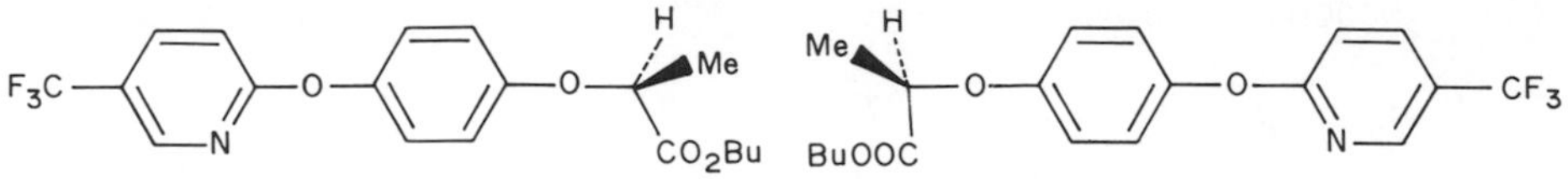

Fluazifop butyl

(*S*) inactive (*R*) herbicide

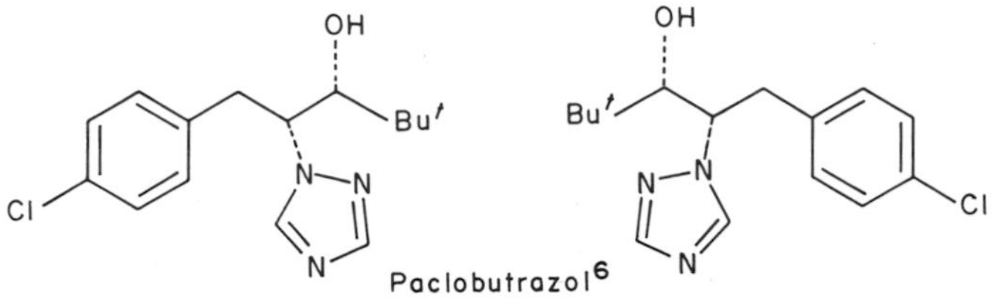

Paclobutrazol[6]

(2*R*,3*R*) fungicide (2*S*,3*S*) plant growth regulator

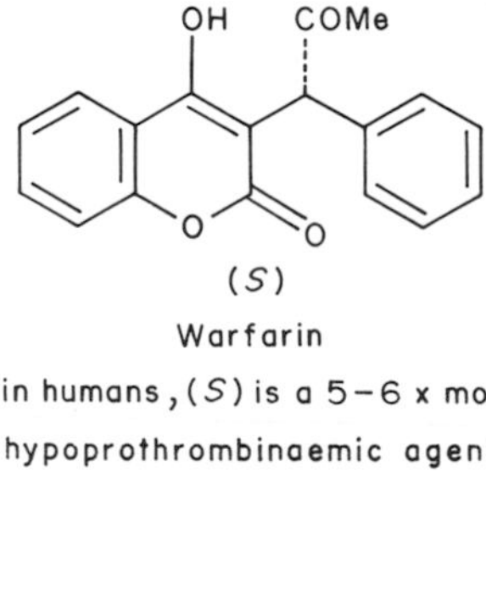

(*S*)

Warfarin

in humans, (*S*) is a 5–6 x more potent hypoprothrombinaemic agent than (*R*)[7]

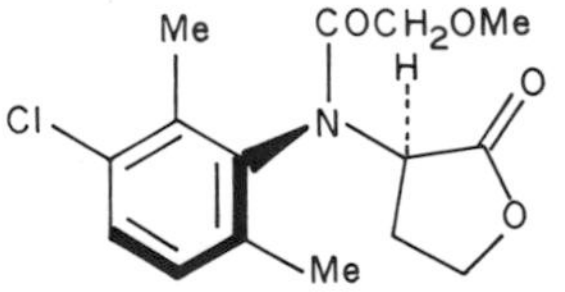

(αS,3*R*) (**3**)

Clozylacon

fungicidal activity arises mainly from (αS, 3*R*) isomer[8]

enantiomerically pure materials by methods applicable to at least multi-kilogram amounts and in many cases to hundreds or thousands of tonnes.

All conceivable methods for the production of optically pure chiral materials are being actively researched. The field is served by a steady stream of monographs, reviews[9] and specialist conferences[10] and new journals dedicated to the topic have appeared.[11] The perspective of this book is industrial and the topic is approached from the standpoint of the person with an interest in producing at least a few kilograms of an enantiomerically pure material for initial field trials or toxicological studies during the development of a new pesticide or therapeutic agent and who may ultimately have to consider how to obtain tonne quantities by the most economic route and within tight time constraints. The aim of this Chapter is to survey the methods available for producing optically active materials, how they have been applied to targets of commercial interest and their suitability for producing, at least, multi-kilogram amounts.

The subject could be addressed either by reference to a selection of target molecules and examining the approaches brought to bear on the problems, or by systematic reference to sources of, or techniques for, the generation of optically active compounds. The latter has been adopted because it allows the merits of the various approaches to be kept in better perspective.

Excluding the isolation of natural products, the production of optically pure materials has generally presented a significant challenge bearing in mind that, to be of practical large-scale use, the enantiomeric excesses (*ee*s) ought to be at least 70% and preferably greater than 80% for the crude material which is initially produced.

Approaches which may be applied are utilization of chiral pool materials, separation of racemates and creation from prochiral precursors.

1.2 METHODS FOR OBTAINING OPTICALLY ACTIVE COMPOUNDS

1.2.1 THE CHIRAL POOL

The chiral pool customarily refers to relatively inexpensive, readily available optically active natural products, substances whose commercial availability generally falls in the range 10^2–10^5 tonnes per annum; respresentative materials are listed in Table 2. However, as a result of the pressures to produce an ever-growing number of commercial products as single enantiomers, there is an increasing and still largely unrecognized and unexploited source of new materials which should justifiably be added to the traditional pool. These materials, industrial end products and process intermediates, are often produced in very significant quantities, 10^2–10^3 tonnes per annum. They have a diversity

Table 2. Representative substances from the chiral pool[a]

Compound	Approx. price (US dollars kg^{-1})
Ascorbic acid	13
(+)-Calcium pantothenate	16
(−)-Carvone	23
Anhydrous dextrose	1.2
Ephedrine hydrochloride	62
(+)-Limonene	3
L-Lysine	3.2
Mannitol	7.5
Monosodium glutamate	2
Norephedrine hydrochloride	24
Quinidine sulphate	130
Quinine sulphate	75
Sorbitol	1.7
L-Threonine	12–50, depending on grade
L-Tryptophan	68

[a] Data from *Chemical Marketing Reporter*, Schnell Publishing, New York, 13 April (1990); reproduced by permission of the Editor.

which must soon exceed that of the natural counterparts if this is not already the case. Although in many instances the output of these substances exactly matches consumption, it is important to recognize that they exist, that necessary technology to produce them exists and that more could be made if the demand was there.

This section is concerned primarily with use of chiral pool substances as building blocks. They are incorporated into the target structure with any necessary modification in order to achieve the desired chiral features.

1.2.1.1 Amino Acids

α-Amino acids are readily available, in bulk, usually with high *ee*s and are one of the oldest sources of optical activity. Crystalline glutamic acid (**4**) was isolated from gluten hydrolysate in 1866.[12] The development in Japan early in this

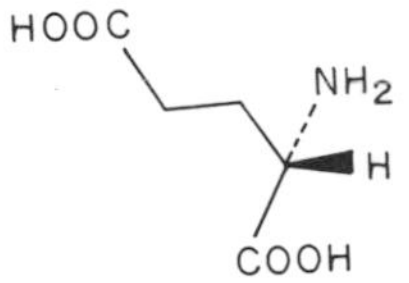

HOOC NH2 H COOH

(**4**)

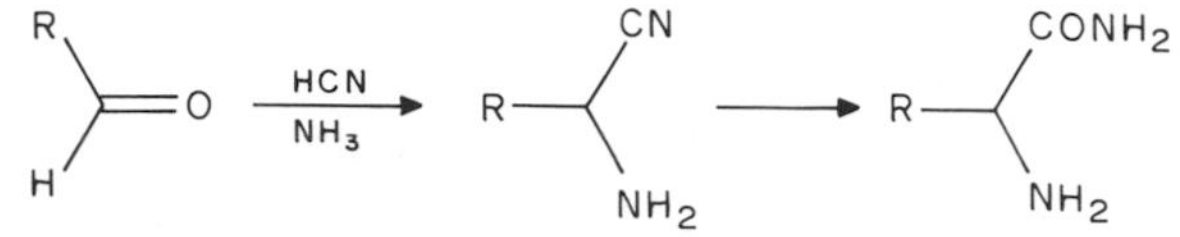

Scheme 1

century of monosodium L-glutamate as a flavour enhancer laid the foundation for an amino acid industry still dominated by Japanese companies. Other major producers tend to have a primary capability in cyanide chemistry and make racemic α-amino amides by the Strecker synthesis[13] (Scheme 1) and then resolve enzymically (Section 1.2.2.3) or, in some instances, classically (Section 1.2.2.1); see Chapters 8 and 2.

All the proteinogenic L-amino acids are available commercially on scales

Amino acid	*Product*	*Ref.*
D-Valine	Fluvalinate: insecticide	17
D-Serine	PyroGlu-His-Trp-Ser-Tyr-D-Ser(Bu^t)-Leu-Arg-Pro-NHEt Buserelin (**5**)·prostatic cancer treatment	18
D-Alanine	Alitame (**6**): synthetic sweetener	19
D-Phenylglycine	Ampicillin: antibiotic	

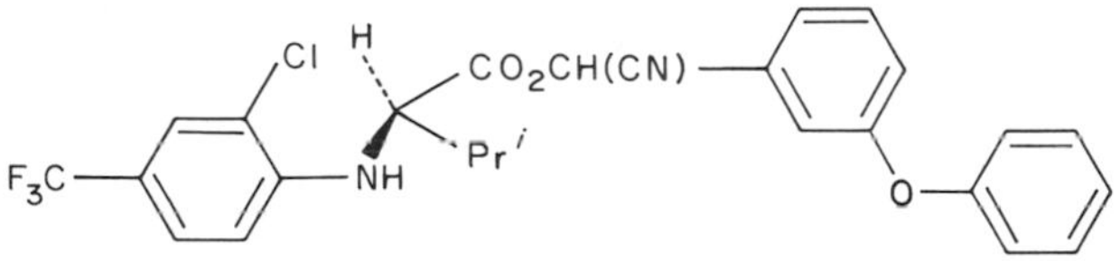

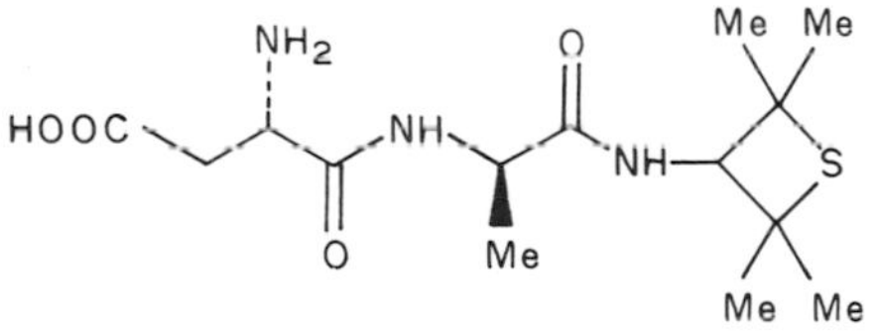

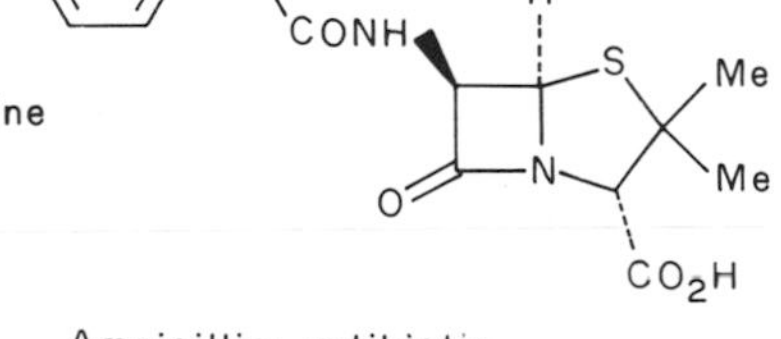

Chart 1

ranging from 10 to 10^5 tonnes per annum (tpa).[14] L-Lysine (*ca* 70 000 tpa) and monosodium L-glutamate (*ca* 350 000 tpa)[15,16] are produced on scales which rank with petrochemicals. Although in the past it has been the natural, L-series, acids which have been available, D-amino acids are now becoming increasingly available as a result of being needed as components of materials of industrial interest (Chart 1). Peptide drugs such as buserelin (**5**) often incorporate D-amino acids to thwart proteolytic cleavage at positions where this is otherwise most likely to occur.

Pfizer's new dipeptide sweetener alitame (**6**) required D-alanine, not hitherto widely available; however, as with any potentially important new product, supply was stimulated. In the same area Searle's sweetener, aspartame (see Chapter 11), had already stimulated sources of L-phenylalanine. D-Alanine is now being advertised by several sources.[20] In the case of L-phenylalanine world demand grew from <50 tpa in 1980 to >3000 tpa by 1985 and more than 30 companies have been involved in process route development.[21]

It should not be overlooked that D-amino acids are in principle, usually, just as cheap and abundant as their L-enantiomers. The method of direct crystallization, which generates L- and D-isomers with equal facility, is applicable to most amino acids or to simple derivatives thereof (Section 1.2.2.2) and the aminopeptidase and hydantoinase routes (Section 1.2.2.3) also furnish D-amino acids.

In addition to the examples in Chart 1, a selection of other products which

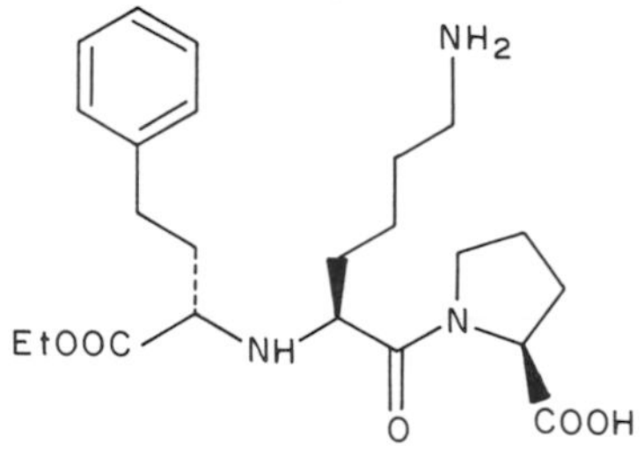

(**7**)

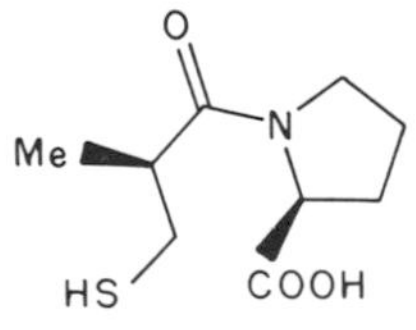

(**9**)

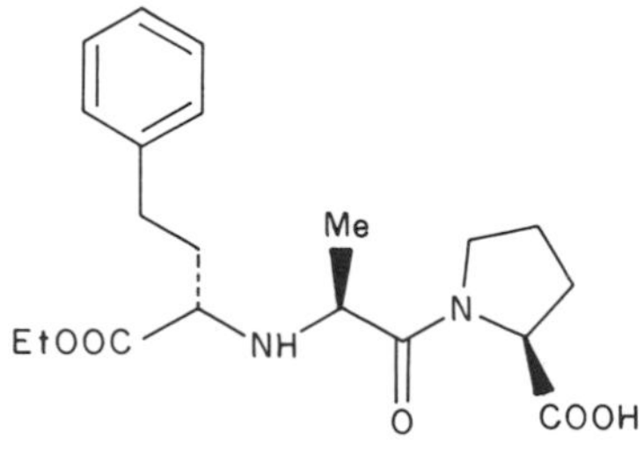

(**8**)

derive optical activity from amino acids is given on the previous page. In most cases the chemistry by which the acids are incorporated is obvious and straightforward.

Synthetic peptides include some important drugs. The angiotensin-converting enzyme (ACE) inhibitors lisinopril (**7**), enalapril (**8**) and captopril (**9**) all

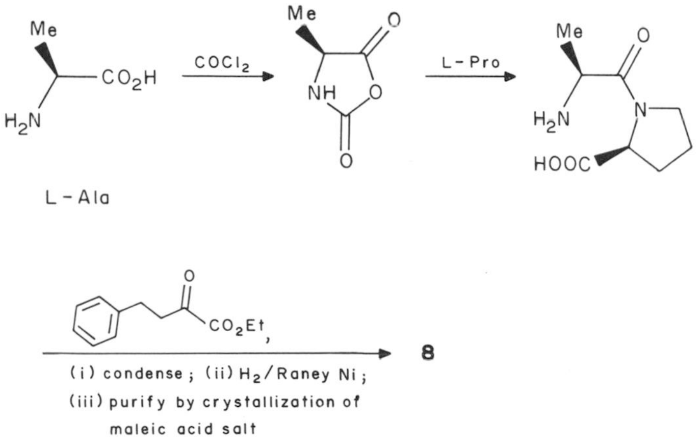

L-Ala $\xrightarrow{COCl_2}$ $\xrightarrow{\text{L-Pro}}$

$\xrightarrow[\text{(i) condense; (ii) } H_2\text{/Raney Ni; (iii) purify by crystallization of maleic acid salt}]{PhCH_2CH_2COCO_2Et,}$ **8**

Scheme 2

incorporate L-proline. A synthesis of **8** is shown in Scheme 2. A key step here is the Raney nickel reduction which gives 87% of the product as the required (*S*, *S*, *S*)-isomer.[9a]

Oxytocin (**10**), used for induction of labour, is representative of other peptide drugs composed of a small number of amino acid residues.

H–Cys–Tyr–Ile–Glu–Asn–Cys–Pro–Leu–Gly–NH_2

(**10**)

Syntheses of carbapenem antibiotics have been developed which, variously, utilize the chirality of L-threonine,[22] L-glutamic acid[23] and L-aspartic acid.[24]

In addition to the proteinogenic amino acids, 6-aminopenicillanic acid (6-APA) (**11**), D-(−)-phenylglycine (**12**) and D-(−)-4-hydroxyphenylglycine (**13**) are all produced in large quantities. 6-APA is produced on a scale of around 5000 tonnes per annum by enzymic cleavage of penicillin-V or G.[25] Compound **12** is also made on a 1000-tonne scale for semi-synthetic β-lactam antibiotics such as ampicillin (Chart 1); see Chapter 20. Compound **13** is a building block for amoxicillin, **14** and other antibiotics.

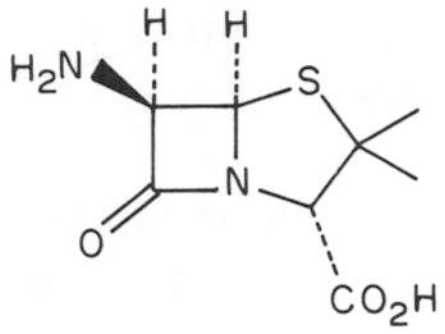

H H
H_2N S
N
O
CO_2H
(11)

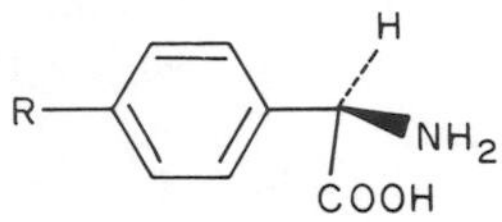

H
R NH_2
COOH
(12) R = H
(13) R = OH

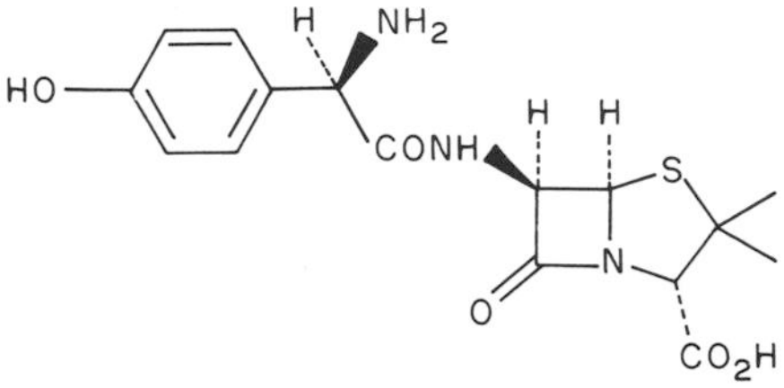

H NH_2
HO
H H
CONH S
N
O
CO_2H
(14)

1.2.1.2 Hydroxy Acids

The common hydroxy acids are shown in Chart 2.

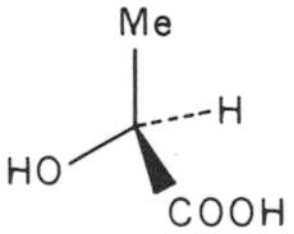

Me
H
HO
COOH
(S)-(+)-Lactic acid

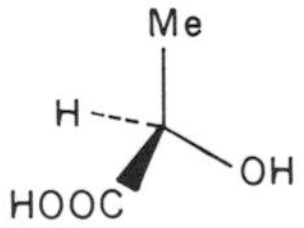

Me
H
OH
HOOC
(R)-(−)-Lactic acid

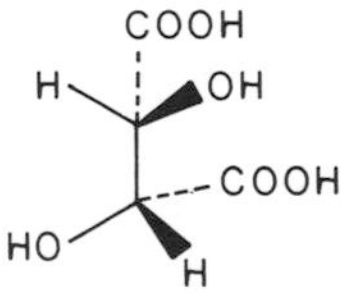

COOH
H OH
COOH
HO
H
(R,R)-(+)-Tartaric acid

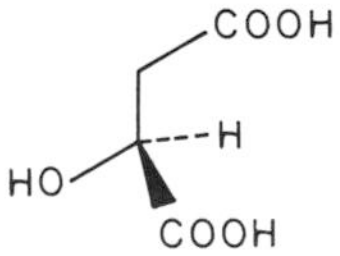

COOH
H
HO
COOH
(S)-(−)-Malic acid

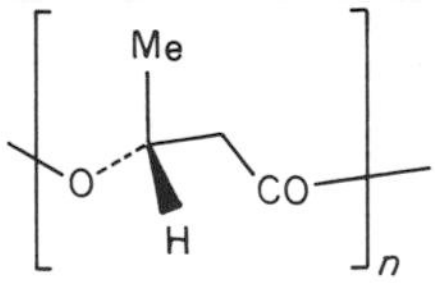

Me
O CO
H
n
(17)
Poly[(R)-3-hydroxybutyrate]

Chart 2

Lactic Acid

Both enantiomers are produced commercially by fermentation and a significant outlet for the 'unnatural'[26] (*R*)-enantiomer has been as a source of esters of (*S*)-2-chloropropionic acid used for the large-tonnage aryloxypropionate herbi-

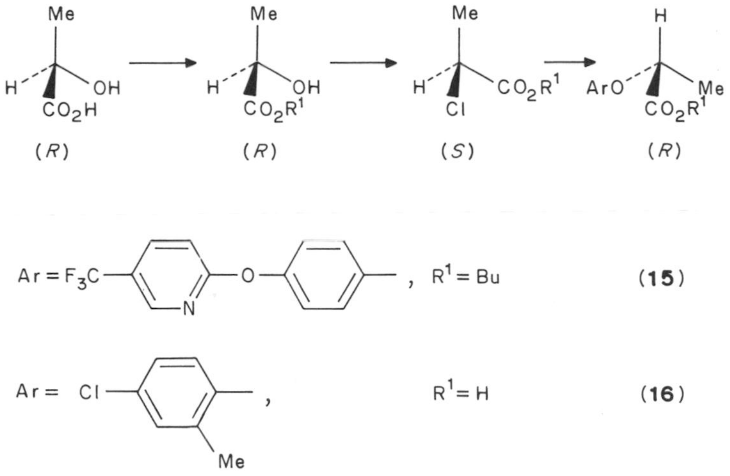

Me H···C(—OH)(CO2H) (R) → Me H···C(—OH)(CO2R1) (R) → Me H···C(—CO2R1)(Cl) (S) → H ArO···C(—Me)(CO2R1) (R)

Ar = F3C–(pyridyl)–O–(phenyl)–, R1 = Bu (15)

Ar = Cl–(phenyl, Me)–, R1 = H (16)

Scheme 3

cides such as fluazifop-butyl (**15**) and mecoprop-P (**16**) (Scheme 3). However, (*S*)-2-chloropropionic acid is now available directly (Section 1.2.2.3 and Chapter 5) without the need for (*R*)-lactic acid.

Tartaric Acid

Tartaric acid has an ancient history within organic stereochemistry. It was isolated by Scheele in 1769, its optical activity was recognized by Biot in 1832[27] and it had been resolved both biologically, using a mould, and chemically, via an alkaloid salt by Pasteur as early as 1858.[28] Despite its abundance and long-standing availability, it has found relatively little application as a chiral building block and industrially finds more use as a resolving agent (Section 1.2.2.1). Of 15 applications cited in *Pharmazeutische Wirkstoffe*,[29] 14 are resolutions. An efficient synthesis of (*R*)-(−)-γ-amino-β-hydroxybutyric acid (GABOB) has been described starting from (+)-tartaric acid.[30]

The other important outlet for optically active tartaric acid is as a source of chirality in catalysts for asymmetric synthesis (Section 1.2.3.1).

Malic Acid

(*S*)-Malic acid is available by fermentation, by hydration of fumaric acid (Section 1.2.3.2 and Chapter 19) and by asymmetric synthesis[31] (Section 1.2.3.1 and Chapter 18) but, like tartaric acid, it has found relatively little use as a chiral building block. A recent application for which (*S*)-malic acid was considered as a building block was to produce an intermediate for the fungicide CGA

80 000 (clozylacon) (**3**).[8] Also, a short and efficient synthesis of both (*R*)- and (*S*)-carnitine from (*R*)- and (*S*)-malic acid, respectively, has been reported.[32]

3-Hydroxybutyric Acid

Poly[(*R*)-3-hydroxybutyrate] (**17**) is a more recent addition to the chiral pool. It is produced in bulk by ICI.[33] Monomeric esters are readily made from it by heating with the corresponding alcohol in the presence of a catalyst.[34] Biological production of the polymer, generated by *Alcaligenes eutrophus* bacteria at up to 80% of its dry weight, can be controlled to give various proportions of (*R*)-3-hydroxyvalerate as copolymer; depolymerization and distillation thus also provide access to esters of (*R*)-3-hydroxyvaleric acid. An example of the utility of **17** is in the synthesis of carbapenem antibiotics.[35] (*S*)-3-Hydroxybutyric acid is also readily obtained by yeast reduction of acetoacetates[34] (Section 1.2.3.2).

More detailed examples of the preparation and elaboration of optically active hydroxy acids are given in Chapters 12 and 13, respectively.

1.2.1.3 Carbohydrates and Derivatives

Carbohydrates are renewable, often cheap and abundantly available, but as chiral building blocks they are generally only available in one enantiomeric form and rarely bear a close structural relationship to a target. They suffer from a profusion of chirality. The carbon chains are generally too long, necessitating costly transformation to smaller, more useful species. Mannitol is an exception; the symmetry of the molecule permits cleavage to two identical, and still chiral, subunits. This is utilized in the synthesis of (*S*)-solketal (**19**)[36] from the D-isomer (**18**) (Scheme 4) and in the synthesis of 2, 3-*O*-isopropylidene-D-glyceric acid in a convenient procedure reported by Emons *et al.*[37]

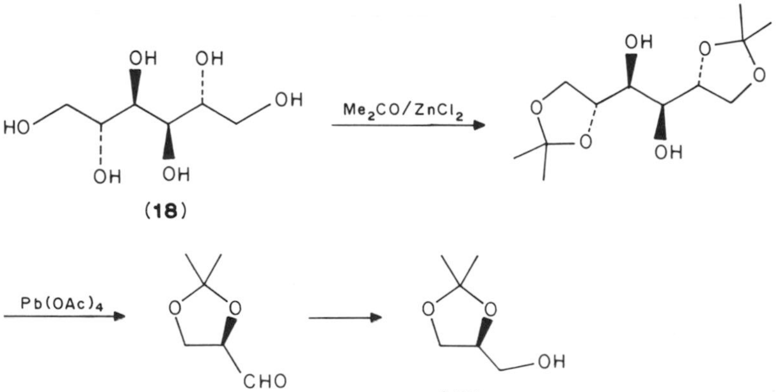

HO, OH, OH, OH, OH, OH (**18**) $\xrightarrow{Me_2CO/ZnCl_2}$ [bis-acetonide, OH, OH] $\xrightarrow{Pb(OAc)_4}$ [acetonide aldehyde, CHO] ⟶ [acetonide, OH] (**19**)

Scheme 4

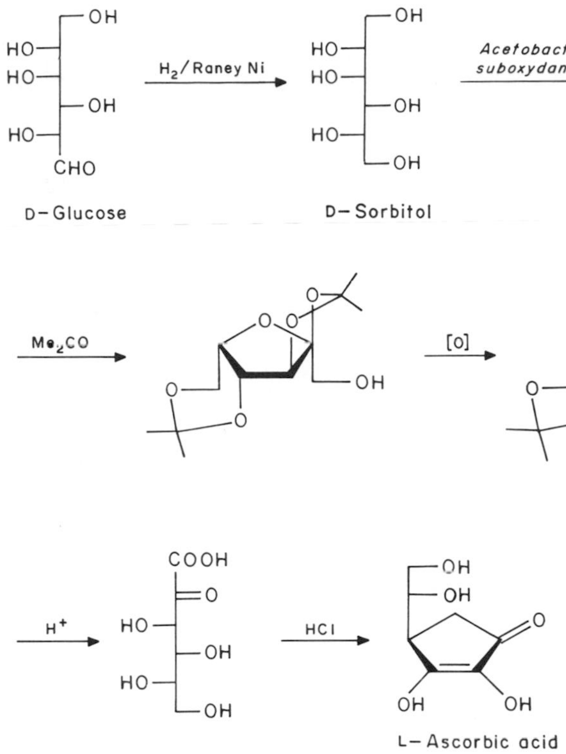

H_2/Raney Ni
Acetobacter suboxydans
D-Glucose D-Sorbitol L-Sorbose
Me_2CO [O]
H^+ HCl
L-Ascorbic acid

Scheme 5

Utilization of D-glucose in the classical Reichstein–Grussner process for ascorbic acid (Scheme 5) is a major, and one of the earliest, industrial examples of a chiral pool substance being used in synthesis, albeit for the production of another 'pool' material. The process was developed in the 1920s and, currently, is used to produce more than 35 000 tonnes per annum. These levels of production place L-ascorbic acid and D-sorbitol in the class of major chiral pool materials, the latter being produced independently for a large number of other industrial applications such as resins and surfactants.

L-Ascorbic acid may be converted (Scheme 6) into the useful C_3 synthon (*R*)-solketal (**20**),[38] which can be employed in the synthesis of (*S*)-β-blockers. An efficient use of D-sorbitol is its conversion into the coronary vasodilator isosorbide dinitrate (**21**) (Scheme 7).[39]

An example of a multi-step conversion, in order to use a carbohydrate material, is shown in Scheme 8.[40] A synthesis of this length can only be contemplated when the target is of high value, in this case a key prostaglandin intermediate.

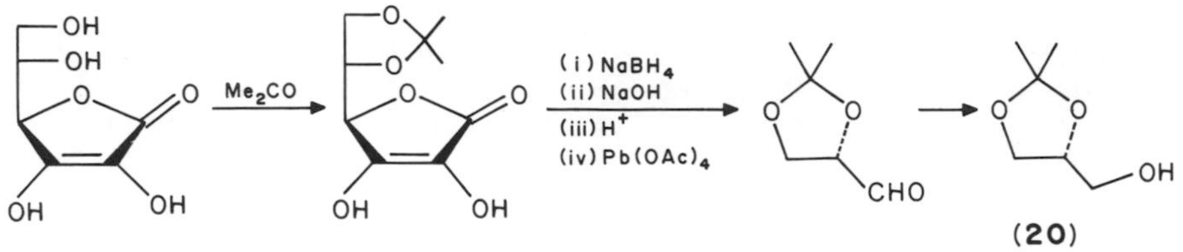

Scheme 6

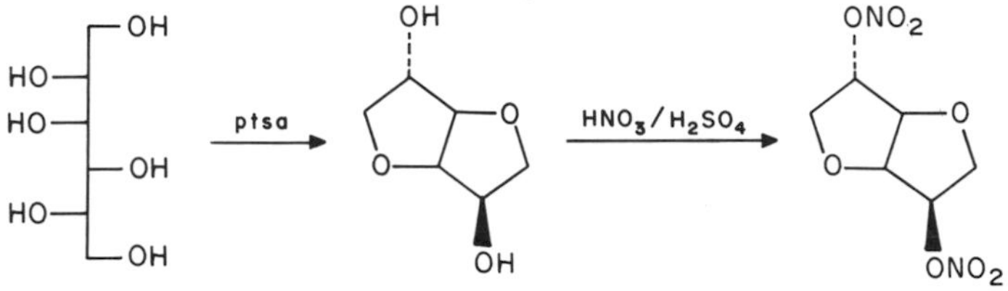

Scheme 7

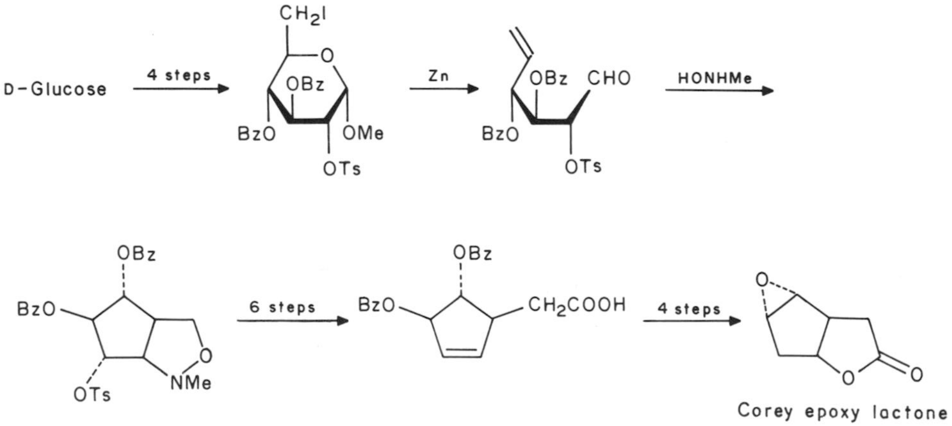

Scheme 8

Tolstikov *et al.*[41] reported the synthesis of **23** in seven steps from levoglucosan (**22**). Compound **23** is an intermediate for high-value HMG–CoA reductase inhibitor drugs (**24**) (Scheme 9).

A recently developed C_4 carbohydrate building block is L-erythrulose (**25**)[42] which can be transformed into the drug γ-amino-β-hydroxybutyric acid (GABOB), biotin or the C_3 synthon (*S*)-1-glyceraldehyde, and thence into β-blockers or prostaglandins.

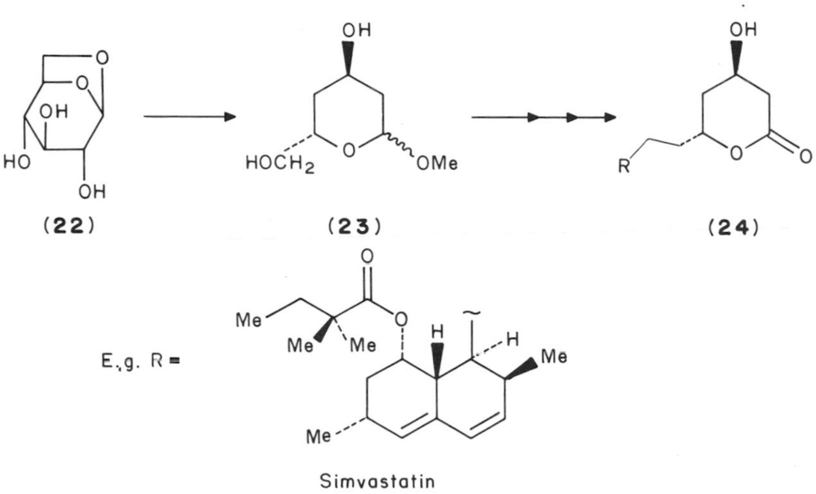

(22) (23) (24)

E.g. R =

Simvastatin

Scheme 9

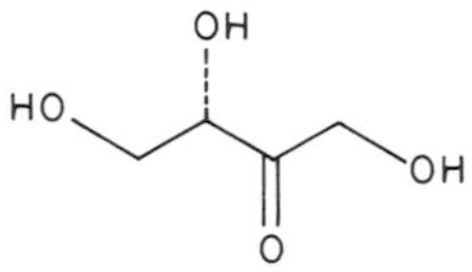

(25)

1.2.1.4 Terpenes

Some of the available optically active terpenes are shown in Chart 3. In general, terpenes do not lend themselves to direct incorporation as building blocks and find their main outlets as precursors for resolving agents and as the source of chirality in catalysts for asymmetric synthesis. Not all are available in both enantiomeric forms and they are not always available with high chemical and optical purities. Purification to high *ee* is not always easy. Most are liquids and are generally devoid of suitable functionality for simple purification via derivatives.

Camphor derivatives, however, generally are crystalline and the issue of non-destructive chirality transfer has been addressed by Oppolzer.[43] Bornane-10, 2-sultam (**26**), and its enantiomer, accessible from (+)- and (−)-camphorsulphonic acids (Scheme 10), serve as versatile chiral auxiliaries and have been available in kilogram quantities.

A structural feature of terpenes which has invited particular attention is the occurrence of geminal dimethyl groups in conjuction with optical activity; the stimulus has been synthetic pyrethroid insecticides, examples of which are given

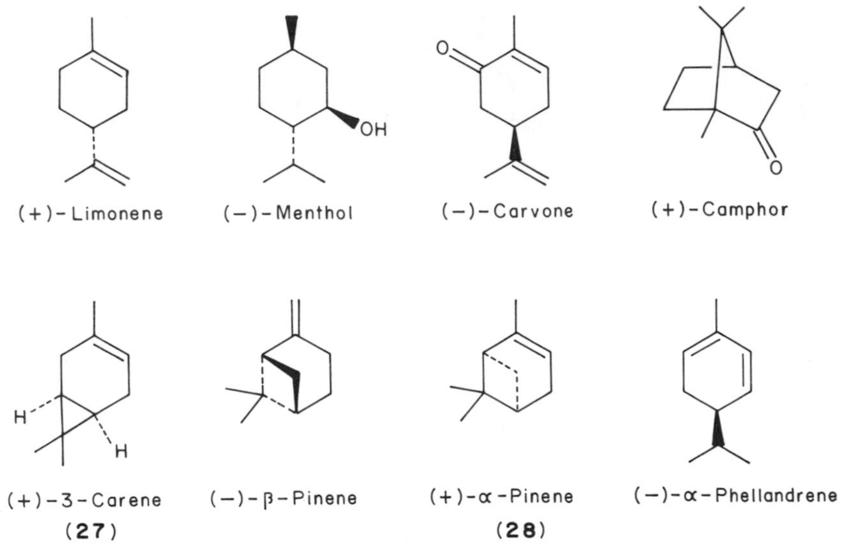

(+)-Limonene (−)-Menthol (−)-Carvone (+)-Camphor

(+)-3-Carene (**27**) (−)-β-Pinene (+)-α-Pinene (**28**) (−)-α-Phellandrene

Chart 3

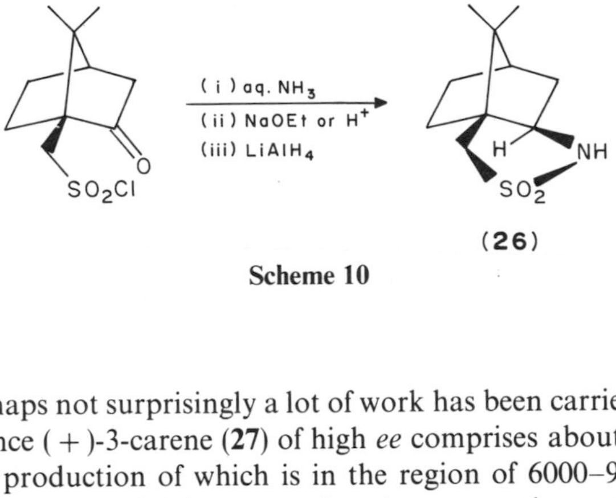

(i) aq. NH_3
(ii) NaOEt or H^+
(iii) $LiAlH_4$

(**26**)

Scheme 10

in Chart 4. Perhaps not surprisingly a lot of work has been carried out in Indian laboratories, since (+)-3-carene (**27**) of high *ee* comprises about 60% of Indian turpentine, the production of which is in the region of 6000–9000 tonnes per annum.[44] (+)-3-Carene (**27**) is present in other turpentines at lower but still commercially interesting levels. Scheme 11 outlines some of the approaches which have been considered for exploiting terpenes in pyrethoid synthesis (see also Chapter 4). Caronaldehyde (**29**) permits the synthesis of the highly active (1*R*, *cis*)-isomers of the halovinyl series (Scheme 12). Terpenes also find use as starting materials for other terpenes. Menthol has been manufactured from (*R*)-(−)-α-phellandrene and (+)-3-carene (Section 1.3.1.3).

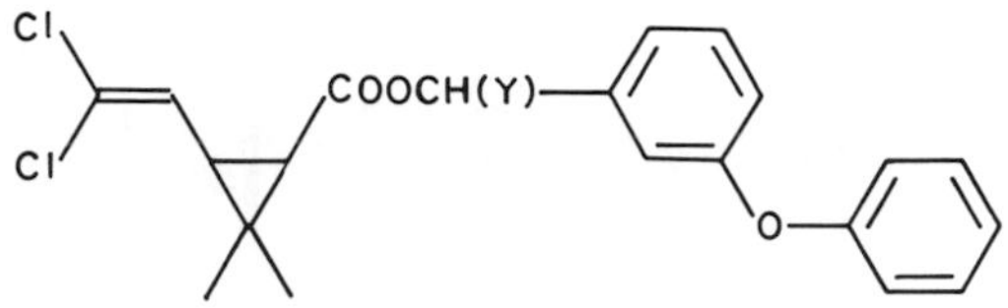

Permethrin Y = H
Cypermethrin Y = CN

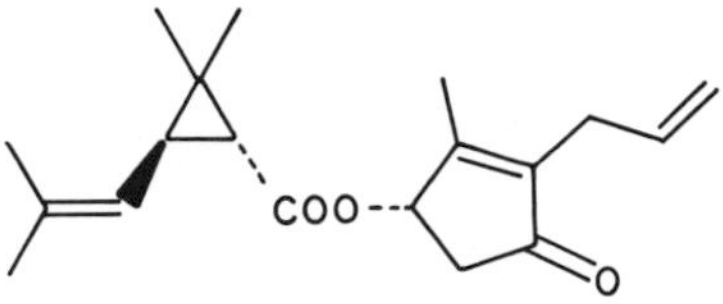

Bioallethrin

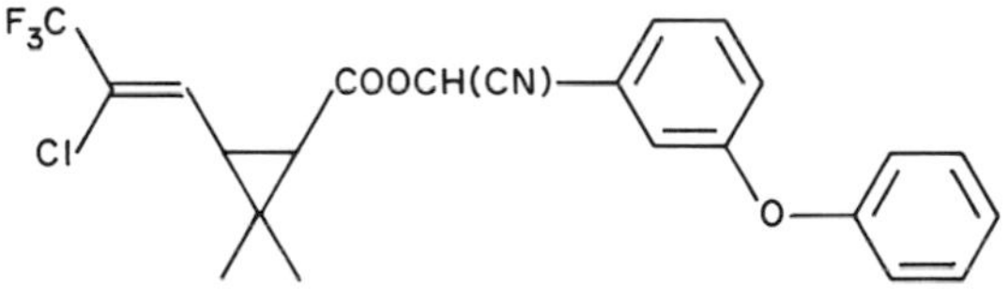

Cyhalothrin

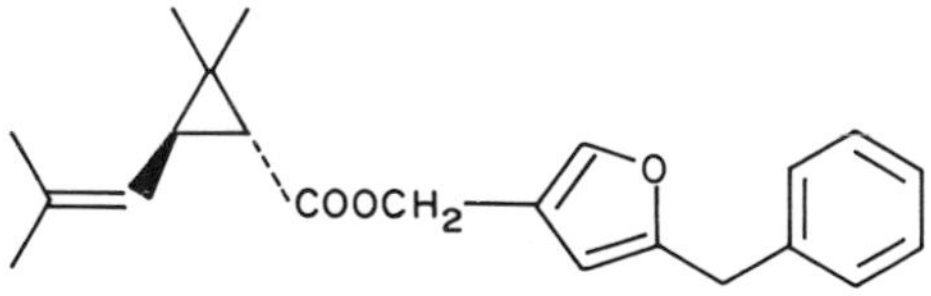

Bioresmethrin

Chart 4

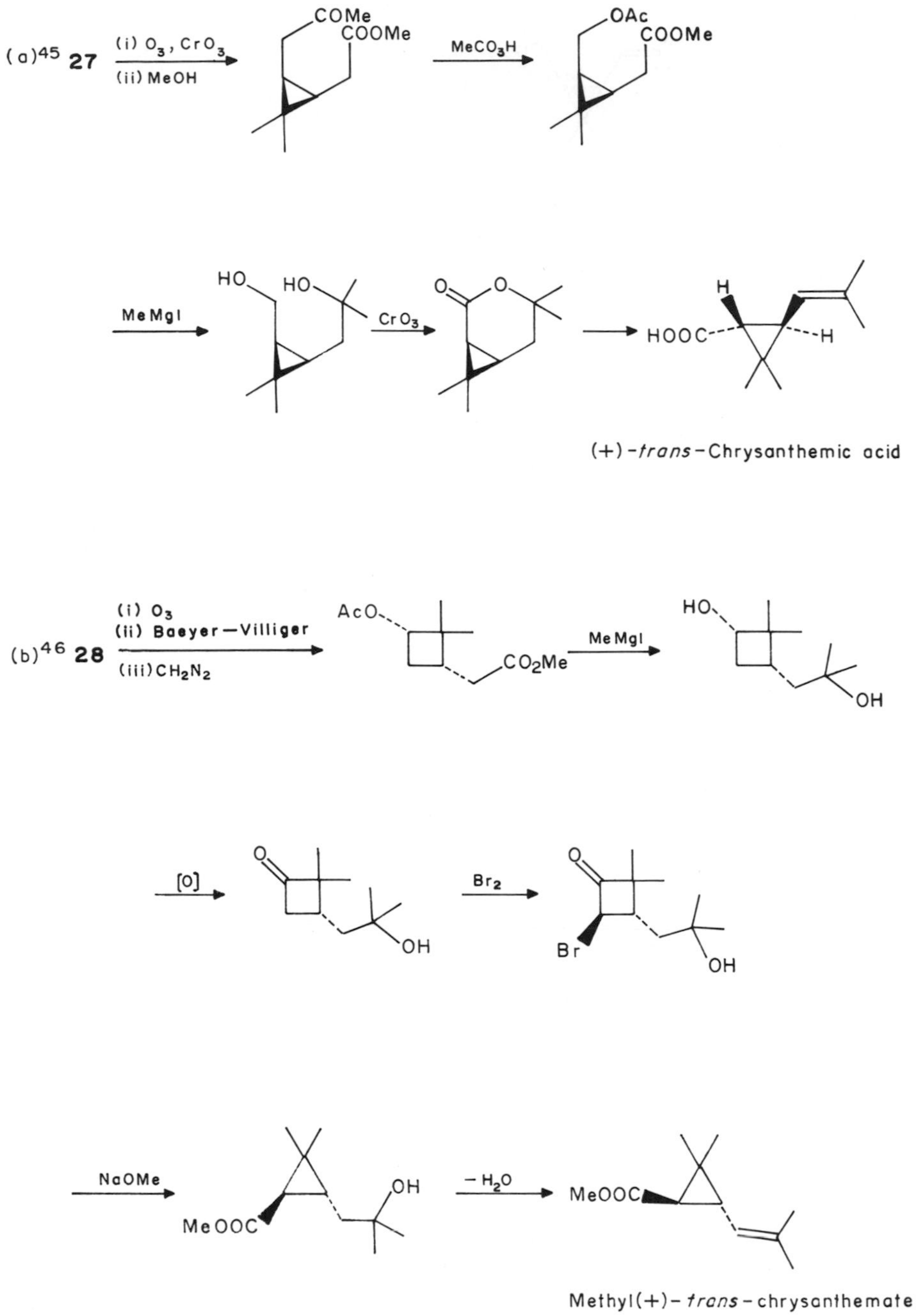

(a)[45] **27** (i) O_3, CrO_3 (ii) MeOH → COMe, COOMe → $MeCO_3H$ → OAc, COOMe → MeMgI → HO, HO → CrO_3 → O, O → HOOC, H, H

(+)-*trans*-Chrysanthemic acid

(b)[46] **28** (i) O_3 (ii) Baeyer—Villiger (iii) CH_2N_2 → AcO, CO_2Me → MeMgI → HO, OH

→ [O] → O, OH → Br_2 → O, Br, OH

→ NaOMe → MeOOC, OH → $-H_2O$ → MeOOC

Methyl(+)-*trans*-chrysanthemate

Scheme 11 (*Continued*)

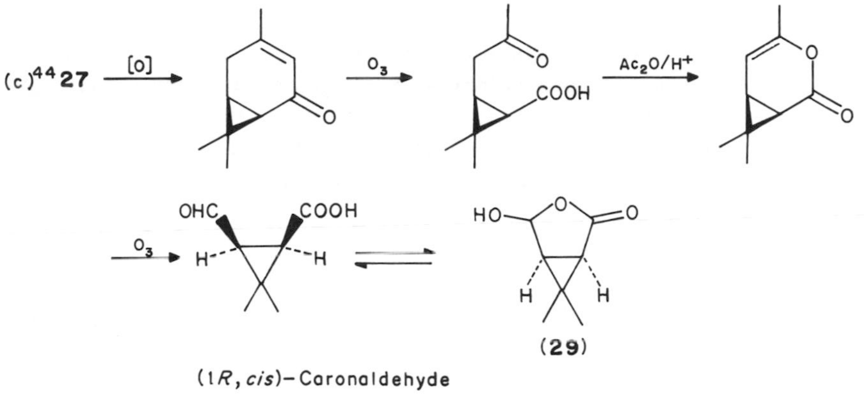

(c)[44] **27** $\xrightarrow{[O]}$ $\xrightarrow{O_3}$ $\xrightarrow{Ac_2O/H^+}$ $\xrightarrow{O_3}$

(1*R*, *cis*)-Caronaldehyde

(**29**)

Scheme 11

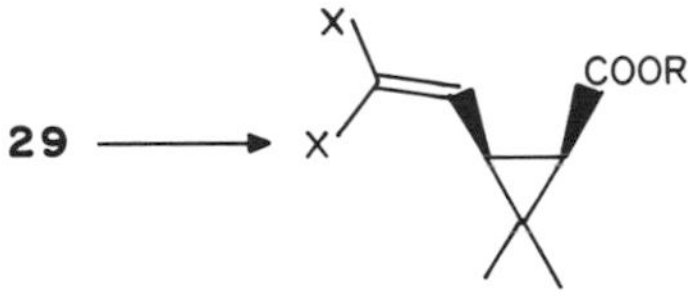

29 ⟶

Scheme 12[47]

1.2.1.5 Alkaloids

By far the most useful alkaloids for the production of optically active materials are the cinchona bases (Chart 5). They are moderately expensive (Table 2) and are very unlikely to find application as building blocks. They are, however, most valuable chiral auxiliaries[48] (Section 1.2.3.1; see also Chapter 18) and much

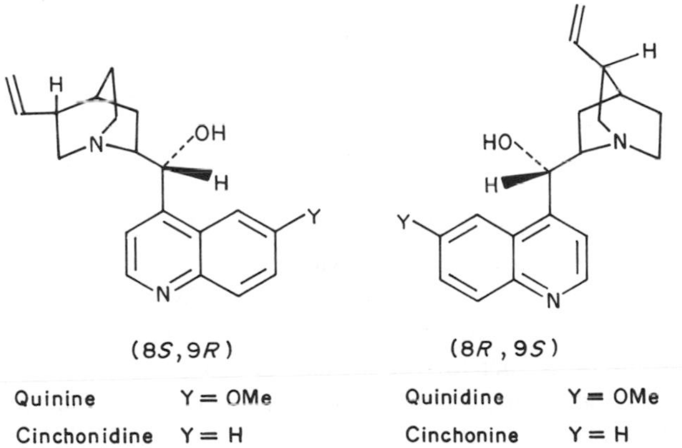

(8*S*,9*R*)

Quinine Y = OMe

Cinchonidine Y = H

(8*R*, 9*S*)

Quinidine Y = OMe

Cinchonine Y = H

Chart 5

used as resolving agents on the small scale. For large-scale resolutions, a highly efficient recycle is imperative. The four cinchona bases comprise two pairs of diastereomers but the critical β-hydroxyamino segments are enantiomeric and they behave as such in many applications. Their toxicity is also low compared with, for example, nicotine, brucine and strychnine, the large-scale use of which could only be considered in closely controlled circumstances.

1.2.1.6 The New Pool

The whole gamut of industrially produced chiral products is available for consideration. Some such as 6-APA (**11**) are not particularly new, having been made on tonnage scales for 20–30 years, but they do not feature in any conventional listing.

Consideration should also be given to by-products which are normally recycled, such as L-phenylglycine in the manufacture of D-phenylglycine[49] and D-(+)-α-amino-ε-caprolactam (**30**) in the manufacture of L-lysine (see Scheme 28).

In addition, many conventional pool materials such as ephedrine (**31**), menthol and L-phenylalanine have had their availability considerably augmented by manufacture.

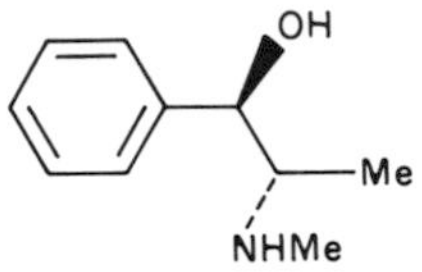

OH

Me

NHMe

(**31**)

1.2.2 SEPARATION OF RACEMATES

1.2.2.1 Classical Resolution

Despite its 'low technology' image, classical resolution via diastereoisomer crystallization is widely used industrially and in particular furnishes a large proportion of those optically active drugs which are not derived from natural products. Examination of a representative group of such drugs[50] shows that 65% owe their optical activity to classical resolution. There are clearly many instances where resolution is both economically viable and the method of choice.

It is a technique which is sometimes viewed as owing too much to empiricism to be worthy of serious consideration. However, Wilen *et al.*[51] have provided guidelines which permit a rational approach with a high probability of success. There is still much on offer for the industrial practitioner or anyone requiring kilograms of a product and where some effort spent in system optimization can be justified. It has been said that success is never guaranteed,[52] but it is probably more guaranteed than, for example, asymmetric synthesis is at present.

This section is not concerned with detailed appraisal of the many and well known reagent–functional group combinations which may be employed to obtain the necessary diastereomers. Likewise, the methodology, principles of the technique and criteria for good resolving agents are described more fully in ref. 48 and Chapter 2.

Attractions of classical resolution include wide applicability, provided there is suitable functionality in the molecule through which to form the diastereomer, and, usually, access to both enantiomers. Classical resolution becomes particularly attractive where it can be combined with *in situ* racemization in a crystallization-induced asymmetric transformation (Scheme 13), a process designated 'deracemization.'[54] It is then possible to obtain almost complete conversion to the required enantiomer; precipitation of one diastereomer drives the equilibrium in favour of that isomer. An elegant example comes from Merck[55] (Scheme 14) in which a catalytic amount of an aldehyde facilitates racemization in solution at ambient temperature via the imine and the desired (*S*)-amine continuously crystallizes as its (+)-camphor-10-sulphonic acid salt. Racemization is presumed to be effected by the small amount of amine present as the free base. This efficient, one-pot resolution–racemization process has been operated on a 6 kg scale to produce an intermediate for a candidate cholecystokinin antagonist.

$$(R)\text{-A}\cdot(R)\text{-B} \rightleftharpoons (R)\text{-B} + (R)\text{-A} \rightleftharpoons (S)\text{-A} + (R)\text{-B} \rightleftharpoons (S)\text{-A}\cdot(R)\text{-B}$$

Soluble ⇓ Precipitate

Scheme 13

Despite the wide applicability of classical resolution, molecules devoid of suitable functionality present difficulties. The problem may in principle be tackled by the formation of inclusion complexes. Compound **32** forms crystalline 1:1 inclusion complexes which allow the resolution of molecules such as **33**–**35** giving products of high *ee*.[56] However, despite the commercial interest[56b,c] which appears directed towards resolved glycidyl compounds, reagents such as **32** are clearly not cheap. The reagent **32** is made by oxidative coupling of pure **36**, which in turn has to be resolved by forming inclusion complexes with alkaloids.

As with the chiral pool, there has been a tendency to think of resolving agents as naturally occurring materials coming from the pool, or derivatives thereof. In reality, these have for some time included materials such as 1-phenylethylamine (**37**), which is now widely used since it was shown to be simply resolved with tartaric acid.[57]

Existing tonnage products, and their intermediates, should not be overlooked as potential resolving agents. Consider, for example, D-(+)-α-amino-ε-caprolactam (**30**), a readily accessible and cheap potential by-product of L-lysine manufacture

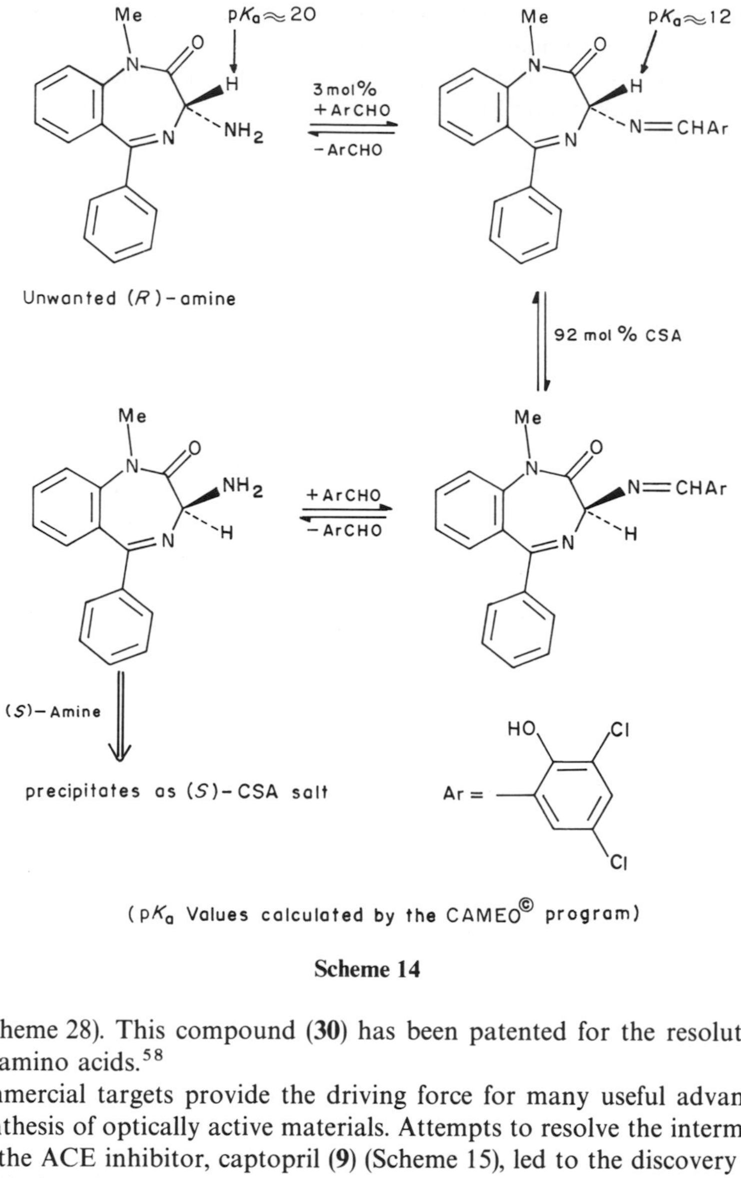

(pK_a Values calculated by the CAMEO© program)

Scheme 14

(see Scheme 28). This compound (**30**) has been patented for the resolution of *N*-acylamino acids.[58]

Commercial targets provide the driving force for many useful advances in the synthesis of optically active materials. Attempts to resolve the intermediate **38** for the ACE inhibitor, captopril (**9**) (Scheme 15), led to the discovery of the new chiral amine, (*S*)-*N*-(isopropyl)phenylalaninol (**39**).[59] This reagent was developed when several naturally occurring alkaloids and other commercially available chiral amines proved ineffective. Its development emphasizes the value of a systematic study of physical parameters. The solubility ternary phase diagram was determined in order to find optimum conditions for purification of the crude salt.

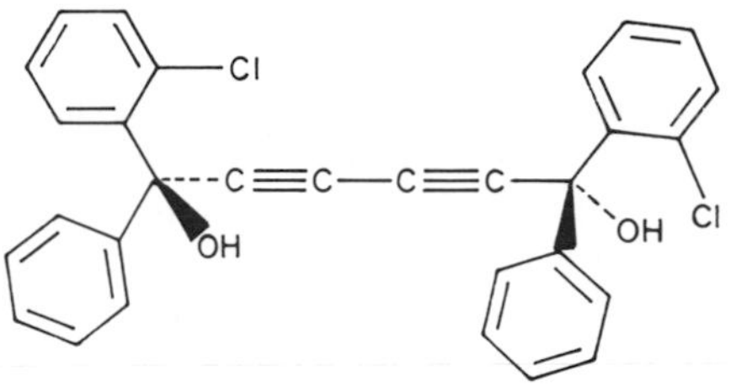

(32) 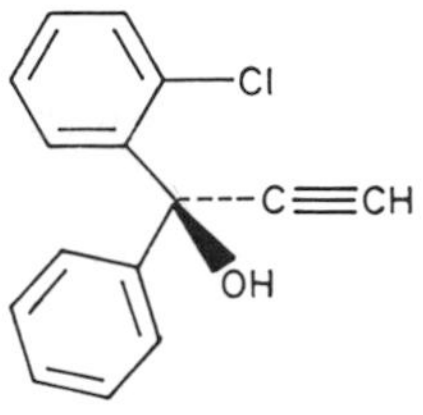 (36)

(33) 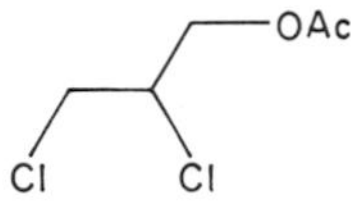 (34) 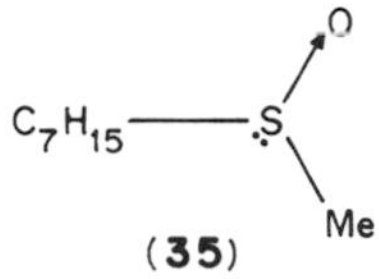 (35)

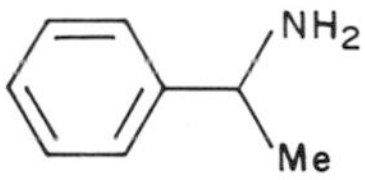

(37) 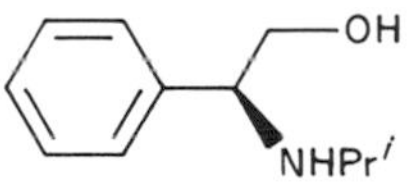 (39) 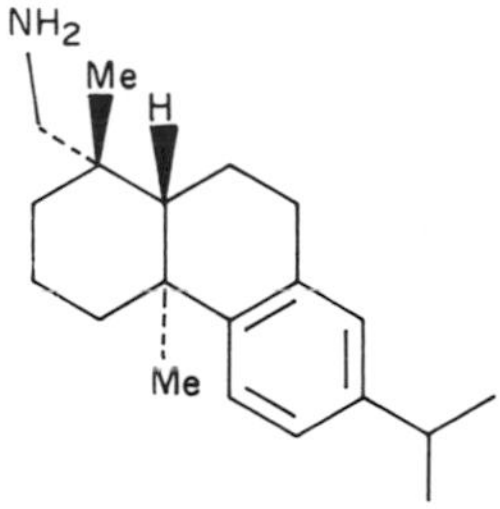 (40)

The captopril intermediate **38** has also been resolved using Amine D™, (+)-dehydroabietylamine (**40**).[60] Derived from wood rosin, it is of interest because of its potential abundance and cheapness; one of its major uses has been to promote adhesion of stone aggregates in road construction! It was introduced in 1964[61] but has not been used very often. Amine D™ is approximately 50% pure with about 20% each of the di- and tetrahydroabietylamine analogues.[62] Whilst it may be readily purified via the acetate,[63] it is the author's experience that it may be used satisfactorily in its crude state,[64] particularly in conjunction with lower alkanols as solvents. It forms salts with carboxylic acids which are soluble in organic solvents. Disadvantages are its availability in only one enantiomeric form and its high molecular weight. However, it is still approximately three times cheaper per mole than resolved **37**, when purchased in bulk.

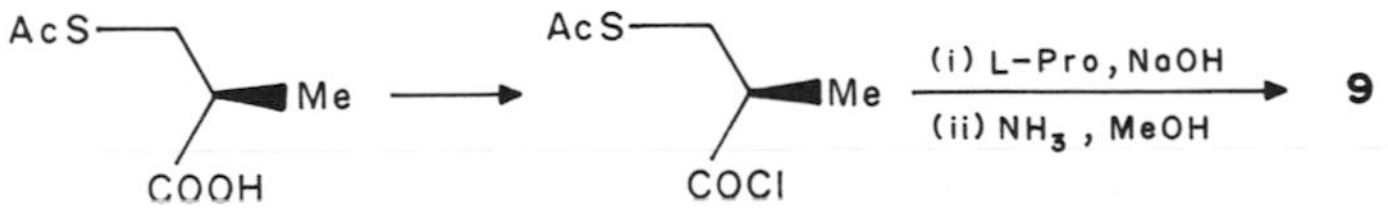

(38) → (i) L-Pro, NaOH; (ii) NH_3, MeOH → 9

Scheme 15

1.2.2.2 Resolution by Direct Crystallization

This is an attractive method; auxiliaries and reagents, other than a solvent are not required. The excellent treatise by Jacques *et al.*[53d] gives a detailed exposition of the theory. In simple terms it depends on the occurrence of some substances as crystalline conglomerates (racemic mixtures) rather than racemic compounds. Although in bulk a conglomerate is optically neutral, individual crystals contain only one enantiomer, whereas in a racemic compound individual crystals contain equal amounts of both enantiomers. Conglomerate formation is a prerequisite for resolution by direct crystallization.

Before the method may be applied it is obviously necessary to establish the existence of the conglomerate; this may be done in a number of ways: (i) by selecting small, discrete crystals and subjecting them to any sensitive method for measuring enantiomer excess (polarimetry, chiral LC or GC, NMR with shift reagent, effect on the nematic phase of a liquid crystal);[65] (ii) determination of binary or ternary phase diagrams (see Chapter 2); (iii) effecting resolution by direct crystallization (more likely to be used as a confirmatory test); (iv) powder X-ray or solid-state IR spectra (enantiomers give spectra identical with those of racemic conglomerate but differ for racemic compounds).

There are a number of variations in the way resolution by direct crystallization may be effected in practice. In the first method, simultaneous crystallization of the two enantiomers is carried out in an apparatus of the type shown schematically in Figure 1. Initially seeds are introduced into the two crystallization chambers and crystallization from the supersaturated solution occurs. The depleted solution is resaturated at a higher temperature in a make-up vessel before recooling to restore the original level of supersaturation required in the

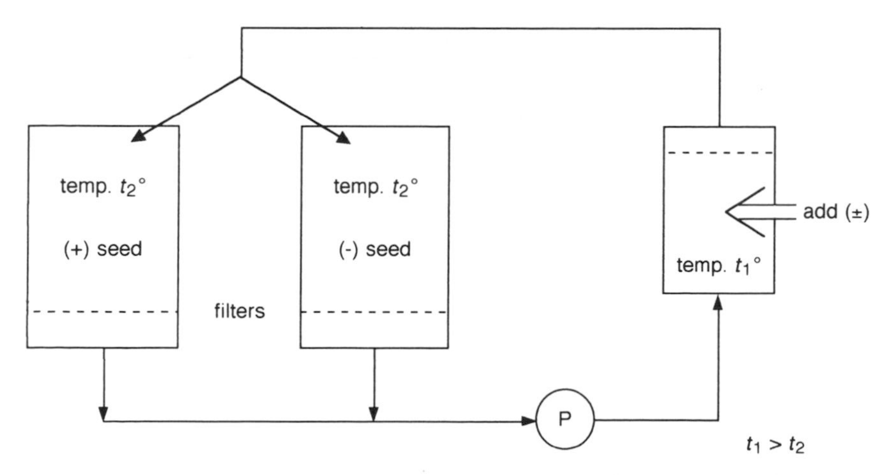

Figure 1. Schematic diagram of apparatus for crystallization of enantiomers

crystallizers. This was in essence the method used in the Merck process for the antihypertensive methyldopa.[66] The technique has also been applied to the useful C_3 synthon glycidyl 3-nitrobenzenesulphonate (**41**).[67]

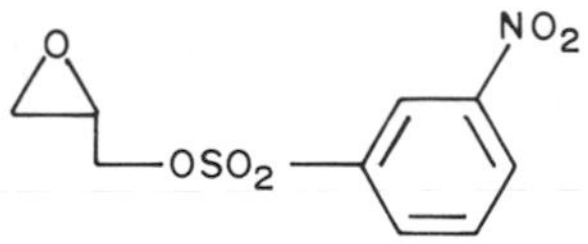

OSO_2 NO_2

(**41**)

A successful large-scale process was developed by Haarmann and Reimer for (−)-menthol, which is separated as an ester.[68] A simple apparatus suitable for laboratory operation has been described by Sato *et al.*,[69] who used it to resolve DL-lysine-3,5-dinitrobenzoate.

Another method consists in taking alternate crops of each of the enantiomers using a single vessel; this is the so-called method of resolution by entrainment.[70] It has its origins in the work of Gernez, who in 1866[71] demonstrated that resolution could occur when a supersaturated solution of racemate was seeded with one of the enantiomers. To a supersaturated solution of the racemate initially artificially enriched with, say, the (+)-enantiomer are added seed crystals of the (+)-enantiomer. A crop of (+)-enriched product is collected equal to approximately twice the amount of material used for the original enrichment. An amount of racemate equal to the weight of the (+)-crop is then dissolved in the filtrates by warming and the solution is cooled to the operating temperature to restore the original degree of supersaturation, but now with the (−)-enantiomer in excess. The solution is then seeded with the (−)-enantiomer and the whole process repeated *ad infinitum*. The main practical limitation on the number of cycles through which such a process can be operated is the build-up of impurities and the tolerance to these of the crystallization. A highly purified starting racemate may be essential if an economic number of cycles is to be achieved. This process has been operated on a large scale to make chloramphenicol.

Other examples of this procedure are indicated in patents to Industria Chimica Profarmaco for the resolution of naproxen as its ethylamine salt[72] and for 2-hydroxy-3-(4-methoxyphenyl)-3-(2-acetylaminophenylthio)propionic acid (**42**), an intermediate for optically active benzothiazepines.[73]

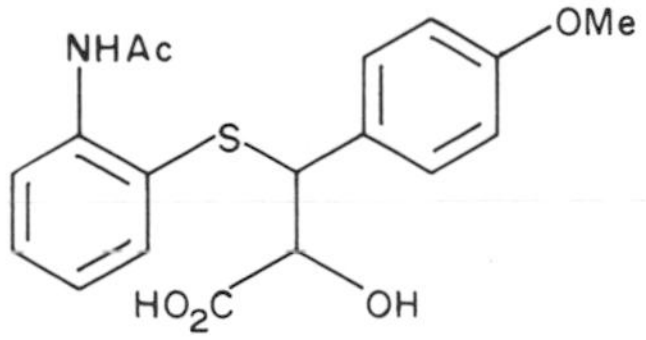

NHAc S OMe HO_2C OH

(**42**)

The method of entrainment may be used to resolve two useful chiral auxiliaries, hydrobenzoin (**43**)[74] and 1,1′-binaphthalene-2,2′-diol, as its dimethyl ether (**44**).[75] The latter material is required for the industrially important Noyori catalyst (Section 1.2.3.1 and Chapter 17), whilst the former can be elaborated into chiral crown ethers which have been used, *inter alia*, to effect enantioselective hydride reductions and cyanohydrin formation.[76]

HO OH

OMe
OMe

(**43**) (**44**)

A further industrially important example of resolution by direct crystallization is Ajinomoto's process for L-glutamic acid (Scheme 16), introduced in the 1960s, operated on a scale in excess of 10 000 tpa, and competitively with the fermentation process for monosodium glutamate (MSG).[77]

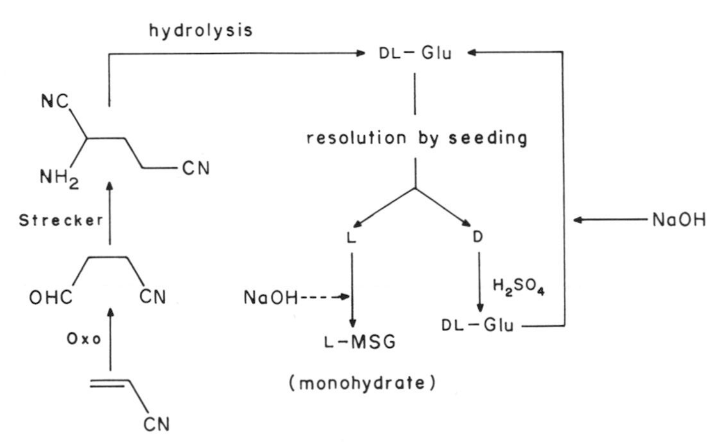

Scheme 16. Ajinomoto L-monosodium glutamate process

A more esoteric variant has been reported[78] in which the supersaturated solution of racemate is seeded simultaneously with large crystals of one enantiomer and very small crystals of the other. After crystallization has occurred the products are separated by sieving.

The ultimate process is a combination of resolution by direct crystallization with facile *in situ* racemization of the unwanted enantiomer. This leads to a so-called[79] second-order asymmetric transformation (Figure 2). Such a technique

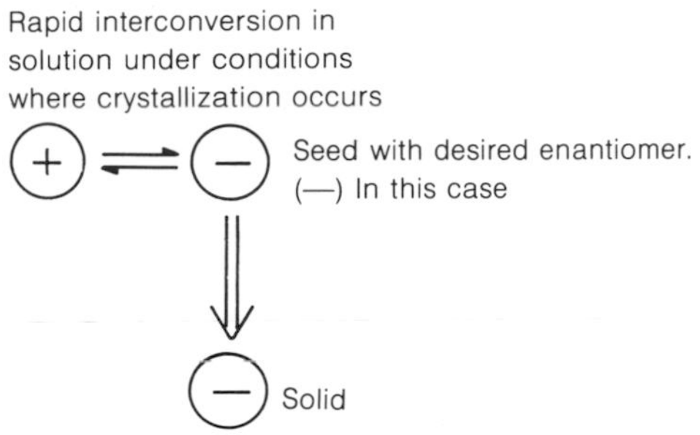

Rapid interconversion in solution under conditions where crystallization occurs

(+) ⇌ (−) Seed with desired enantiomer. (—) In this case

⇓

(−) Solid

Figure 2. The ultimate resolution process

has been applied to compound **45**, the ketone precursor of paclobutrazol; the unwanted enantiomer is racemized *in situ* by treatment with base.[80] The lysine precursor α-amino-ε-caprolactam has also been resolved in this manner (Chapter 3).

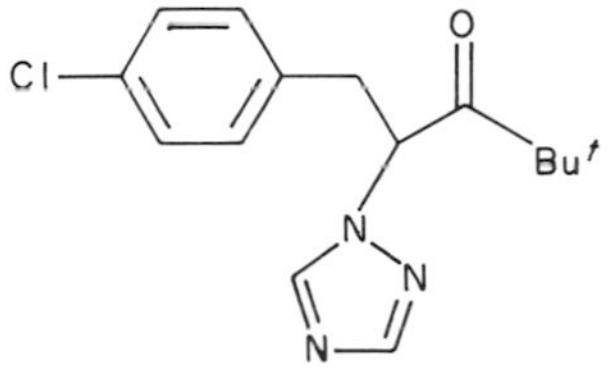

Cl — Bu^t — O — N

(**45**)

Where applicable, a second-order asymmetric transformation would be expected to be economically competitive with any other procedure for generating optical activity. Such a process, proceeding in the absence of deliberate seeding, is a 'spontaneous resolution' and the enantiomer obtained becomes a matter of chance.

Another attraction of direct crystallization is that unlike classical resolution it is not necessary for substrates to possess any particular functionality for it to work. However, it is of limited, and unpredictable, applicability. The occurrence of conglomerates has been estimated at perhaps less than 10% of all crystalline racemates.[81] However, the frequency amongst salts has been estimated to be two or three times that for covalent compounds,[82] and this provides a basis for increasing the chance of discovering a conglomerate. Note, for example, that of the naturally occurring α-amino acids virtually all are resolvable either directly or as derivatives;[83] cf. alanine monomaleate.[84] The technique is clearly amenable to large-scale operation but may require very fine temperature control.[68] Uniform quality of feedstock, perferably of high chemical purity, will be required in order to achieve reproducible crystallizations.

1.2.2.3 Kinetic Resolution[85]

This is a process in which one of the enantiomers (A) of a racemate (AB) is more readily converted to product than the other:

$$A \xrightarrow{k_A} y$$

$$B \xrightarrow{k_B} z$$

$$E = k_A/k_B$$

The enantiomer ratio, E, dictates the efficiency of resolution and is related to the enantiomeric excess of the recovered reactant (ee_R) and of the product (ee_P), at a given degree of conversion (c), by the equations:

$$E = \ln[(1-c)(1-ee_R)]/\ln[(1-c)(1+ee_R)]$$

$$E = \ln[1-c(1+ee_P)]/\ln[1-c(1-ee_P)]$$

For a derivation of these equations see ref. 86a. A useful graphical representation of the dependence of ee_R as a function of c and E has been given by Martin *et al.*[86b] The attraction of kinetic resolution is that the *ee* of the residual substrate improves with the degree of conversion and with only modest selectivity it is still possible to recover the substrate with high *ee*. In practice, an E value > 20 would be sought for a commercially attractive process such that a high *ee* (*ca* 98%) can be attained at 60% and preferably closer to 50% conversion. The converse is that when E is high (*ca* 50 or greater) the product, if chiral, is simultaneously obtained with high *ee*; product *ee* deteriorates as c increases and is poor for low E values at the point at which reactant has a satisfactory *ee*. For an economic process it is important to be able to utilize the resolution product; this is more likely to be the case when it also has a high *ee*.

Kinetic resolution may be realized by chemical or enzymic methods; in the former case the reaction may be either catalytic or stoichiometric with respect to the optically active auxiliary; from an economic standpoint catalysis is obviously preferred. Kinetic resolutions and high E values are more commonly found with enzymic than chemical processes, this being reflected in the number of commercially relevant examples given below.

Stoichiometric Chemical Reaction

The work of Sharpless and co-workers[86b,87] using the original, stoichiometric, version of the epoxidation reagent focused attention on chemical kinetic resolution and the very high *ee*s which are attainable (cf. examples in Scheme 17). In a different example (Scheme 18), stoichiometric use of an auxiliary has been applied to naproxen synthesis.[88] The anhydride of racemic naproxen is reacted with optically active 1-(4-pyridyl)ethanol giving a product of modest optical

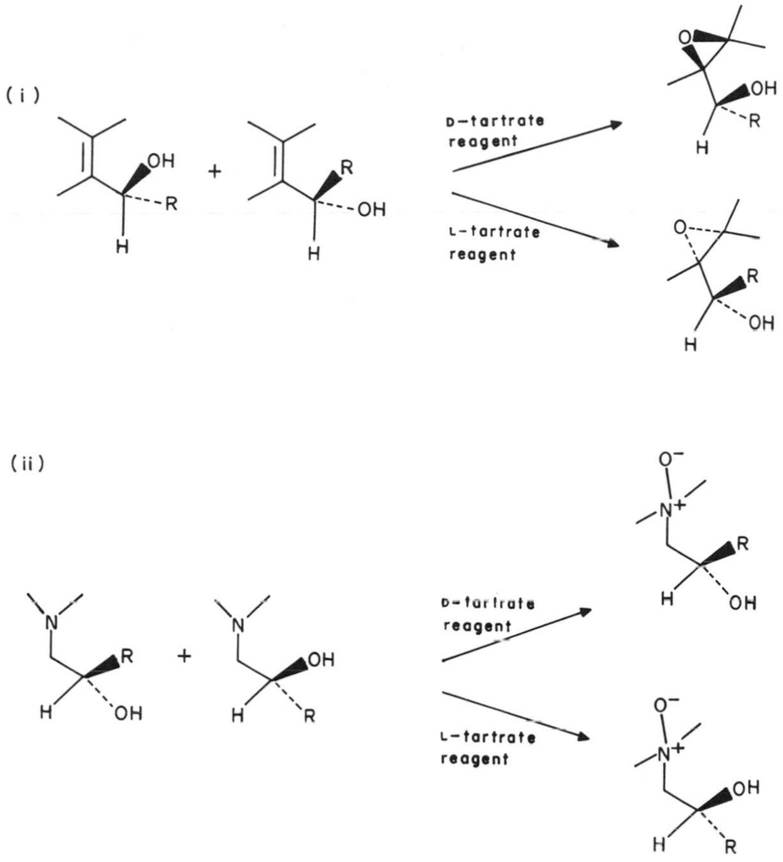

(i)

D-tartrate reagent

L-tartrate reagent

(ii)

D-tartrate reagent

L-tartrate reagent

Scheme 17

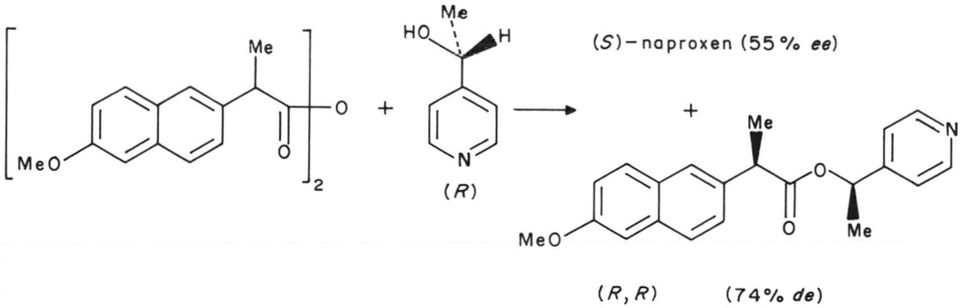

(S)-naproxen (55% *ee*)

(R) (R,R) (74% *de*)

Scheme 18

purity. The same approach has been used for the resolution of pyrethroid acids[89] and aryloxypropionic acids.[90]

Chemical Catalysis

The area is still relatively new and applications will go hand in hand with the general development of new catalysts for asymmetric synthesis where these can be applied to chiral substrates. Its industrial potential has yet to be realized.

For pointers one can return here to the Sharpless epoxidation (catalytic version).[91] Another major area of asymmetric catalysis, hydrogenation, also provides examples of kinetic resolution. Both rhodium-based catalysts and the Ru–BINAP catalysts of Noyori *et al.* (Chapter 17) have been used (Scheme 19).[92,93]

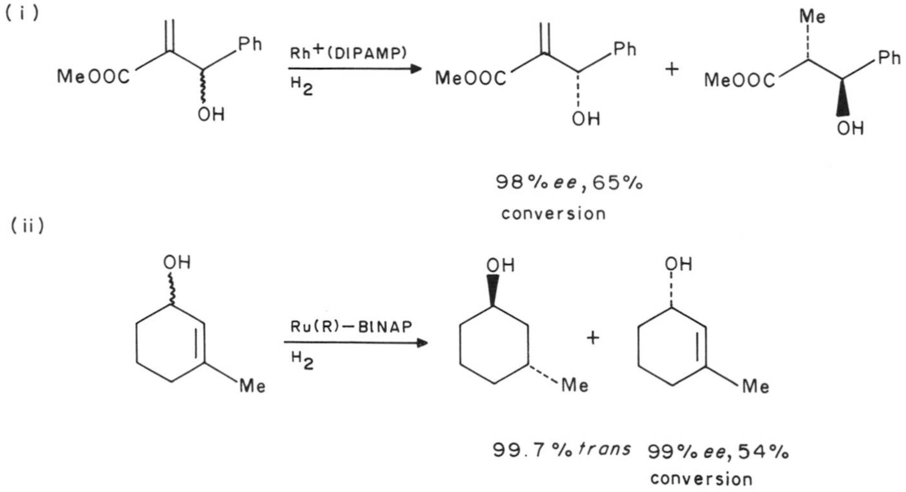

(i)
MeOOC Ph OH → Rh+(DIPAMP), H_2 → MeOOC Ph OH + MeOOC Me Ph OH
98% *ee*, 65% conversion

(ii)
OH Me → Ru(R)–BINAP, H_2 → OH Me + OH Me
99.7% *trans* 99% *ee*, 54% conversion

Scheme 19

Enzymic Catalysis

There are many industrially relevant examples under this heading. Consider first the various approaches to β-blockers (Scheme 20).

(R)-*Glycidyl butyrate* (**46**). There is much industrial interest in this compound.[94] Processes for large-scale synthesis have been developed using a lipase-catalysed hydrolysis (Scheme 21). The reaction was first reported by Ladner and Whitesides.[95] It has been refined and is now operated on a multi-tonne scale,[94a] despite the relatively modest enantioselectivity of porcine pancreatic lipase, which requires the reaction to be run well beyond 50% conversion to ensure high *ee*. If the process could be operated with higher selectivity, such that the

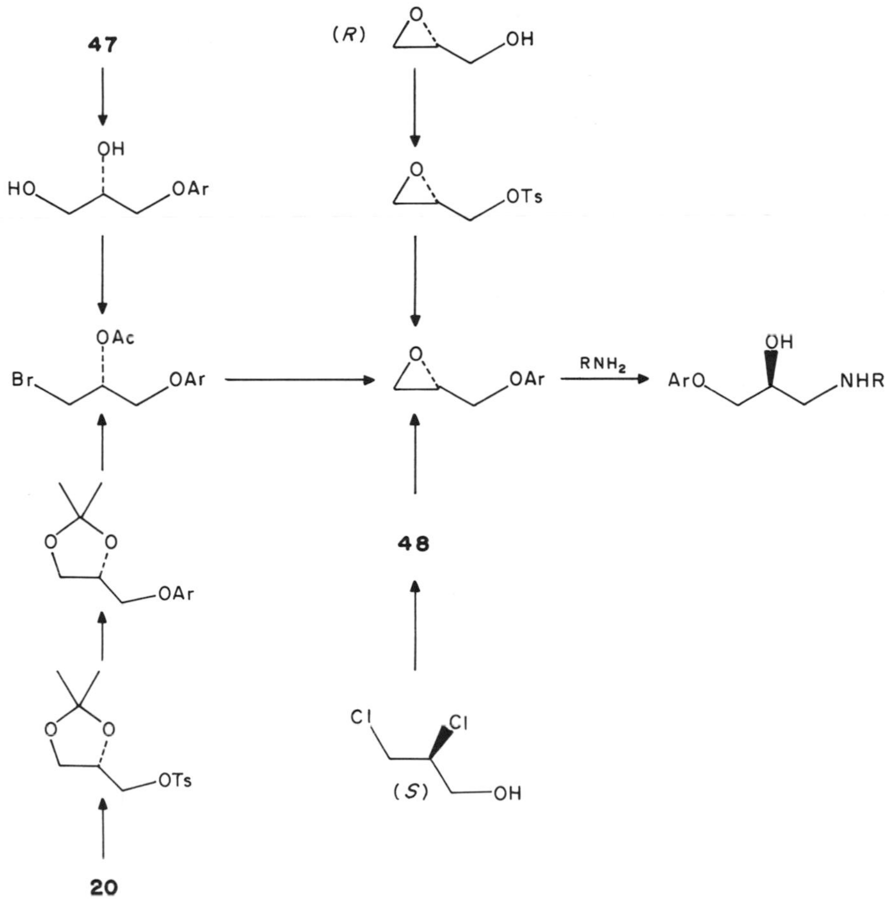

Scheme 20

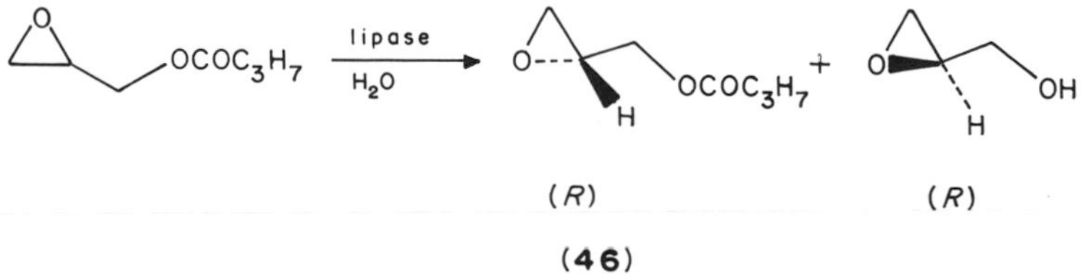

Scheme 21. Enzymic resolution of glycidyl butyrate

by-product (*R*)-glycidol was also produced with a comparably high *ee* (96–98%), both products could in principle be converted into the desired (*S*)-β-blockers (Scheme 22).[96]

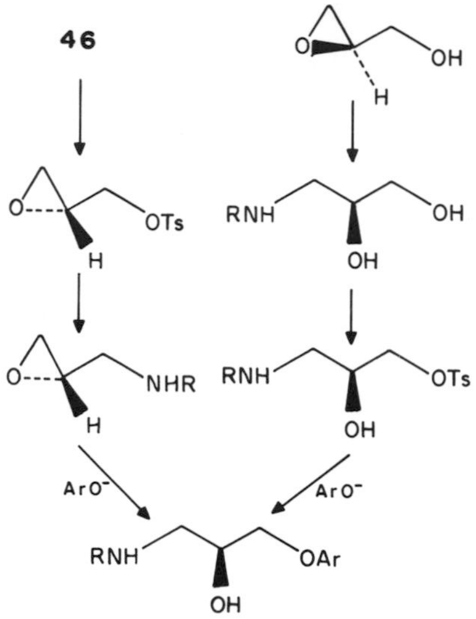

Scheme 22

(R)-*Isopropylideneglycerol* (**20**). This synthon is difficult and expensive to obtain by direct synthesis (Schemes 6 and 23).[38,97] International Bio-Synthetics developed a route in which the resolution is effected by a selective microbial oxidation of the (*S*)-enantiomer into (*R*)-isopropylideneglyceric acid (**47**) (Scheme 24).[98] The *ee* of **20** exceeds 98% and that of the product **47** is greater than 90%.[99]

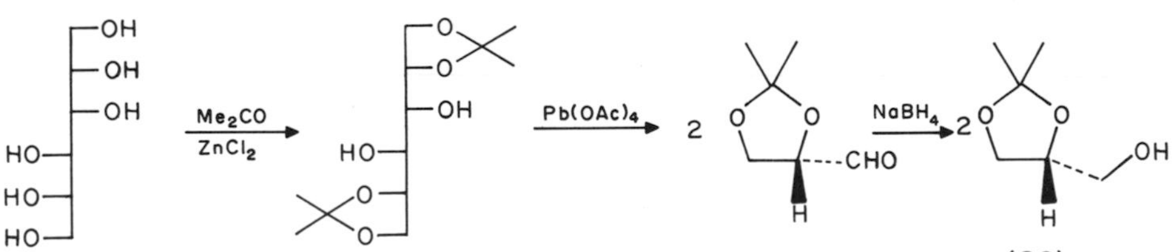

Scheme 23. (*R*)-Isopropylideneglycerol from unnatural L-mannitol

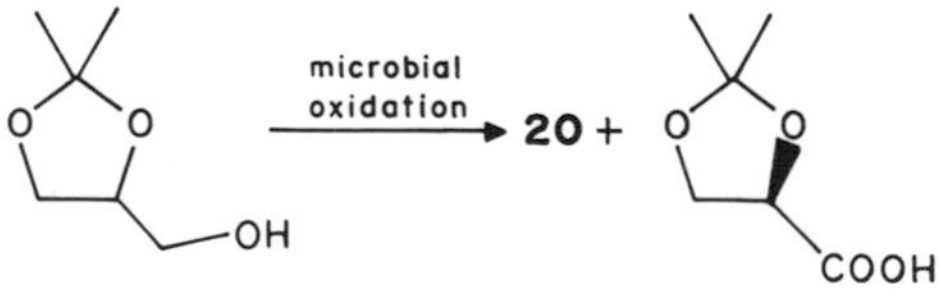

Scheme 24. Resolution of IPG

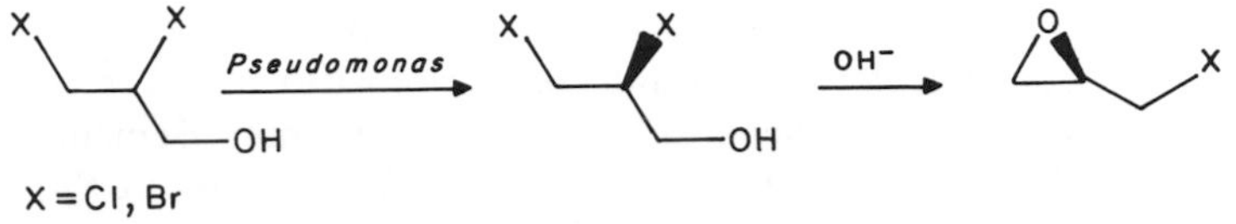

X = Cl, Br

Pseudomonas

OH^-

Scheme 25. Generation of (*S*)-2,3-dihalopropan-1-ols through selective microbial degradation of (*R*)-enantiomers

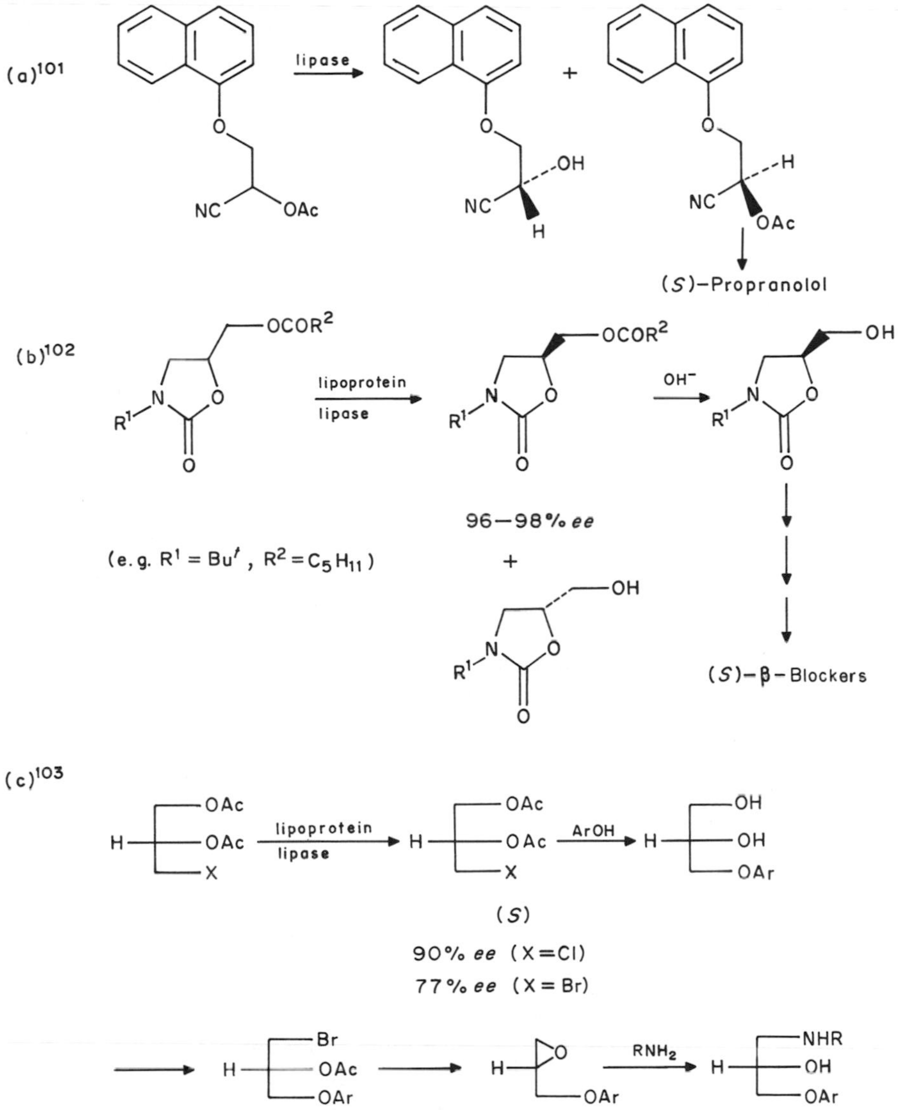

(a)[101]

lipase

(*S*)-Propranolol

(b)[102]

lipoprotein lipase

OH^-

(e.g. $R^1 = Bu^t$, $R^2 = C_5H_{11}$)

96–98% *ee*

(*S*)-β-Blockers

(c)[103]

lipoprotein lipase

ArOH

(*S*)

90% *ee* (X=Cl)

77% *ee* (X=Br)

RNH_2

Scheme 26

(R)-*Epichlorohydrin* (***48***). Processes have been patented for the production of (*R*)-enantiomers of both epichlorohydrin and epibromohydrin via the corresponding (*S*)-2,3-dihalopropan-1-ols (Scheme 25).[100] Both enantiomers of epichlorohydrin are available from Osaka Soda. Other, lipase-based, approaches to (*S*)-β-blocker intermediates are shown in Scheme 26.

Intermediates for ACE inhibitors have also been obtained using enzyme-mediated kinetic resolution (Scheme 27).[104,105] In this case yields are more than 50%, which indicates that the substrate enantiomers are continuously equilibrated via the enol form, then the microorganism removes one by reduction.

Industrial production of α-amino acids is another area where enzymic resolutions have commanded attention and several large-scale processes are being operated.

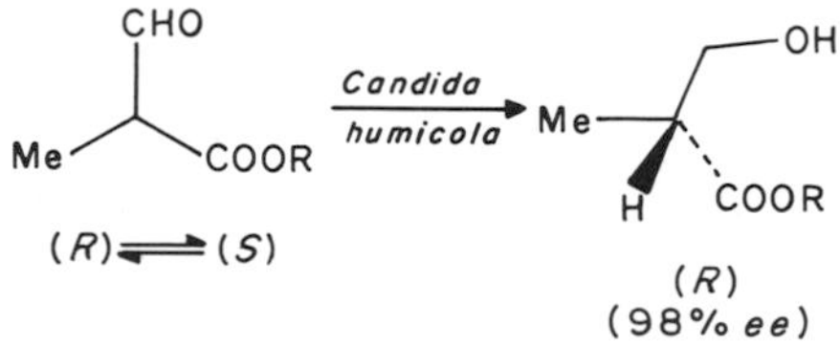

Scheme 27

As described in Chapter 19, Tanabe have operated a process using an amino acylase immobilized on DEAE-Sephadex.[106] This was one of the earliest examples using an immobilized enzyme in commercial production and has been used, *inter alia*, for the production of L-methionine, L-valine and L-phenylalanine.

Another example comes from DSM, who developed an amino acid process based on resolution of amino acids using an aminopeptidase; this is detailed in Chapter 8. The process accepts a range of substrates and produces either the D- or L-acid, as required. These are valuable attributes for a commercial route. More recently the same company has been developing a much simpler amidase-based process with *in situ* racemization of the unwanted enantiomer.[107]

D-4-Hydroxyphenylglycine is made via enzymic kinetic resolution with hydantoinases by both SNAM Progetti[108] and Kanegafuchi[109] companies. Since the hydantoins racemize readily under the conditions of the enzymic hydrolysis, this allows virtually quantitative conversion of the racemic hydantoin into the D-amino acid. The process is described further in Chapter 20.

The Toray process for L-lysine (Scheme 28) is a further case of an industrial enzymic kinetic resolution process in which total conversion of the substrate may be achieved. DL-α-Amino-ε-caprolactam (**49**) is hydrolysed with simultaneous racemization of the unwanted D-isomer (**30**). The latter transformation may also be effected enzymically.[110]

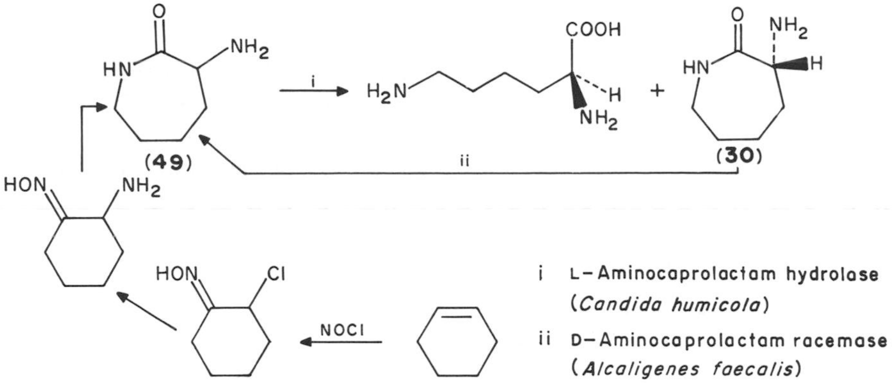

i L-Aminocaprolactam hydrolase (*Candida humicola*)

ii D-Aminocaprolactam racemase (*Alcaligenes faecalis*)

Scheme 28. Toray process for L-lysine

Synthesis of the important optically active 2-halopropionic acids (cf. Section 1.2.1.2) has also seen the industrial development of enzymic resolutions. Stauffer Chemical investigated the lipase-catalysed route shown in Scheme 29[111] whilst ICI have commercialized a dehalogenase-based process on a very large scale (Scheme 30)[112] (see Chapter 5). In the Stauffer procedure it is essential that excess, insoluble ester is present. In aqueous solution non-stereoselective hydrolysis occurs; this process requirement is further aided by the presence of added organic solvent.

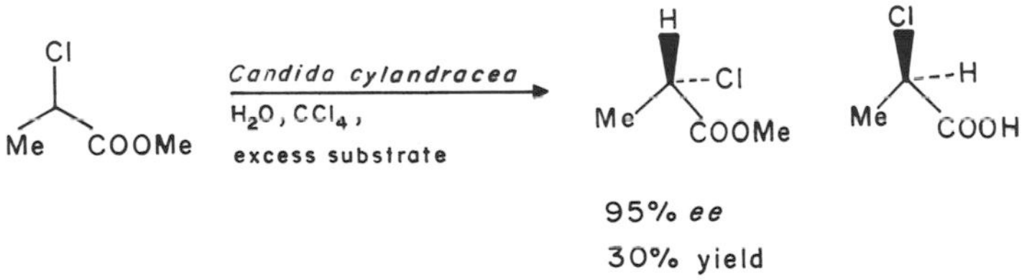

Candida cylandracea, H_2O, CCl_4, excess substrate

95% *ee*, 30% yield

Scheme 29

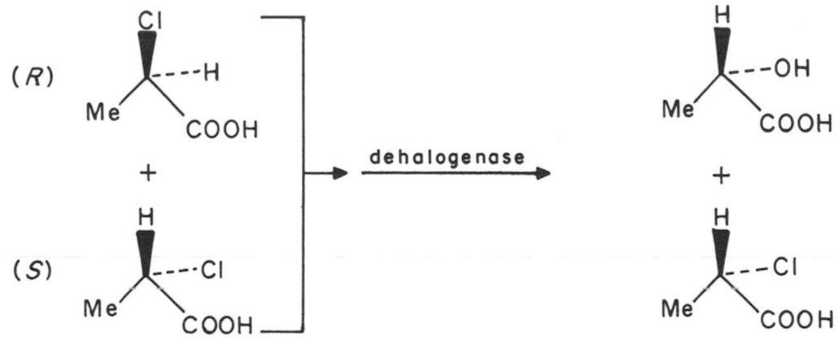

Scheme 30. ICI dehalogenase process for (*S*)-2-chloropropionic acid

Other enzymic resolution approaches to 2-chloropropionic acid include esterification,[113] transesterification[114] and aminolysis.[115]

Finally, an excellent example in which enzymic resolution is combined with chemical inversion of the unwanted enantiomer has been applied by workers at Sumitomo to the synthesis of, *inter alia*, pyrethroid alcohols (Scheme 31).[116] The crude mixture of (*R*)-alcohol and (*S*)-acetate is esterified either with fuming nitric acid or methanesulphonyl chloride (MsCl) and the crude mixture is then hydrolysed, with inversion of the nitrate or mesylate, to give a product which is predominantly the required (*S*)-alcohol.

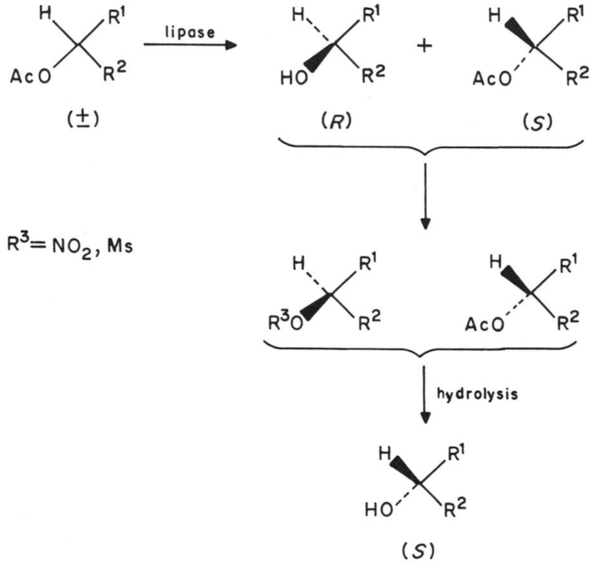

$R^3 = NO_2$, Ms

Scheme 31

1.2.3 FROM PROCHIRAL COMPOUNDS: ASYMMETRIC SYNTHESIS

Acquisition of enantiomerically pure materials through transformation of prochiral substrates (Scheme 32) necessitates the intervention of an optically

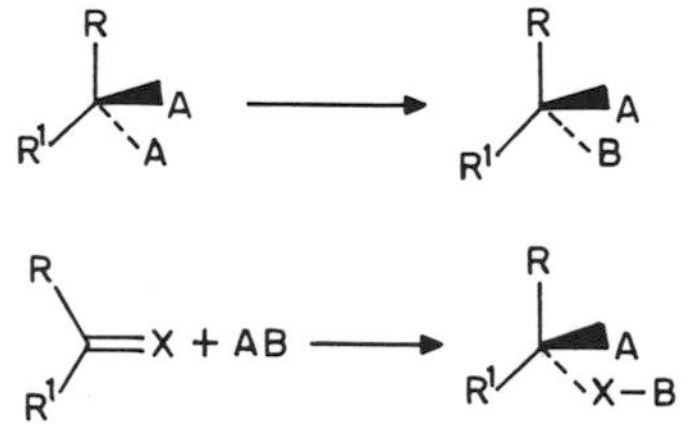

Scheme 32

active agent, used either stoichiometrically or catalytically, which expresses its chirality. We shall be concerned here with the latter case, which is particularly challenging when non-enzymic catalysts are used.

1.2.3.1 Non-enzymic Methods[117]

Asymmetric hydrogenation has its origins in the soluble Wilkinson catalyst modified with chiral phosphine ligands; this led to the Monsanto process for levodopa (**50**) (Scheme 33), commercialized in the early 1970s. This was a landmark in industrial asymmetric synthesis[118] and a spur for a huge amount of other industry-based research. A similar levodopa process has been developed by VEB Isis-Chemie[119] but uses a catalyst in which **51** is the chiral ligand. Another commercial asymmetric hydrogenation is the Enichem synthesis of (*S*)-phenylalanine.[117b]

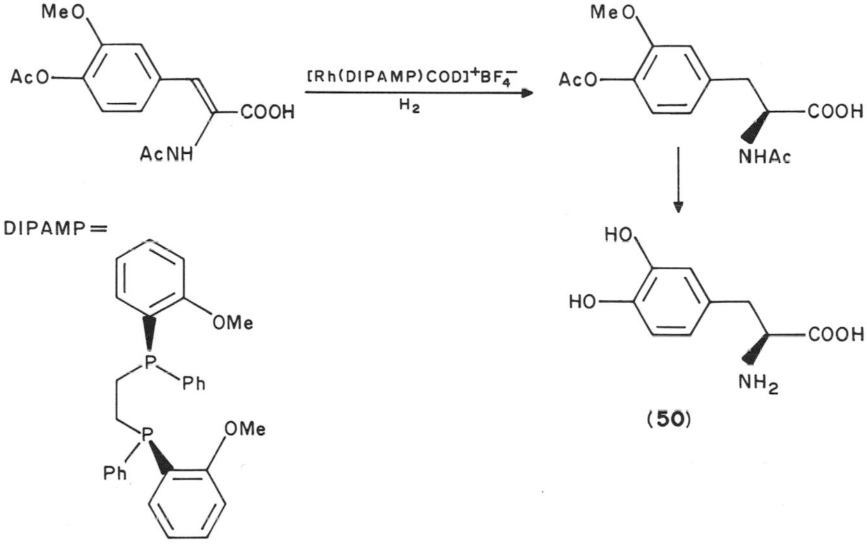

Scheme 33. Monsanto process for L-DOPA

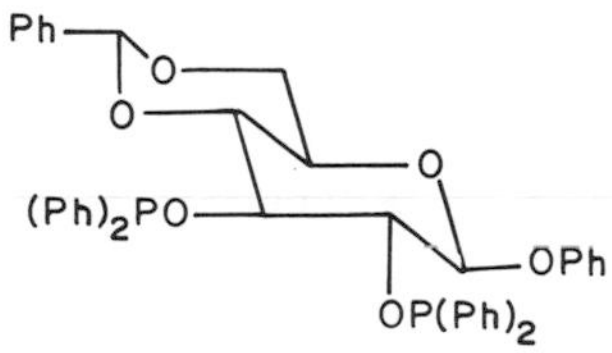

(**51**)

The important non-steroidal anti-inflammatory group of drugs has stimulated almost every conceivable approach to their synthesis as single enantiomers (Section 1.3.1.4).[120] One of particular interest under this heading is shown in Scheme 34.[121]

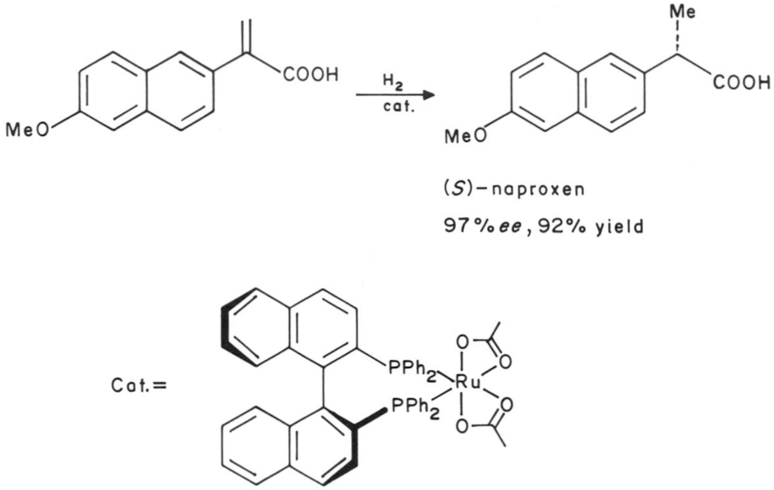

MeO COOH $\xrightarrow[\text{cat.}]{H_2}$ MeO COOH Me

(*S*)-naproxen

97% *ee*, 92% yield

Cat. = PPh_2 PPh_2 Ru O O O O

Scheme 34

Although asymmetric reductions of ketones are probably still better served by enzymic methods with respect to the diversity of substrates accepted (see Table 3), there have been considerable advances in recent years by Noyori and co-workers[122] through the use of Ru–BINAP systems (see Chapter 17). Tartaric acid-modified Raney nickel was one of the earliest chemical catalysts, for the reduction of *β*-keto esters and *β*-diketones, and has been applied on a commercial scale.[9b] It was used by Hoffmann-La Roche for the transformation shown in Scheme 35;[9b] the product (**52**) was made on a 6–100 kg scale as an intermediate for a pancreatic lipase inhibitor.

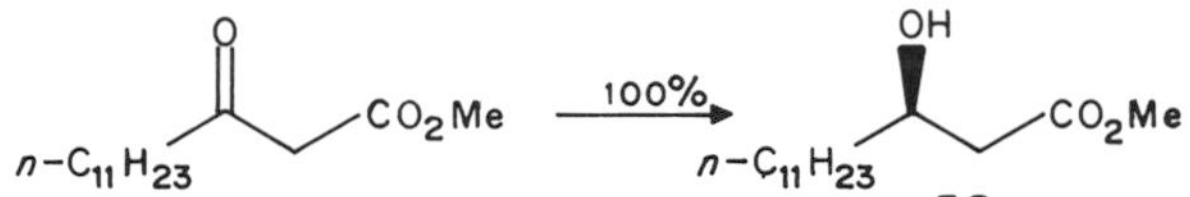

$n\text{-}C_{11}H_{23}$ O CO_2Me $\xrightarrow{100\%}$ $n\text{-}C_{11}H_{23}$ OH CO_2Me **52**

90–92% *ee*

>99% *ee* after recryst.

Scheme 35

Enantioselective hydrogenation has been investigated by Blaser and co-workers[123] of Ciba-Geigy for the synthesis of the benazepril intermediate

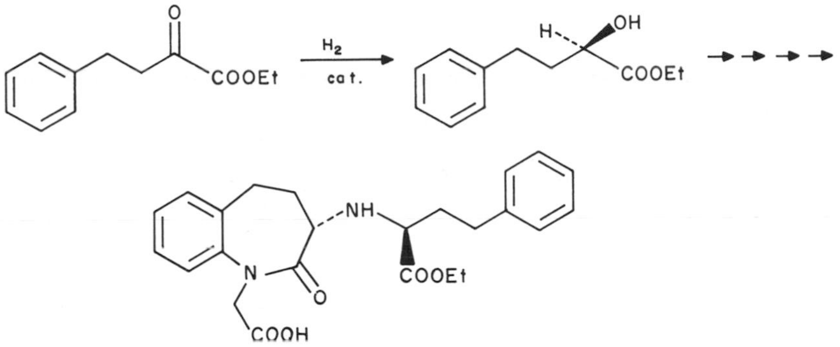

Scheme 36

ethyl (*R*)-2-hydroxy-4-phenylbutyrate (Scheme 36). The reaction can be achieved with 96% *ee* using $[Rh(NBD)Cl]_2$–(2*S*, 3*S*)-NORPHOS as catalyst or *ca* 80% *ee* using the heterogeneous catalyst Pt–Al_2O_3–10,11-dihydrocinchonidine,[124] the latter reaction having been operated on a 10–200 kg scale.

Syntheses of epinephrine (Scheme 37)[125] and of pantolactone (Scheme 38)[126] are further examples of asymmetric carbonyl group reductions applied to

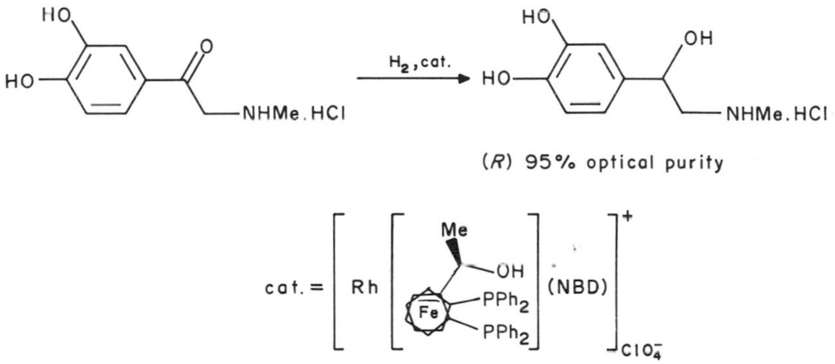

(*R*) 95% optical purity

Scheme 37

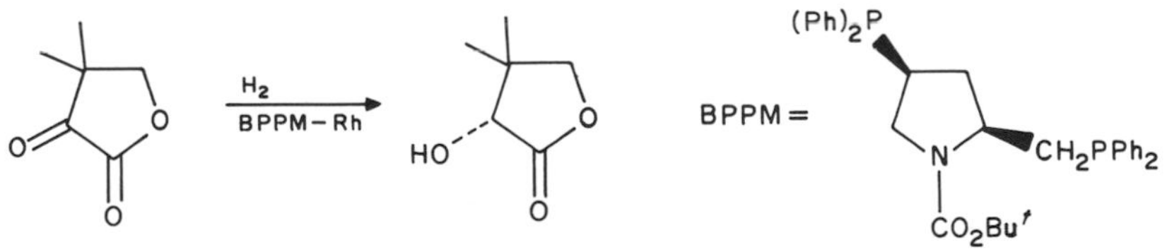

(*R*)–Pantolactone
100% yield, 86% *ee*
70% yield, 100% *ee* after recryst.

Scheme 38

biologically important compounds. In these cases the activities of the catalysts were too low to be the basis for commercial processes.

Asymmetric hydroformylation (Scheme 39) was demonstrated in 1972, but has not been as thoroughly studied as hydrogenation or exploited industrially; catalysts are typically based on platinum or rhodium with chiral phosphine ligands. Enantiomeric excesses are often mediocre and racemization of the optically active aldehyde can be a problem. The potential of this reaction for the synthesis of pharmaceuticals has recently been reviewed.[127]

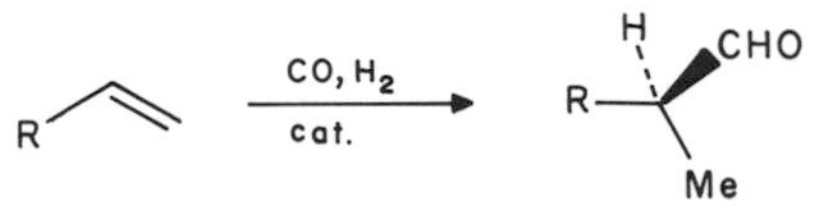

R CO, H_2 cat. R H CHO Me

Scheme 39

One of the major applications of asymmetric synthesis in industry is the Takasago process for (−)-menthol,[128] which is based on a highly efficient *asymmetric isomersization* of diethylgeranylamine (**53**) to citronellal diethylenamine (**54**) (Scheme 40). The development of this process is discussed in Chapter 16.

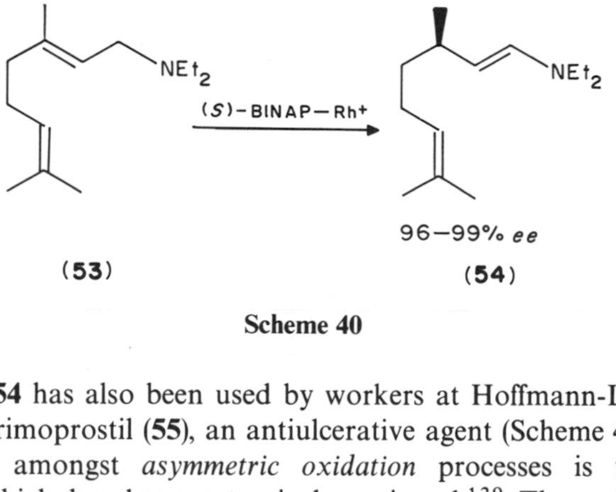

NEt_2 (*S*)−BINAP−Rh^+ NEt_2

96−99% *ee*

(**53**) (**54**)

Scheme 40

Compound **54** has also been used by workers at Hoffmann-La Roche as a precursor for trimoprostil (**55**), an antiulcerative agent (Scheme 41).[117d]

Pre-eminent amongst *asymmetric oxidation* processes is the Sharpless epoxidation, which has been extensively reviewed.[129] The catalytic version of the process[91,130] has been scaled up by ARCO for production of both enantiomers of glycidol (Scheme 42)[131] on at least a 600 g mol scale.[132] The process has also been operated on a multi-kilogram scale by Upjohn[133] to give the epoxide **56** (Scheme 43). The catalytic version of the process has an appreciably better volume productivity than the stoichiometric version (10% vs < 2%).

Jacobsen and co-workers[134] have disclosed details of a catalytic asymmetric epoxidation using manganese complexes of chiral Schiff bases which permit the

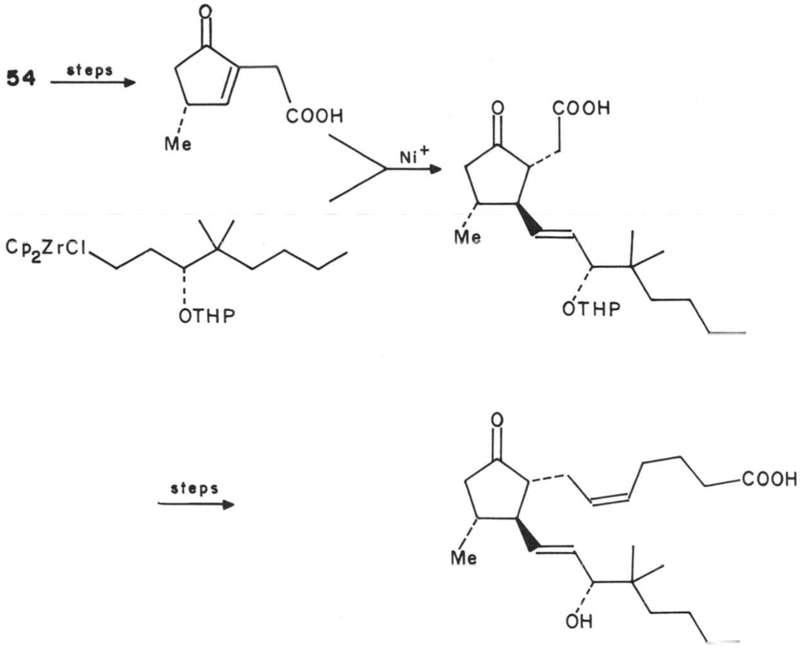

54 steps

COOH

Me

Cp_2ZrCl

OTHP

Ni^+

COOH

Me

OTHP

steps

COOH

Me

OH

(55)

Scheme 41

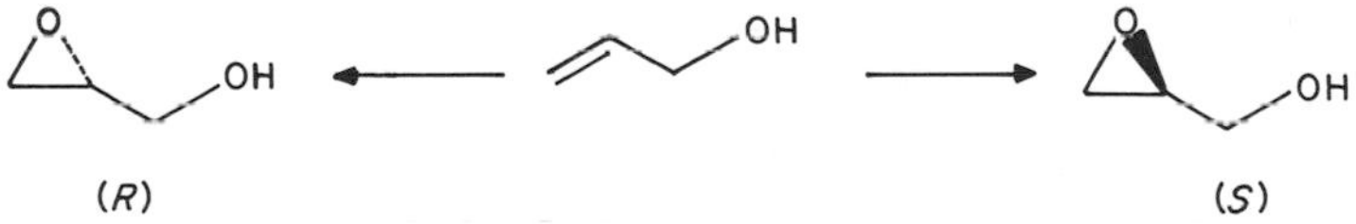

OH

OH

(R)

OH

(S)

Scheme 42

OH

CH_2Cl_2, −15 °C

$Ti(OPr^i)_4$, (+)−DET, TBHP

OH

DET = diethyl tartrate

TBHP = *tert* − butyl hydroperoxide

> 98% *ee* (after recryst.)

(56)

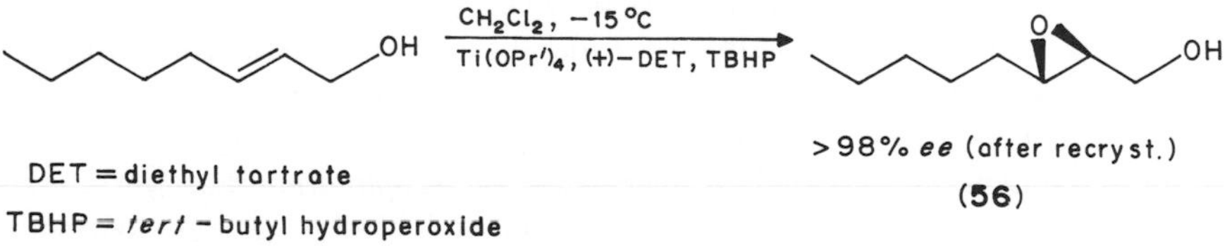

Scheme 43

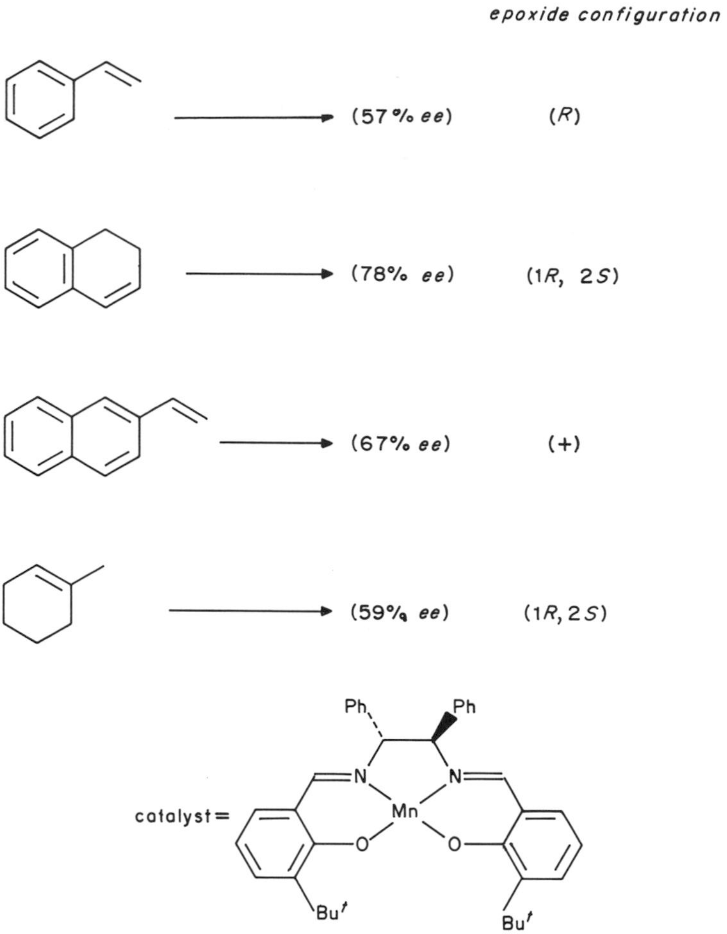

		epoxide configuration
styrene →	(57% ee)	(R)
dihydronaphthalene →	(78% ee)	(1R, 2S)
2-vinylnaphthalene →	(67% ee)	(+)
1-methylcyclohexene →	(59% ee)	(1R, 2S)

catalyst = [Mn salen complex: Ph, Ph, N, N, Mn, O, O, Bu^t, Bu^t]

Scheme 44

epoxidation of a range of alkyl- and aryl-substituted olefins (examples are given in Scheme 44). Preliminary studies indicate that NaOCl is an effective oxidant. Although the reported *ee*s are still only modest in some cases, these results break significant new ground and must be seen as having industrial potential. Significantly, it has proved possible to 'tune' these catalysts electronically in a very straightforward manner, allowing *ee*s to be varied over wide ranges. Electron-donating groups lead to higher enantioselectivities.[135]

Although much investigated, *asymmetric cyclopropanation* has yet to be applied on a large scale to potentially the most lucrative targets, synthetic pyrethroids. One process that has been scaled up is for (+)-ethyl (1*S*)-2,2-dimethylcyclopropanecarboxylate (**57**) (Scheme 45).[136] Compound **57** is obtained

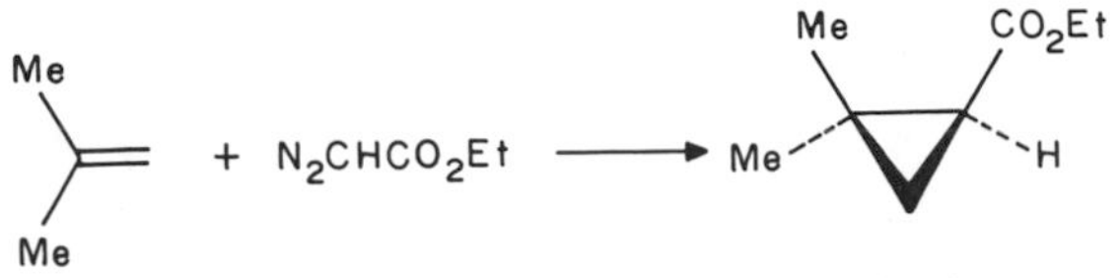

(57)

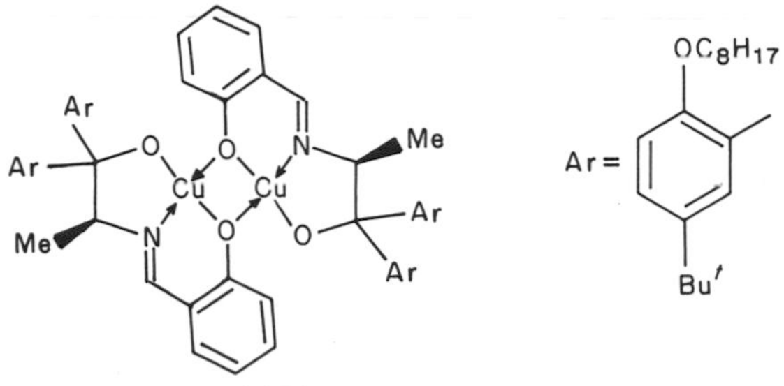

(58)

Scheme 45

by reaction of isobutene with ethyl diazoacetate using the catalyst **58**; **57** is a key intermediate for cilastatin (**59**), used in conjunction with imipenem (**60**) to suppress hydrolysis of the latter by renal enzymes.

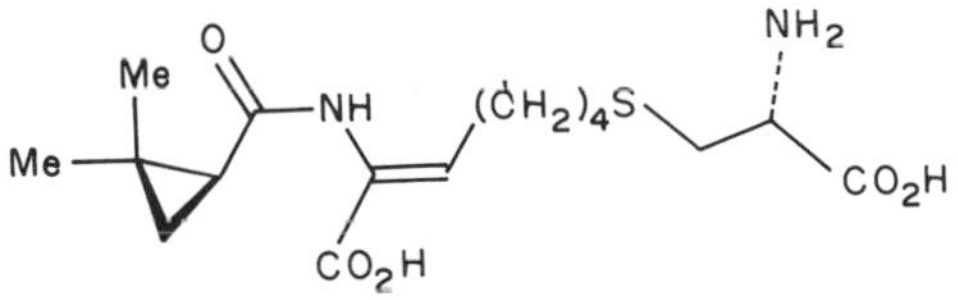

(59)

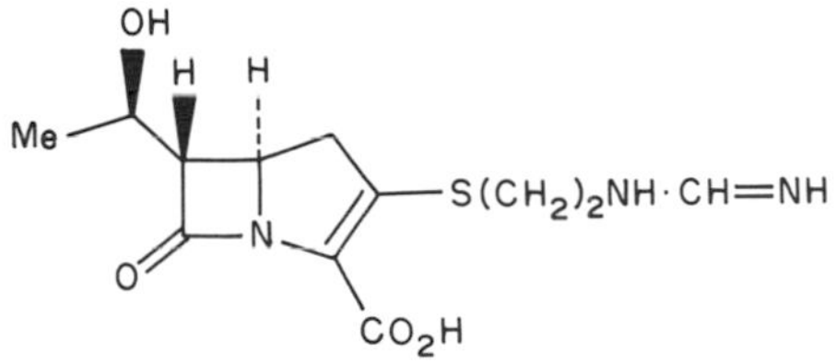

(60)

Asymmetric phase transfer catalysis had proved a difficult area with false dawns[137] before the signal achievement by the Merck Group of Dolling *et al.*[138] who obtained the alkylated indanone **61** (Scheme 46) in 92% *ee* and 95% yield. Compound **61**, an intermediate in a proposed commercial synthesis of (*S*)-(+)-indacrinone (**62**) (MK-0197), was only obtained with high *ee* after painstaking

Scheme 46

Scheme 47

and methodical development work[139] which followed an initial observation of only 6% *ee*. More recently, O'Donnell *et al.*,[140] also using a quaternized cinchona alkaloid as the catalyst, alkylated the Schiffs base ester **63** (Scheme 47) to achieve an α-amino acid synthesis. Although by most standards the *ee* of the product of the alkylation step would be considered low for a practical synthesis (66%), recrystallization leads in one step to an optically pure product (in the filtrate); to date this process has only been run on a multi-gram scale.

Using *N*-(*p*-trifluoromethylbenzyl)cinchonidinium bromide as the catalyst, Nerinckx and Vandewalle[141] have applied phase-transfer catalysis to the synthesis of the potent analgesic (−)-Wy-16 225 (**64**) and to (+)-podocarp-8(14)-en-13-one (**65**) (Scheme 48); the latter synthesis employed an asymmetric Robinson annulation.

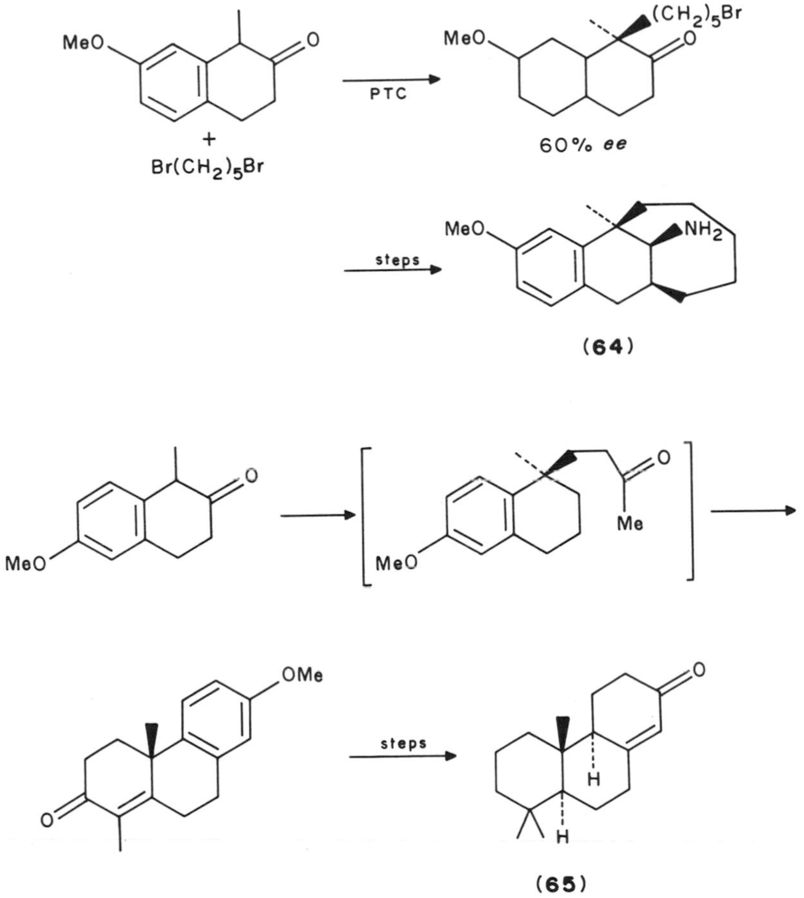

Scheme 48

Cycloaddition is another of the select and disparate asymmetric syntheses which have been commercialized[142] (see Chapter 18) and highlights again the value of cinchona alkaloids.[27] The cycloaddition of ketene and chloral at low temperature in the presence of 1 mol% of the alkaloid proceeds quantitatively and in high *ee* to the lactone **66**. Hydrolysis of lactone **66** affords (*S*)-malic acid; the unnatural (*R*)-acid results if quinine is the catalyst. Use of 1, 1, 1-trichloroacetone in place of chloral leads to citramalic acid (**67**),[143] a promising synthon for natural products owing to its isoprenoid structure, but not hitherto easily made in optically active form or extensively exploited. The reaction has been extended to a range of other aldehydes and ketones containing 1, 1-dichloro substituents.[144] Products from this chemistry are being commercialized by Lonza and are detailed in Chapter 18.

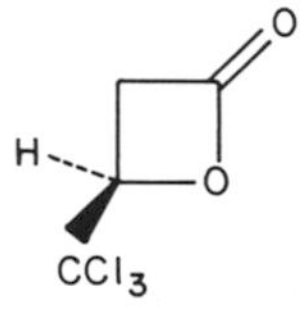

(**66**)

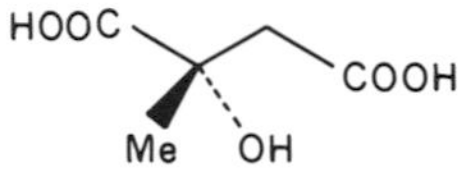

(*S*) (when catalyst is quinidine)

(**67**)

In the quest for a more efficient route to benazepril, Blaser *et al.*[145] achieved catalytic *enantioselective hydrodehalogenation* of α, α-dichlorobenzazepin-2-one with up to 50% *ee* (Scheme 49); this was effected with a 5% Pd/BaSO$_4$ catalyst modified with cinchonine. Disappointingly, attempts to extend this very useful

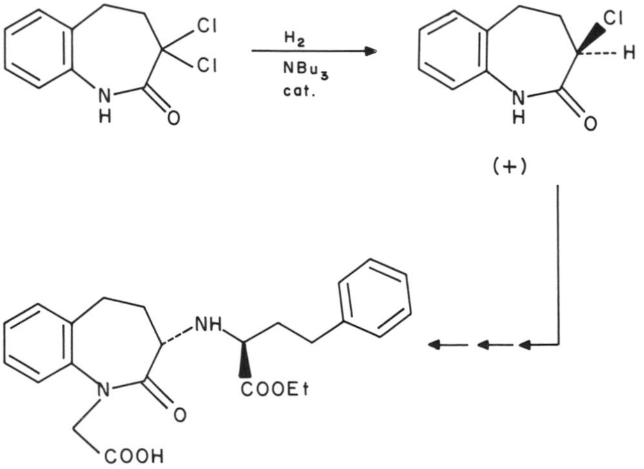

Benazepril

Scheme 49

reaction to other α,α-dihalogen-substituted acid derivatives were unsuccessful. Prior to this work the only other successful approach had been enantioselective electrochemical dehalogenation on mercury electrodes modified with alkaloids.[146]

1.2.3.2 Enzymic Methods

A rising proportion of syntheses of homochiral materials include enzymic steps (Figure 3), showing an increasing recognition of the contribution that enzymes can make and a willingness by the chemical community to employ them.

This section is concerned with use of microorganisms or isolated enzymes to catalyse single transformations, rather than multi-step sequences from basic feedstocks such as carbohydrates. Those biological systems are included which may be complex in themselves but which can be regarded as sources of, for example, reducing or oxidizing power.

Oxidation

Chiral epoxides. Microbial epoxidation dates from 1963,[147] and it was in 1973 that the reaction was found to be stereoselective.[148] Much of the pioneering work was carried out by Nippon Mining, as will be described in Chapter 7. By 1992 twelve 1,2-epoxyalkanes, C_7–C_{18}, were commercially available from this technology, as well as several polyfunctional epoxides, such as aryl glycidyl ethers with the (*S*)-configuration required for the pharmacologically active β-blockers (cf. Scheme 50).

Shell and Gist-Brocades have also reported a route to single enantiomer β-blockers using stereoselective microbial epoxidation (Scheme 50).[149,150] These epoxides are produced with a stereochemistry which is consistent with the observed *si* side stereoselectivity for linear olefins.[151] Such routes, at the time,

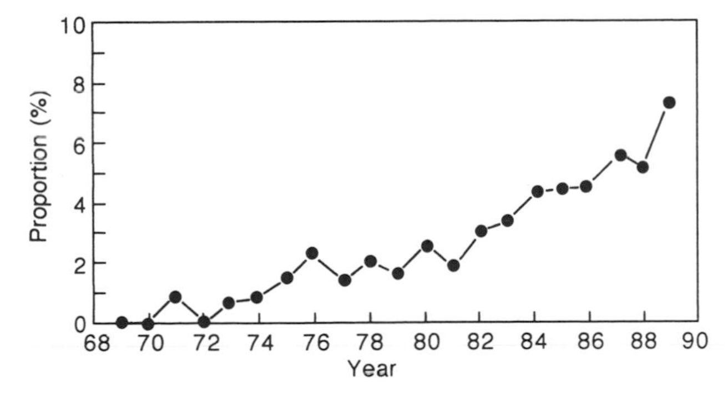

Figure 3. Proportion of papers on asymmetric synthesis cited by *Chemical Abstracts* which employ enzymic methods

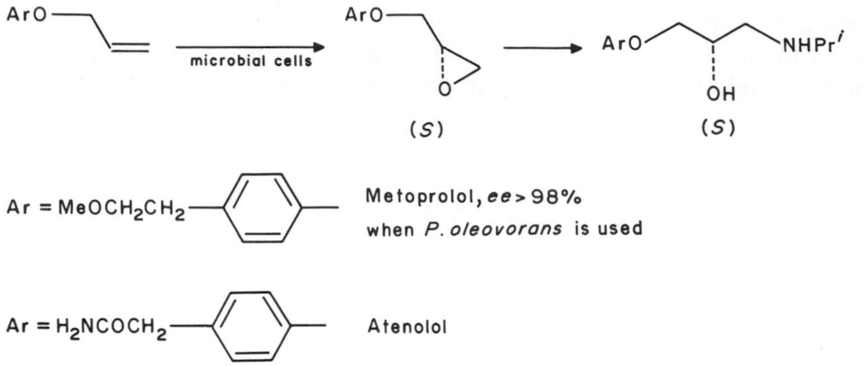

Scheme 50

were obviously attractive over earlier lengthy chemical synthesis, involving pre-synthesis of chiral synthons such as (*R*)-isopropylideneglycerol.

Dihydrocatechols. By the use of a genetically manipulated microorganism, ICI has developed the process depicted in Scheme 51 which proceeds in high yield and with total selectivity. Although originally developed with benzene as substrate to produce polymerization monomers,[152] the process may be applied to a range of substituted benzenes to give the corresponding (*cis*, 1*S*)-1,2-dihydrocatechols. The process was developed in the face of challenging technical hurdles. It employs a catalyst which can be used in standard, non-sterile chemical plant and oxygen is reacted with flammable volatile cytotoxic water-immiscible substrates. These are problems which will be encountered in many of the potentially most useful biotransformations. The technology, which has been proved at the tonnage scale,[153] is described in Chapter 6.

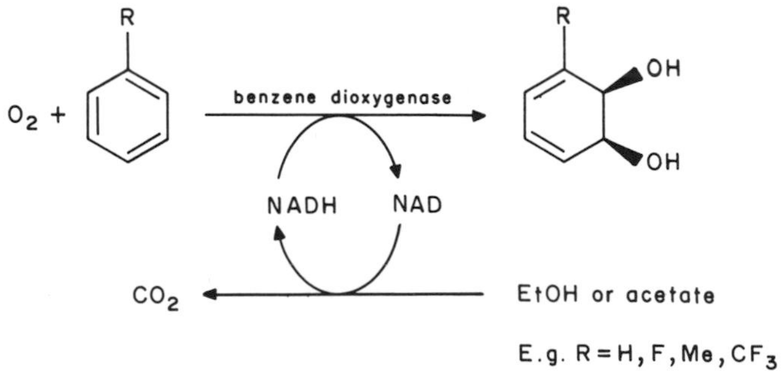

Scheme 51

Stereoselective side-chain oxidation. Examples are shown in Scheme 52. Methods have been patented[154] for the production of (*R*)-(−)-3-hydroxyisobutyric acid required for the Squibb cardiovascular drug captopril (**9**). Shell–Gist-Brocades have investigated similar oxidation processes to (*S*)-naproxen and the active (*R*)-isomers of the aryloxypropionate herbicides[155,156] [(b) and (c) in Scheme 52].

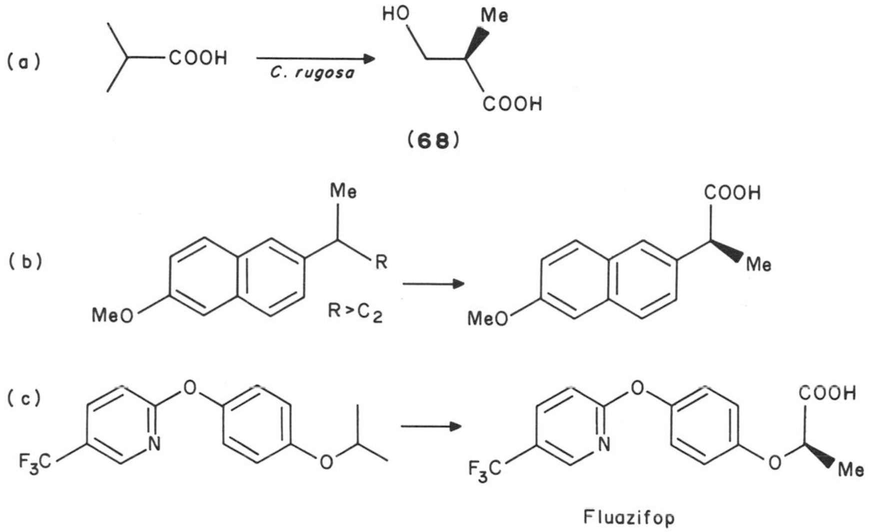

(a) COOH → *C. rugosa* → HO Me COOH (**68**)

(b) MeO, Me, R, R>C$_2$ → MeO, COOH, Me

(c) F_3C, N, O, O → F_3C, N, O, O, COOH, Me

Fluazifop

Scheme 52

Reduction (Scheme 53)

As far as the bench chemist is concerned, applications are dominated by the use of baker's yeast.[157] Its use is largely empirical. Attractions are cheapness, it does not demand elaborate equipment or sterile technique and it is applicable

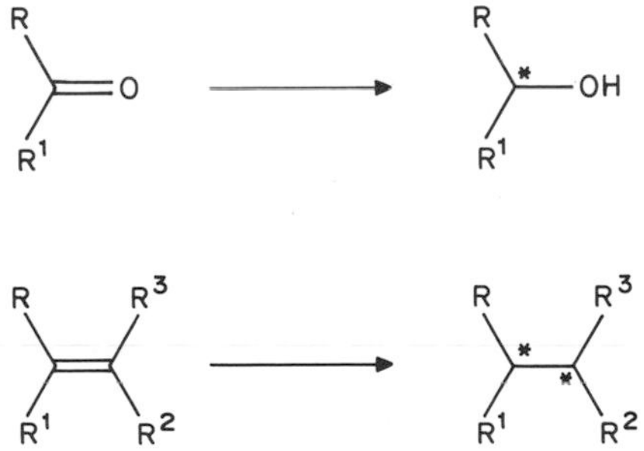

R, R^1, O → R, R^1, *, OH

R, R^3, R^1, R^2 → R, R^3, *, *, R^1, R^2

Scheme 53

Table 3. Structural diversity of substrates accepted in yeast reductions

Substrate	Product	*ee*(%)	Ref.
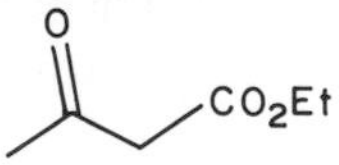 O, CO_2Et	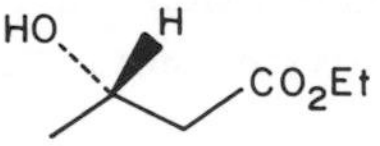 HO, H, CO_2Et	97	158
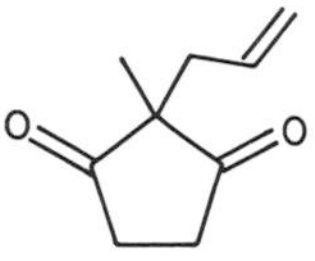 O, O	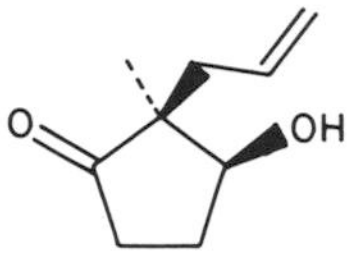 O, OH	>98	159
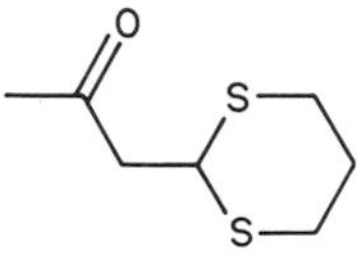 O, S, S	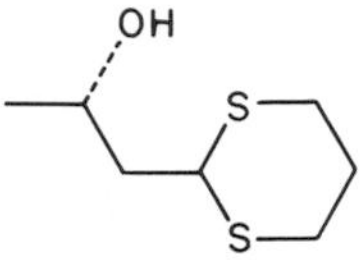 OH, S, S	99	160
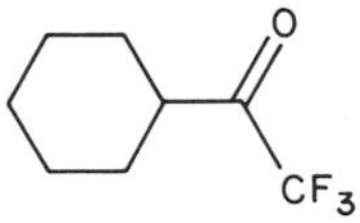 O, CF_3	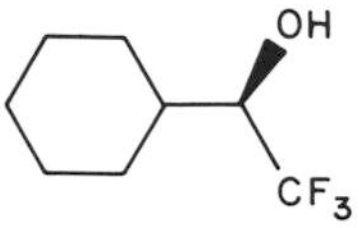 OH, CF_3	99	161
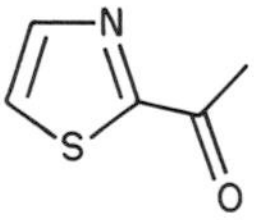 N, S, O	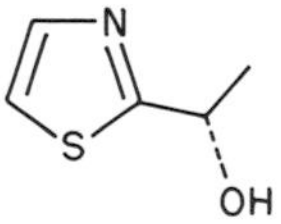 N, S, OH	95	162
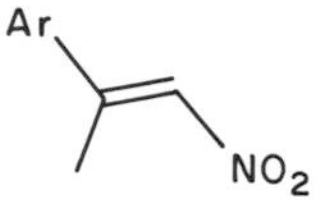 Ar, NO_2	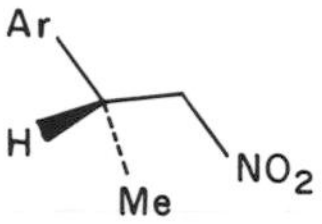 Ar, H, Me, NO_2	89–98	163
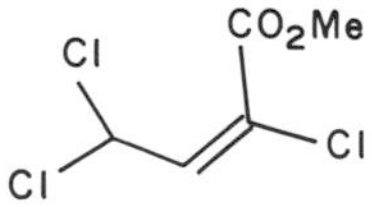 Cl, Cl, CO_2Me, Cl	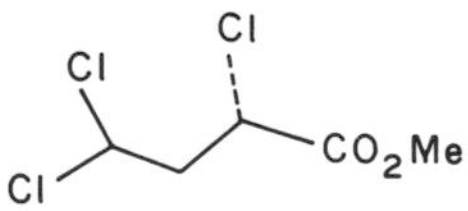 Cl, Cl, Cl, CO_2Me	84–98	164
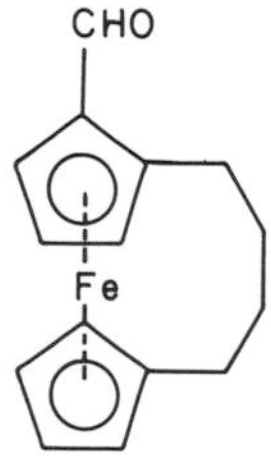 (±) CHO, Fe	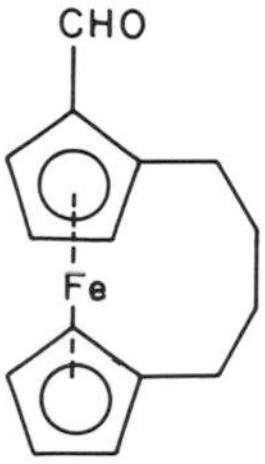 CHO, Fe (89% *ee*) + Fe, CH_2OH (80% *ee*)	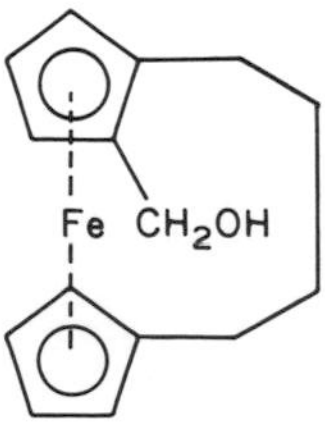	165

to an extremely diverse range of substrates (Table 3), often giving very high *ees*. Little attention has been paid to problems relevant to very large-scale operation. Some of the more obvious disadvantages are the need for high biomass: substrate ratios and dilution which, for the commercial operator, translate into difficult isolations, serious effluent problems and poor productivities. For products required in kilograms rather than tonnes it is a viable technique.[158b] The transformation shown in the first entry of Table 3 is one which has been demonstrated on a multi-kilogram scale; this complements PHB (Section 2.1.2) and makes both enantiomeric forms of 3-hydroxybutyrates easily accessible.

Sih and co-workers[166] have reported an interesting compact synthesis of L-carnitine (**69**) (Scheme 54) using *Saccharomyces cerevisiae*. Key to the success of this route was 'manipulating' Prelog's rule[167] (Scheme 55), by making the R group sufficiently large (C_8H_{17}) so that the required (*R*)-stereochemistry was obtained. In contrast, use of R = Et leads to the (*S*)-product.

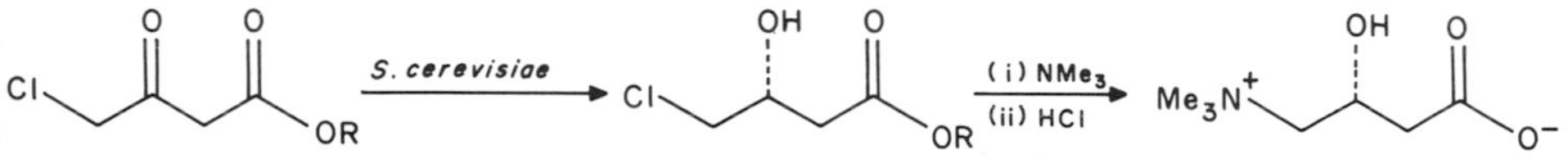

Scheme 54

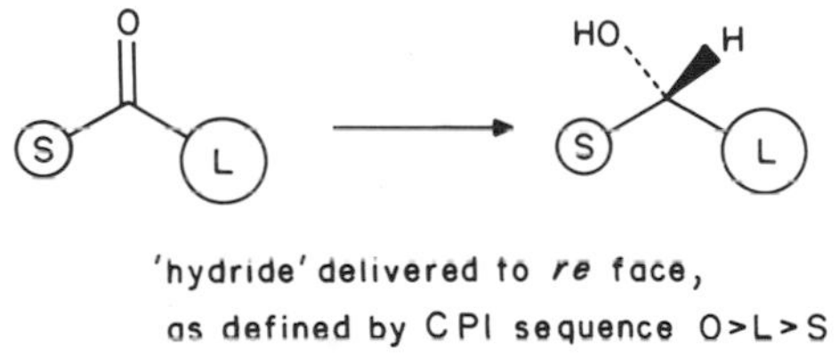

Scheme 55

Hydrogenation of activated carbon—carbon double bonds has been less extensively exploited but, again, yeasts will carry out such reductions and usually with high stereospecificity (Table 3 and ref. 168). A group at Hoffmann-La Roche have devised a technical synthesis of (4*R*,6*R*)-4-hydroxy-2,2,6-trimethyl-cyclohexanone (**70**) (Scheme 56),[169] using baker's yeast reduction of a carbon—carbon double bond; the starting material is readily available oxoisophorone. The intermediate diketone is a solid which may be strained off whilst continuing to pump-feed the liquid starting material.[169b] Compound **70** is a building block for the synthesis of optically active hydroxy carotenoids and other terpenoid compounds.

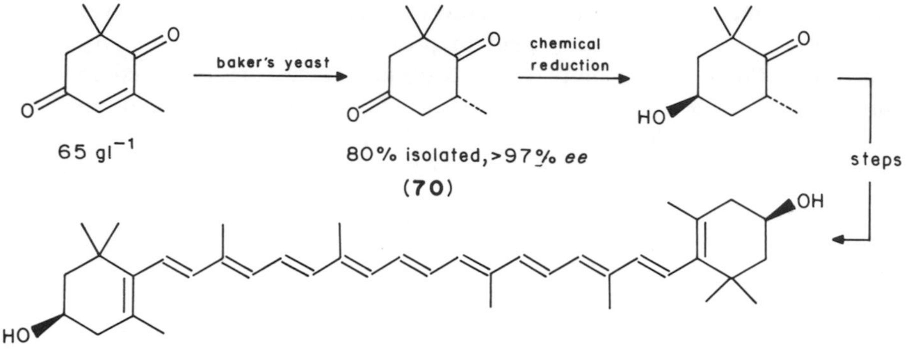

Scheme 56

Reductive Amination

Scheme 57 shows an interesting example developed by Wandrey at the Institute of Biotechnology, Nuclear Research Centre, Jülich: the process has been commercialized by Degussa (see Chapter 20).

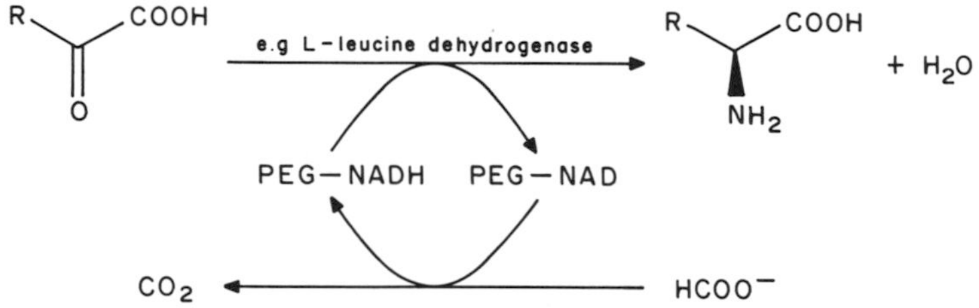

Scheme 57

Ammonia Addition

Production of L-aspartic acid (Scheme 58a) is an industrially important example using enzymic addition of ammonia to a prochiral substrate, fumaric acid. The production of pharmaceutical-grade material was about 4000 tonnes per annum in 1987.[170] It is possible to immobilize the enzyme and still retain virtually all (97%) of the aspartas activity; the half-life of such an immobilized catalyst was estimated at 3–4 years.[170] Commercialization of a corresponding process to L-phenylalanine (Scheme 58b) was achieved by Genex.[171] This is a more difficult reaction because of low conversion, poor stability of the enzyme and substrate inhibition.

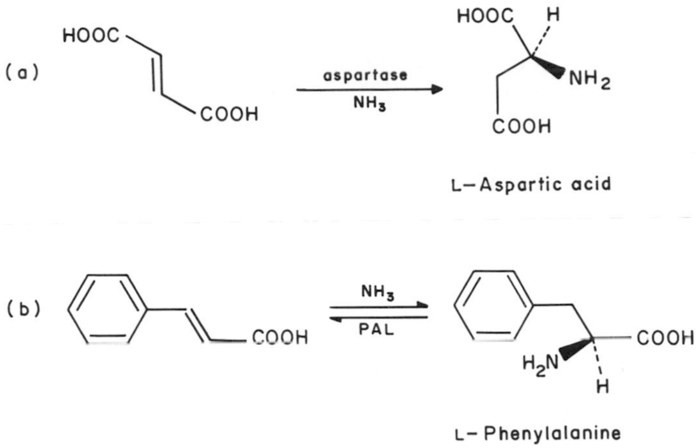

(a) HOOC–CH=CH–COOH —aspartase, NH_3→ L-Aspartic acid

(b) PhCH=CH–COOH ⇌ (NH_3 / PAL) L-Phenylalanine

PAL = L-Phenylalanine ammonia lyase

Scheme 58

Transamination

Another large-scale (600 tonnes per year) approach to L-phenylalanine, developed by Purification Engineering,[172] is transamination from aspartic acid to phenylpyruvic acid (**71**) (Scheme 59). The initial by-product, **72**, decarboxylates yielding pyruvic acid. The use of aminotransferases is discussed in detail in Chapter 9.

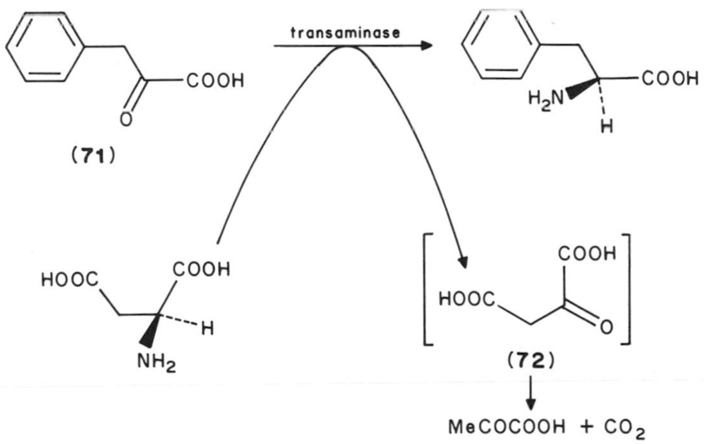

(71) —transaminase→ L-phenylalanine; aspartic acid → (72) HOOC–CH₂–CO–COOH → MeCOCOOH + CO_2

Scheme 59

Hydration

A process for the production of (*S*)-malic acid (**73**) (Scheme 60; see also Chapter 19) has been operated by Tanabe since 1974.[173] Citramalic acid (**67**) could be similarly obtained using *Clostridium tetanomorphum*.[174]

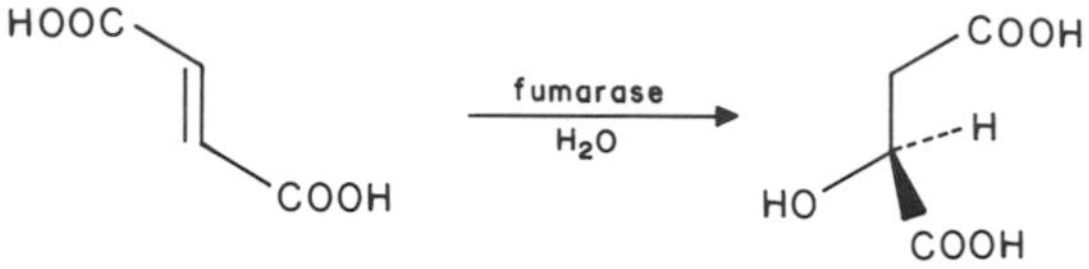

HOOC–CH=CH–COOH —(fumarase, H_2O)→ HO–CH(COOH)–CH_2–COOH (**73**)

Scheme 60

Cyanohydrin Formation

This is a reaction which has attracted much attention[175a–d] but has only recently been realized on a practical scale. It is reviewed in detail in Chapter 14.

1.3 CONCLUDING REMARKS

The task of the industrial process development chemist, who may in the early stages of definition of a route for the manufacture of an optically active material have at least 10 route options of perhaps 6–10 stages each, is clearly far from easy. Where and by what means in the overall synthetic scheme should optical activity be introduced? Chirality is an additional, constraining dimension which has to be added to the other chemical and physical considerations, not to mention issues such as registration, patents, effluents, toxicity and economics.

It is instructive to consider a number of points: (i) when and how optical activity is generated in a synthesis; (ii) pros and cons of biological versus chemical methods; (iii) historically, how approaches to a commercially important chiral product have evolved; and (iv) how the advent of a commercially important new structure stimulates the range of approaches.

1.3.1 WHERE AND HOW TO INTRODUCE OPTICAL ACTIVITY IN A MULTI-STAGE SYNTHESIS

Any attempt to draw up guidelines is fraught with exceptions. There is no best point to insert optical activity into a synthesis which is predictable; in many instances it is only after experimentation and with a knowledge of physical parameters that this may be determined. Solubility and crystallization behaviour can be crucial. For example, does a conglomerate exist (Section 1.2.2.2) or is easy enantiomer purification possible?

On a small laboratory scale, it may be advantageous to carry larger amounts of material through a synthesis for ease of manipulation. Large-scale process economics favour processing the minimum amount of material.

If a resolution is involved, the stage which allows most efficient recycle of unwanted material is likely to dictate where the optical activity is produced. If an *in situ* racemization can be coupled with the resolution, allowing a 'second-order' process, then it will be immaterial where it takes place. Similarly, when a catalytic asymmetric synthesis is employed, any point in the synthetic sequence is probably acceptable.

All methods are obviously constrained by the requirement that there should be no significant risk of racemization following the creation of optical activity. It is generally beneficial to have the opportunity for crystallization after the 'chiral' step to permit *ee* enhancement.

Different route options to a given target may, individually, only be amenable to a particular approach, and be dictated by the means by which the optical activity is created. Comparison of these options becomes part of the wider issue of overall process economics, encompassing capital costs, operating costs, materials costs, environmental costs, etc. All these parameters have to be optimized and the 'chiral' part of the equation may be subordinated to other factors when there are two or more 'chiral' options. Conversely, if there is only one credible approach to the target in optically active form this will dictate the route. However, there are usually several possible synthetic routes (see Section 1.3.4).

1.3.2 THE PLACE OF BIOLOGICAL METHODS

Increasingly, reported syntheses of optically active compounds include biological steps (Figure 3). The number and variety of large-scale enzymic processes is also increasing (see Sections 1.2.3.2 and 1.2.2.3), demonstrating that in many situations biology offers the most economic solution.

Although many large-scale applications employ enzyme reactions which do not require cofactors, cofactor recycle has been successfully addressed as, for example, in the Wandrey–Degussa membrane reactor or, more simply, through the use of intact cells in conjunction with another substrate, to drive the cofactor recycle (cf. Scheme 51).

Enzymes are catalytic proteins. This dictates both their capabilities and drawbacks. Advantageous features of biological systems are:

(i) Specificity. The corollary being perhaps limited substrate range, but it is disputable whether enzymes are any worse than chemical catalysts in this respect; witness the substrate limitations of the Sharpless epoxidation catalyst and chiral hydrogenation catalysts. Many enzymes are at least as catholic in their acceptance of substrate variation.

(ii) They generally conform to well worked out kinetics. As Whitesides and Wong have pointed out,[176] the time spent in measurement of kinetic constants is well rewarded when there is a need to optimize a process by rational control of enzyme activity.
(iii) Most enzymes operate under similar conditions of pH and temperature. This allows consecutive reactions to be run in the same vessel without intermediate product isolation.
(iv) Catalytic properties may be manipulated through the application of protein engineering[177] and the catalyst concentration can be increased through the use of cloning techniques.
(v) Safety. It is possible to avoid the use of reagents which are costly and difficult to handle on a large scale (e.g. highly toxic metal catalysts or pyrophoric reagents).
(vi) Catalyst discovery. Although it can be an empirical process, statistically the chances of success are good because of the large population of organisms from which to choose and the further ability to produce mutants.
(vii) Mild conditions of use. These can permit the isolation of products which would not survive traditional chemical processing and also allow for cheaper engineering.
(viii) Biocatalysts are able to perform the equivalent of most known organic reactions. Notable exceptions include the Cope rearrangement and Diels–Alder reaction.[178]

Negative features are:

(i) It is not always possible to access either enantiomer as easily as with chemical methods.
(ii) Low volumetric productivity is often a problem but should not be assumed. Some bio-processes operate at concentrations which match or exceed those commonly encountered in conventional chemical manufacture. A spectacular example is a process being developed by Novo–Nordisk for sugar-based surfactants.[179] Using a lipase as catalyst, a long-chain fatty acid is esterified with a sugar under solventless conditions to achieve almost 100% product formation.
(iii) The need for cofactors, but this is increasingly amenable to solution.
(iv) Chemical and thermal instability.
(v) Incompatibility with organic solvents, particularly polar solvents.
(vi) Often lacking in properties sought by the process engineer: mechanical strength and rheological properties, where application, whole cells or supported enzymes, leads to heterogeneous systems.
(vii) Reactions are often very slow in comparison with chemical processes but this can be a consequence of enzyme concentration rather than reaction rate.
(viii) Subject to inhibition.

(ix) Phase separation problems during work-up.
(x) Poor substrate: biomass ratios leading to serious effluent problems if whole cells are used (e.g. yeast reduction).

It is debatable whether an ability to operate in an aqueous medium is a plus point. It can often be easier to recycle organic solvents and disposal of aqueous biological waste streams is a problem. Many substrates which are of interest are water soluble, which makes their isolation more difficult.

Reasons for disfavouring biological processes are more concerned with operational features; they score highly as a means of effecting chiral synthesis. A balanced view must be kept of the bio-option. It does not offer a magical solution any more than do other types of asymmetric catalysis. It may introduce additional unit processes into the flowsheet which add to the cost of an apparently straightforward transformation. Also, successful operation of enzyme processes in the manufacture of fine chemicals depends crucially on an integrated skill base spanning biotechnology–classical organic chemistry–physical chemistry–process engineering. Biocatalyst manufacture will not often be a commercially attractive exercise in its own right.[178]

1.3.3 EVOLUTION OF METHODS FOR A COMMERCIAL PRODUCT: (−)-MENTHOL

World demand (1988) for (−)-menthol (**74**) is approximately 4000 tonnes.[44] Historically it has been produced from *Mentha arvensis* (Japanese mint) oil by

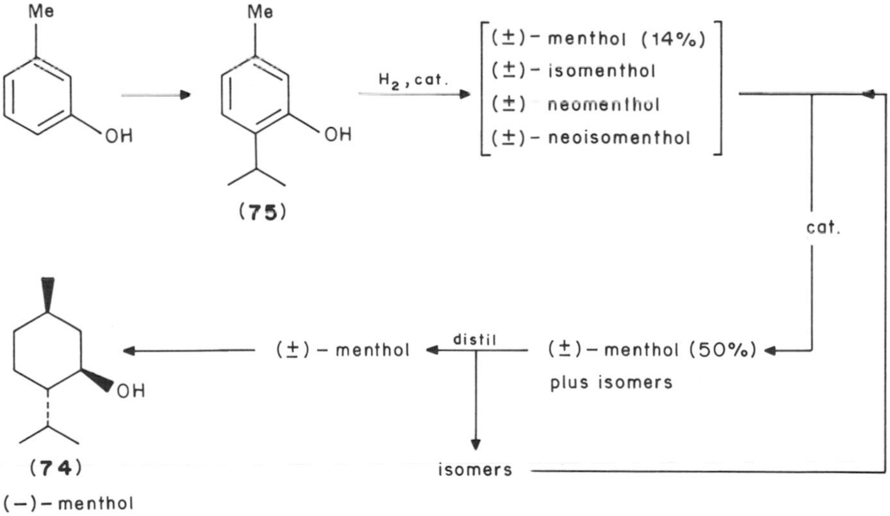

Scheme 61

cooling and separating the crystals by centrifugation. This is an inherently unpredictable source, owing to climatic variations and competing cash crops, and this has led to the development of many synthetic and semi-synthetic methods. The Haarmann and Reimer synthesis (Scheme 61)[68] exemplifies the former. 3-Cresol is converted into thymol (**75**), which is hydrogenated to an eight-isomer mixture; its (±)-menthol content is enhanced by passing it over a heterogeneous catalyst. After separation of racemic menthol by distillation, the residue is re-isomerized and recycled. The racemic menthol is then resolved by the method of preferential crystallization (Section 1.2.2.2). An esterase method has also been investigated for the resolution of (±)-menthol.[180] A technical process from phellandrene (Scheme 62)[181] illustrates a semi-synthetic approach. Another semi-synthetic process was disclosed in 1978 for the manufacture of (−)-menthol starting from (+)-3-carene and this has since been commercialized.[182] Details of the Malti-Chem process, from (+)-3-carene, are given in Scheme 63. Based upon this technology, a 120 tonne plant was built in 1982.[44] Most recent has been the emergence of the Takasago asymmetric synthesis, which is described in Chapter 16.

Production of (−)-menthol parallels nicely the development of methods for producing optically active materials. It evolved from the isolation from natural

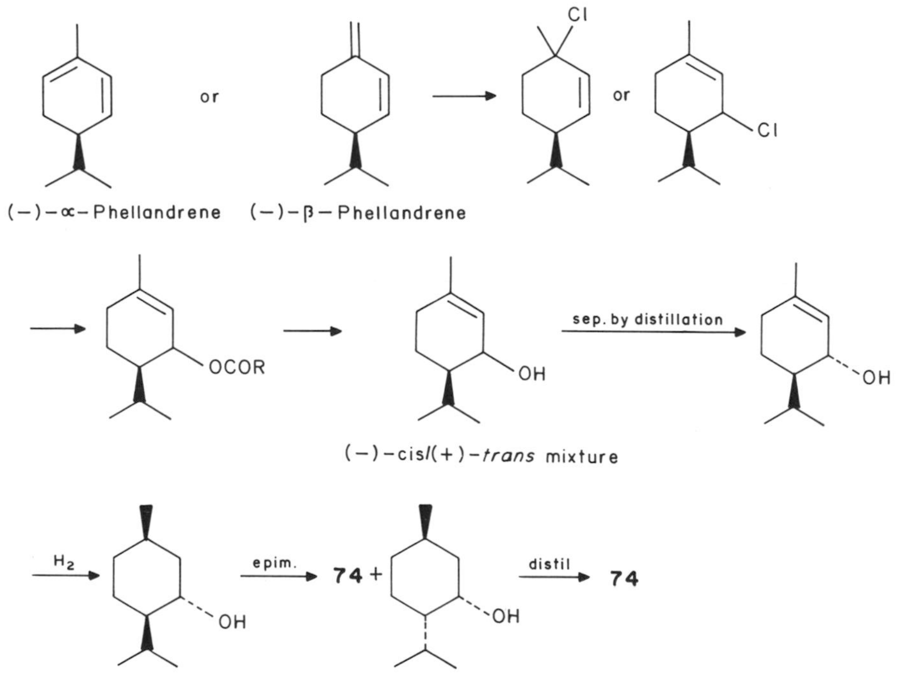

(−)-α-Phellandrene or (−)-β-Phellandrene → [chlorides] → [acetate, OCOR] → [alcohol, OH] (−)-*cis*/(+)-*trans* mixture —sep. by distillation→ [OH] —H_2→ [OH] —epim.→ **74** + [OH] —distil→ **74**

Scheme 62

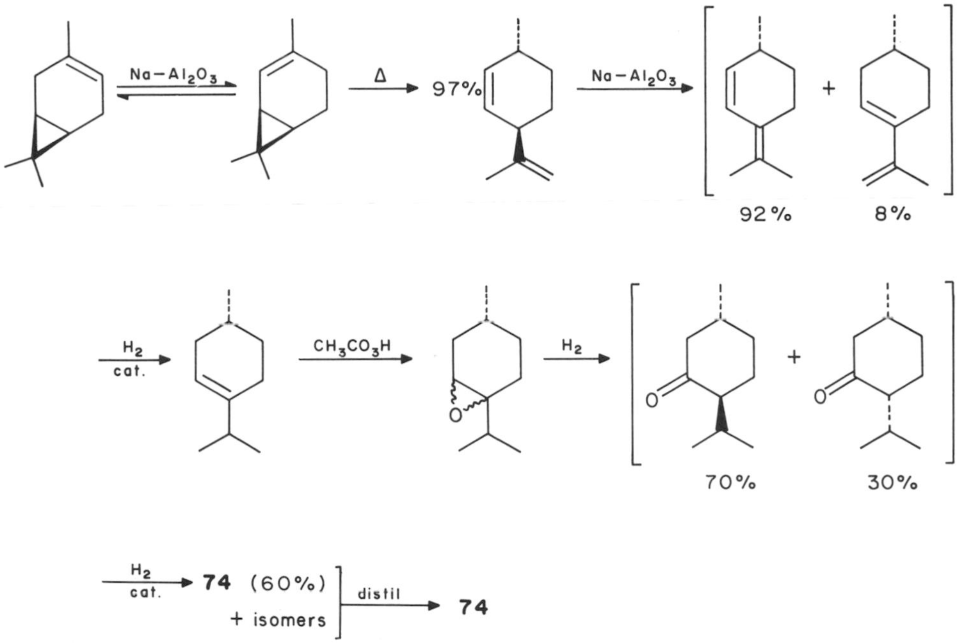

$Na-Al_2O_3$ Δ 97% $Na-Al_2O_3$ 92% 8%

H_2 cat. CH_3CO_3H H_2 70% 30%

H_2 cat. **74** (60%) + isomers distil **74**

Scheme 63

materials to the use of one of the most challenging of techniques, catalytic asymmetric synthesis. It has also encompassed chiral pool methods and the resolution of totally synthetic material.

1.3.4 THE STIMULUS OF A COMMERCIALLY IMPORTANT OPTICALLY ACTIVE STRUCTURE

Alongside the evolution of methods for production of (−)-menthol, which adopted technologies and techniques as they became available and offered improved methods, it is interesting to review the response to a structural target of more recent origin and which from the outset had a wider range of developed techniques and more advanced chemistry on which to build. This is well illustrated by the non-steroidal, anti-inflammatory, α-arylalkanoic acids (examples are given in Chart 6). The average daily dose for this class of drugs can be from several hundred milligrams to 1–2 g and this results in multi-hundred ton requirements.

All the approaches discussed in this chapter have been exemplified at least on the research scale and several have been investigated on a multi-kilogram scale.[9b,72,88,155,183–190] Many of these are discussed in Chapter 15, which details the elegant Zambon process.

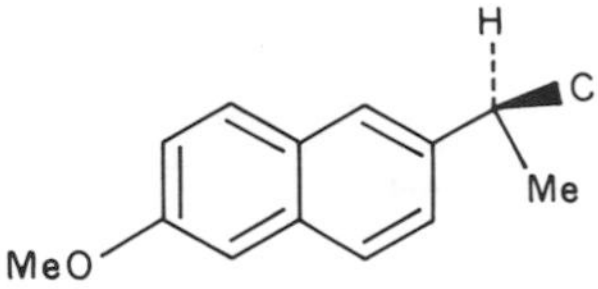

(*S*)-Naproxen 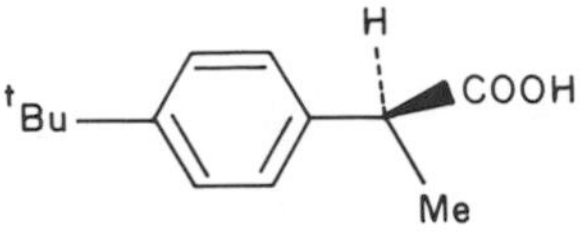 (*S*)-Ibuprofen

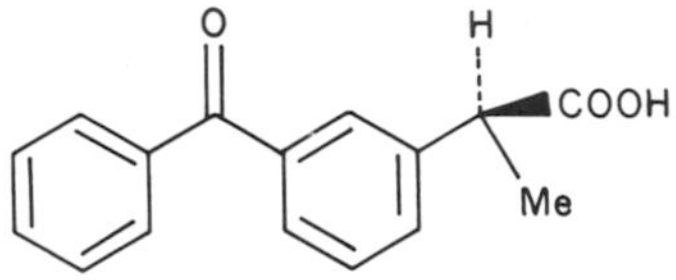

(*S*)-Ketoprofen 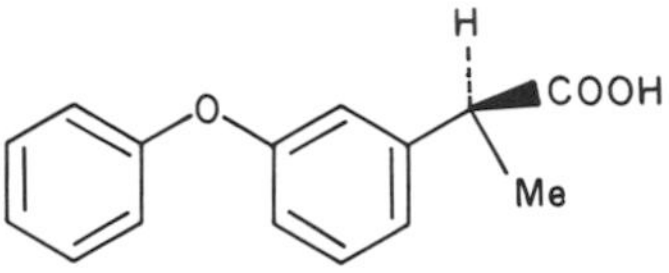 (*S*)-Fenoprofen

Chart 6

This is by no means the only target which could be cited: the aryloxypropionate herbicides, pyrethroids, β-blockers, carbapenems, HMG–CoA reductase inhibitors, etc., could all serve to illustrate the breadth of synthetic and technological ingenuity which has been applied to the problems of obtaining single enantiomers. This underscores the importance of commercial targets as catalysts of progress in all approaches to the synthesis of optically active materials.

1.4 NOTES AND REFERENCES

1. Ariëns E. J., *Eur. J. Clin. Pharmacol.*, **26**, 663 (1984).
2. Midland, M. M., and Nguyen, N. H., *J. Org. Chem.*, **46**, 4107 (1981).
3. De Camp, W. H., *Chirality*, **1**, 2 (1989); Shindo, H., and Caldwell, J., *Chirality*, **3**, 91 (1991).
4. Stewart, A. W., *Stereochemistry*, 2nd edn, Longmans Green, London, 1919, p. 248.
5. Cushny, A. R., *J. Physiol.*, **30**, 193 (1904).
6. Sugavanam, B., *Pestic. Sci.*, **15**, 296 (1984).
7. Testa, B., and Trager, W. F., *Chirality*, **2**, 129 (1990), and references cited therein.
8. Buser, H. P., Pugin, B., Spindler, F., and Sutter, M., *Tetrahedron*, **47**, 5709 (1991).
9. (a) Sheldon, R. A., *Speciality Chem.*, Feb., 31 (1990); (b) Kagan, H. B., *Bull. Soc. chim. Fr.*, 846 (1988); (c) Ariëns, E. J., van Rensen, J. J. S., and Welling, W. (Eds), Stereoselectivity of Pesticides: Biological and Chemical Problems, Elsevier, Amsterdam, 1988; (d) Sheldon, R. A., Porskamp, P. A., and ten Hoeve, W., in *Biocatalysts in Organic Synthesis* (ed. J. Tramper, H. C. van der Plas and P. Linko), Elsevier, Amsterdam, 1985 p. 59; (e) Morrison, J. D., and Scott, J. W. (Eds), *Asymmetric Synthesis*, Academic Press, Orlando, 1983–85, Vols. 1–5; (f) Sheldon, R., *Chem, Ind. (London)*, 212 (1990); (g) Eliel, E. L., Wilen, S. H., and Allinger, N. L. (Eds) *Topics in Stereochemistry*, Wiley, New York, 1967–89, Vols. 1–19; (h) Blaser, H.-U., *Tetrahedron: Asymmetry*, **2**, 843 (1991).
10. Some recent conferences dedicated to chiral topics: (a) St. Mary's Discussion Forum, London, 5–6 Dec., 1988; (b) Chiral Synthesis Symposium and Workshop, Spring

Innovations, Manchester, UK, 18 April 1989; (c) Smith Kline and French Research Symposium: Chirality in Drug Design and Synthesis, Cambridge, UK, 27–28 March 1990; (d) The International Conference on Chirality, Cancun, Mexico, 6–9 June 1990; (e) Chiral 90, Spring Innovations, Manchester, UK, 18–19 Sept., 1990; (f) Second International Symposium on Chiral Discrimination, Rome, 27–31 May 1991.
11. (a) *Chirality*, 1989, Vol. 1; (b) *Tetrahedron: Asymmetry*, 1990, Vol. 1.
12. *J. Prakt. Chem.* **99**, 6 (1866).
13. Strecker, A., *Justus Liebigs Ann. Chem.*, **75**, 27 (1850).
14. (a) Izumi, Y., Chibata, I., and Itoh, T., *Angew. Chem., Int. Ed. Engl.*, **17**, 176 (1978); (b) Martens, J., in *Topics in Current Chemistry* (ed. F. L. Boschke), Springer, Berlin, 1984, Vol. 125, p. 165; (c) Layman, P. L., *Chem. Eng. News*, Jan. 3, 18 (1983).
15. Roth, H. J., Kleemann, A., and Beisswenger, T., *Pharmaceutical Chemistry*, Ellis Horwood, Chichester, 1988, Vol. 1, p. 23.
16. Harper, D., *Manuf. Chem. Aerosol News*, July, 35 (1983).
17. *Farm Chemicals Handbook*, 72nd edn, Meister, Willoughby, 1986.
18. Ref. 15, p. 381.
19. Budavari, S. (Ed.), *The Merck Index*, 11th edn, Merck, Rahway, NJ, 1989, entry 235, p. 42.
20. Suppliers of D-alanine include: (i) S.S.T. Corp. Clifton, NJ, USA; (ii) Nagase, Tokyo, Japan; (iii) Toray Industries, Tokyo, Japan.
21. Jones J. L., Fong, W. S., Hall, P., and Cometta, S., *Chemtech*, 304 (1988).
22. Shiozaki, M., Ishida, N., Hiraoka, T., and Yanagisawa, H., *Tetrahedron Lett.*, **22**, 5205 (1981).
23. (a) Ohta, T., Kimura, T., Sato, N., and Nozoe, S., *Tetrahedron Lett.*, **29**, 4303 (1988); (b) Ohta, T., Sato, N. Kimura, T., Nozoe, S., and Izawa, K., *Tetrahedron Lett.*, **29**, 4305 (1988).
24. Salzmann, T. N., Ratcliffe, R. W., Christensen, B. G., and Bouffard, F. A., *J. Am. Chem. Soc.*, **102**, 6161 (1980).
25. Gupta, N., and Eisberg, N., *Performance Chem.* Aug.–Sept. 19 (1991).
26. It is, patently, natural.
27. For historical aspects, see Ramsay, O. B., *Stereochemistry*, Heyden, London, 1981.
28. Ref. 27, p. 76.
29. Kleemann, A., and Engel, J., *Pharmazeutische Wirkstoffe*, Georg Thieme, Stuttgart, 1982.
30. Bose, D. S., and Gurjar, M. K., *Synth. Commun.*, **19**, 3313 (1989).
31. Wynberg, H., and Staring, E. G. J., *J. Am. Chem. Soc.*, **104**, 166 (1982).
32. Bellamy, F. D., Bondoux, M. and Dody, P., *Tetrahedron Lett.*, **31**, 7323. (1990).
33. *Chem. Ind. (London)*, 274 (1990).
34. Seebach, D., and Züger, M., *Helv. Chim. Acta*, **65**, 495 (1982).
35. Iimori, T., and Shibasaki, M., *Tetrahedron Lett.*, **26**, 1523 (1985).
36. Baer, E., *Biochem. Prep.*, **2**, 31 (1952).
37. Emons, C. H. H., Kuster, B. F. M., Vekemans, J. A. J. M., and Sheldon, R. A., *Tetrahedron: Asymmetry*, **2**, 359 (1991).
38. Jung, M. E., and Shaw, T. J., *J. Am. Chem. Soc.*, **102**, 6304 (1980).
39. Goldberg, L., *Acta Physiol. Scand.*, **15**, 173 (1948).
40. Ferrier, R. J., and Prasit, P., *J. Chem. Soc., Chem. Commun.*, 983 (1981).
41. Tolstikov, G. A., Miftakhov, M. S., Valeev, F. A., Ibragimova, I. R., and Gareev, A. A., *Zh. Org. Khim.*, **27**, 415 (1991).
42. (a) Berkovitch, I. *Manuf. Chem.*, March, 41 (1989); (b) De Wilde, H., De Clercq, P., and Vandewalle, M., *Tetrahedron Lett.*, **28**, 4757 (1987); (c) Marco, J. L., *J. Chem. Res. (S)*, 276 (1988).

43. Oppolzer, W., *Pure Appl. Chem.* **62**, 1241 (1990).
44. Dev, S., *Proc. Indian. Natl. Sci. Acad., Part A*, **54**, 745 (1988).
45. Sobti, R., and Dev, S., *Tetrahedron*, **30**, 2927 (1974).
46. Mitra, R. B., and Khanra, A. S., *Synth. Commun.*, **7**, 245 (1977).
47. Roussel-ULCAF, *Ger. Pat., Ger. Offen.*, 1 935 320.
48. Wynberg, H., in *Topics in Stereochemistry*, (ed. E. L. Eliel, S. H. Wilen, N. L. Allinger), Interscience, New York, 1986, Vol. 16, p. 87.
49. Meijer, E. M., Boesten, W. H. J., Schoemaker, H. E., and van Balken, J. A. M., in *Biocatalysts in Organic Synthesis* (ed. J. Tramper, H. C. van der Plas, and P. Linka), Elsevier, Amsterdam, 1985, p. 142.
50. Roth. H. J., Kleeman, A., and Beisswenger, T., *Pharmaceutical Chemistry*, Ellis Horwood, Chichester, 1988, Vol. 1.
51. Wilen, S. H., Collet, A., and Jacques, J., *Tetrahedron*, **33**, 2725 (1977).
52. Davies, S. G., Brown, J. M., Pratt. A. J., and Fleet, G. W. J., *Chem. Br.*, 259 (1989).
53. (a)Newman, P., *Optical Resolution Procedures for Chemical Compounds*, Optical Resolution Information Center, New York, 1978, Vol. 1, *Amines and Related Compounds*; 1981, Vol. 2, *Acids*; 1984, Vol. 3, *Alcohols, Phenols, Thiols, Aldehydes and Ketones*; (b) Wilen, S. H., in *Topics in Stereochemistry*, (ed. N. L. Allinger, and E. L. Eliel), Interscience, New York, 1971, Vol. 6, p. 107; (c) Boyle, P. H., *Chem. Soc. Rev.*, **25**, 323 (1971); (d) Jacques, J., Collet, A., and Wilen, S. H., *Enantiomers, Racemates and Resolutions*, Wiley-Interscience, New York, 1981.
54. Stinson, S. C., *Chem. Eng. News*, 19 Oct., 31 (1987).
55. Reider, P. J., Davis, P., Hughes, L. D., and Grabowski, E. J. S., *J. Org. Chem.*, **52**, 955 (1987).
56. (a) Toda, F., in *Topics in Current Chemistry*, (ed. E. Weber), Springer, Berlin, 1987, Vol. 140, p. 43; (b) *Jpn. Chem. Week*, 20 March, 1 (1986); (c) *Eur. Pat.*, 220 887.
57. (a) Theilacker, W., and Winkler, H.-G., *Chem. Ber.*, **87**, 690 (1954); (b) *Org. Synth., Coll. Vol.*, **5**, 932; (c) Cope, A. C., Moore, W. R., Bach, R. D., and Winkler, H. J. S., *J. Am. Chem. Soc.*, **92**, 1243 (1970).
58. *Jpn. Pat.*, J6 2 114 945.
59. Sawayama, T., Tsukamoto, M., Sasagawa, T., Naruto, S., Matsumoto, J., and Uno, H., *Chem. Pharm. Bull.*, **37**, 1382 (1989).
60. *US Pat.*, 4 294 775.
61. Sjoberg, B., and Sjoberg, S., *Ark. Kemi.*, **22**, 447 (1964).
62. Gottstein, W. J., and Cheney, L. C., *J. Org. Chem.*, **30**, 2072 (1965).
63. Corey, E. J., and Hashimoto, S.-i., *Tetrahedron Lett.* **22**, 299 (1981), and references cited therein.
64. Crosby, J., Unpublished results.
65. Ref. 53d, p. 54.
66. *Chem. Eng.*, 8 Nov., 247 (1965).
67. *Eur. Pat.*, 441 471.
68. Davis, C. J., *Chem. Eng.*, 22 May, 62 (1978).
69. Sato, N., Uzuki, T., Toi, K., and Akashi, T., *Agric. Biol. Chem.*, **33**, 1107 (1969).
70. Amiard, G., *Bull. Soc. chim. Fr.*, 447 (1956).
71. Gernez, D., *C. R. Acad. Sci.*, **63**, 843 (1866).
72. *Eur. Pat.*, 298 395.
73. *Eur. Pat.*, 325 965.
74. Collet, A., Brienne, M. J., and Jacques, J., *Chem. Rev.*, **80**, 215 (1980).
75. Gottarelli, G., and Spada, P. G., *J. Org. Chem.*, **56**, 2096 (1991).
76. Crosby, J., Fakley, M. E., Gammell, C., Martin, K., Quick, A., Slawin, A. M.,

S-Zavareh, H., Stoddart, J. F., and Williams, D. J., *Tetrahedron Lett.*, **30**, 3849 (1989), and references cited therein.
77. Wakamatsu, H., *Food Eng.*, Nov., 92 (1968).
78. *US Pat.*, 2 898 558.
79. Ref. 53d, p. 371.
80. Black, S. N., Williams, L. J., Davey, R. J., Moffatt, F., Jones, R. V. H., McEwan, D. M., and Sadler, D. E., *Tetrahedron*, **45**, 2677 (1989).
81. Ref. 53d, p. 81.
82. Jacques, J., Le Clercq, M., and Brienne, M.-J., *Tetrahedron*, **37**, 1727 (1981).
83. Ref. 53d, p. 76.
84. *Br. Pat.*, 1 345 113.
85. For a recent review, see Kagen, H. B.; Fiaud, J. C., In *Topics in Stereochemistry*; Eliel, E. L., Wilen, S. H., Eds; Interscience: New York (1988), Vol. 18, p. 249.
86. (a) Chen, C. S., and Sih, C. J., *Angew. Chem., Int. Ed. Engl.*, **28**, 695 (1989); (b) Martin, V. S., Woodard, S. S., Katsuki, T., Yamada, Y., Ikeda, M., and Sharpless, K. B., *J. Am. Chem. Soc.*, **103**, 6237 (1981).
87. (a) Katsuki, T., and Sharpless, K. B., *J. Am. Chem. Soc.*, **102**, 5974 (1980); (b) Miyano, S., Lu, L. D.-L., Viti, S. M., and Sharpless, K. B., *J. Org. Chem.*, **48**, 3608 (1983).
88. Franck, A., and Rüchardt, C., *Chem. Lett.*, 1431 (1984).
89. Rüchardt, C., Gärtner, H., and Saltz, U., *Angew. Chem., Int. Ed. Engl.*, **23**, 162 (1984).
90. Salz, U., and Rüchardt, C., *Chem. Ber.*, **117**, 3457 (1984) .
91. Gao, Y., Hanson, R. M., Klunder, J. M., Ko., S. Y., Masamune, H., and Sharpless, K. B., *J. Am. Chem. Soc.*, **109**, 5765 (1987).
92. Kitamura, M., Kasahara, I., Manabe, K., and Noyori, R., *J. Org. Chem.*, **53**, 708 (1988).
93. Brown, J. M., *Angew. Chem., Int. Ed. Engl.*, **26**, 190 (1987), and references cited therein.
94. Amongst interested companies are Andeno (a), BASF (b), Lonza (a), Genzyme (c) and Sepracor (d): (a) Proceedings of the Chiral Synthesis Symposium and Workshop; Manchester, UK, 18 April 1989; (b) Schneider, M., *Performance Chem.* April, 28 (1989); (c) Richards, A., and Roach, D., *Performance Chem.*, April, 14 (1989); (d) Young, J. W., *Speciality Chem.*, Feb., 18 (1990).
95. Ladner, W. E., and Whitesides, G. M., *J. Am. Chem. Soc.*, **106**, 7250 (1984).
96. Sheldon, R. A., *P T-Procestech*, **43**, 32 (1988).
97. (a) Baer, E., and Fischer, H. O. L., *J. Am. Chem. Soc.*, **70**, 609 (1948); (b) Hirth, G., and Walther, W., *Helv. Chim. Acta*, **68**, 1863 (1985).
98. *Eur. Pat.*, 244 912.
99. Kooreman, H. J., Van Nistelrooij, H., and Pryce, R. J., in ref. 94a, p. 37.
100. (a) *Jpn. Pat.*, 86 129 797; (b) *Eur. Pat.*, 207 636.
101. Matsuo, N., and Ohno, N., *Tetrahedron Lett.*, **26**, 5533 (1985).
102. Hamaguchi, S., Asada. M., Hasegawa, J., and Watanabe, K., *Agric. Biol. Chem.*, **49**, 1661 (1985).
103. Iruchijima, S., and Kojima, N., *Agric. Biol. Chem.*, **46**, 1153 (1982).
104. Matzinger, P. K., and Leuenberger, H. G. W., *Appl. Microbiol. Biotechnol.*, **22**, 208 (1985).
105. *Eur. Pat.*, 144 832.
106. Chibata, I., in *Asymmetric Reactions and Processes in Chemistry* (ed. E. L. Eliel, and S. Otsuka), ACS Symposium Series, Vol. 185, American Chemical Society, Washington, DC 1982, p. 195.
107. Schoemaker, H. E., Boesten, W. H. J., Kaptein, B., Hermes, H. F. M., Sonke, T., Broxterman, Q. B. , Van den Tweel, W. J. J., and Kamphuis, J., paper presented at

IUPAC–NOST International Symposium, Enzymes in Organic Synthesis, New Delhi, 6–9 Jan., 1992.
108. Olivieri, R., Fascetti, E., Angelini, L., and Degen, L., *Biotechnol, Bioeng.*, **23**, 2173 (1981).
109. Takahashi, S., Ohashi, T., Kii, Y., Kumagai, H., and Yamada, H., *J. Ferment. Technol.*, **57**, 328 (1979).
110. Fufkumura, T., *Agric. Biol. Chem.*, **40**, 1687 (1976).
111. Dahod, S. K., and Siuta-Mangano, P., *Biotechnol. Bioeng.*, **30**, 995 (1987).
112. Taylor, S. C., paper presented at SCI Meeting, *Opportunities in Biotransformations*, Cambridge, UK, 3–5 April, 1990.
113. Bodnár, J., Gubicza, L., and Szabó, L.-P., *J. Mol. Catal.* **61**, 353 (1990).
114. *Eur. Pat.*, 396 447.
115. Brieva, R., Rebolledo, F., and Gotor, V., *J. Chem. Soc., Chem. Commun.*, 1386 (1990).
116. Danda, H., Nagatomi, T., Maehara, A., and Umemura, T., *Tetrahedron*, **47**, 8701 (1991).
117. For recent articles containing references to applications of industrial relevance, see (a) Scott, J. W., in *Topics in Stereochemistry* (ed. E. L. Eliel, and S. H. Wilen), Wiley, New York, 1989, Vol. 19, p. 209; (b) Ojima, I., Clos, N., and Bastos, C., *Tetrahedron*, **45**, 6901 (1989); (c) Parshall, G. W., and Nujent, W. A., *Chemtech.*, **184**, 376 (1988); (d) Refs 9b and 9f.
118. Knowles, W. S., *Acc. Chem. Res.*, **16**, 106 (1983).
119. Vocke, W., Hänel, R., and Flöther, F.-U., *Chem. Techn.*, **39**, 123 (1987).
120. For a recent review, see Sonawane, H. R., Bellur, N. S. Ahuja, J. R., and Kulkarni, D. G., *Tetrahedron: Asymmetry*, **3**, 163 (1992).
121. Ohta, T., Takaya, H., Kitamura, M., Nagai, K., and Noyori, R., *J. Org. Chem.*, **52**, 3174 (1987).
122. Noyori, R., *Chem. Soc. Rev.*, **18**, 187 (1989), and references cited therein.
123. Spindler, F., Pittelkow, U., and Blaser, H.-U., *Chirality*, **3**, 370 (1991).
124. Blaser, H.-U., *Tetrahedron: Asymmetry*, **2**, 843 (1991).
125. Hayashi, T., Katsumura, A., Konishi, M., and Kumada, M., *Tetrahedron Lett.*, **20**, 425 (1979).
126. Ojima, I., Kogure, T., Terasaki, T., and Achiwa, K., *J. Org. Chem.*, **43**, 3444 (1978).
127. Botteghi, C., Paganelli, S., Schionato, A., and Marchetti, M., *Chirality*, **3**, 355 (1991).
128. Tani, K., Yamagata, T., Akutagawa, S., Kumobayashi, H., Taketomi, T., Takaya, H., Miyashita, A., Noyori, R., and Otsuka, S., *J. Am. Chem. Soc.*, **106**, 5208 (1984).
129. (a) Rossiter, B. E., in *Asymmetric Synthesis*, (ed. J. D. Morrison), Academic Press, Orlando, 1985, Vol. 5, p. 194; (b) Finn, M. G., and Sharpless, K. B., as (a), p. 247.
130. Hanson, R. M., and Sharpless, K. B., *J. Org. Chem.*, **51**, 1922 (1986).
131. Kagen, H. B., in *Stereochemistry of Organic and Bioorganic Transformations* (ed. W. Bartmann and K. B. Sharpless), VCH, Weinheim, 1987, p. 31.
132. Sharpless, K. B., paper presented at the Smith Kline and French Research Symposium: Chirality in Drug Design and Synthesis, Cambridge, UK, 27–28 March 1990.
133. Sharpless, K. B., *Janssen Chim. Acta*, **1**, 3 (1988).
134. Zhang, W., Loebach, J. L., Wilson, S. R., and Jacobsen, E. N., *J. Am. Chem. Soc.*, **112**, 2801 (1990).
135. Jacobsen, E. N., Zhang, W., and Güler, M. L., *J. Am. Chem. Soc.*, **113**, 6703 (1991).
136. (a) Aratani, T., *Pure Appl. Chem.*, **57**, 1839 (1985); (b) *US Pat.*, 4 552 972.
137. Dehmlow, E. V., Singh, P., and Heider, J., *J. Chem. Res. (S)*, 292 (1981); this paper highlights the unreliability of conclusions based on optical rotation data; it re-examines earlier work and concludes that some must be viewed with suspicion.

138. Dolling, U. H., Davis, P., and Grabowski, E. J. J., *J. Am. Chem. Soc.*, **106**, 446 (1984).
139. Reider, P., paper presented at SCI Meeting: Chirality Recognition in Synthesis, London, 17 March 1988.
140. O'Donnel, M. J., Bennett, W. D., and Wu, S., *J. Am. Chem. Soc.*, **111**, 2353 (1989).
141. Neriackx, W., and Vandewalle, M., *Tetrahedron: Asymmetry*, **1**, 265 (1990).
142. Lonza Ltd, Organic Chemicals, CH-4002, Basle, Switzerland.
143. (a) Wynberg, H., and Staring, E. G. J., *J. Chem. Soc., Chem. Commun.*, 1181 (1984); (b) Staring, E. G. J., Moorlag, H., and Wynberg, H., *Recl. Trav. Chim. Pays-Bas*, **105**, 374 (1986).
144. Wynberg, H., and Staring, E. G. J., *J. Org. Chem.*, **50**, 1977 (1985).
145. Blaser, H.-U., Boyer, S. K., and Pittelkow, U., *Tetrahedron: Asymmetry*, **2**, 721 (1991).
146. Hazard, R., Jaouannet, S., and Tallec, A., *Tetrahedron*, **18**, 93 (1982); Tallec, A., Hazard, R., Le Bouc, A., and J. Grimshaw, *J. Chem. Res. (S)*, 342 (1986).
147. van der Linden, A. C., *Biochim. Biophys. Acta*, **77**, 157 (1963).
148. May, S. W., and Schwartz, R. D., *J. Am. Chem. Soc.*, **96**, 4031 (1974).
149. Johnstone, S. L., Phillips, G. T., Robertson, B. W., Watts, P. D., Bertola, M. A., Koger, H. S., and Marx, A. F., in *Biocatalysts in Organic Media*, (ed. C. Laane, J. Tramper, and M. D. Lilly), Elsevier, Amsterdam, 1987, p. 387.
150. *Eur. Pat.*, 193 228.
151. (a) May, S. W., and Swartz, R. D., *J. Am. Chem. Soc.*, **96**, 4031 (1974); (b) Ohta, H., and Tetsukawa, H., *J. Chem. Soc., Chem. Commun.*, 849 (1978). (c) de Smet, M. J., Witholt, B., and Wynberg, H., *J. Org. Chem.*, **46**, 3128 (1981); (d) H-Crutzen, A. Q. H., Carlier, S. J. N., de Bout, J. A. M., Wistabon, D., Schurig., V., Hartmans, S., and Tramper, J., *Enzyme Microb. Technol.*, **7** 17 (1985).
152. (a) Taylor, S. C., in *Enzymes in Organic Synthesis*, Ciba Foundation Symposium 111, Pitman, London, 1985, p. 71; (b) Ballard, D. G. H., Courtis, A., Shirley, I. M., and Taylor, S. C., *J. Chem. Soc., Chem. Commun.*, 954 (1983).
153. Taylor, S. C., *Speciality Chem.*, **8**, 236 (1988).
154. (a) *Jpn. Pat.*, 58 158188; (b) *Ger. Pat.*, 3041224; (c) Hasegawa, J., *Hakko Kogako Kaishi*, **62**, 341 (1984); *Chem. Abstr.*, **101**, 226524 (1984); *US Pat.*, 4981 794.
155. *Gist-Brocades/Shell Eur. Pat.*, EP 274146.
156. E.g. Shell, *Eur. Pat.*, EP 319100.
157. (a) Davies, H. G., Green, R. H., Kelly, D. R., and Roberts, S. M., *Biotransformations in Preparative Organic Chemistry*, Academic Press, London, 1989, p. 99; (b) Servi, S., *Synthesis*, 1 (1990), and references cited therein; (c) Gramatica, P., *Chim. oggi*, 17 (1988), and references cited therein.
158. (a) Seebach, D., Sutter, M. A., Weber, R. H., and Zuger, M. F., *Org. Synth.*, **63**, 1 (1985); (b) Wipf, B., Kupfer, E., Bertazzi, R., and Leuenberger, H. G. W., *Helv. Chim. Acta*, **66**, 485 (1983).
159. Brooks, D. W., Mazdiyasni, H., and Grothaus, P. G., *J. Org. Chem.*, **52**, 3223 (1987).
160. Ghiringhelli, D., *Tetrahedron Lett.*, **24**, 287 (1983).
161. Bucciarelli, M., Forni, A., Moretti, I., and Torre, G., *J. Chem. Soc., Chem. Commun.*, 456 (1978).
162. Dondoni, A., Fantin, G., Fogagnolo, M., Mastellari, A., Medici, A., Nefrini, E., and Pedrini, P., *Gazz. Chim. Ital.*, **118**, 211 (1988).
163. Ohta, H., Ozaki, K., and Tsuchihashi, G.-I., *Chem. Lett.*, 191 (1987).
164. Utaka, M., Konishi, S., Okubo, T., Tsuboi, S., and Takeda, A., *Tetrahedron Lett.*, **28**, 1447 (1987).
165. Izumi, T., Murakami, S., and Kasahara, A., *Chem. Ind.* (*London*), 79 (1990).
166. Zhou, B-n., Gopalan, A. S., VanMiddlesworth, F., Shieh, W.-R., and Sih, C. J., *J. Am. Chem. Soc.*, **105**, 5925 (1983).

167. Prelog, V., *Pure Appl. Chem.*, **9**, 119 (1964).
168. Crout, D. H. G., and Christen, M., in *Modern Synthetic Methods* (ed. R. Scheffold), Springer, Berlin, 1989, Vol. 5, p. 64.
169. (a) Leuenberger, H. G. W., Boguth, W., Widmer, E., and Zell, R., *Helv. Chim. Acta*, **59**, 1832 (1976); (b) Leuenberger, H. G. W., paper presented at Warwick Workshop, Enzymes as Reagents, Warwick University, April 1987.
170. Calton, G. J., in *Biotechnology in Agricultural Chemistry*, ACS Symposium Series, Vol. 334. American Chemical Society, Washington, DC, 1987, p. 181.
171. Evans., C., *Performance Chem.*, July–Aug., 58 (1989).
172. (a) Purification Engineering, *Eur. Pat.*, EP 132999; (b) *Biotechnol. News*, **4**, No. 19, 7 (1984).
173. Cheetham, P. S. J., in *Handbook of Enzyme Biotechnology*, 2nd edn. (ed. A. Wiseman), Ellis Horwood, Chichester, 1985, p. 274.
174. Barner, R., and Schmid, M., *Helv. Chim. Acta*, **62**, 2384 (1979).
175. (a) Sumitomo, *Eur. Pat.*, EP 80827; (b) ICI *Eur. Pat.*, EP 132392; (c) Shell *Eur. Pat.*, EP 109681; (d) Asada, S., Kobayashi, Y., and Inoue, S., *Makromol. Chem.*, **186**, 1755 (1985); (e) Joint work by Leiden University–Duphar BV–PEBOC leading to development-scale production of optically active cyanohydrins was reported by Kruse, C. G., Brussee, J., and van der Gen, A., at Chiral 90, Spring Innovations, Manchester, UK, 18–19 Sept., 1990.
176. Whitesides, G. M., and Wong, C.-H., *Angew. Chem., Int. Ed. Engl.* **24**, 617 (1985).
177. Fersht, A. R., paper presented at the International Conference on Enzyme Engineering, Cambridge, UK, 25–26 Sept., 1986.
178. Teeuwen, H., and Polastro, E., *Performance Chem.*, Sept.–Oct., 20 (1989).
179. Godtfredsen, S. E., paper presented at SCI meeting, Opportunities in Biotransformations, Cambridge, UK, 3–5 April, 1990.
180. Takasago Perfumery, *Ger. Pat.*, 1815845.
181. Bauer, K., and Garbe, D., *Common Fragrance and Flavor Materials*, VCH, Weinheim 1985, p. 40.
182. (a) Dev, S., in *Proceedings of 11th IUPAC International Symposium on Chemistry of Natural Products*, Bulgarian Academy of Sciences, Sofia, 1978, Vol. 4, Part. 1, p. 433; (b) Bedoukian, P. Z., *Perfum. Flavorist*, **8**, 5 (1983).
183. Ref. 15p. 101.
184. Young. J. W., *Speciality Chem.*, Feb., 18 (1990).
185. Larsen, R. D., Corley, E. G., Davis, P., Reider, P. J., and Grabowski, E. J. J., *J. Am. Chem. Soc.*, **111**, 7650 (1989).
186. Parrinello, G., and Stille, J. K., *J. Am. Chem. Soc.*, **109**, 7122 (1987).
187. Hayashi, T., Konishi, M., Fukushima, M., Kanehira, K., Hioki, T., and Kumada, M., *J. Org. Chem.*, **48**, 2195 (1983).
188. (a) Piccolo, O., Spreafico, F., and Visentin, G., *J. Org. Chem.*, **52**, 10 (1987); (b) Honda, Y., Ori, A., and Tsuchihashi, G., *Bull. Chem. Soc. Jpn.*, **60**, 1027 (1987).
189. (a) Giordano, C., Castaldi, G., Cavicchioli, S., and Villa, M., *Tetrahedron*, **45**, 4243 (1989); (b) Castaldi, G., Cavicchioli, S., Giordano, C., and Uggeri, F., *J. Org. Chem.*, **52**, 3018 (1987); (c) *US Pat.*, 4697036, 4734507, 4810819, 4855464.
190. Sonawane, H. R., Nanjudiah, B. S., Kulkarni, D. G., and Ahuja, J. R., *Tetrahedron: Asymmetry*, **2**, 251 (1991).

NON-BIOLOGICAL RESOLUTIONS

2 Resolution of Racemates by Diastereomeric Salt Formation

C. R. BAYLEY and N. A. VAIDYA
Norse Laboratories, Newbury Park, CA, USA

2.1 INTRODUCTION

There can be no doubt that the production of pure enantiomers of drugs is of growing importance to the pharmaceuticals industry. It is now accepted that in some circumstances the introduction of a pure enantiomer, rather than a racemic pharmaceutical, can provide the benefit of a smaller dose and possibly an enhancement of the therapeutic effect.[1] However, 95% of the drugs with an asymmetric carbon atom introduced between 1983 and 1985 were in a racemic form, and even today about 25% of drugs are marketed as racemates.[2] It is anticipated that where practical, enantiomerically pure drugs will be marketed in the future, and this has led to the current demand for economic methods for the preparation of pure enantiomers.

To date, the resolution of racemic mixtures via diastereomeric salt formation has been the most commonly used industrial technique, and this chapter reviews the factors that can affect the success of this approach.[3–6] As noted in Chapter 1, an alternative to diastereomeric salt formation is direct, preferential crystallization of the desired enantiomer. If applicable, preferential crystallization of enantiomers is a highly economic approach, and it has been used by Merck with great success in the manufacture of α-methyl-DOPA.[7] However, in practice the method has limited application because it can be applied only to a conglo-

Chirality in Industry. Edited by A. N. Collins, G. N. Sheldrake and J. Crosby

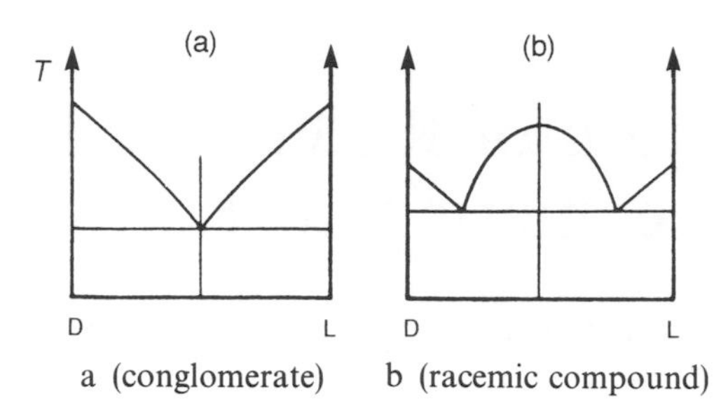

Figure 1. Binary (melting-point) phase diagrams a and b. Reproduced by permission from ref. 8)

merate, i.e. a mechanical mixture of crystals of the two enantiomers. In contrast in the much commoner case of a racemic compound where both enantiomers exist in a unit cell resolution by preferential crystallization is not possible and separation must be via diastereomers. Further examples of industrial applications of direct crystallization are given in Chapter 1 (Section 1.2.2.2), and a comprehensive description of conglomerates and their properties has been given by Jacques *et al.*[8]

In designing a resolution process, the possibility of using preferential crystallization should be checked first. This can be done by studying the melting-point characteristics of the subject using a differential scanning calorimeter—a conglomerate has a sharp melting point which is lower than that of the enantiomers, whereas a true racemic compound has a higher melting point (Figure 1).

2.2 THE INDUSTRIAL APPLICATION OF DIASTEREOMERIC CRYSTALLIZATION

Diastereomeric crystallization is used so widely that it provides a measure for judging alternative processes. It is based on the interaction of a racemic product with an optically active material (resolving agent), to give two diastereomeric derivatives (usually salts):

(DL)-A	+	(L)-B	⟶	(D)-A·(L)-B	+	(L)-A·(L)-B
racemate		resolving agent		n-salt		p-salt

The salts formed are diastereomers with different physical properties and may be separated in a number of ways, for example by chromatography, but the most efficient method of separating such diastereomers is by crystallization. Many significant pharmaceuticals are resolved using diastereomeric crystallization (see Table 1 for some examples).

In many of these examples, for economic or other process reasons, an intermediate is resolved rather than the end product. The technology utilizing

Table 1. Examples of pharmaceuticals resolved using diastereomeric crystallization in the process

Pharmaceutical	Resolving agent
Ampicillin	D-Camphorsulphonic acid
Ethambutol	L-(+)-Tartaric acid
Chloramphenicol	D-Camphorsulphonic acid
Dextropropoxyphene	D-Camphorsulphonic acid
Dexbrompheniramine	D-Phenylsuccinic acid
Fosfomycin	R-(+)-Phenethylamine
Thiamphenicol	D-(−)-Tartaric acid (unnatural)
Naproxen	Cinchonidine
Diltiazem	*R*(+)-Phenethylamine

diastereomeric crystallization has the advantage of relative simplicity and requires only standard production equipment. From the practical point of view, the method is flexible and suited to intermittent batch production, which is often the norm in pharmaceutical manufacture. However, in spite of this simplicity, the procedure has some disadvantages on a large scale, including the need for a great deal of process equipment—reactors, holding tanks, etc. The storage of the various mother liquors, and second and third crops held pending future re-work, takes up considerable space in the plant and can create a bottleneck in plant utilization. Recovery of the resolving agent is necessary for environmental and economic reasons, but the cost of recovery, particularly for inexpensive resolving agents, can be relatively high. The feasibility of racemization of the 'unwanted' enantiomer has a major effect on the economics. This is illustrated in the resolution of racemic 2-aminobutanol, where the (*S*)-(+)-enantiomer is required for the manufacture of ethambutol. No convenient method exists for the racemization of the (*R*)-(−)-enantiomer byproduct, and this creates a disposal problem. It is in these situations that asymmetric synthesis can appear attractive.

2.3 THE VARIABLES AFFECTING THE SEPARATION OF ENANTIOMERS

Two equilibria exist in diastereomer resolutions: first, the formation of the diastereomeric salts, where the resolving agent is a controlling variable, and second, the separation of the salts by crystallization, where the solvent plays a vital role. In general the temperature is not a variable (of course, temperature control is important during the crystallization process and the application of programmed cooling to ensure reproducible results is becoming more common in industrial practice). There are therefore two factors which directly affect the success of the resolution by diastereomer crystallization: (i) choice of resolving agent and (ii) solvent composition. Although an experienced investigator can develop a 'feel' for the resolution of a racemic mixture, in practice working out the experimental conditions is still a process of trial and error. Nevertheless, there are approaches which can rationalize the work and these are described in a later section on the selection of solvents. They consist of a study of the resolution system through construction and analysis of a phase diagram and a systematic screening procedure for the resolving agent and solvent.

2.3.1 SELECTION OF A RESOLVING AGENT

Ideally, a good resolving agent should have the following properties relating to the formation of diastereomeric salts and subsequent crystallization.

i. The diastereomeric salt must crystallize well and there must be an appreciable difference in solubility between (+)-A·(−)-B and (−)-A·(−)-B, where (±)-A is the racemic mixture and (−)-B is the resolving agent. This difference depends on both the nature of A and B and the solvent chosen. Selection of a resolution system is still a matter of trial and error and some resolutions have defied solution even after trying many different resolving agents and solvents. This may be due to the formation of a double salt (+)-A·(−)-B·(−)-A·(−)B in a particular solvent.

ii. The compound between the resolving agent and the substance to be resolved should be easily formed, and the resolving agent should be easily recoverable in a pure state from the salt following the crystallization step. The diastereomeric salt is split by treatment with mineral acid or alkali and the resolving agent is recovered by an appropriate method. If the resolving agent can be recovered by distillation, e.g in the case of α-methylbenzylamine, this is a significant advantage compared with other procedures. For example, the recovery of tartaric acid by ion exchange is a relatively expensive process for a low cost material.

In general, the diastereomeric method is used to resolve acids (carboxylic or sulphonic) and amines, and in practice strong acids and bases give better results than their weaker equivalents. Other functional groups which are not ionic can be converted into derivatives which in turn will form salts. For example, octan-2-ol may be converted into a half-ester with phthalic acid and then resolved with α-methylbenzylamine.[9]

iii. For industrial purposes, a resolving agent should be relatively inexpensive and readily recoverable in high yield after completion of the resolution. It is worth noting that the agents listed in Table 1, used in the resolution of commercially important pharmaceuticals, are mostly relatively inexpensive. In industrial practice the quantity of resolving agent is often reduced to half the stochiometric amount. This reduction usually allows for a better separation of the desired enantiomer and a lower cost. A fuller listing of commonly used resolving agents is given in Table 2.

Table 2. Commonly used resolving agents

For acids	For bases
α-Methylbenzylamine	1-Camphor-10-sulphonic acid
α-Methyl-*p*-nitrobenzylamine	Malic acid
α-Methyl-*p*-bromobenzylamine	Mandelic acid
2-Aminobutane	α-Methoxyphenylacetic acid
N-Methylglucamine	α-Methoxy-α-trifluoromethyl-phenylacctic acid
Dehydroabietylamine	2-Pyrrolidone-5-carboxylic acid
α-(1-Naphthyl)ethylamine	Tartaric acid
threo-2-Amino-1-(*p*-nitrophenyl)-propane-1,3-diol	
Cinchonine	
Cinchonidine	
Quinine	
Ephedrine	

Comprehensive reviews of applications have been given by Newman[10] and Wilen.[11]

As can be seen from Table 2, there are more bases readily available for the resolution of acids than vice versa. Among the bases the resolving agent of choicc is often α-methylbenzylamine (α-phenethylamine) ($35 kg^{-1}), which is recoverable in high yields and may be purified by distillation. α-Methylbenzylamine, one of the largest volume resolving agents, is produced synthetically, by reductivc amination of acetophenone and resolution with tartaric acid. Derivatives of α-methylbenzylamine, c.g. 4-bromo-, 4-nitro- and 4-methoxy-, and other aromatic amines are also effective in some resolutions.

A useful resolving amine is α-(1-naphthyl)ethylamine, which gives good recoveries and nicely crystalline derivatives. Its use has been limited by its price ($135 kg^{-1}), but a major bulk use would probably allow for more competitive pricing. A naturally occurring and inexpensive base, dehydroabietylamine, was used extensively at one time (in the resolution of calcium pantothenate), but its use has declined owing to problems with the supply of raw material. Alkaloids such as brucine, strychnine and ephedrine have been used for a long time as resolving agents and have the advantage of being relatively inexpensive (price range $45–105 kg^{-1}). As natural products they are available in a high degree of chiral purity. Alkaloids may be recovered in high yield after the resolution by treating the diastereomeric salt in alcoholic solution with acid and filtering, or by extracting the resolved acid. The alkaloid remains in the aqueous, alcoholic layer as hydrochloride and may be liberated and recovered by addition of ammonia.

(+)-Tartaric acid, which occurs naturally, is inexpensive ($5 kg^{-1}) and is the first acid to be tried in any screening to find a resolving agent. It may be recovered by the use of ion exchange although the economics for such a low-priced substance are problematic. The other enantiomer, (*S*)-(−)-tartaric acid, is fairly expensive ($150 kg^{-1}) but nevertheless has found some industrial uses. The dibenzoyl and ditolyl derivatives of (+)-tartaric acid (about $40 kg^{-1}), are excellent resolving agents and are easier to recover than tartaric acid itself because of their solubility properties. D-Mandelic acid is also an excellent resolving agent (*ca* $40 kg^{-1}) and it also has the advantage of ease of recovery in high yield.

Among the other acids that may be considered, (+)-10-camphorsulphonic acid, made from natural camphor, is readily available and relatively inexpensive ($15 kg^{-1}), and has found large-scale industrial use in the manufacture of D-phenylglycine and dextropropoxyphene. 2-Ketogulonic acid, an industrial intermediate in the manufacture of ascorbic acid, has considerable promise as a resolving agent but can be unstable on long-term storage.

iv. In general, a resolving agent should be available in an optically pure form because a substance to be resolved cannot be obtained in a higher state of optical purity than the resolving agent by mere crystallization of diastereoisomers. Naturally occurring resolving agents such as the alkaloids and tartaric acid exist in an optically pure form as opposed to resolving agents prepared synthetically, e.g. α-methylbenzylamine, which can be obtained optically pure only with some difficulty because of the technological limitations of their preparation. It will be interesting to see how resolving agents prepared in high optical purity by enzymatic methods[12] find use in diastereomeric crystallizations.

v. The chiral centre should be as close as possible to the functional group responsible for salt formation. An extreme example of the problems which may arise if this condition is not met was the resolution of 4-(2′-pentyl)benzoic acid, which required 24 crystallizations to achieve purification of the diastereomeric salt![13] The resolving agent used should give a diastereomeric salt with a tight and rigid structure. This usually results from the presence of a number of polar functional groups forming hydrogen bonds. Interaction by hydrogen bonding, charge transfer and π-complex formation can be as important as the closeness of the chiral centre to the functional group of the resolving agent.

vi. An agent must be chemically stable and not racemize under the conditions of the resolution process. In the authors' experience, even D-(−)-mandelic acid, which one would have imagined to be stable, eventually shows some racemization on many recoveries. It should therefore withstand strongly acidic or alkaline conditions and the temperatures encountered during the enantiomer liberation step.

vii. Resolving agents should be available as both enantiomers so that both isomers of the substrate can be prepared. This is necessary in any investigation of biological activity. A drawback with the use of naturally occurring agents is their existence, usually, as only one enantiomer; this is an advantage of synthetic agents. A major problem with the diastereomer separation method is that generally only one diastereomer (the less soluble one) can be obtained from solution in the pure state. The other diastereomer is apt to stay in the mother liquor, contaminated with a residue of the less soluble material, and decomposition of such a mixture will evidently not give enantiomerically pure material. Thus, the common situation is that one enantiomer may be obtained pure or nearly so, whereas the other is recovered in a far from optically pure state. From the practical point of view, this is not always too serious a drawback, since it is often not necessary to have both enantiomers pure in a manufacturing situation. However, for research purposes this is usually a requirement and it may be achieved by the use of the antipode of the original agent to complete the resolution of the impure second enantiomer or, of course, to carry out the resolution of the racemic product. It should be noted that resolutions may not always be reciprocal.[14] In other words, if a given resolving agent (−)-A serves to resolve a particular mixture of (±)-B enantiomers, one of these enantiomers, e.g. (+)-B, will not necessarily serve as resolving agent for (±)-A. However, in practice, reciprocity seems to apply to most situations.

2.3.2 SELECTION OF SOLVENT

The nature and composition of the solvent used are important in determining the success of the resolution, and trying a range is important in finding the optimum conditions for a resolution. Crystallization of diastereomers usually depends on solubility differences between the diastereomers which are formed in equal quantities, and the solubilities of a three-component system (two diastereomers and the solvent) obtained in such a resolution can be expressed in a ternary phase diagram. After the first crystallization, an examination of the enantiomeric composition of the less soluble diastereomer together with that of the mother liquor, taken over a range of concentrations, enables a phase diagram to be constructed. From this one can calculate the maximum theoretical yield of the desired diastereomer. If this yield is not workable, other solvents can be tried. An excellent description of this most useful approach was given by Jacques *et al.*[8]

The importance of solvent composition is demonstrated by the work of Sjoberg[15] on the resolution of α-(2-thianaphthenyl)propionic acid (Table 3). It can be seen that, apart from the success achieved with only two resolving agents, selection of the solvent played a decisive role.

The described experiments typify the setting up of a screening procedure where a range of resolving agents and substrates are dissolved in various solvents and left to stand under identical conditions and the onset of crystallization is observed. This preliminary screening can identify systems to be investigated in

Table 3. A typical screening experiment to determine resolution conditions

Base	Specific rotation (°)			
	From methanol	From ethanol	From acetone	From ethyl acetate
Cinchonine	Oil	Oil	Oil	Oil
Cinchonidine	−2	+13	+4	+4
Quinine	0	0	0	0
Quinidine	Oil	Oil	Oil	Oil
Brucine	−2	0	−2	0
Strychnine	Oil	Oil	Oil	Oil
Morphine	+23	0	+20	+26
Ephedrine	Oil	Oil	Oil	Oil
(+)-α-Phenylethylamine	−2	−3	0	0
(+)-α-(2-Naphthyl)ethylamine	−21	−8	−5	−28

Specific rotation of α-(2-thianaphthenyl)propionic acid in absolute ethanol with different resolving solvents.

more detail. Carrying out the screening work on a small scale enables expensive materials to be conserved and results to be obtained quickly.

Solvent composition is also important, and in particular small quantities of water can have a large influence. Examples from the authors' own experience can confirm this; the resolution of racemic 2-phenylpropionic acid using α-methylbenzylamine is best carried out with propan-2-ol containing 10% water as solvent. The use of dry propan-2-ol is less successful, owing to the decreased solubility of the (+)-phenylpropionic acid–(−)-methylbenzylamine salt. Similarly, in the preparation of the important intermediate (+)-2-aminobutanol (used to manufacture ethambutol), resolution may be carried out with tartaric acid in water. The desired (+)-2-aminobutanol-*d*-tartrate salt is extremely soluble in water, whereas the diastereomeric (−)-aminobutanol–*d*-tartrate is much less soluble and crystallizes from water in good yield. The (+)-enantiomer may therefore be produced by filtering off the (−)-amino butanol–*d*-tartrate and recovering the relatively pure (+)-2-aminobutanol salt from the filtrate. This product may be purified by treatment with ethanol, where the solubilities are reversed and the salt of the (−)-enantiomer is now more soluble.[16]

In conclusion, a wide variety of resolution methods are available and resolution through diastereomeric salt crystallization continues to be a practical approach to the separation of enantiomers. Because of this it is important to optimize the resolving agent–solvent combination to develop the most economic industrial process.

2.4 REFERENCES

1. Williams, K., and Lee, E., *Drugs*, **30**, 333 (1985).
2. Gross, M., *Annu. Rep. Med. Chem.*, **25**, 323 (1989). See also Ariens, E. J., and Wuis, E., *Clin. Pharmacol. Therap.* **42**, 361 (1987).
3. Eliel, E. L., *Stereochemistry of Carbon Compounds*, McGraw-Hill, New York, 1962.
4. Sheldon, R. A., Leusen, F. J. J., van der Haest, A. D., and Wijnberg, H., *Proceedings of the Chiral 90, Symposium*, Spring Innovations, Manchester, 1990 p. 101.
5. Secor, R., *Chem. Rev.*, **63**, 297 (1963).
6. Boyle, P. H., *Q. Rev. Chem. Soc.*, **25**, 323 (1971).
7. US Pat., 3 158 648, 1964 Jones, R. T., Kreiger, K. H. and Lago, J., assigned to Merck & Co; Reinhold, D. F., Firestone, R. A., Gaines, W. A., Chemerda, J. M., and Sletzinger, M., *J. Org. Chem.*, **33**, 1209 (1968).
8. Jacques, J., Collet, A., and Wilen, S. H., *Enantiomers, Racemates and Resolutions*, Wiley, New York, 1981.
9. Kenyon, J., *J. Chem. Soc.*, **121**, 2540 (1922).
10. Newman, P., *Optical Resolution Procedures*, Optical Resolution Information Center, Manhattan College, NY, 1981.
11. Wilen, S. H., *Tables of Resolving Agents*, University of Notre Dame Press, 1972.
12. Stirling, D. I., Zeitlin, A. L., and Matcham, G. W., *US Pat.*, 4 950 606, 1990.
13. Wilen, S. H., in *Topics in Stereochemistry* (ed. E. L. Eliel and N. L. Allinger), Wiley–Interscience, New York, 1971, vol. 6, p. 107.
14. Markwald, W., *Chem. Ber.*, **29**, 42 (1896).
15. Sjoberg, B., *Ark. Kemi*, **12**, 588 (1958).
16. Pitre, D., and Grabitz, E. B., *Chimia*, **23**, 399 (1969).

3 L-Lysine via Asymmetric Transformation of α-Amino-ε-caprolactam

S. SIFNIADES
Allied-Signal, Inc., Morristown, NJ, USA

3.1 INTRODUCTION

L-Lysine is an essential amino acid that is used commercially as a feed supplement for poultry and, to a lesser extent, for swine. At the time when this work was carried out, L-lysine was manufactured by fermentation of sugars.

Several synthetic routes to L-lysine had been described, and most of the routes directed towards commercial production utilized DL-α-amino-ε-caprolactam (DL-ACL) as an intermediate. The resolution of DL-ACL is a crucial step in these processes.[1] To be commercially viable, resolution must be coupled with racemization so that the undesired D-enantiomer may also be utilized, and in all processes reported prior to this work this was accomplished as a separate step on isolated D-ACL.[2] Some years ago, the author and co-workers discovered and developed at Allied-Signal a synthesis of L-lysine that relied on the second order asymmetric transformation of DL-ACL to L-ACL via the complex (L-ACL)$_3$NiCl$_2$.[3] This chapter describes this unique process, and its scale-up in a mini-pilot plant.

Chirality in Industry. Edited by A. N. Collins, G. N. Sheldrake and J. Crosby

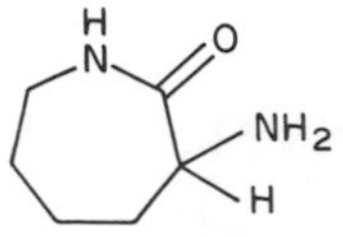

α-Amino-ε-caprolactam (ACL)

3.2 UNDERLYING PRINCIPLES

A second order asymmetric transformation involves the resolution of a pair of enantiomers by crystallization of one enantiomer with simultaneous racemization of the other.[4] For a pair to be amenable to such a transformation two conditions must be met: (a) the pair must exist as a racemic mixture (conglomerate), rather than as a racemic compound or solid solution, under the conditions of the resolution;[5] and (b) an asymmetric center must be present that is labile under the conditions of the resolution.

Only a few examples of such second-order asymmetric transformations had been reported at the time that this work was completed. Two involved restricted rotation,[6,7] the third involved conformational isomerism,[8] the fourth was a labile ammonium salt[9] and the fifth an amino acid–metal complex.[10] In all cases the asymmetric center was naturally labile under the conditions of resolution. None involved racemization at a tetrahedral carbon atom.[11]

Earlier work at Allied-Signal had shown that the second-order asymmetric transformation of ACL might be feasible. L-ACL had been racemized at 95°C in methanol solution in the presence of catalytic amounts of $NiCl_2$[12] and the $NiCl_2$ complex of DL-ACL had been resolved from ethanol solution at 50°C by seeding with crystals of $(L\text{-}ACL)_3NiCl_2 \cdot EtOH$.[13] The aim was to find conditions that allowed these two processes to take place simultaneously. This would imply an increase in the rate of racemization of ACL so that it might be reasonably rapid at temperatures and in media compatible with the crystallization.

3.3 SIMULTANEOUS RESOLUTION AND RACEMIZATION OF ACL

The resolution of DL-ACL via the nickel complex is kinetically controlled. The driving force for crystallization is the supersaturation of the complex in solution, a non-discriminating, thermodynamic force, which, on its own, would result in the formation of racemic crystals. The selectivity of the process relies on the kinetic control provided by the chiral surface of the seed, and if the process is allowed to proceed without addition of fresh seed, the enantiomeric excess (*ee*) of the crystalline product will deteriorate to unacceptably low levels. Fresh seed of high *ee* is produced by partial dissolution of product.

The racemization of ACL in the presence of $NiCl_2$ may be achieved by general base catalysis. The base abstracts the hydrogen from the α-position of a chelated ACL, which results in destruction of the asymmetry. In previous work[12] this task was accomplished by excess of ACL. The author and co-workers found that the racemization rate could be dramatically increased by adding catalytic amounts of ethoxide ion along with $NiCl_2$ to an ethanol solution of ACL. An excess of ACL over the stoichiometry of the complex $(ACL)_3NiCl_2$ was essential to achieve the full catalytic effectiveness of ethoxide. This is due to the need to keep the nickel ion coordinatively saturated so that it will not bind ethoxide and thus render it ineffective as a base. When the ACL/$NiCl_2$ ratio was 4 or higher the reaction was first order in ethoxide,[3c] and racemization half-lives of the order of 2–4 min were obtained at *ca* 80°C, which is the boiling point of a typical reaction mixture. Crystallization of (L-ACL)$_3$NiCl$_2\cdot$EtOH on seed crystals proceeded smoothly at that temperature also. Thus, all the elements necessary for a continuous process were in place.

The process was demonstrated in the laboratory using a stirred evaporative crystallizer (100–120 ml effective volume) with continuous addition of DL-ACL, $NiCl_2$ and Ni(OMe)Cl in ethanol, and removal of solvent overhead. Seed crystals of (L-ACL)$_3$NiCl$_2\cdot$EtOH (average diameter $< 5\ \mu m$) were added at the beginning of the run and every hour thereafter; product crystals were removed at the same time. At the end of a 28 h operation, 214 g of (L-ACL)$_3$NiCl$_2\cdot$EtOH crystals of 97% *ee* in L-ACL had been collected; 62 g of seed crystals of 99% *ee* were used during the process. The net gain of 152 g represented an 86% yield based on $NiCl_2$ and a 76% conversion of DL-ACL to L-ACL for a net crystallization rate of $54\ g\ l^{-1}\ h^{-1}$. The concentration of $(ACL)_3NiCl_2\cdot$EtOH in solution during the run averaged 22 ± 2 wt% with an *ee* of D-ACL of $5 \pm 2\%$. Approximately 1% of the ACL in solution was hydrolysed to lysine owing to the presence of adventitious amounts of water.[3c]

3.4 ANCILLARY STEPS

The ethoxide required for the racemization of D-ACL is conveniently produced from a weak base by a series of equilibria. In the first step (reaction 1), a methanolic solution of nickel(II) chloride is treated with a weakly basic ion-exchange resin such as Dowex-MWA1 (represented by R_3N). The formation of methoxide is favored because it becomes a weaker base than the resin when strongly bound to the nickel. The equilibrium is prevented from returning to the left in the subsequent steps because the HCl is physically removed with the resin.

In the second step (reaction 2), methoxide is displaced from the nickel complex when the mixture is added to excess of ACL. Three molecules of the bidentate

ACL coordinatively saturate the nickel, ensuring that this equilibrium lies to the right.

In the final step (reaction 3), the equilibrium between methoxide and ethoxide is forced to the right by the physical removal of methanol by distillation.

The resin is regenerated by treatment with ammonia and so the net process is the generation of a strong base (ethoxide) from a much weaker base (ammonia).

$$NiCl_2 + R_3N + MeOH \rightleftharpoons Ni(OMe)Cl + R_3NH^+Cl^- \quad (1)$$

$$Ni(OMe)Cl + 3ACL \rightleftharpoons (ACL)_3NiCl^+MeO^- \quad (2)$$

$$MeO^- + EtOH \rightleftharpoons MeOH + EtO^- \quad (3)$$

The product of the resolution/racemization, $(\text{L-ACL})_3NiCl_2 \cdot EtOH$, is converted into L-ACL·HCl by dissolution in methanol and decomposition with anhydrous hydrogen chloride (reaction 4). L-ACL·HCl of 99–100% *ee* crystallizes, although the original complex has a lower enantiomeric purity. This optical enrichment is due to the fact that DL-ACL·HCl is a racemic mixture (conglomerate) in methanol, and remains in solution while L-ACL·HCl crystallizes. Small amounts of nickel remaining in the product are removed by treatment with a chelating resin.[3c]

$$(\text{L-ACL})_3NiCl_2(l) + 3HCl \longrightarrow 3\text{L-ACL}\cdot HCl(s) + NiCl_2 \quad (4)$$

L-ACL·HCl is converted into L-lysine by hydrolysis with hydrochloric acid, racemization during hydrolysis being limited to less than 0.2% by operating at 115°C or lower. L-Lysine hydrochloride is recovered by crystallization after evaporation of excess of hydrochloric acid, addition of methanol and neutralization with ammonia.[3c] A closed-loop scheme has been devised for the recovery of residual lysine from the mother liquors.[14]

3.5 MICRO-PILOT PLANT STUDIES

The process and ancillary steps were further studied in a micro-pilot plant (Figure 1). The resolution/racemization reactor was a 5 l continuous-flow, stirred, evaporative crystallizer, and all other pieces of equipment were operated in batch fashion.

DL-ACL, $NiCl_2$, Ni(OMe)Cl and ethanol were fed to the reactor along with a slurry of $(\text{L-ACL})_3NiCl_2 \cdot EtOH$ seed crystals. The grown crystals were filtered, washed with ethanol and partly dissolved in methanol. The undissolved portion had an improved *ee* and was returned to the reactor as seed. Next, the solution was treated with hydrogen chloride in a stirred vessel to decompose the $(\text{L-ACL})_3$-$NiCl_2$ complex, and the L-ACL·HCl crystals formed were filtered, washed with methanol and sent to the hydrolysis section. The mother liquor and wash liquors were sent to a weakly basic ion-exchange column to produce Ni(OMe)Cl (reaction 1). The small portion of nickel ion that was held on the resin was

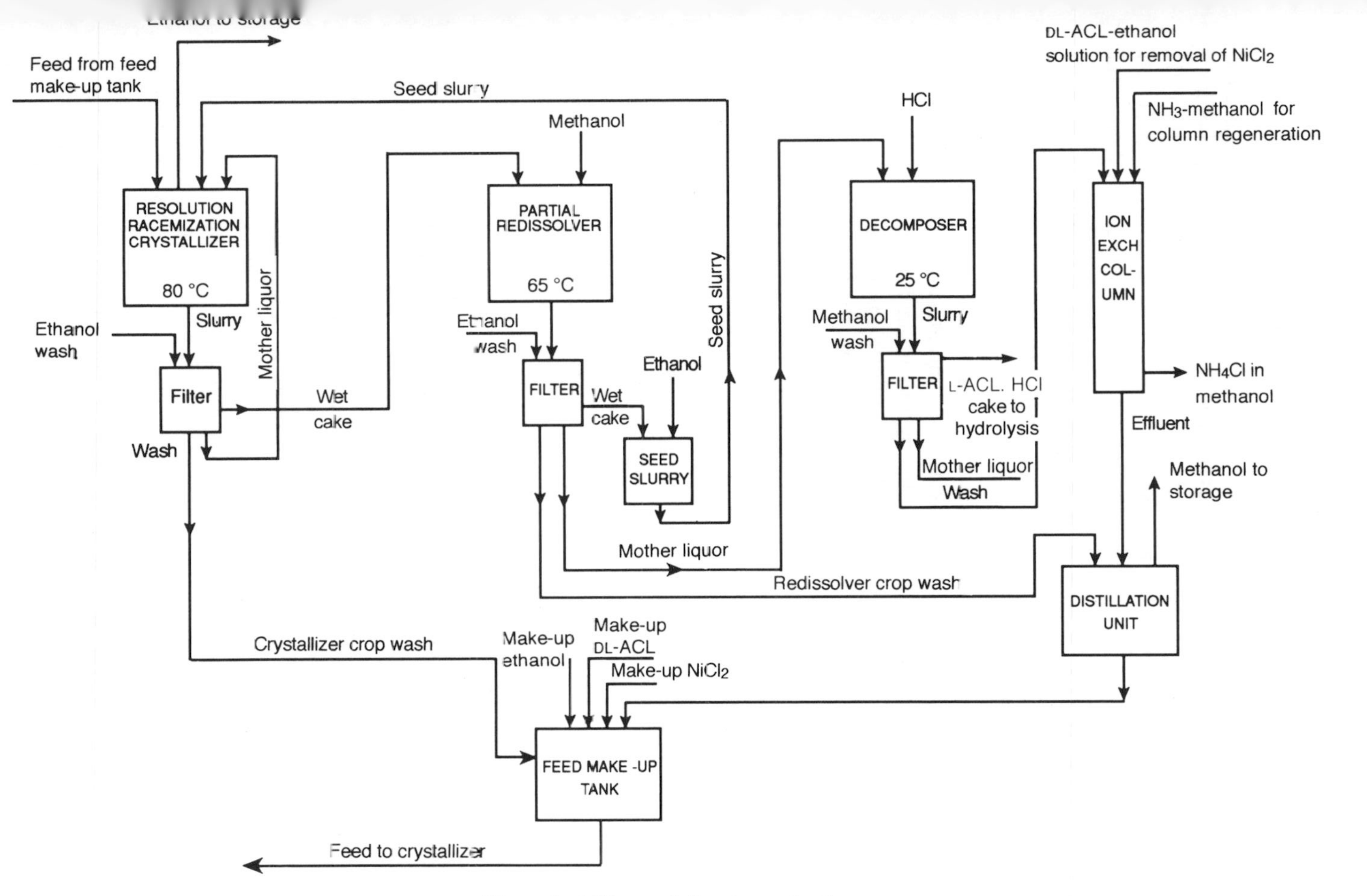

Figure 1. ACL resolution/racemization

removed by treatment with a DL-ACL–ethanol solution. The combined effluents were submitted to distillation to remove methanol and returned to the resolution/racemization reactor, and the resin was regenerated with methanol–ammonia.

The most important factors affecting the resolution/racemization were the size of the seed crystals, the ratio of net product to seed crystals (net growth factor), the crystal slurry density, the ratio of ACL to nickel in solution and the ratio of ethoxide to nickel. The water concentration was kept below certain limits. These factors will be briefly examined.

With steady-state operation, the rate of (L-ACL)$_3$NiCl$_2$·EtOH crystallization and removal must equal the feed rate of ACL and $NiCl_2$. The rate of (L-ACL)$_3$NiCl$_2$·EtOH crystallization increases with the surface area of the crystals present in the reactor and with the concentration of $(ACL)_3NiCl_2$ in solution. Spontaneous nucleation, which results in the formation of crystals of the D-enantiomer, is triggered once that concentration exceeds certain limits. The higher the concentration of D-ACL in solution, the lower are those limits. Note that the D-ACL concentration is governed by the rate of racemization. The concentration of $(ACL)_3NiCl_2$ in solution increases as the surface area of the crystals decreases.

To obtain a product of high *ee* at a given feed rate, it was found to be convenient to use small seed crystals at a high slurry density. This provided a large surface area for crystallization, and resulted in a relatively low steady-state concentration of $(ACL)_3NiCl_2$ in solution that forestalled spontaneous nucleation. It was also important to maintain a high rate of racemization.

In laboratory experiments, seed crystals as small as 3 μm average diameter were used to obtain excellent enantiomeric selectivity at relatively high rates of crystallization, and net growth factor. In the micro-pilot plant studies several disadvantages of such small crystals became apparent: the product was difficult to filter, an excessive amount of washing was required to remove adhering mother liquor and the crystal slurry had high viscosity and tended to plug transfer lines (plugging may be less of a problem in a commercial-scale plant).

A compromise was struck between the demand for high *ee* and mechanical limitations, by coordinating seed crystal average size, slurry density and net growth factor to arrive at conditions under which a product *ee* of 92% could be maintained at a net crystallization rate of up to 60 g l^{-1} h^{-1}.

The rate of racemization of D-ACL depends on the ACL/nickel and ethoxide/nickel molar ratios in solution. High ACL and ethoxide concentrations favor racemization but they hinder crystallization by increasing the solubility of the ACL–$NiCl_2$ complex. The optimum ratios are *ca* 4.5 and 0.15, respectively. These ratios were established by the initial charge to the reactor, and maintained by using ratios of 3.2 and 0.004, respectively, in the feed to the reactor to compensate for losses due to entrainment of mother liquor with the product.

Water has deleterious effects on the process because it inhibits the racemization of D-ACL. A primary effect is the rapid reaction with ethoxide to produce ethanol and hydroxide, which is a weaker base, and a secondary effect is slow hydrolysis of ACL to lysine, which neutralizes an equivalent of ethoxide. Additionally, lysine inhibits the crystallization (see the next section). Water content is controlled by using dry reagents and avoiding undue exposure to the atmosphere. It is also important to use a non-aqueous, non-hydroxylic base for the regeneration of the resin that is used to produce Ni(OMe)Cl, (methanol-ammonia serves that purpose well). Up to 0.2 wt% of water can be tolerated in the resolution/racemization at steady state.

3.6 SOURCE OF SEED CRYSTALS FOR START-UP

In a commercial resolution/racemization plant, a large quantity of L-ACL must be available to prepare seed crystals for start-up. A ready source is commercially available L-lysine hydrochloride, which can be cyclized to L-ACL in a two step process.[3c] This, however, is a laborious expedient that would require additional equipment.

The author and co-workers discovered a variant of the process that produces $(L\text{-}ACL)_3NiCl_2 \cdot EtOH$ crystals of modest *ee* without the need for optically active seed. When a supersaturated solution of DL-ACL, $NiCl_2$ and ethoxide ion is allowed to crystallize in the presence of racemic seed crystals, the product is invariably racemic. If, however, a catalytic amount of L-lysine is added to the solution (*ca* 4 mol% on ACL), the crystals formed are predominantly $(L\text{-}ACL)_3NiCl_2 \cdot EtOH$ (the enantiomeric complex prevails if D-lysine is used).[15] The *ee* of the product is *ca* 60%, but it can be upgraded easily by partial dissolution and/or recrystallization.

In the presence of $(L\text{-}ACL)_3NiCl_2 \cdot EtOH$ seed crystals of high *ee*, no additional advantage is gained by doping with L-lysine, which in fact inhibits crystallization. Doping such crystals with D-lysine inhibits crystallization to an even greater extent. These effects cast some light on the mechanism of the doping: racemic seed crystals are a conglomerate of D and L enantiomers. Doping with L-lysine inhibits crystallization on the D crystals more that it does on the L crystals, and therefore the latter grow preferentially. If L crystals of high purity are used as seed, no D crystals are present and doping serves no useful purpose.

3.7 CONCLUSION

The refinement of the second order asymmetric transformation outlined above led to a viable process, operated on a pilot-plant scale, for a commercially useful product. The method could be used, at least in principle, with equal ease for the production of D-ACL and D-lysine, and there appears to be some demand for the latter.[16]

Second order asymmetric transformations have been used in the manufacture of several other amino acids, as discussed in Chapters 8 and 20.

3.8 ACKNOWLEDGEMENTS

The author gratefully acknowledges contributions by Drs W. J. Boyle, Jr. and J. F. van Peppen in the discovery of the second-order asymmetric transformation of ACL; A. S. Bogeatzes, S. E. Belsky, V. H. Chudamani and E. A. Egbuniwe in the design of the micro-pilot plant; and Dr. M. Lahav of the Weizmann Institute of Science, Rehovot, Israel, for the discovery of a process for the preparation of $(L\text{-}ACL)_3NiCl_2{\cdot}EtOH$ crystals in the absence of optically active seed while on sabbatical leave at Allied-Signal.

3.9 NOTES AND REFERENCES

1. See bibliography in ref. 3c for references to ACL resolution.
2. An ingenious process has been devised for converting DL-ACL into L-lysine with simultaneous racemization of D-ACL using microorganisms; Toray Industries, *Jpn. Pat.*, 73 28 679, 1973.
3. (a) Sifniades, S., Boyle, W. J., Jr, and van Peppen, J. F., to Allied Chemical, *US Pat.*, 3 988 320, 1976; (b) Sifniades, S., Boyle, W. J., Jr, and van Peppen, J. F., *J. Am. Chem. Soc.*, **98**, 3738 (1976); (c) Boyle, W. J., Jr, Sifniades, S., and van Peppen, J. F., *J. Org. Chem.*, **44**, 4841 (1979).
4. Eliel, E. L., *Stereochemistry of Carbon Compounds*, McGraw-Hill, New York, 1962, p 42.
5. Secor, R. M., *Chem. Rev.*, **63**, 297 (1963).
6. Pincock, R. E., and Wilson, K. R., *J. Am. Chem. Soc.*, **93**, 1291 (1971).
7. Wynberg, H., and Groen, M. B., *J. Am. Chem. Soc.*, **90**, 533, (1968).
8. Baker, W., Gilbert, B., and Ollis, W. D., *J. Chem. Soc.*, 1443 (1952); Powell, H. M., *Nature (London)*, **170**, 155 (1952).
9. Havinga, E., *Biochim. Biophys. Acta*, **13**, 171 (1954).
10. Job, R. C., *J. Chem. Soc., Chem. Commun.*, 258 (1977). Although an auxiliary chiral agent was used, it was present in only catalytic (1%) amounts.
11. Several examples of second order asymmetric transformations involving tetrahedral carbon have been shown in recent years, e.g. Hongo, C., Tohyama, M., Yoshioka, R., Yamada, S., and Chibata, I., *Bull. Chem. Soc. Jpn.*, **58**, 433 (1985).
12. Weidler-Kubanek, A.-M., Kim, Y. C., and Fuhrmann, R., *Inorg. Chem.*, **9**, 1282 (1970); Allied Chemical, *US. Pat.*, 3 692 775, 1972.
13. Kubanek, A.-M., Sifniades, S., and Fuhrmann, R., to Allied Chemical, *US Pat.*, 3 824 231, 1974.
14. Sifniades, S., to Allied Chemical, *US Pat.*, 3 917 684, 1975.
15. Sifniades, S., Lahav, M., and Boyle, W. J., Jr, to Allied Chemical, *US Pat.*, 4 259 239, 1981; van Mill, J., Addadi, L., Lahav, M., Boyle, W. J., Jr, and Sifniades, S., *Tetrahedron*, **43**, 1281 (1987).
16. *Chem. Econ. Eng. Rev.*, **15**, No. 5 (No. 168), 50 (1983).

4 The Development and Manufacture of Pyrethroid Insecticides

J. MARTEL
Roussel-Uclaf, Romainville, France

4.1 INTRODUCTION

The development of the extremely effective and successful family of modern pyrethroid insecticides represents a major achievement by the fine chemicals industry. Pyrethrins, which are natural insecticides found in some *Chrysanthemum* species, provided the model from which ever more active synthetic analogues were devised. Biological properties were systematically refined until compounds were obtained which represent some of the most potent insecticides known to man, e.g. deltamethrin (**1**).[1] Many hundreds of tonnes of these compounds

Chirality in Industry. Edited by A. N. Collins, G. N. Sheldrake and J. Crosby

are now produced worldwide, and the quest for higher activity and optimum cost efficacy has prompted great efforts by industrial and academic groups towards the development of practical, stereoselective routes to the most active stereoisomers. Indeed, the diversity of synthetic strategies which have been developed in various companies is a tribute to the ingenuity of the research groups involved.

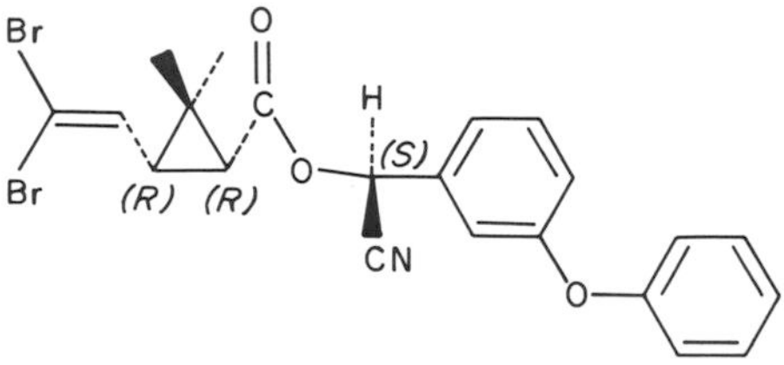

(1)
Deltamethrin

It is not possible in this short chapter to survey the whole field of pyrethroid synthesis and, in any case, several reviews have covered the topic.[2,3] Some reference has been made already to the use of terpenoid natural products in pyrethroid synthesis (see Section 1.2.1.4). The aim of this chapter is to review the pioneering work of Roussel–Uclaf on the development of new pyrethroids, with special emphasis on the resolution processes which have been adopted for their manufacture.

4.2 THE PYRETHRINS

The insecticidal properties of pyrethrum, a petroleum extract of the dried flowers of *Chrysanthemum cinerariaefolium*, have been recognized for many centuries. This product, which is still produced commercially in East Africa, has a potent 'knock-down' effect on flying insects, and is particularly suitable for use in a domestic environment. The knock-down action is reversible, however, and to overcome this, synergists such as piperonyl butoxide (**2**) and piperonal (**3**) may be added to pyrethrum preparations. These compounds potentiate the action of pyrethrum by inhibiting detoxifying esterase enzymes. The combination of pyrethrum with synergists results in a very safe insecticide

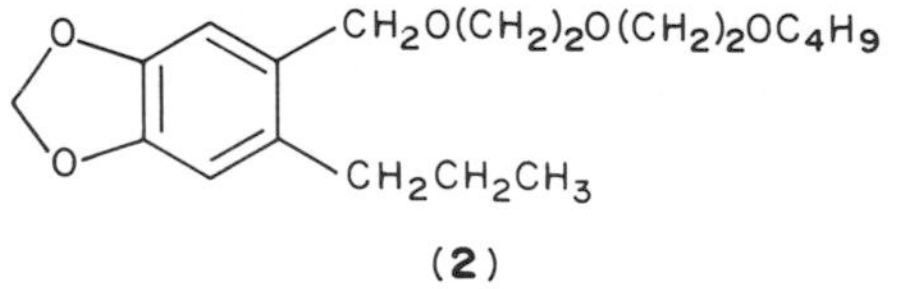

$CH_2O(CH_2)_2O(CH_2)_2OC_4H_9$ $CH_2CH_2CH_3$ CHO

(2)
Piperonyl butoxide

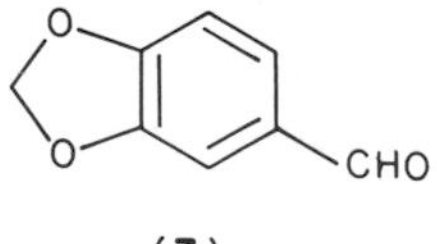

(3)
Piperonal

which is not toxic to mammals. Pyrethrum preparations are, however, labile to heat, light and air oxidation, which limits their utility for crop protection.

The biological properties of pyrethrum are now known to be associated with six esters, known collectively as 'pyrethrins,' which constitute about 20% of the dried flower extract.[4] Three of these are esters of the cyclopropanecarboxylic acid **4**, termed chrysanthemic acid, while the others are derivatives of the closely related compound pyrethric acid (**5**). Owing to the presence of two chiral centers in the cyclopropane ring, $2^2 = 4$ stereoisomers of chrysanthemic acid are possible. As is the case for many compounds with a biological function, the differing spatial arrangement of functional groups in each of the stereoisomers has a profound effect on the ease with which the molecule can interact with the target site. Chrysanthemic acid derived from the natural source contains only the (1*R*, 3*R*) stereoisomer [also designated (1*R*, *trans*)].

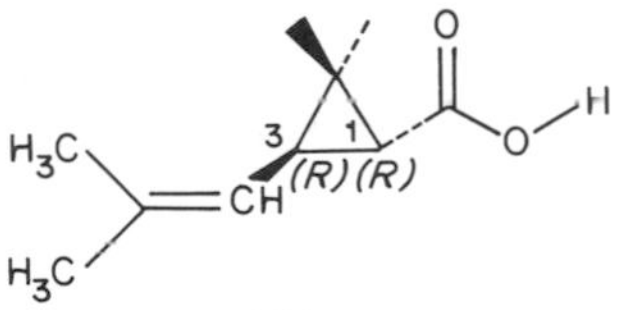

(**4**)
(1*R*)-*trans*

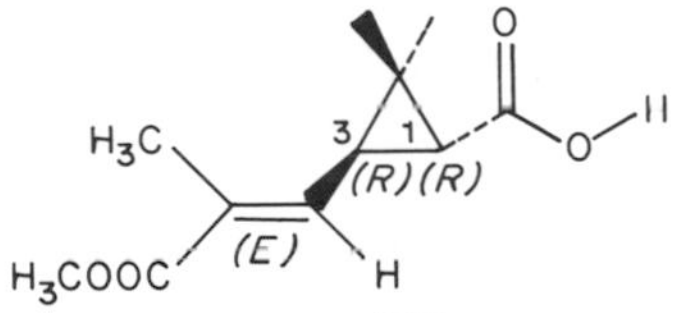

(**5**)
(1*R*)-*trans*

Pyrethric acid differs from chrysanthemic acid by the presence of a carbomethoxy group on the side-chain which confers on the molecule an additional source of isomerism, (*E* or *Z* at the double bond). Thus, eight geometric and optical isomers are possible, although the form which occurs in nature has only the *E* configuration at the double bond, and (1*R*, 3*R*) configurations at the cyclopropane ring. Three alcohols, known collectively as rethrolones (**6**), give the six constituents of pyrethrum noted above when esterified with the two natural forms of chrysanthemic and pyrethric acids. The chiral centre at C-4 of the cyclopentenolone ring, together with possible geometric isomerism of the

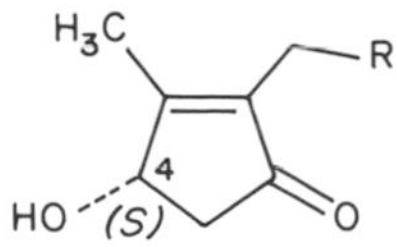

(**6**)
Rethrolones

Natural:

R = —CH$\overset{(Z)}{=}$CHCH=CH$_2$ Pyrethrolone

R = —CH$\overset{(Z)}{=}$CHCH$_3$ Cinerolone

R = —CH$\overset{(Z)}{=}$CHCH$_2$CH$_3$ Jasmolone

Table 1. Esters of the pyrethrins mixture

Alcohol component	Acid component: Chrysanthemic acid (**4**)		Acid component: Pyrethric acid (**5**)	
Pyrethrolone	Pyrethrin I	Pyrethrins I	Pyrethrin II	Pyrethrins II
Cinerolone	Cinerin I		Cinerin II	
Jasmolone	Jasmolin I		Jasmolin II	

side-chain double bond, gives rise to four possible stereoisomers of each rethrolone. The rethrolones pyrethrolone, jasmolone and cinerolone are closely related molecules in which the stereochemical configurations are *Z* for the double bond in the side-chain and *S* for the C-4 carbon of the cyclopentenolone ring. The esters of chrysanthemic acid with the three rethrolones are collectively termed 'pyrethrins I' and the corresponding esters of pyrethric acid 'pyrethrins II' (see Table 1).

Thus, while the natural pyrethrum extract contains six compounds (pyrethrins I and II), the reconstitution of this mixture by chemical synthesis (without control of stereochemistry) would produce a mixture of 144 compounds (Table 2). In fact, many of these have been prepared separately, and it has been shown that none matches the insecticidal potency of the naturally occurring representatives, the majority being much inferior.

In order to develop synthetic variants of the pyrethrins, methods for the construction of generic acid and alcohol structures and techniques for the isolation of individual stereoisomers had to be developed. The enormity of these tasks is apparent when one considers that in 1964 when research in the field of pyrethrin synthesis began at Roussel-Uclaf, the total synthesis of natural rethrolones, even as racemates, was unknown, as was a practical and unambiguous route to pyrethric acid.[5]

Table 2. Possible stereoisomers in reconstituted synthetic pyrethrum

Compound	Number of asymmetric factors	Number of obtainable isomers
Pyrethrin I	4	$2^4 = 16$
Pyrethrin II	5	$2^5 = 32$
Cinerin I	4	$2^4 = 16$
Cinerin II	5	$2^5 = 32$
Jasmolin I	4	$2^4 = 16$
Jasmolin II	5	$2^5 = 32$
		Total 144

4.3 THE ALLETHRIN SERIES

4.3.1 FROM ALLETHRIN TO (*S*)-BIOALLETHRIN[R]

Allethrin (**7**) was the first synthetic pyrethroid to be produced in large quantities.[6] Chemically, it is closely related to the natural product, although a simplification of the side-chain of the alcohol component [allethrolone, (B) in Scheme 1] removes one possibility of geometric isomerism. Both the acid and alcohol components of allethrin were comparatively easy to synthesize. A copper-catalysed carbene insertion reaction was used to produce racemic chrysanthemic acid (A) as a mixture of *cis* and *trans* isomers. Racemic allethrolone (B) was also prepared in a one-step process as shown in Scheme 1.

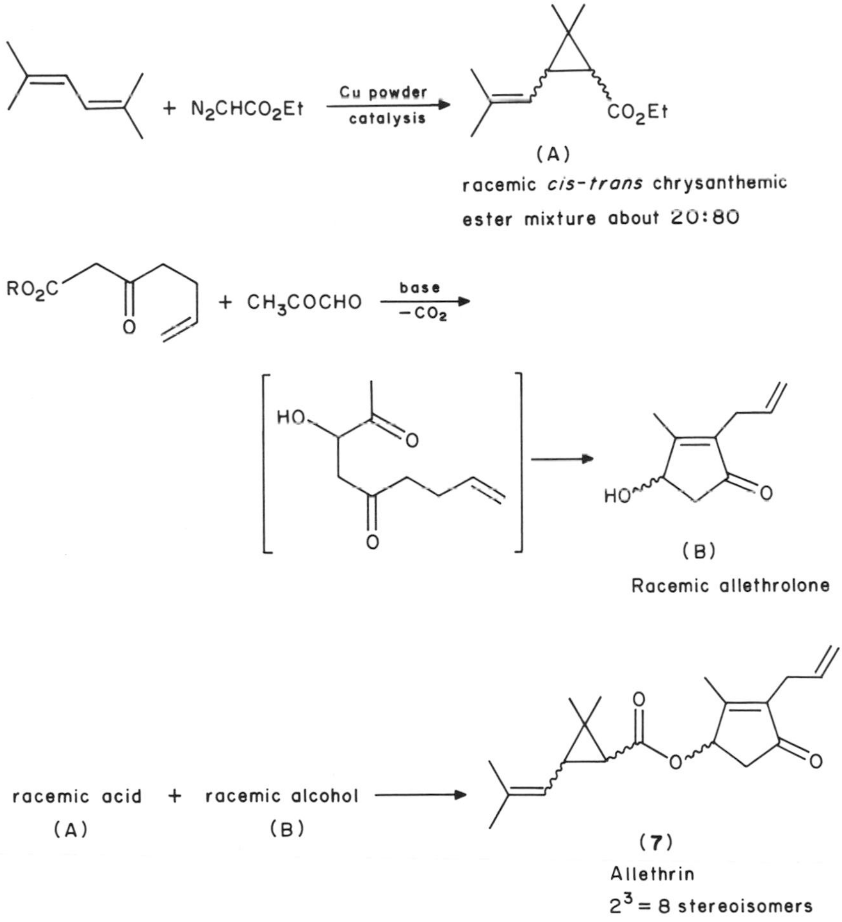

+ N_2CHCO_2Et —Cu powder catalysis→ CO_2Et

(A)

racemic *cis-trans* chrysanthemic ester mixture about 20:80

RO_2C … O + CH_3COCHO —base, $-CO_2$→

[HO … O … O] → HO … O

(B)

Racemic allethrolone

racemic acid (A) + racemic alcohol (B) → O … O … O

(**7**)

Allethrin

$2^3 = 8$ stereoisomers

Scheme 1. Synthesis of allethrin

Although allethrin reproduces the pyrethrins' mode of action, it is considerably less potent. Only one of the mixture of eight stereoisomers obtained by the synthetic route possesses useful levels of activity, this being the isomer with the stereochemical characteristics of the natural products of the pyrethrins I family, i.e. (1*R*, *trans*) for the acid moiety (4*S*) for the alcohol. The content of this desirable isomer in allethrin prepared by the original route did not exceed 20%. As a fourfold increase in potency might be expected by esterifying (1*R*, *trans*)-chrysanthemic acid with racemic allethrolone (a reduction in the number of isomers of allethrin from eight to two), a robust, stereoselective route to this acid moiety became a primary research target.

4.3.2 ACCESS TO (1*R*, *trans*)-CHRYSANTHEMIC ACID

At the time of this research, asymmetric synthesis had been little used on an industrial scale. A more feasible approach might have been chemical resolution of the *trans* racemate, although efficient isolation of one enantiomer from the four constituents of the commercially available *cis–trans* mixture seemed unlikely. Another problem with this approach would have been the toxic and physical hazards associated with the industrial operation of the diazoacetate reaction (see Scheme 1). For these reasons, a safer synthesis which was selective for the *trans* geometric isomer of chrysanthemic acid was sought.

A rational approach which takes into account the monoterpenoid nature of the target molecule is shown in Scheme 2. The proposed condensation of two five-carbon fragments reflects the biogenesis of (1*R*, *trans*)-chrysanthemic acid precursors in plants, which occurs by 'middle to tail' fusion of two isoprenylpyrophosphate units.[7] Although this unusual combination of isoprenoid units had at the time (1964) no equivalent in chemical synthesis, the availability of two potential starting materials, i.e. 3,3-dimethylacrylate esters and prenyl halides, was recognized. To effect the α, β- and α, γ-fusions, it would be necessary first to convert the halogen (X) of the prenyl halide into an electron-withdrawing group (Y), to allow the generation of a carbanion at the α-carbon and a possible Michael addition to the conjugated ester (α, γ-fusion). If the activating group (Y) were also a leaving group, then cyclopropane ring formation could follow as a result of an intramolecular nucleophilic displacement.

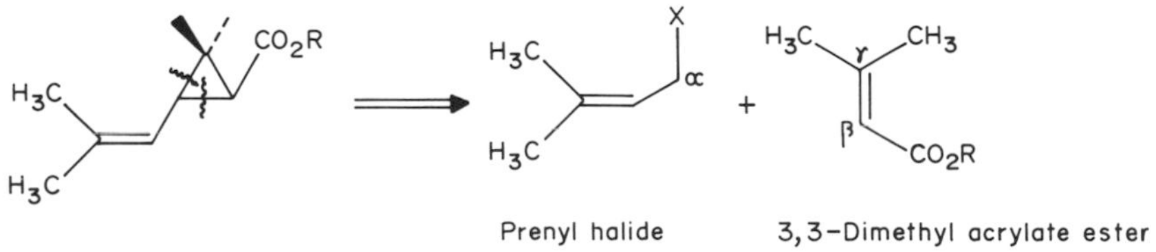

Prenyl halide

3,3-Dimethyl acrylate ester

Scheme 2. Retrosynthetic analysis of *trans*-chrysanthemic acid

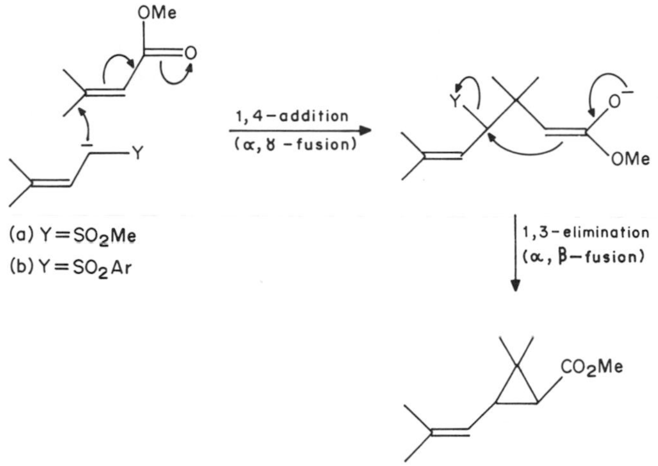

Scheme 3. Synthesis of *trans*-chrysanthemic acid

While the sulphone group ($Y = SO_2R$ in Scheme 3) would fulfil the criteria above, the Michael addition of carbanions to 3,3-disubstituted acrylate esters was known to be difficult to achieve. Indeed, the first trials with prenyl methyl sulphone [(a) in Scheme 3] and methyl 3,3-dimethyl acrylate did not lead to the expected cyclopropane, but gave instead the six-membered cyclic sulphone **9** in good yield. This result did confirm, however, the feasibility of the conjugate addition of carbanions stabilized by α-sulphones to dimethylacrylate esters. Formation of the six-membered ring was probably the result first of an equilibrium between the initially formed ester carbanion **8a** and the methyl sulphone anion **8b**, which can attack the ester in an intramolecular fashion as shown in

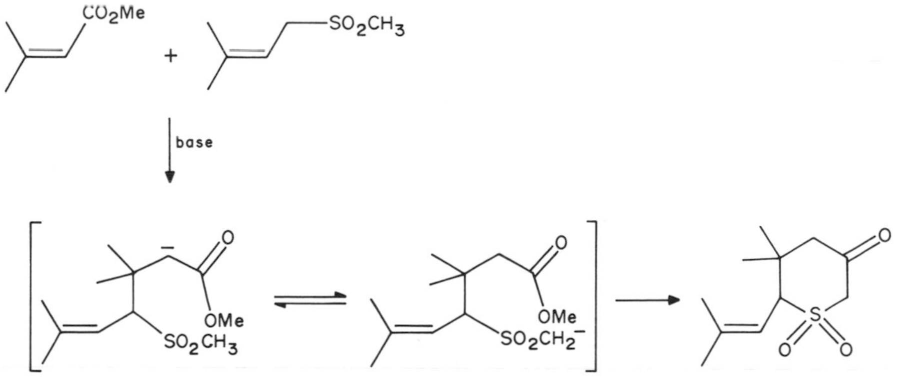

Scheme 4. Formation of cyclic sulphone **9**

Scheme 4. When the alkyl group of the sulphone was replaced with an aryl group [(b) in Scheme 3] and the reaction was carried out in dimethyl sulphoxide or dimethylformamide with sodium methylate as base, chrysanthemic ester was obtained in an excellent yield. These conditions favoured the production of the more stable *trans* diastereoisomer to an extent of more than 95%. Saponification of this chrysanthemic ester provided the desired, racemic *trans* acid.

Resolution of racemates on an industrial scale is often a delicate process (see Chapter 2). In this case, the choice of the optically active base was critical; a clear crystallization of one of the resultant diastereomeric salts was required. Nearly ideal conditions for the resolution of *trans*-chrysanthemic acid were obtained with the amine **10** from the Roussel-Uclaf process for the manufacture of chloramphenicol.

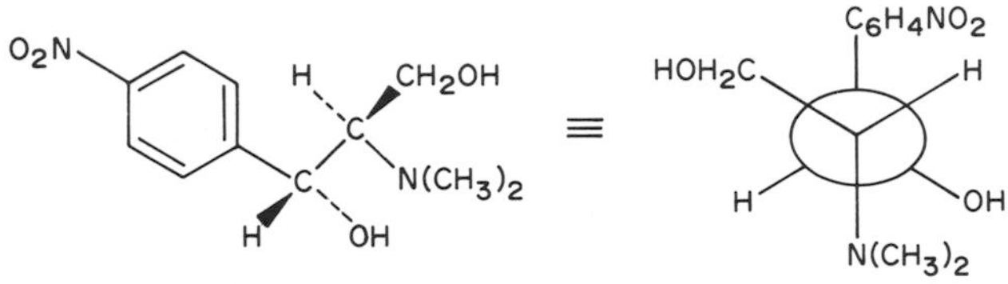

(10)

Treatment of *trans*-chrysanthemic acid with this D-(−)-*threo*-amine resulted in the clean separation of the salt of the (1*R*, *trans*) acid. Isolated by a simple filtration, this salt yielded, after acidification, pure chrysanthemic acid which was identical with the natural product (an efficient recovery and recycle of the base was also possible). The product of the esterification of (1*R*, *trans*)-chrysanthemic acid with racemic allethrolone was launched in 1967 as Bioallethrin^{R}. The content of the potent stereoisomer with the (4*S*) configuration in the alcohol component was, of course, only 50%. In order to achieve the goal of a selective industrial process for the (4*S*)-allethronyl-(1*R*, *trans*)-chrysanthemate, a good resolution of allethrolone was first required.

4.3.3 ACCESS TO (4*S*)-ALLETHROLONE

The resolution of allethrolone was achieved via the hemisuccinate esters as shown in Scheme 5. These formed separable diastereoisomeric salts with (+)-ephedrine, although the conditions for the crystallization process[8] were more critical than those for chrysanthemic acid. Subsequent production of the free enantiomers was achieved by careful hydrolysis of the hemiester under acidic conditions. Great care was necessary in the latter reaction to avoid degradation of the fragile allethrolone moiety or loss of optical purity.

This circuitous sequence of chemical reactions was expensive and difficult to implement on a large scale. Among the alternative methods which were developed for the resolution of the unstable racemic alcohols, one is worthy of mention

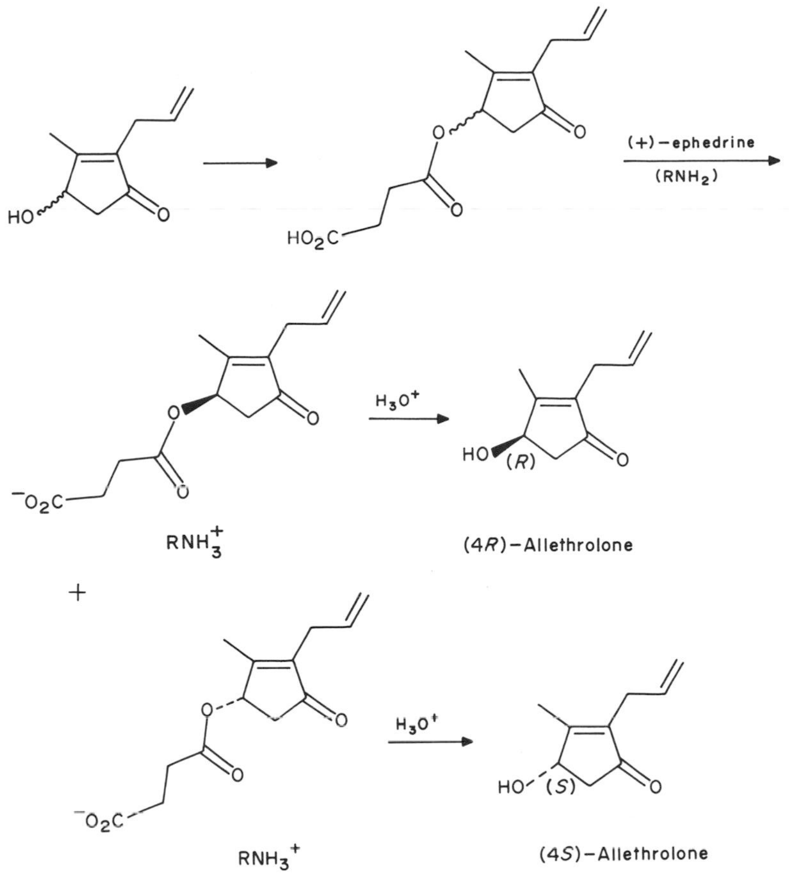

(+)-ephedrine (RNH_2)

H_3O^+

(4*R*)-Allethrolone

RNH_3^+

+

H_3O^+

(4*S*)-Allethrolone

RNH_3^+

Scheme 5

here, as it was also used to resolve cyanohydrins of type **11**, which are components of more recent photostable pyrethroids (see Section 4.4.3).

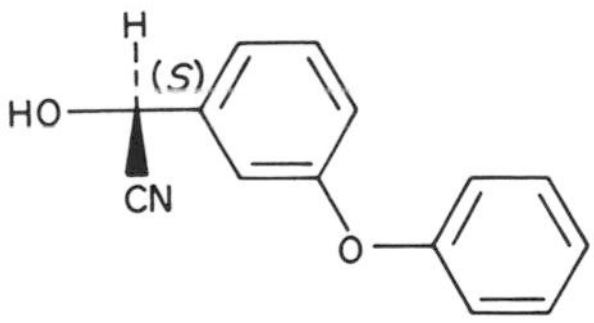

(11)

This technique[9] involved the formation of ketals of racemic alcohols with optically active *cis*-lactols. Among these reagents, (1*R*, *cis*)-caronaldehyde (**12**) was to become a cornerstone in pyrethroid chemistry. The diastereoisomeric ketals of racemic allethrolone (Scheme 6) were generated smoothly under acidic

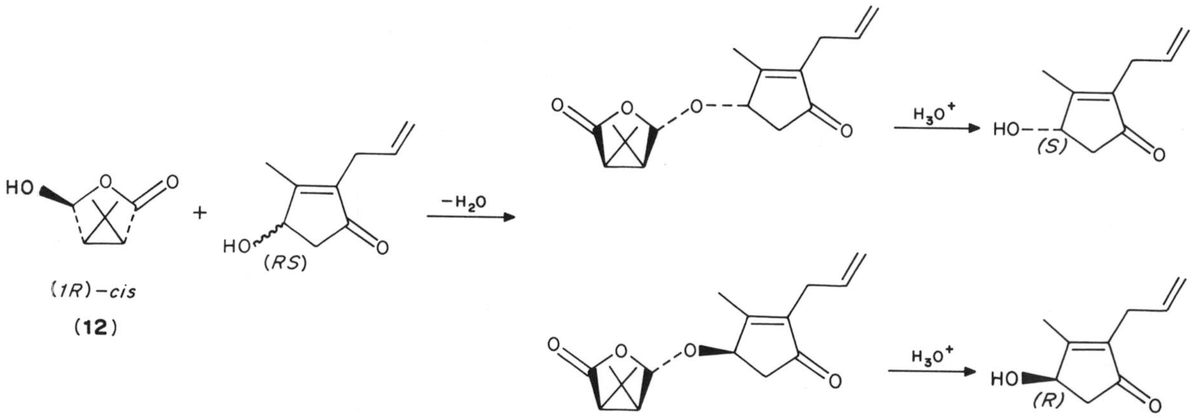

HO

(1*R*)−*cis*

(**12**)

+

HO

(*RS*)

$-H_2O$

H_3O^+

HO

(*S*)

H_3O^+

HO

(*R*)

Scheme 6

conditions with elimination of water. These compounds could be separated readily by crystallization, and were isolated from the medium with their chemical and optical purities intact.

A kinetic resolution of allethrolone via its acetate esters was also achieved by the use of various lipases.[10] These enzymes were used to cleave selectively the (*R*)-ester from the racemic mixture (Scheme 7). The reverse reaction, stereoselective acylation of the (*S*)-allethrolone enantiomer from the racemate, was also found to be feasible[11] (Scheme 8). In both cases, however, the separation of free alcohol from its acetate was difficult, and high chemical and optical yields could not be obtained simultaneously.

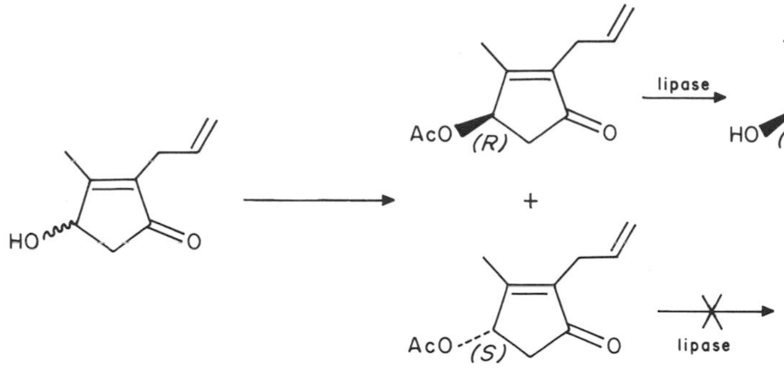

Scheme 7

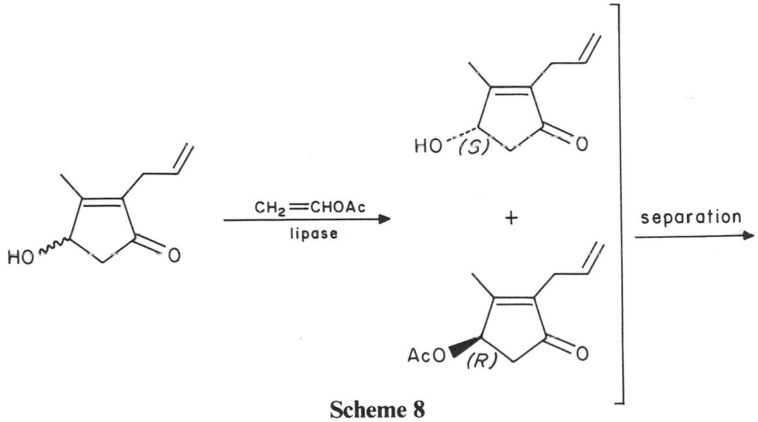

Scheme 8

The esterification of (*S*)-allethrolone with (1*R*, *trans*)-chrysanthemic acid gave (4*S*)-allethronyl-(1*R*, *trans*)-chrysanthemate, which was launched in 1973 under the trade name (*S*)-Bioallethrin^R^.

Although successful on a large scale, the resolution methodology chosen for the manufacture of Bioallethrin^R^ would not have been cost effective without an efficient means to recover and recycle both the residual (1*S*, *trans*)-chrysanthemic acid and the (4*R*)-allethrolone. This is discussed in the next section.

4.3.4 RECOVERY OF (1*S*, *trans*)-CHRYSANTHEMIC ACID

The recycle of unwanted enantiomers from resolution processes is generally achieved by conversion back to the initial racemate. In the case of *trans*-chrysanthemic acid, one method for racemization might involve a specific cleavage of the cyclopropane bond between the C-1 and C-3 chiral centres, followed by recombination. Processes which involved the homolytic cleavage of the cyclopropane were found to be difficult to control, however. Attempts to form the diradical by means of light or heat produced only 3,3-dimethylacrylate derivatives as a result of bond breakage between C-2 and C-3 in the initially formed 1,3-diradical (Scheme 9), and attempts to produce an ionic cleavage also met with only partial success.[12] Ultimately, the best conditions for racemization of chrysanthemic acid gave only 50% recovery from each run. For this reason, alternative procedures involving inversion of configuration at the asymmetric centres were investigated.

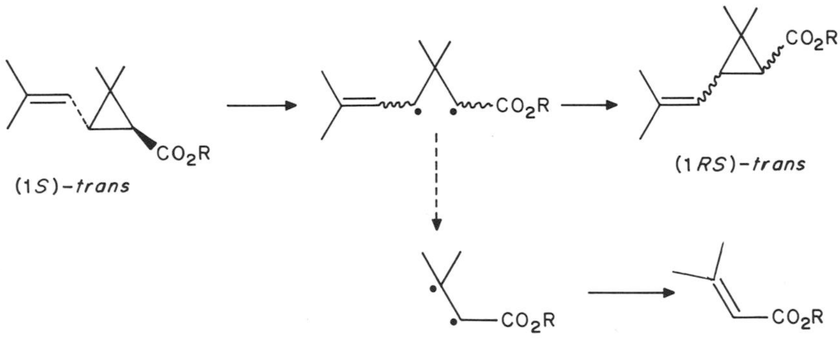

CO_2R (1*S*)-*trans* → CO_2R → CO_2R (1*RS*)-*trans*

CO_2R → CO_2R

Scheme 9

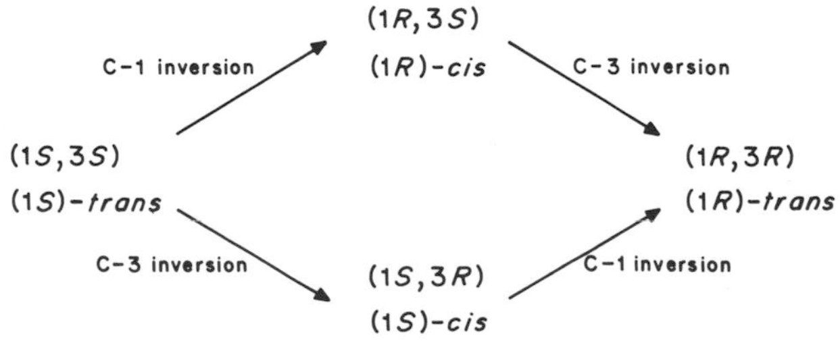

(1*S*,3*S*) (1*S*)-*trans* —C-1 inversion→ (1*R*,3*S*) (1*R*)-*cis* —C-3 inversion→ (1*R*,3*R*) (1*R*)-*trans*

(1*S*,3*S*) (1*S*)-*trans* —C-3 inversion→ (1*S*,3*R*) (1*S*)-*cis* —C-1 inversion→ (1*R*,3*R*) (1*R*)-*trans*

Scheme 10

Depending on the order of inversion of C-1 and C-3, two general strategies are possible (see Scheme 10). In either case, the first inversion would lead to a thermodynamically disfavoured *cis* derivative. This problem was solved by the formation of a bicyclic lactone (**13**) in which the *cis* geometry at the cyclopropane is forced (Scheme 11). Thus, hydration of (1*S*, 3*S*)-chrysanthemic acid followed by esterification produced the *trans*-alcohol ester. Base-catalysed epimerization of the C-1 centre in the presence of methoxide in methanol led to an equilibrium amount of the *cis* isomer, which lactonized under the reaction conditions.[13] The cleavage of lactone **13** with acid catalysis then gave (1*R*, *cis*)-chrysanthemic acid in good yield.

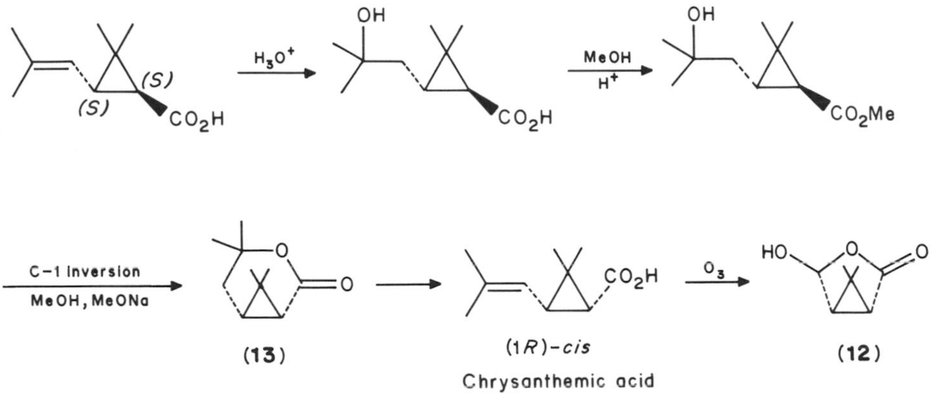

(**13**) (1*R*)-*cis* Chrysanthemic acid (**12**)

Scheme 11

The inversion of configuration at C-3 was achieved in the following manner. Ozonolysis of (1*R*, *cis*)-chrysanthemic acid and reduction *in situ* of the transient methoxy hydroperoxide produced caronaldehyde (**12**), a stable, crystalline compound which exhibits a preference for the bicyclic hemiacetal form. On treatment with aqueous base, **12** reverted to the monocyclic form, which possesses an enolizable hydrogen at C-3. Under these conditions, caronaldehyde equilibrates to its more stable *trans* configuration (Scheme 12). Finally, a Wittig olefination reconstituted the chrysanthemic acid side-chain and, at the same time, rendered the configuration at C-3 stable to further base-catalysed epimerization.

This versatile sequence could be operated in the reverse order, i.e. C-3 inversion followed by inversion at C-1 (Scheme 10). Thus, if (1*S*, *trans*)-chrysanthemic acid was subjected to ozonolysis followed by base-catalysed epimerization at C-3 and *cis*-lactol formation, (1*S*, *cis*)-caronaldehyde resulted (Scheme 12). A sequence of Wittig olefination followed by epimerization at C-1 produced (1*R*, *trans*)-chrysanthemic acid.

In this manner, two efficient methods for the recycle of (1*S*, *trans*)-chrysanthemic acid to its (1*R*, *trans*) isomer were achieved. A still more valuable

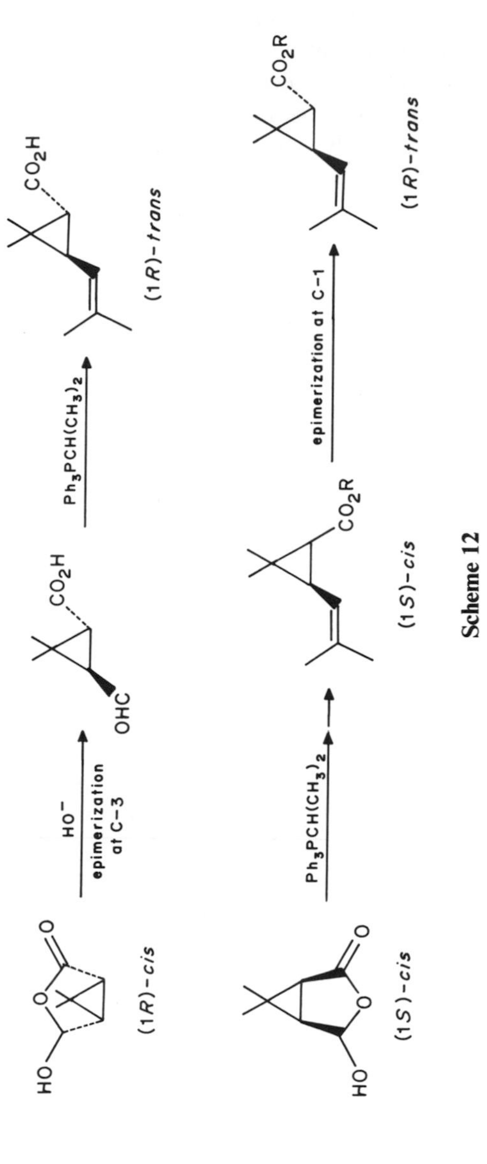
HO
O
O
(1R)-cis
HO⁻
epimerization
at C-3
OHC
CO2H
Ph3PCH(CH3)2
CO2H
(1R)-trans
HO
O
O
(1S)-cis
Ph3PCH(CH3)2
CO2R
(1S)-cis
epimerization at C-1
CO2R
(1R)-trans

Scheme 12

outcome of these studies, however, was an efficient route to (1*R*, *cis*)-caronaldehyde starting from either enantiomer of *trans*-chrysanthemic acid. (1*R*, *cis*)-Caronaldehyde is perfectly adapted to the production of artificial chrysanthemates belonging to the (1*R*, *cis*) series from which modern pyrethroids have evolved. Kadethrinic acid[14] (**14a**), for example, was accessible from caronaldehyde (**12**) either by Horner–Wittig olefination with a phosphonium ylid or by a three-step aldocrotonization route, as shown in Scheme 13. The ester of kadethrinic acid and 5-benzyl-3-furanylmethyl alcohol[15] is known as Kadethrin^R (**14b**), one of the most active 'knock-down' agents over produced on an industrial scale.

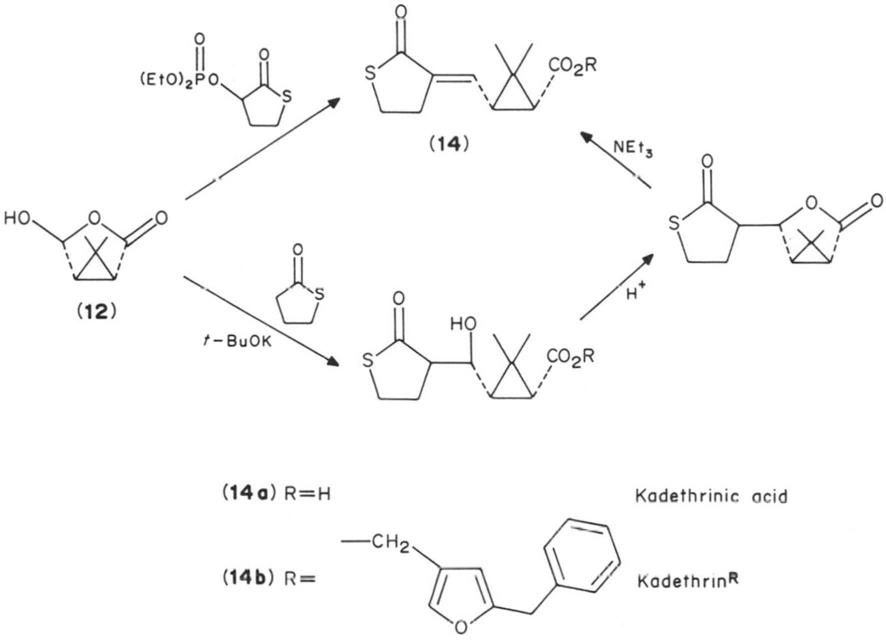

(**14a**) R=H Kadethrinic acid

(**14b**) R= Kadethrin^R

Scheme 13

4.3.5 RECOVERY OF (4*R*)-ALLETHROLONE

A general method for epimerization of an alcoholic chiral centre is to prepare the sulphonic ester, then to displace the sulphonate group in an S_N2 process with inversion of configuration. Initial attempts to carry out this sequence with the unwanted (*R*)-allethrolone enantiomer failed to give the expected sulphonic ester, however. Instead, treatment of (*R*)-allethrolone with an alkylsulphonyl chloride in pyridine produced the (*S*)-4-chlorocyclopentenone derivative as the major product, which resulted from displacement of the sulphonate by chloride.

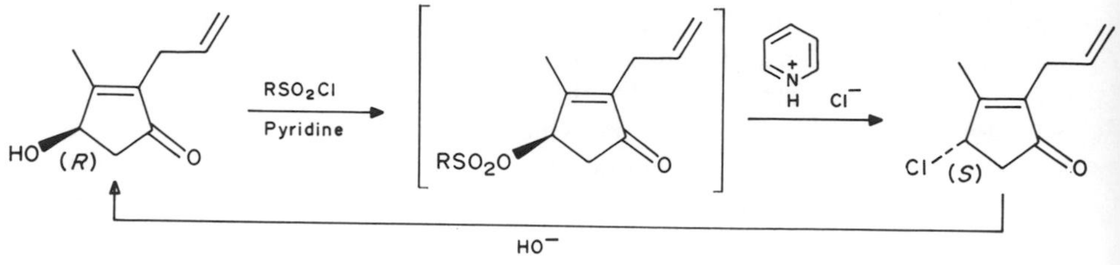

HO (R) O
RSO_2Cl
Pyridine
RSO_2O O
H Cl^-
Cl (S) O
HO^-

Scheme 14

Treatment of this prouct with aqueous base simply reformed the starting material (R)-allethrolone (Scheme 14). This problem was overcome by preparing instead the methanesulphonate ester via the sulphene reagent derived from methanesulphonyl chloride and triethylamine. In the subsequent displacement of the methanesulphonate ester, it proved difficult to find conditions which favoured the desired S_N2 reaction over the competing $E2$ and S_N1 processes (which led to a series of undesirable side-products as shown in Scheme 15).

Although nucleophiles such as hydroxide and monochloroacetate could be

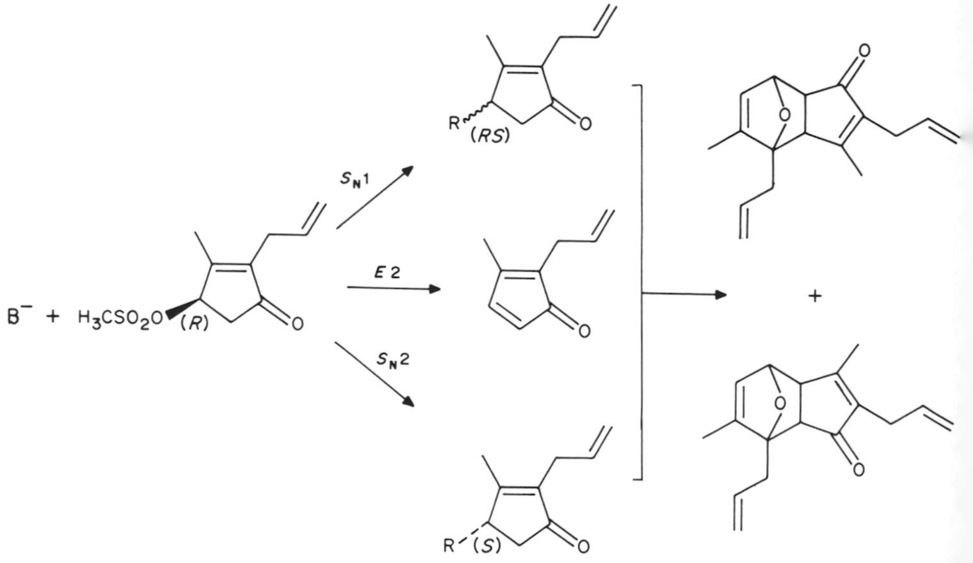

R (RS) O
S_N1
$E2$
S_N2
$B^- + H_3CSO_2O$ (R) O
R (S) O
+

Scheme 15

made to react in an S_N2 fashion with the methanesulphonate ester of allethrolone, the best results were obtained with the (1R, *trans*)-chrysanthemate ion (Scheme 16).

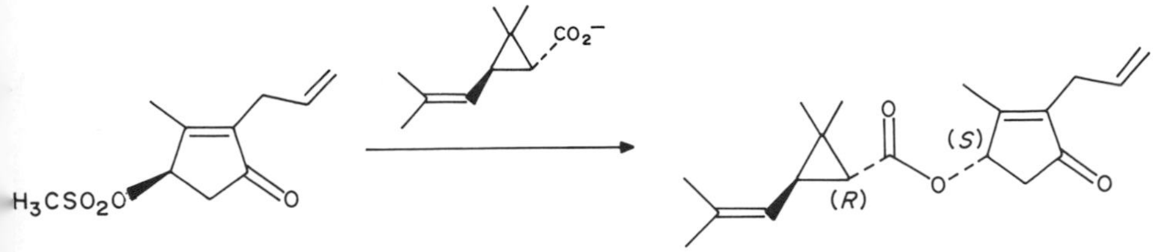

H$_3$CSO$_2$O (S) (R) CO_2^-

Scheme 16

The latter reaction gave pure (*S*)-Bioallethrin directly in excellent yield, and with no loss of optical purity.[16]

4.4 PHOTOSTABLE PYRETHROIDS

4.4.1 DELTAMETHRIN

Although successful in their own right, Bioallethrin^R and Kadethrin^R suffered from the same drawbacks as their natural precursors, i.e. instability to heat and light and susceptibility to oxidative breakdown by detoxifying enzymes in insects. As a consequence, the use of these products in agriculture was also limited. In the chrysanthemate moiety, susceptibility to light- or heat-induced degradation was found to be related to the presence of the vinylic *gem*-dimethyl group. Replacement of the methyl groups in this position by halogens (either chlorine in the case of **15**[17] or bromine in **16**[18]) led to a significant improvement in stability.

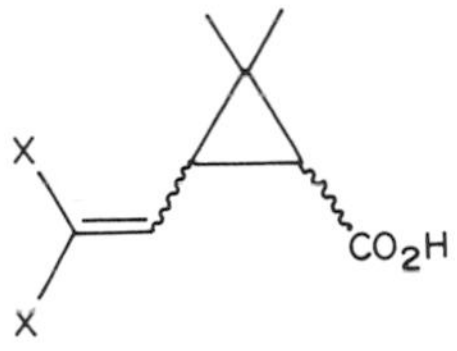

X X CO_2H

Racemic *cis–trans*

(**15**) X=Cl
(**16**) X=Br

The alcohol moieties **17–19** also contain functional groups which are susceptible to oxidation, i.e. terminal double bonds in alcohols **17** and **19**, the conjugated keto-alcohol system in **17**, the activated methylene group in **18** and **19** and the furan ring of **18**. These structural features are absent in alcohols **20**[19] and **21**[20] on which a new generation of pyrethroids was based. Esterification of benzylic alcohol **20** with the racemic *cis–trans* acid **15** (permethric acid) gave

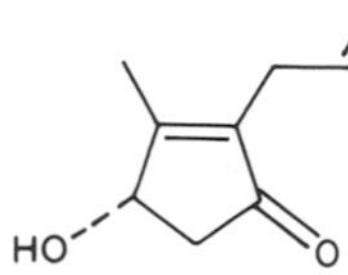

(17) 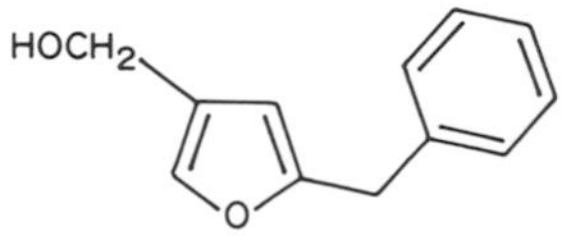 (18) 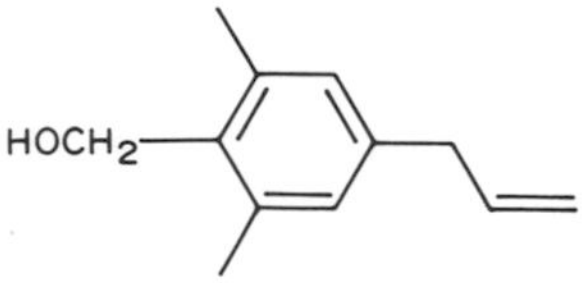 (19)

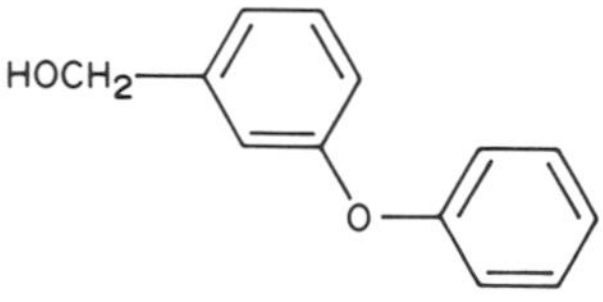

(20) 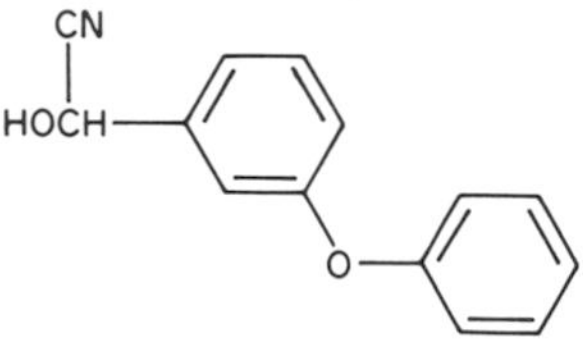 (21)

permethrin[18] (**22**), the first photostable pyrethroid with good insecticidal properties to be offered for crop protection. More active still was the ester cypermethrin[19] (**23**) derived from the racemic cyanohydrin **21** and permethric acid (**15**), despite the presence in this product of eight stereoisomers.

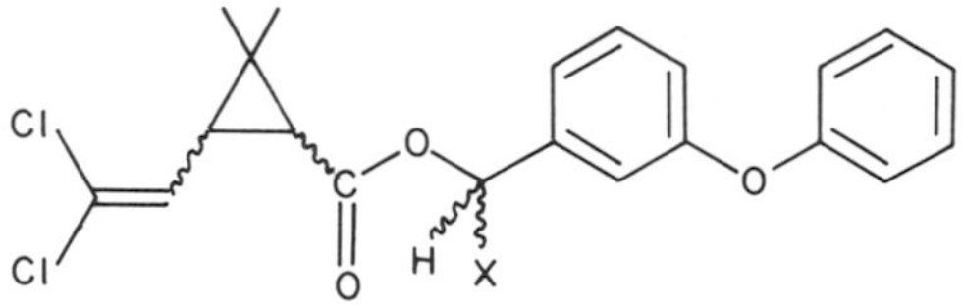

(**22**) X = H Permethrin
(**23**) X = CN Cypermethrin

Starting from (1*R*, *cis*)-chrysanthemic acid, available from Roussel-Uclaf, (1*R*, *cis*)-dibromovinyl acid (**16**) was prepared as a single stereoisomer at Rothamsted Experimental Station and combined with the racemic cyanohydrin **21**. The ester thus obtained as a mixture of two diastereoisomers was considerably more active than cypermethrin. More interesting still was the fact that from this two-component mixture, the crystallization of one diastereoisomer took place spontaneously. The crystalline stereoisomer was shown to be wholly responsible for the insecticidal effect. Initially, this product was assigned the trivial name decamethrin (reflecting the tenfold increase in activity over permethrin), although the name deltamethrin has been in use since 1980.

4.4.2 INDUSTRIAL ACCESS TO DELTAMETHRIN

Two possible industrial routes to the acidic moiety of deltamethrin from the available (1*R*, *cis*)-caronaldehyde (**12**) were investigated. Route A (Scheme 17)

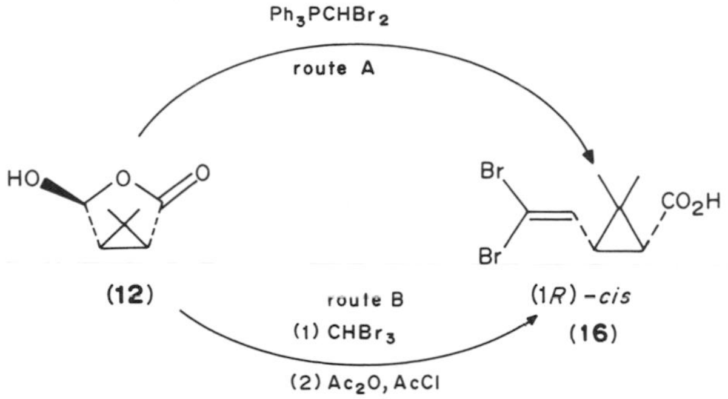

Scheme 17

involved a Wittig reaction with a phosphonium ylid containing the dibromomethylene group. In practice, a two-step synthesis (route B), involving condensation with bromoform followed by elimination of hydrogen bromide, was the more efficient, large-scale route.

4.4.3 RESOLUTION OF THE CYANOHYDRIN COMPONENT

The racemic cyanohydrin **21** can be prepared by various routes based on Ullmann coupling reactions (Scheme 18). As the crystallization of one diastereoisomer of deltamethrin from a mixture according to the discoverers' method was not felt to be practical on a large scale, methods to resolve this cyanohydrin precursor were investigated. The resolution of cyanohydrins, especially those with a particularly acidic α-hydrogen such as **21**, is a delicate process (see Section 14.2.1). Nevertheless, the method which had been applied previously[9] to the resolution of allethrolone (Scheme 6) was found to be adequate for the cyanohydrin **21**. Thus, treatment of the cyanohydrin with (1*R*, *cis*)-caronaldehyde (**12**)

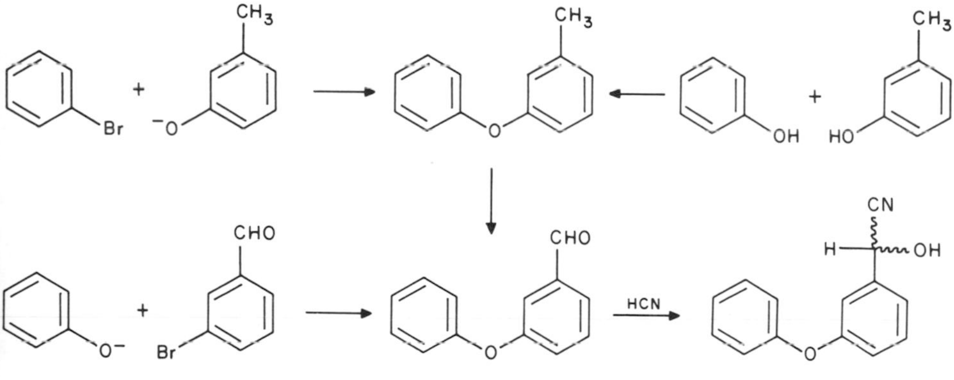

Scheme 18

gave a mixture of diastereoisomeric ketals. These were separated by fractional crystallization, the desired (*S*)-cyanohydrin being obtained by hydrolysis of the appropriate diastereoisomer. Conversion of (1*R*, *cis*)-deltamethrinic acid (**16**) to its acid chloride and subsequent esterification with the (*S*)-cyanohydrin **11** gave deltamethrin (**1**) with all three chiral centres in the correct configurations for optimum activity.

4.4.4 SECOND-ORDER ASYMMETRIC TRANSFORMATIONS

For the manufacture of deltamethrin on an industrial scale, a process was developed[21] which avoided the need to resolve the cyanohydrin **21**. This important simplification was made possible by three favourable properties of the product:

(a) the acidity of the benzylic proton which is activated by the adjacent cyano and benzyl groups;
(b) the crystalline nature of the desired diastereoisomer of deltamethrin;
(c) the insolubility of this diastereoisomer relative to its epimer at the α-carbon.

These factors were capitalized upon in the design of the following second-order asymmetric transformation. A solvent system for the final esterification was chosen in which the desired (1*R*, αS) diastereoisomer of deltamethrin had a very low solubility (Scheme 19). The addition of a weak organic base to the medium allows interconversion of the two diastereoisomers to take place via epimerization at the α-carbon (a careful choice of base and reaction conditions avoids hydrolysis of the cyano or ester functional groups). As the reaction proceeds, pure (αS)-deltamethrin crystallizes from solution leading to a deficiency of this diastereoisomer in the liquid phase, and a net conversion of the (αR) to the (αS) diastereoisomer then occurs in solution to restore the equilibrium between the diastereoisomers. The crystallization of the required (αS) stereoisomer thus drives the liquid-phase (αR) to (αS) transformation to an extent limited only by the solubility of the (αS) diastereoisomer in the reaction medium. In fact, this conversion is virtually complete in the industrial process.

Similar second-order asymmetric transformations have been applied to other pyrethroids. The technique can even be used in cases where the substrate is a mixture of the four stereoisomers which result from the use of racemic acid and alcohol moieties. An industrial route[22] to *cis*-cypermethrin, for example, involves a combination of the racemic cyanohydrin **21** with racemic *cis*-permethrinic acid (**15**) to give a mixture of four stereoisomers represented by structure **23**, namely:

(A) 1*R*, *cis*, αS
(B) 1*S*, *cis*, αR
(C) 1*S*, *cis*, αS
(D) 1*R*, *cis*, αR

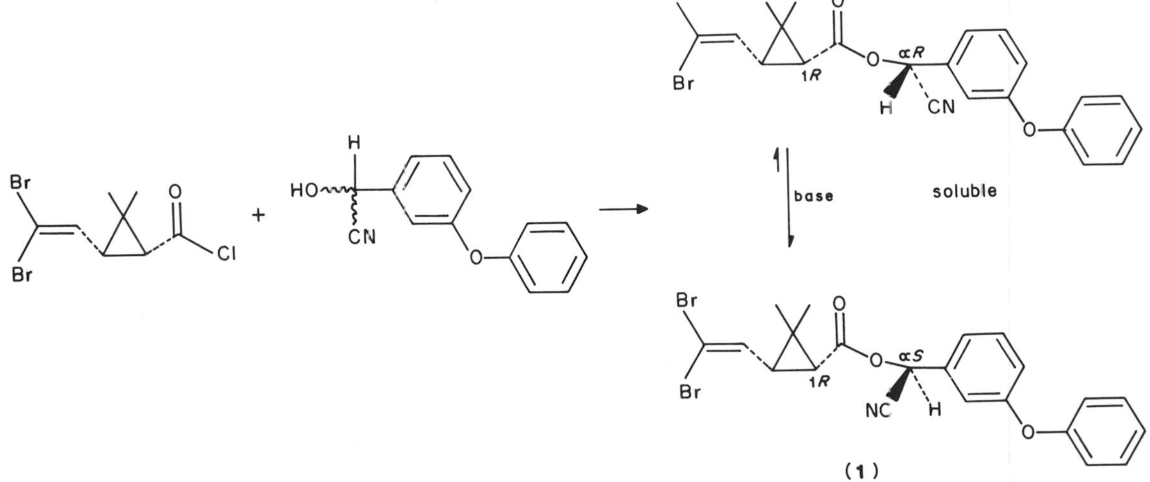

soluble

base

(1)
m.p. 101–102 °C
insoluble

Scheme 19

Of these, only diastereoisomer A possesses the desired activity. Fortunately, isomer A and its enantiomer B co-crystallize in the form of a racemic compound, whereas the other enantiomer pair, C and D, are more soluble. If the crystallization is performed in an appropriate basic medium, epimerization at the α-carbon can take place to transform isomer C (1*S*, *cis*, αS) into B (1*S*, *cis*, αR), and D into A. This reduction in the number of stereoisomers from four to two effectively doubles the concentration of A in the product. A similar process is operated by ICI to produce cyhalothrin,[23] the product of esterification of the racemic *cis*-acid **24** with racemic alcohol **21**, as a mixture of only two stereoisomers.

Cl

CO_2H

F_3C

(**24**)

4.5 CONCLUSION

The aim of this chapter has been to illustrate how the development of new pyrethroids at Roussel-Uclaf was paralleled by advances in methods for the production of active stereoisomers. Beginning with classical resolutions, and progressing to second-order asymmetric transformations, ever more sophisticated procedures have become feasible on an industrial scale. Selective production of the most active stereoisomers has brought benefits both to the manufacturer in terms of a greater economy in the use of raw materials and to the environment as much less insecticide need be applied to crops.

The manufacture of pure enantiomers, which was relatively rare 25 years ago, has increasingly become a yardstick of progress in chemical process development. The large-scale production of some of the pyrethroids mentioned in this chapter has been, for the agrochemical industry, the flagship of such progress, and the area will, no doubt, continue to produce exciting developments in manufacturing technology.

4.6 REFERENCES

1. Elliott, M., Farnham, A. W., Janes, N. F., Needham, P. H., and Pulman, D. A., *Nature* (*London*), **248**, 710 (1974).
2. Bowers, W. S., Ebing, W., Martin, D., and Wegler, R. (Eds), *Chemistry of Plant Protection*, Vol. 4: *Synthetic Pyrethroid Insecticides: Structures and Properties*, Springer, Berlin, 1990.

3. Naumann K., *Chemistry of Plant Protection*, Vol. 5: *Chemistry and Patents*, Springer, Berlin, 1990.
4. Staudinger, H., and Ruzicka, L., *Helv. Chim. Acta*, 456 (1924).
5. First industrial route described by Roussel-Uclaf, *Ger. Pat.*, 2 005 489, 1970.
6. Schechter, M. S., Green, N., and LaForge, F. B., *J. Am. Chem. Soc.*, **71**, 3165 (1949).
7. Crowley, M. P., Godin, P. J., Inglis, H. S., Snarey, M., and Thain, E. M., *Biochim. Biophys. Acta*, **60**, 312 (1962).
8. Roussel-Uclaf, *Fr. Pat.*, 2 166 503, 1971.
9. Martel, J. J., Demoute, J. P., Tèche, A. P., and Tessier, J. R., *Pestic. Sci.*, **11**, 188 (1980).
10. Sumitomo, *Jpn. Pat.*, 50 013 365, 1973.
11. Hoechst, *Eur. Pat.*, 321 918, 1988.
12. Sumitomo, *Ger. Pat.*, 2 723 383, 1977.
13. Roussel-Uclaf, *Fr. Pat.*, 2 031 793, 1970.
14. Roussel-Uclaf, *Ger. Pat.*, 2 029 043, 1970.
15. Elliott, M.,: *Br. Pat.*, 52406/65, 1965; *Nature* (*London*) **213**, 493 (1967). See also *Br. Pat.*, 6552406.
16. Roussel-Uclaf, *Fr. Pat.*, 2 364 199, 1976.
17. Farkas, J., Sorm, F., and Kourim, P., *Collect. Czech. Chem. Commun.*, **24**, 2230 (1959).
18. NRDC, PR 72GB-024809, 1972; *US Pat.*, 4 024 163, 1977.
19. Elliott, M., Farnham, A. W., Janes, N. F., Needham, D. H., Pulman, D. A., and Stevenson, J. H., *Nature* (*London*), **246**, 169 (1973); NRDC, *Br. Pat.*, 1 413 491, 1972.
20. Sumitomo, *Jpn. Pat.*, 4 882 034, 1972.
21. Roussel-Uclaf, *Belg. Pat.*, 853 867, 1977.
22. Shell, *Fr. Pat.*, 2 470 117, 1980; *US Pat.*, 4 427 598, 1984; *Ger. Pat.*, 3 044 391, 1980.
23. ICI, *Eur. Pat.*, 106 469, 1983.

BIOLOGICAL METHODS

5 Isolation and Improvement of Biological Catalysts for the Synthesis of Optically Active Chemicals

R. A. HOLT
ICI Bio Products and Fine Chemicals, Billingham, UK

Lord, I fall upon my knees
And pray that all my syntheses
May no longer be inferior
To those conducted by bacteria.[1]

5.1 INTRODUCTION

The capabilities of microorganisms as synthetic organic chemists are remarkable. Many are able to synthesize all their proteins, lipids, nucleic acids, vitamins and other minor cell components from an aqueous solution of inorganic ions and a single source of carbon such as glucose or pyruvate. The rate at which these reactions occur is equally remarkable, with many microorganisms able to double their cell number and mass in less than 1 h in what amounts to an

Chirality in Industry. Edited by A. N. Collins, G. N. Sheldrake and J. Crosby

autocatalytic reaction occurring at an exponential rate. An important aspect of biological systems is their ability to carry out these reactions in a regio- and enantiospecific manner. Most biological reactions are catalysed by proteinaceous catalysts (enzymes) built up exclusively from amino acids of the L-stereochemical series. The consequent asymmetry of the enzyme confers on it the capacity to catalyse reactions stereoselectively. The synthesis of the natural toxin palytoxin by *Palythoa toxica*, a primitive Pacific marine coral, serves to illustrate the tremendous capabilities of biological systems in stereospecific synthesis. Palytoxin has the empirical formula $C_{129}H_{223}N_3O_{54}$. Its structure, which was first deduced in 1981, contains 64 defined asymmetric carbon atoms and six double bonds. This Herculean feat of asymmetric synthesis has now been equalled by a team of chemists from Harvard,[2] but only through the efforts of 22 scientists over a period of 7 years.

To harness the capabilities of biological catalysts by transferring them from the status of bench-scale curiosities into the context of large-scale manufacture is a major challenge for industry. This challenge has already been overcome in many cases where the desired compound is naturally formed by the organism, penicillin being a noteworthy example. However, different problems lie in store for those who wish to use 'Nature's catalysts' to perform reactions on unnatural substrates.

The first problem encountered by the chemist who wishes to use a biocatalyst to carry out an unnatural reaction is the selection of an appropriate organism. This decision is aided by a detailed knowledge of microbiology and biochemistry, but in practice the problem may be circumvented by performing a screen of a wide range of microorganisms. Yeast and bacteria possess many characteristics which render them suitable for use in biocatalytic processes. Each of these classes of organism exhibits a rapid rate of growth and catalyses a wide range of metabolic processes. The metabolic diversity within these two classes is broad, with different species from either class being able to grow, for example, on sugars, methanol, alkanes and aromatic hydrocarbons at high or low pH, at high osmolarity or at high temperatures. The bacteria, however, are the more diverse of the two classes of organisms. A further positive feature of bacteria is their relatively simple and well characterized genetic make-up that has allowed biochemists to develop, over the past 20 years, methodologies for controlling the production of certain enzymes or modifying their structure in either a random or specific manner. The application of these methods to the development of biocatalysts will be exemplified later. Although similar techniques can be applied to yeasts, their more complex genetic make-up and life cycles make their modification more difficult.

Having identified an organism which possesses the required catalytic activity, it is often desirable to enhance the expression of this activity. This can only be done systematically when the microorganism's biochemical, genetic and growth characteristics are understood. In the following example such knowledge was applied by J. G. Morris *et al.* to the development of a microorganism enhanced in its ability to reduce ketones alien to its normal metabolism.[3]

5.2 USE OF CHEMOSTAT CULTURE FOR THE SELECTION OF A MUTANT STRAIN OF *CLOSTRIDIUM TYROBUTYRICUM* ENHANCED IN ITS ABILITY TO REDUCE KETONES

Clostridium tyrobutyricum belongs to a group of bacteria known as obligate anaerobes which grow only in the total absence of molecular oxygen. Energy for growth is normally derived from the dissimilation of glucose as shown in Figure 1, with the production of acetic and butyric acids, hydrogen and carbon dioxide as end products of fermentation. The redox balance of the organism's nicotinamide cofactors is maintained by careful balance between production of oxidized and reduced end products—in particular the ratio of acetate to butyrate—as the flux through the hydrogen-evolving pathway is limited.

This constraint on the organism's ability to redistribute reducing equivalents prevents its growth on sugars such as mannitol which are more reduced than glucose. However, *C. tyrobutyricum* is able to grow on mannitol if acetate is additionally supplied to the growth medium. Under these conditions acetate acts as an electron acceptor, being converted into butyrate by the reversal of the normal acetate-generating pathway. Tidswell *et al.*[3] demonstrated that the effect of acetate is due to relief of 'redox stress' by showing that growth on mannitol could be achieved if the culture was supplied with other reducible compounds such as acetoin (3-hydroxybutan-2-one), or even hexacyanoferrate(III) (in the presence of methylviologen as an electron-carrying mediator). Although cell suspensions of *C. tyrobutyricum* were able slowly to reduce pentan-2-one to pentan-2-ol, pentan-2-one would not support growth on mannitol in batch culture. In order to select an organism with an increased capability to reduce pentan-2-one, the technique of chemostat growth was used to exert a powerful selection pressure on the population of microorganisms.

In a chemostat, the growth medium in the culture vessel is continually replaced by feeding fresh medium whilst excess culture (spent medium and cells) is removed by overflow. The medium composition is adjusted so that a single nutrient is present at a limiting concentration. When the culture system reaches a steady state, the rate of growth of the microbial population is equal to the rate of addition of fresh medium. In this example, growth was initiated with mannitol as the substrate, with acetate present as the growth-limiting nutrient. Once a stable culture had been obtained, pentan-2-one (20 mM) was added to the inflowing medium in an attempt to relieve 'redox stress.' Any mutation leading to the enhanced expression or improved catalytic activity of pentan-2-one reductase would give the mutant cell a selective advantage, as it would relieve

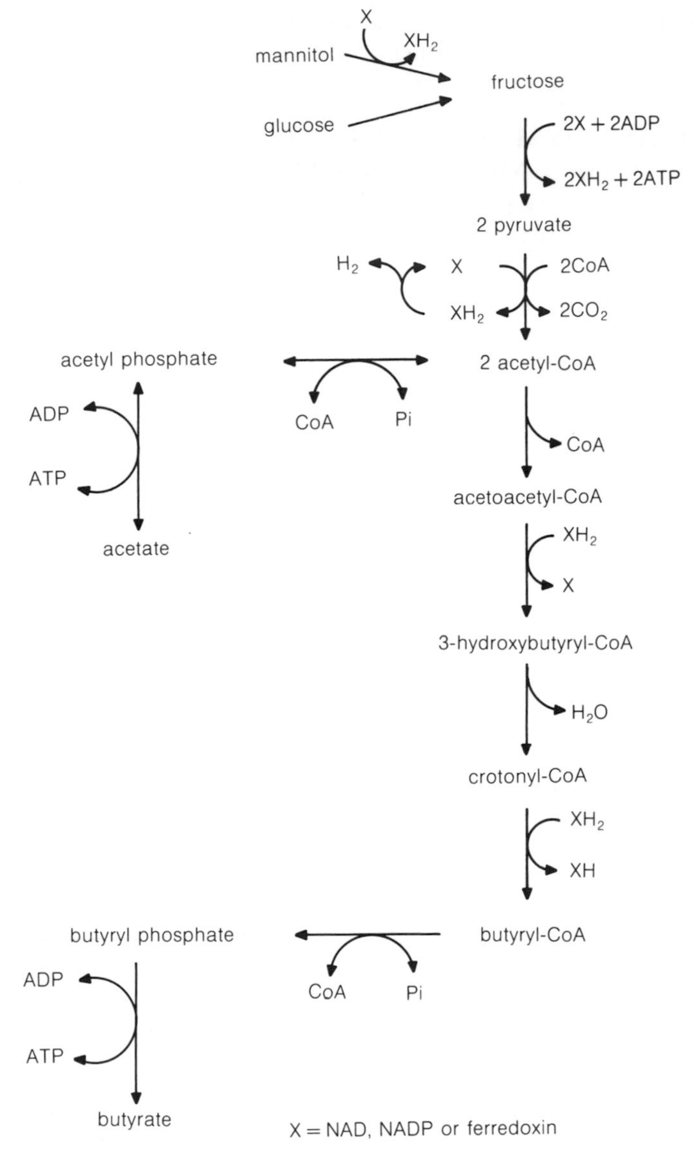

Figure 1. Metabolism of *Clostridium tyrobutyricum*

the restriction on growth imposed by the applied acetate limitation. The growth advantage would rapidly lead to the establishment of the mutant as the predominant organism in the culture.

Following a lag of 150 h, the consumption of mannitol began to increase along with the production of butyric acid and the onset of reduction of pentan-2-one

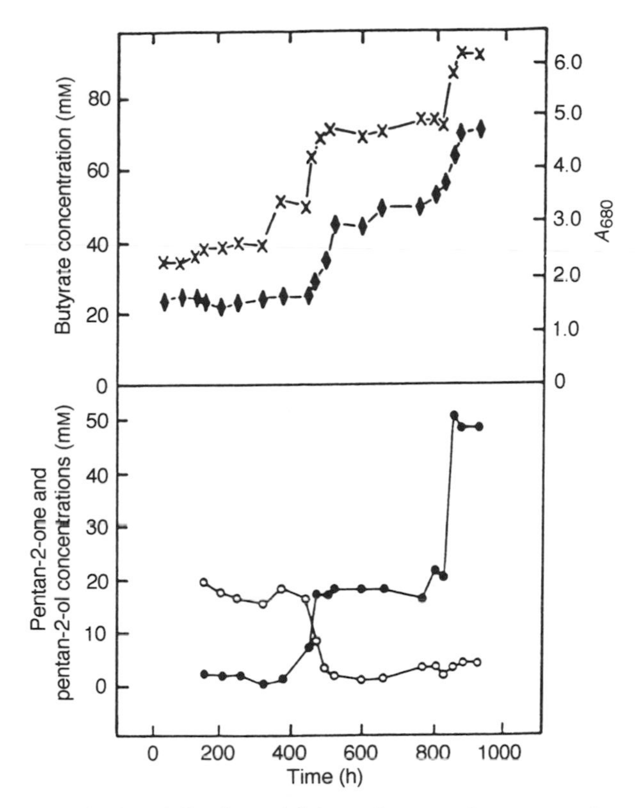

Figure 2. Mutant selection following addition of pentan-2-one to a chemostat culture of *Clostridium tyrobutyricum* growing in a minimal medium containing excess of mannitol (166 mM) but limiting acetate (10 mM), at a dilution rate of $0.15\,h^{-1}$, 35 °C and pH 6.0: x, absorbance at 680 nm (A_{680}); ◆, butyric acid; ●, pentan-2-ol; ○, pentan-2-one. The initial concentration of pentan-2-one was 20 mM and a further 30 mM were added at 808 h. (Adapted from ref. 3 with permission of Springer Verlag).

to pentan-2-ol. These events were accompanied by the doubling of the cell density within the culture vessel. When a further increase in the concentration of added pentan-2-one was made (to 50 mM) there was a further step increase in biomass, butyric acid and pentan-2-ol production (Figure 2).

The dominant organism isolated from the culture after 1000 h was found to be essentially identical with the original inoculum, except for its enhanced ability to reduce pentan-2-one. Further examination revealed that this selected mutant was also improved in its ability to reduce other ketones to secondary alcohols (Table 1).

To those unfamiliar with microbiological systems, the rapid selection of this mutant strain may be surprising. The culture in question, however, contained approximately 1.5×10^{12} cells, each of which would have a spontaneous

Table 1. Reduction of various ketones by parental and mutant strains of *Clostridium tyrobutyricum* (adapted from ref. 3, with permission of Springer Verlag)

Strain of *C. tyrobutyricum*	Growth medium	Primary reductant	Specific rate of reduction[a]		
			Pentan-2-one	6-Methylhept-5-en-2-one	Methyl 4-(4-chloro-phenylthio)-3-oxobutanoate
Parent	Glucose	Glucose	143	13	76
	Mannitol plus acetate	Mannitol	258	ND[b]	ND
Mutant	Glucose	Glucose	2032	114	228
	Mannitol plus acetate	Mannitol	2746	157	329
	Mannitol plus pentan-2-one	Mannitol	4873	ND	ND

[a]mmoles reduced/kg cell dry weight × hour.
[b]ND = not determined.

mutation frequency of approximately 10^{-8} per cell, per generation, i.e. 15 000 mutations per generation (4.62 h), or 3.25×10^6 mutations during the period of culture. Assuming that the bacterial genome contains approximately 3000 genes, it is clear that there is ample opportunity for an individual gene to undergo mutation during a prolonged period of growth. The use of chemostats for selection of microorganisms is a complex subject and the above work is only one example of its application. A more detailed review is given by Dykhuizen and Hartl.[4]

5.3 APPLICATION OF RECOMBINANT DNA TECHNIQUES TO THE PRODUCTION OF AN ENANTIOSELECTIVE ESTERASE

An alternative approach to enhancing gene expression was taken by scientists at International Biosynthetics when they began to develop a process for the resolution of the enantiomers of naproxen and other non-steroidal anti-inflammatories (see Chapter 15 for a full discussion of routes to naproxen). Their approach was to attempt to identify an enzyme capable of enantioselective hydrolysis of the methyl ester of naproxen. An extensive screen of microorganisms eventually produced an organism, *Bacillus subtilis* Thai 1–8, capable of selectively hydrolysing (*S*)-naproxen methyl ester (Figure 3). However, despite the presence of an appropriate enzyme in this organism, the level of enzyme expression was far too low to be of use in a commercial process. To overcome

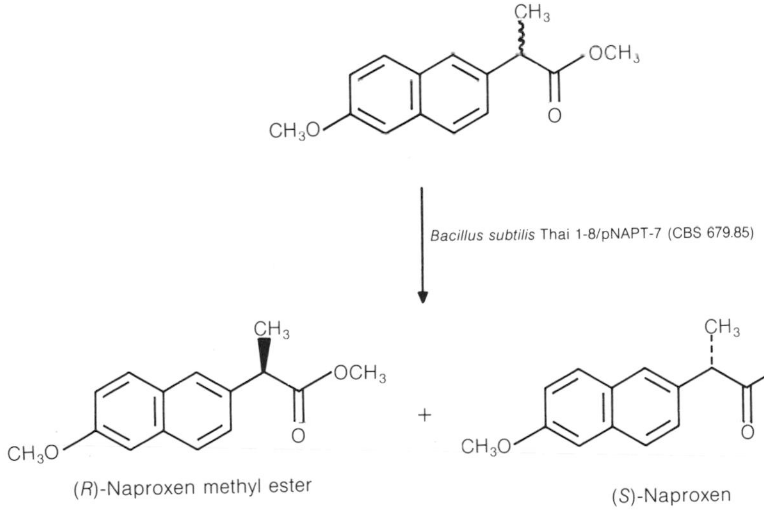

Bacillus subtilis Thai 1-8/pNAPT-7 (CBS 679.85)

(*R*)-Naproxen methyl ester + (*S*)-Naproxen

Figure 3. Enantioselective hydrolysis of naproxen methyl ester

this problem, the team at International Biosynthetics took the direct approach of isolating the gene and cloning it.[5]

A gene library was produced by extracting DNA from *B. subtilis* Thai 1–8 and cutting it into fragments with restriction enzymes. The DNA fragments were then incorporated into a plasmid, pUN121, and transferred into the bacterium *Escherichia coli.* The plasmid pUN121 contains a gene coding for resistance to the antibiotic tetracycline, so organisms carrying the plasmid were easily selected by their ability to grow on agar plates containing tetracycline. Of the organisms which contained the recombinant plasmid, only a very small proportion carried the esterase gene from *B. subtilis* Thai 1–8. These organisms were selected using a colorimetric growth assay. Cultures of *E. coli* containing the plasmid were plated onto agar and incubated. Following growth, the plate was overlaid with an agarose gel containing β-naphthyl acetate and Fast Blue reactive dye. Only organisms producing the esterase were able to hydrolyse β-naphthyl acetate to produce β-naphthol, which reacted spontaneously with Fast Blue and stained the colony. In this way *E. coli* colonies containing the plasmid-borne esterase gene were identified and isolated. From such colonies the plasmids were isolated and examined by standard genetic techniques. An examination of these revealed that, in addition to the desired gene, other unwanted genetic information had been incorporated into the plasmid. This unwanted material was removed and, following further genetic manipulations, the plasmid was transferred back into *B. subtilis* to provide the organism designated *B. subtilis* 1–85/pNAPT-7. This cloning of the esterase resulted in a dramatic increase in esterase production, such that the enzyme became the major protein within the cell, with an 800-fold increase in activity compared with the original organism. Prior to the break-up of International Biosynthetics this enzyme was offered for sale for commercial-scale application.

5.4 DEVELOPMENT OF AN ENANTIOSPECIFIC DEHALOGENASE ENZYME

In addition to enhancing activity, it is also sometimes desirable to remove unwanted enzyme activities which interfere with the desired reaction. Such an approach was taken during the early development of ICI's dehalogenase technology which is now operated commercially for the production of (*S*)-2-chloropropionic acid.

The main use of (*S*)-2-chloropropionic acid is in the synthesis of phenoxypropionic acid herbicides. An interesting feature of this class of molecules is that the herbicidal activity resides predominantly in the (*R*)-enantiomer.

Historically, these herbicides were produced as racemates but the current trend is towards production as single enantiomers, both to reduce the chemical load on the environment and also to decrease the cost of manufacture—particu-

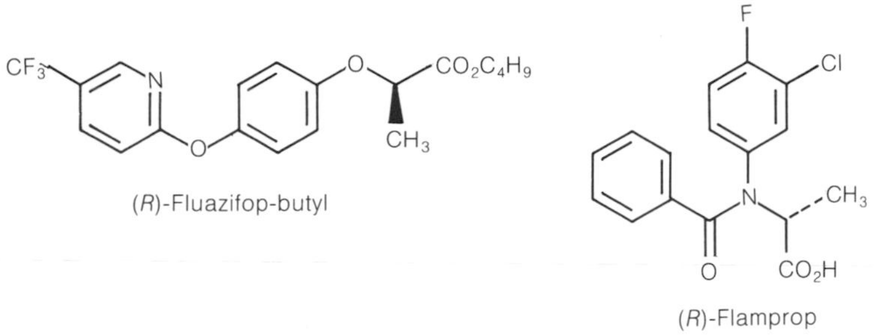

(*R*)-Fluazifop-butyl

(*R*)-Flamprop

Figure 4. Phenoxypropionic acid herbicides

larly important for the newer, more complex compounds such as fluazifop-butyl and flamprop (Figure 4).

ICI's approach to the production of (*S*)-2-chloropropionic acid was to identify a microorganism capable of the enantioselective dehalogenation of racemic 2-chloropropionate. Halogenated compounds occur widely in the environment and it has long been known that they are degraded by microorganisms possessing dehalogenase enzymes.[6] Several studies of microorganisms capable of growth on halogenated compounds had demonstrated a variety of specificities. At the time ICI began its investigation two enzyme specificities had been described.* One class of enzymes was specific for the dehalogenation of (*S*)-2-chloropropionate, whereas the second class was equally active with both enantiomers.[7,8]

ICI initiated a search for novel microorganisms with an (*R*)-specific dehalogenase possessing characteristics suitable for cost-effective application on a plant scale. A large number of organisms were isolated from environmental samples and cultured on racemic 2-chloropropionate. From these isolates, thirteen were found to be capable of growth only on (*R*)-chloropropionate and not the (*S*)-enantiomer. The conclusions to be drawn from these results might have been that these microorganisms possessed only (*R*)-specific dehalogenase, but closer examination revealed this not to be the case. Cell-free extracts of all thirteen isolates were able to dehalogenate both (*R*)- and (*S*)-2-chloropropionate, but were able to *grow* only on (*S*)-lactate [the product of (*R*)-2-chloropropionate dehalogenation] and not on (*R*)-lactate [the product of (*S*)-2-chloropropionate dehalogenation].

The dubious reliability of growth specificity as a tool for the identification of an appropriate microorganism complicated the task of finding an (*R*)-specific dehalogenase. Consequently, an alternative approach was adopted in which those organisms able to dehalogenate both enantiomers of 2-chloropropionate were examined in more detail. One isolate (identified as a strain of *Pseudomonas putida*) was found to contain two separate dehalogenases. One was a low

*Subsequent work did report the existence of an (*R*)-2-chloropropionate dehalogenase.[9]

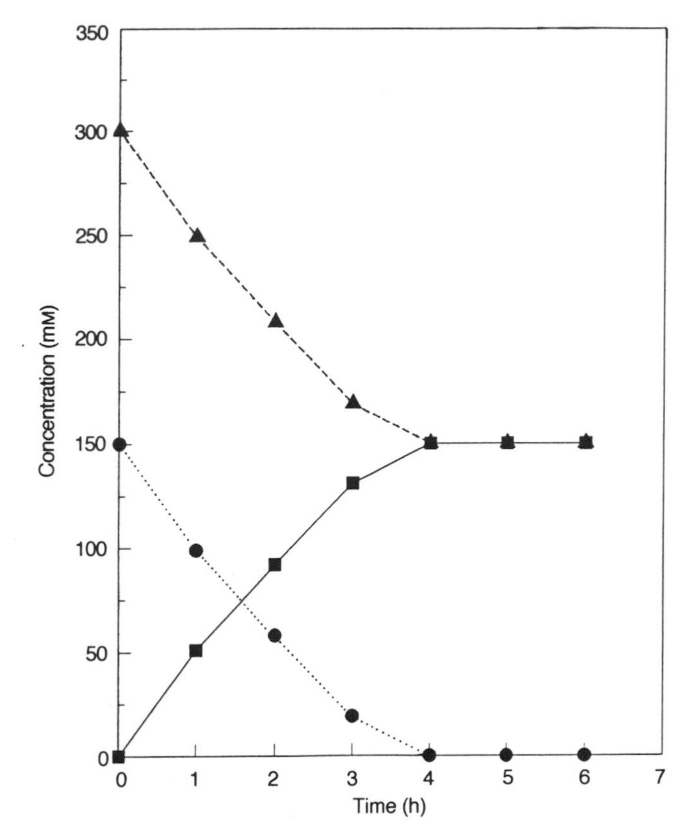

Figure 5. Resolution of racemic 2-chloropropionate by (*R*)-specific *Pseudomonas putida* dehalogenase. Initial CPA concentration, 0.3 M; pH maintained at 7.2 by titration with sodium hydroxide; temperature, 30 °C. ■, (*S*)-Lactate; ▲, total CPA; ●, (*R*)-CPA

molecular weight enzyme showing complete specificity for (*S*)-chloropropionate, whereas the other enzyme was of higher molecular weight and was specific for (*R*)-chloropropionate. *In vitro* experiments with the partially purified enzymes indicated that these enzymes could be used to produce either the (*R*)- or the (*S*)-enantiomer of 2-chloropropionate (Figure 5).

An attractive alternative to the use of purified enzymes to perform the resolution would be to use whole cells, although this would require the removal of the (*S*)-chloropropionate dehalogenase from the organism. To facilitate this, the wild-type organism was exposed to mutagenic agents and then tested for its ability to grow on (*S*)-chloropropionate as the sole source of carbon. Those organisms which were no longer able to grow on (*S*)-chloropropionate were examined, and from these was isolated an organism which retained the (*R*)-chloropropionate dehalogenase and demonstrated other features which were desirable for a process catalyst. The characteristics of the (*R*)-chloropropionate dehalogenase were identical whether the enzyme was isolated from the mutant or the wild type.

Using this (*R*)-specific dehalogenation catalyst, a process for the manufacture of (*S*)-2-chloropropionic acid has been developed which yields product of high optical purity ($>96\%$ enantiomeric excess) in high overall yield. It is believed that this patented process[10] is the first of its kind, and it is currently operated by ICI on a scale in excess of 1000 tonnes per annum.

Further development of both the (*R*)- and (*S*)-dehalogenases is currently under way using the techniques of molecular biology to clone the genes and study the structure and function of the enzymes.[11,12] This has produced catalysts with even higher productivities.

5.5 ALTERATION OF ENZYME SPECIFICITY BY SITE-DIRECTED MUTAGENESIS

A common problem with biocatalysts is that in addition to high enantioselectivity, most also possess high substrate selectivity, a factor which often limits their use to the transformation of a single molecule or at best a small number of substrates. It has been argued[13,14] that an understanding of enzyme mechanism and structure should allow the rational modification of an enzyme to improve or alter its substrate selectivity whilst retaining its enantioselectivity. This hypothesis has been tested by modifying the enzyme L-lactate dehydrogenase from *Bacillus stearothermophilus*. This enzyme was chosen for several reasons: its X-ray crystal structure is known, the enzyme mechanism is known and the primary amino acid sequence is known.

Two regions of the enzyme are thought to play an important role in determining the enzyme's specificity, namely a mobile loop of polypeptide (the coenzyme loop), which folds over the active site, and an α-helix on to which the coenzyme loop folds, enclosing the substrate in a catalytic vacuole.[14] The size, shape and charge of the catalytic vacuole are such that only small, single, negatively charged keto acids allow the loop to fold correctly on to the α-helix such that reduction is catalysed.

J. J. Holbrook *et al.* employed the technique of site-directed mutagenesis to redesign the catalytic vacuole of lactate dehydrogenase such that it could accept the dicarboxylic acid oxaloacetate in preference to pyruvate. Three changes were necessary to achieve this. First, the vacuole size had to be enlarged to accept the larger substrate, this being achieved by replacing threonine with glycine at position 246. Replacement of either aspartate 197 or glutamate 107 with their respective amides removed a negative charge from the periphery of the catalytic vacuole, and decreased repulsion of the second carboxylate group of oxaloacetate/malate. The final change involved insertion of a positively

Table 2. Steady-state kinetic properties of wild-type and mutant lactate dehydrogenases (adapted from ref. 14, with permission. Copyright 1990, American Chemical Society)

Substrate	Constant	Wild type	Mutant 1[a]	Mutant 2[b]	Mutant 3[c]
Pyruvate	K_{cat} (s^{-1})	250	66	167	32
	K_m (mM)	0.06	0.16	4	4
	K_{cat}/K_m ($M^{-1}s^{-1}$)	4.2×10^6	4.1×10^5	4.2×10^4	8×10^3
α-Keto-isocaproate	K_{cat} (s^{-1})	0.33	0.67	1.74	18.5
	K_m (mM)	6.7	1.9	15.4	14.3
	K_{cat}/K_m ($M^{-1}s^{-1}$)	50	353	110	1.3×10^3

[a] Mutant 1: glutamine 102 ⟶ methionine; lysine 103 ⟶ valine; proline 105 ⟶ serine.
[b] Mutant 2: alanine 235 and 236 ⟶ glycine.
[c] Mutant 3: changes to mutant 1 and mutant 2 combined.

charged residue, arginine, in place of glutamine at position 102, the expected position of the second carboxylate group. The result was a switch in specificity against pyruvate and in favour of oxaloacetate by nearly 10^7-fold. The new enzyme displayed a turnover number of $250\,s^{-1}$ and a Michaelis constant of 60 μM.[15]

An unfortunate feature of *B. stearothermophilus* lactate dehydrogenase for those wishing to synthesize optically active α-hydroxy acids is its poor activity with substrates containing large or branched side-chains. In order to accommodate such substrates, the catalytic vacuole had to be enlarged and made more hydrophobic. Alanine residues 235 and 236 on the α-helix were replaced with glycines to increase the vacuole volume and to provide conformational flexibility by destabilizing the helix. The inner surface of the vacuole was made more hydrophobic by substituting glutamine 102 and lysine 103 (on the inner surface of the loop) by methionine and valine, respectively, and proline 105 was replaced with serine to increase flexibility. Changes to residues 235 and 236 alone or residues 102, 103 and 105 alone had little effect on the enzymes' activity against α-ketoisocaproate. However, when both sets of changes were made together, an α-ketoisocaproate dehydrogenase of reasonable activity was obtained[15] (Table 2).

5.6 CONCLUSION

The examples cited above represent only a small sample of the total number of reactions which can be catalysed by biological systems, but demonstrate some of the different approaches that have been used to subvert the natural metabolism of microorganisms to access key commercial materials.

In any attempt to employ biological catalysts, it is important to be aware of their complexity and to understand the fundamentals of the system. Failure to do so can lead to irreproducibility (as is often seen in reductions catalysed by baker's yeast) or even a total lack of reaction. The development of biological catalysts can only be carried out effectively by an integrated multidisciplinary team of microbiologists, biochemists, geneticists and organic chemists, and where development to a large scale is required the involvement of chemical engineers with experience of the specific problems of processing biological materials is essential.

It is clear that biological catalysts will become increasingly important in the synthesis of chemicals on an industrial scale, but in meeting ever more complex technical challenges, the ultimate source of these catalysts, the multi-talented microorganism, should not be forgotten.

Microbes can and will do anything: microbes are smarter, wiser and more energetic than microbiologists, engineers and others.[16]

5.7 REFERENCES

1. Eveleigh, D. E., *Sci. Am.*, **245**, 155 (1981).
2. Armstrong, R. W., Beau, J.-M., Cheon, S. H., Christ, W. J., Fujioka, H., Ham, W.-H., Hawkins, L. D., Jin, H., Kang, S. H., Kishi, Y., Martinelli, M. J., McWhorter, W. W., Mizuno, M., Nakata, M., Stutz, A. E., Talamas, F. X., Taniguchi, M., Tino, J. A., Ueda, K., Uenishi, J.-I., White, J. B., and Yonaga, M., *J. Am. Chem. Soc.*, **111**, 7530 (1989).
3. Tidswell, E. C., Thompson, A. N., and Morris, J. G., *Appl. Microbiol. Biotechnol.*, **35**, 317 (1991).
4. Dykhuizen, D. E., and Hartl, D. L., *Microbiol. Rev.* **47**, 150 (1983).
5. Quax, W. J., Mencke, H. H., de Swaaf, M. P. M. and van der Laken, C. J., in *Proceedings of the 4th European Congress on Biotechnology* (ed. O. M. Neijssel, R. R. van der Meer and K. Ch. A. M. Luyben), Elsevier, Amsterdam, 1987, Vol. 1, p. 519.
6. Senior, E., Bull, A. T., and Slater, J. H., *Nature (London)*, **263**, 476 (1976).
7. Hardman, D. J., and Slater, J. H., *J. Gen. Microbiol.*, **123**, 117 (1981).
8. Weightman, A. J., Weightman, A. L., and Slater, J. H., *J. Gen. Microbiol.*, **128**, 1755 (1982).
9. Leigh, J. A., Skinner, A. J., and Cooper, R. A., *FEMS Microbiol. Lett.*, **49**, 353 (1988).
10. Taylor, S. C., *US Pat.*, 4 758 518, 1988.
11. Barth, P. T., Bolton, L., and Thomson, J. C., *J. Bacteriol.*, **174**, 2612 (1992).
12. Jones, D. H. A., Barth, P. T., Byrom, D., and Thomas, C. M., *J. Gen. Microbiol.*, **138**, 675 (1992).
13. Mutter, M., *Angew. Chem., Int. Ed. Engl.*, **24**, 639 (1985).
14. Wilks, H. M., Halsall, D. J., Atkinson, T., Chia, W. N., Clarke, A. R., and Holbrook, J. J., *Biochemistry*, **29**, 8587 (1990).
15. Dunn, C. R., Wilks, H. M., Halsall, D. J., Atkinson, T., Clarke, A. R., Muirhead, H., and Holbrook, J. J., *Philos. Trans. R. Soc. London, Ser. B*, **332**, 177 (1991).
16. Perlman, D., *Dev. Ind. Microbiol.*, **21**, XV-XXIII (1980).

6 Biologically Derived Arene *cis*-Dihydrodiols as Synthetic Building Blocks

G. N. SHELDRAKE
ICI Specialties, Manchester, UK

6.1 INTRODUCTION

Of the increasing range of biological transformations which have become available to synthetic organic chemists, one of the most intriguing is the *cis*-dihydroxylation of aromatic compounds to give arene *cis*-dihydrodiols. Although no commercial products have yet appeared, several factors give the *cis*-dihydrodiols considerable potential as raw materials for industrial targets:

Chirality in Industry. Edited by A. N. Collins, G. N. Sheldrake and J. Crosby

(a) the biological reaction has no simple chemical equivalent;
(b) there is a wide substrate tolerance;
(c) the substrates are generally inexpensive and readily available;
(d) the products are (usually) optically pure and have predictable stereochemistry;
(e) the stereochemistry is generated from an achiral precursor and so all of the substrate is utilized;
(f) an increasing number of the products are now available commercially.[1]

These factors have combined to allow short and efficient syntheses of biologically active target compounds which had previously been accessible only with difficulty or with poor stereo- and regiospecificity.

The published chemistry of *cis*-dihydrodiols up to 1989 has been comprehensively reviewed by Brown.[2] This chapter, which draws on patents and literature published up to December 1991, describes only the synthetic chemistry which utilizes the stereochemistry of the *cis*-diol centres.

6.1.1 BACKGROUND

The identification of benzene-*cis*-dihydrodiol (**1**) as a product from a mutant strain of the bacterium *Pseudomonas putida* was first made by Gibson *et al.*[3] in 1968 during studies into the mechanism of the degradation of benzene in bacteria. The mutation interrupts the microorganism's normal pathway for the catabolism of aromatic compounds (Scheme 1).

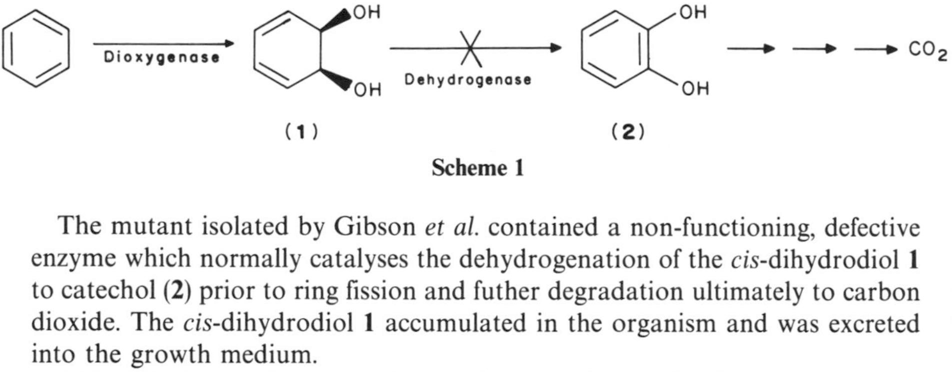

Scheme 1

The mutant isolated by Gibson *et al.* contained a non-functioning, defective enzyme which normally catalyses the dehydrogenation of the *cis*-dihydrodiol **1** to catechol (**2**) prior to ring fission and futher degradation ultimately to carbon dioxide. The *cis*-dihydrodiol **1** accumulated in the organism and was excreted into the growth medium.

Although the precise mechanism of the enzymic reaction is still not certain, it has been found [4–6] that a multi-component iron–sulphur protein is involved and that both atoms of a dioxygen molecule are incorporated into the product. The reaction requires NADH as cofactor via an electron transfer chain. The direct products from bacterial enzymes are invariably *cis*, in contrast to mammalian systems where *trans*-dihydrodiols are found to be products (Scheme 2). *Pseudomonas* species are used most frequently for effecting this

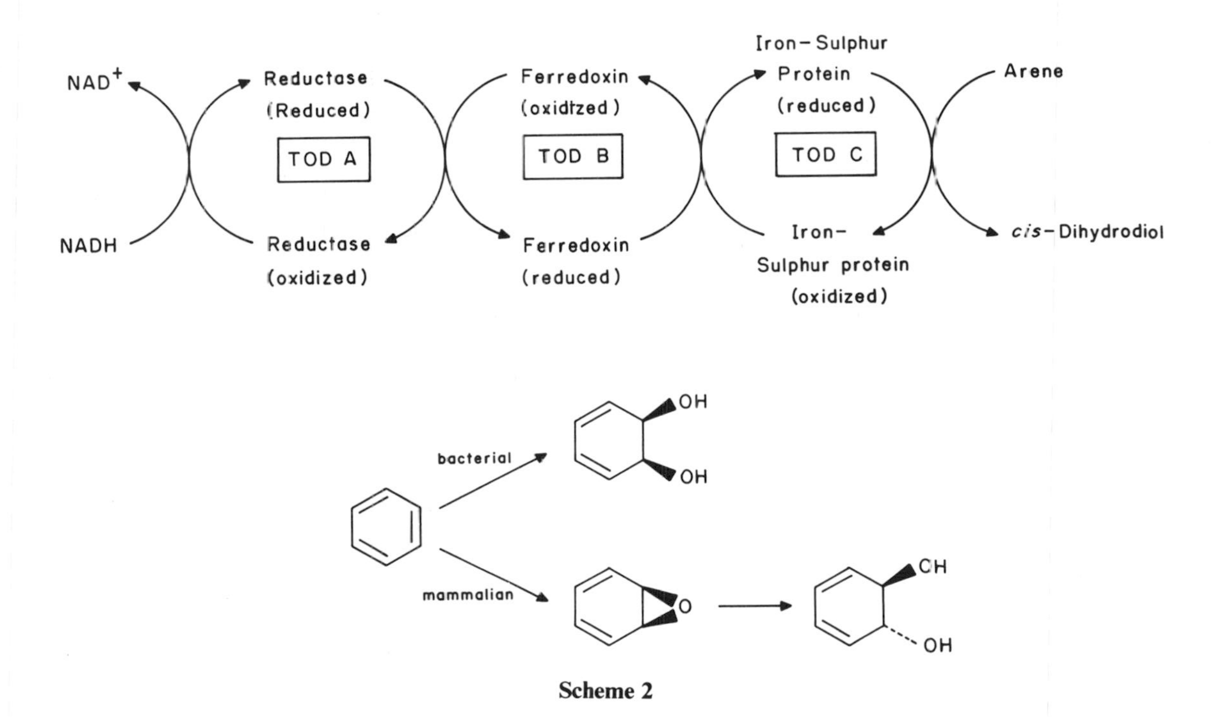
NAD+
NADH
Reductase
(Reduced)
TOD A
Reductase
(oxidized)
Ferredoxin
(oxidtzed)
TOD B
Ferredoxin
(reduced)
Iron-Sulphur
Protein
(reduced)
TOD C
Iron-
Sulphur protein
(oxidized)
Arene
cis-Dihydrodiol
bacterial
mammalian
OH
OH
O
CH
OH

Scheme 2

Table 1. Some arene substrates and *cis*-dihydrodiol products from bacterial *cis*-dihydroxylation reactions

Substrate	Product	Organism	Ref.
	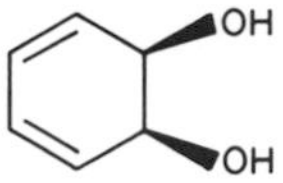 OH OH	*Pseudomonas* spp. *Moraxella* sp.	3 7
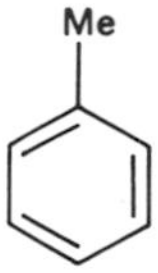 Me	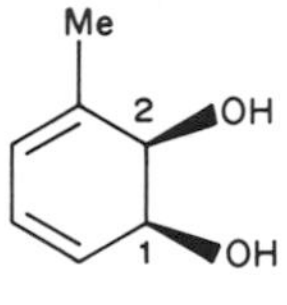 Me 2 OH 1 OH 1*S*, 2*R*	*P. putida*	8,9
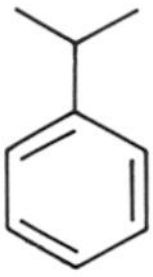	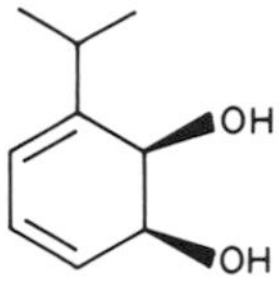 OH OH	*P. desmolytica* *P. convexa*	10
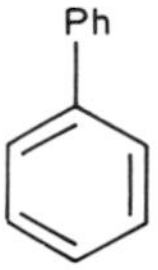 Ph	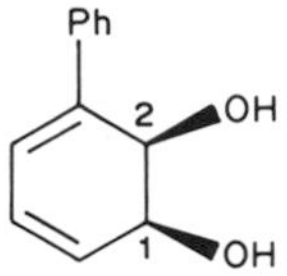 Ph 2 OH 1 OH 1*S*, 2*R*	*P. putida* *Beijerinckia* sp.	11,12 13
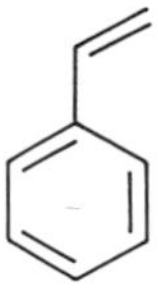	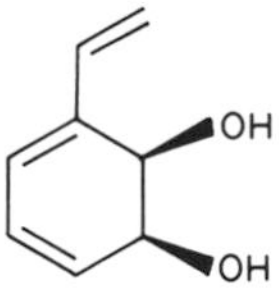 OH OH	*P. putida*	14
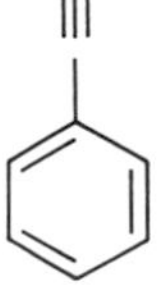	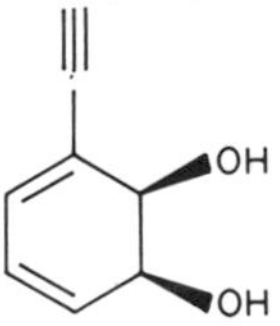 OH OH	*P. putida*	14
	OH OH		15
	HO OH OH	*P. putida*	

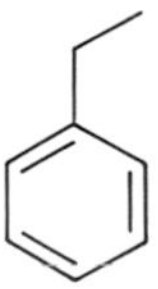

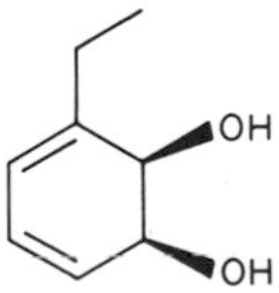

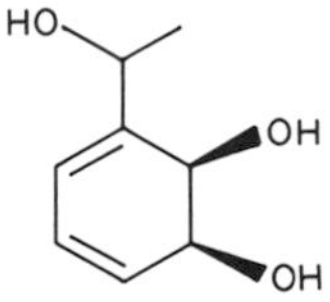

Table 1. (*continued*)

Substrate	Product	Organism	Ref.
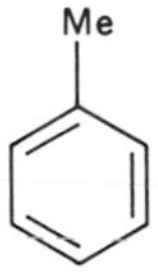 Me	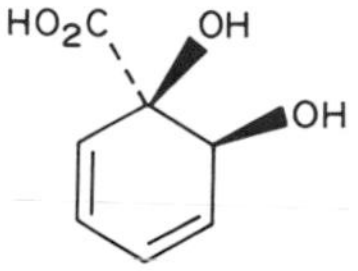 HO_2C OH OH	*P. putida*	16
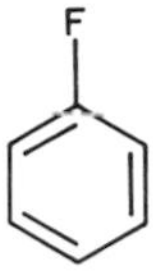 F	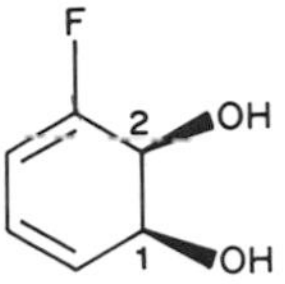 F 2 OH 1 OH 1*S*,2*S*	*P. putida*	17
	1*S*,2*S* 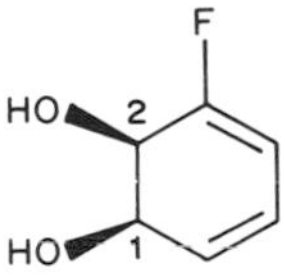 + HO 2 HO 1 F 1*R*,2*R*		18
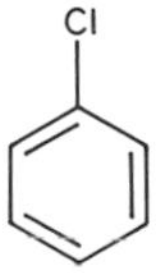 Cl	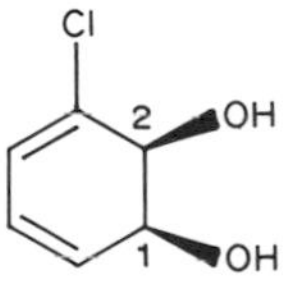 Cl 2 OH 1 OH 1*S*,2*S*	*P. putida*	17
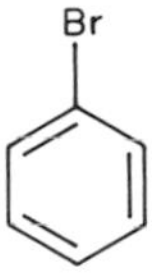 Br	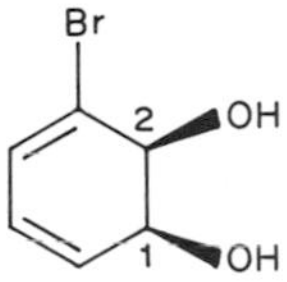 Br 2 OH 1 OH 1*S*,2*S*	*P. putida*	19
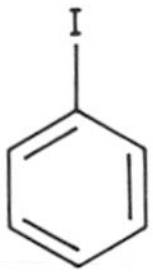 I	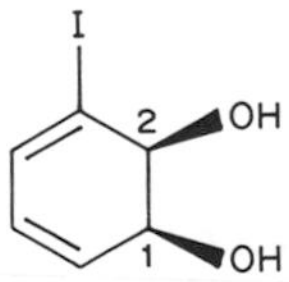 I 2 OH 1 OH 1*S*,2*S*	*P. putida*	20

(*continued*)

Table 1. (*continued*)

Substrate	Product	Organism	Ref.
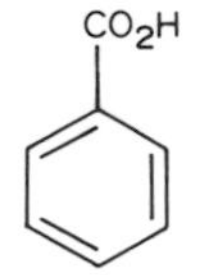 CO_2H	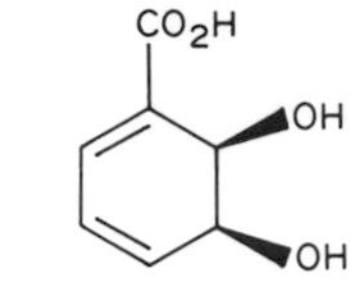 CO_2H, OH, OH	*P. putida*	21
	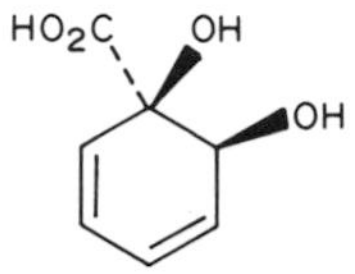 HO_2C, OH, OH	*Alcaligenes eutrophus*	22
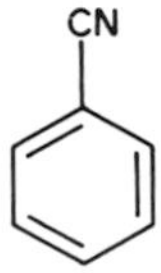 CN	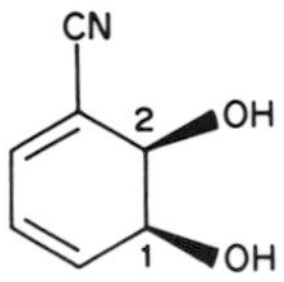 CN, 2, OH, 1, OH 1*S*, 2*R*	*P. putida*	20
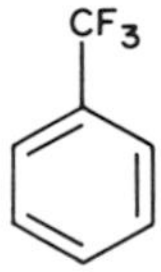 CF_3	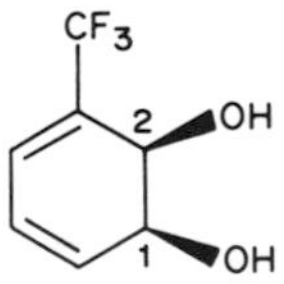 CF_3, 2, OH, 1, OH 1*S*, 2*R*	*P. putida*	23, 18
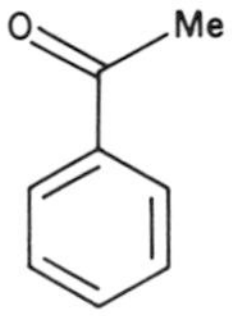 O, Me	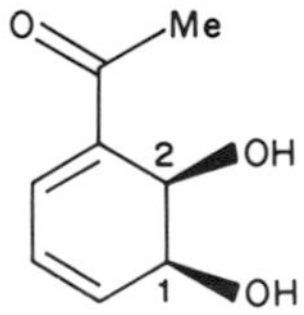 O, Me, 2, OH, 1, OH 1*S*, 2*R*	*P. putida*	15
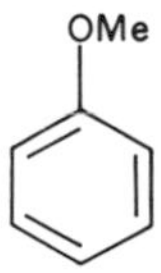 OMe	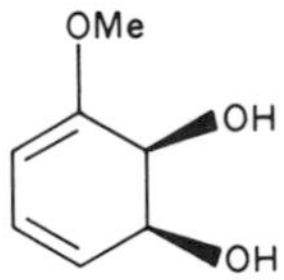 OMe, OH, OH	*P. putida*	17

Table 1. (*continued*)

Substrate	Product	Organism	Ref.
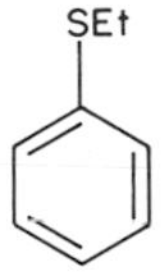 SEt	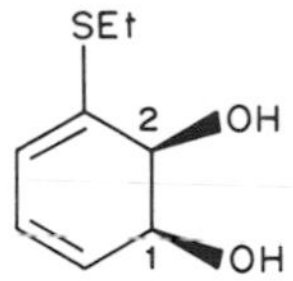 SEt, OH, OH, 2, 1; 1*S*,2*R*	*P. putida*	18
 Me, Br	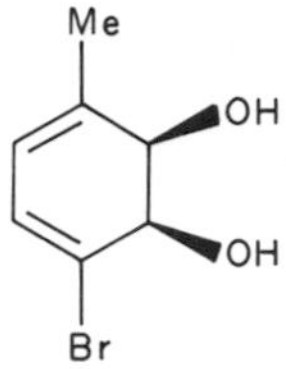 Me, OH, OH, Br	*P. putida*	11
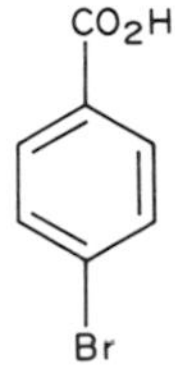 CO_2H, Br	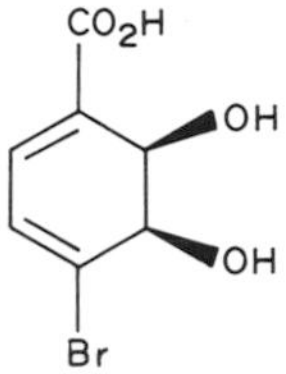 CO_2H, OH, OH, Br	*P. putida*	24
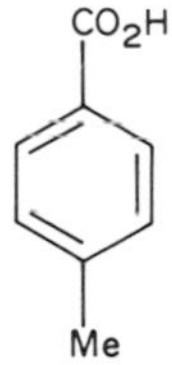 CO_2H, Me	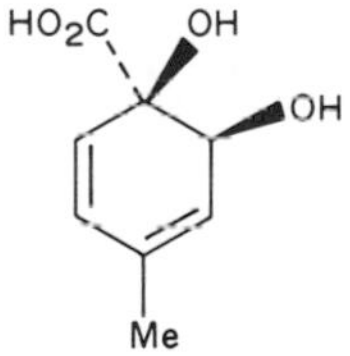 HO_2C, OH, OH, Me	*A. eutrophus*	25
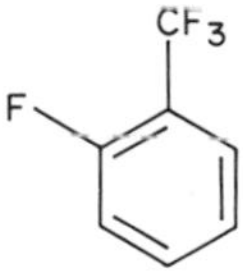 CF_3, F	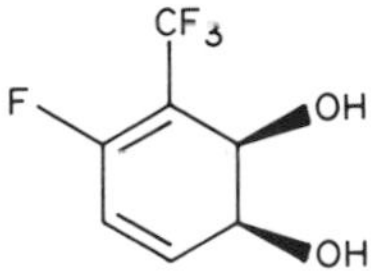 CF_3, F, OH, OH	*P. putida*	26
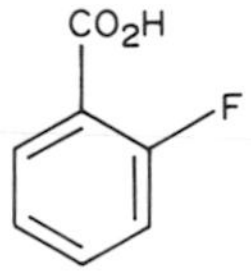 CO_2H, F	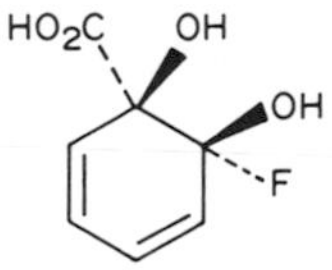 HO_2C, OH, OH, F	*A. eutrophus*	25

(*continued*)

Table 1. *(continued)*

Substrate	Product	Organism	Ref.
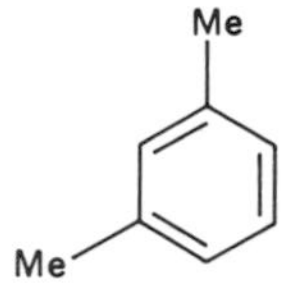	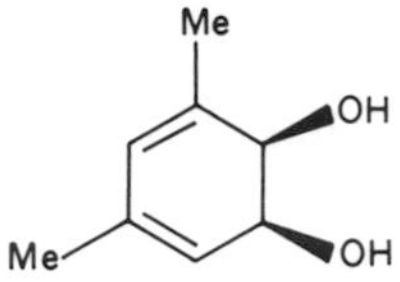	*P. putida*	27
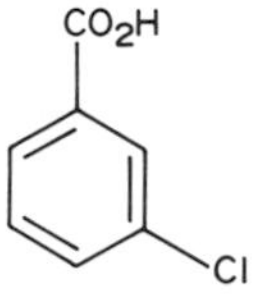	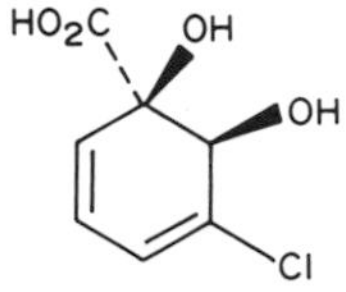	*A. eutrophus*	25
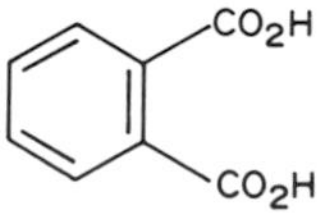	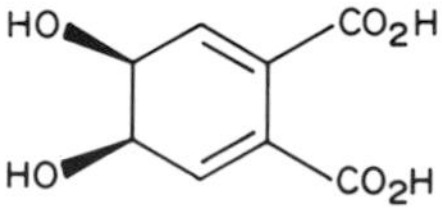	*P. testosteroni*	28
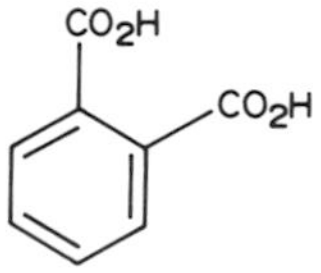	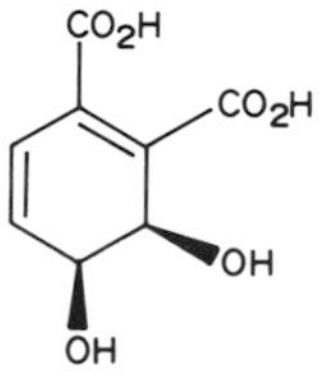	*Micrococcus* sp. *P. testosteroni*	29
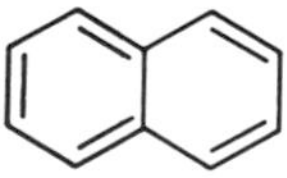	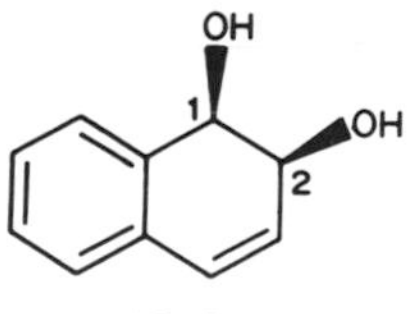 1*R*, 2*S*	*P. putida* *Oscillatoria* sp. *Agmenellum quadruplaticum*	30 31 31
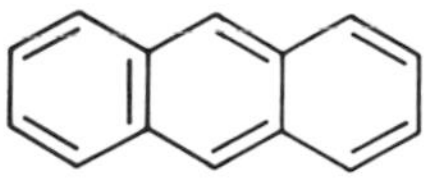	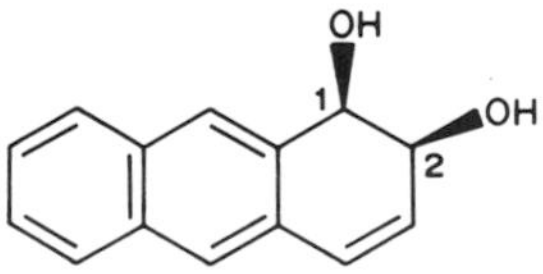 1*R*, 2*S*	*P. putida*	32

Table 1. (*continued*)

Substrate	Product	Organism	Ref.
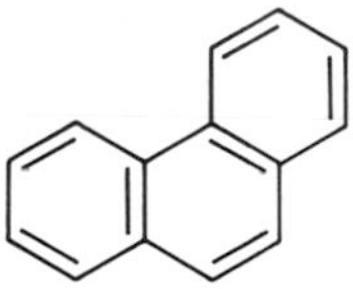	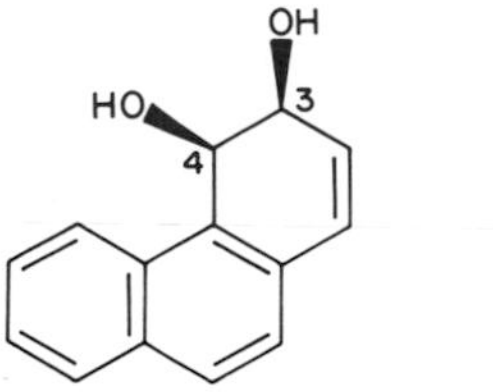 3*S*,4*R*	*P. putida*	32
	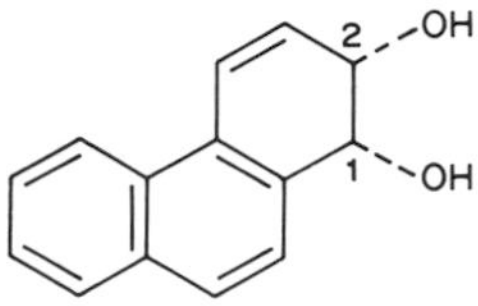 1*R*,2*S*	*Beijerinckia* sp.	32
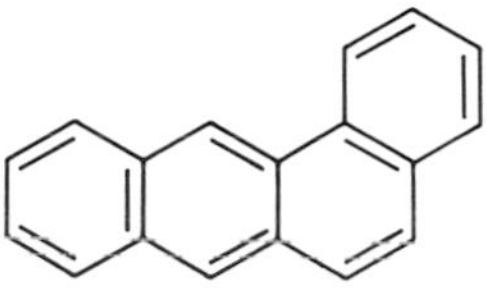	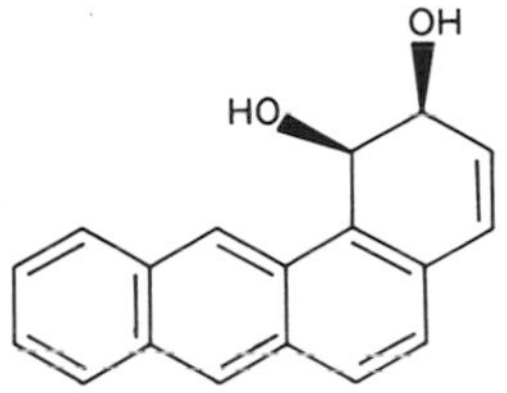 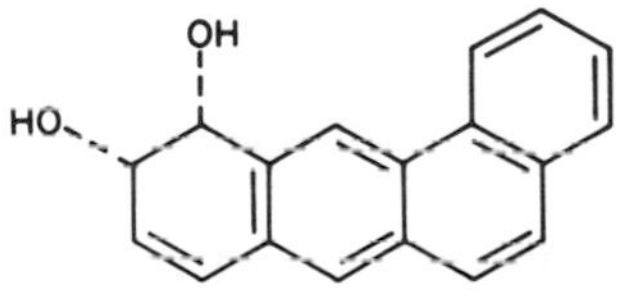 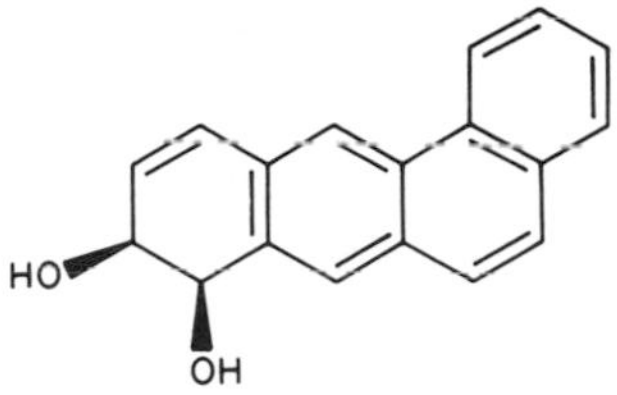 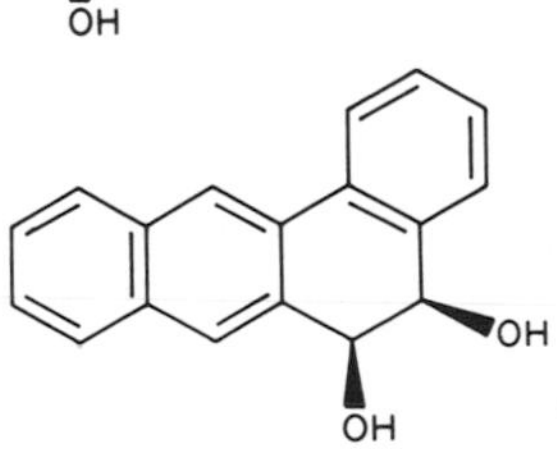	*Beijerinckia* sp.	33

biotransformation but many other bacteria have been shown to possess the enzymes capable of aromatic *cis*-dihydroxylation, as can be seen in Table 1.

Organisms may be mutated to give either constitutive or inducible enzymes. In a constitutive organism, the gene for the dioxygenase enzyme is always expressed and the organism produces the enzyme under all conditions. In an inducible organism, the gene is only 'switched on' in the presence of an aromatic substrate and the enzyme is only produced as a response to this particular stimulus.

6.1.2 SUBSTRATE RANGE AND REGIO- AND STEREOSPECIFICITY

The limits of the substrate range of bacterial aromatic dioxygenase systems have not yet been fully defined and new *cis*-dihydrodiols are reported each year. The regiochemistry, substrate specificity and activity may vary considerably from species to species and even strain to strain. Table 1 illustrates the diverse range of compounds which are accepted by these enzymes.

Within certain limits, the electronic nature of a substituent on the aromatic ring is less important than the size of the atom or group. Strong electron-withdrawing groups, such as trifluoromethyl, and electron donors, such as methoxy, are acceptable. Nitrogen-containing groups, such as amino or nitro, have not been found to be tolerated but benzonitrile is a good substrate. All four monohalobenzenes are good substrates and oxygen and sulphur are acceptable as ethers and thioethers. Alkenes and alkynes attached directly to the aromatic ring tend to come through the biological oxidation unscathed. To date, monocyclic heterocycles have not been successfully used as substrates, although the bicyclic quinolines and related compounds have given *cis*-dihydrodiols.

The size of a group is important and activity generally falls away rapidly as the steric bulk of the substituent increases. Aliphatic groups sometimes suffer side-chain oxidation, in addition to aromatic *cis*-dihydroxylation, to give alcohols or even carboxylic acids.

For most monosubstituted arenes, enzymic attack is at the two carbons *ortho* and *meta* to the carbon bearing the substitutent. There are some examples, particularly with benzoic acid as the substrate for certain *Alcaligenes* species, of attack at the *ipso* and *ortho* positions. For di- and polysubstituted arenes the regiochemistry is less clear and a systematic study of the priority order of substituent groups for directing the point of attack has not yet been carried out.

Until fairly recently, the absolute configuration and even optical purity of only a handful of the *cis*-dihydrodiols were known with certainty.[9,11,15,34,35] During 1991, two academic groups collaborating with ICI each published a general method for the assignment of absolute configuration, one of which also gives a measure of enantiomeric purity.

One of the main problems in developing a general method for the *cis*-dihydrodiols is that the racemic mixtures needed to test the validity of the method are not easily accessible by chemical or biological synthesis. Published chemical syntheses of *cis*-dihydrodiols have been limited to a couple of lengthy routes to unsubstituted (and therefore achiral) benzene-*cis*-dihydrodiol (**1**).[36,37] This problem was circumvented by Boyd *et al.*[18] in the manner outlined in Scheme 3.

First, the *cis*-dihydrodiol of toluene (**3a**), which was known to be optically pure and of the absolute stereochemistry indicated, was treated with Cookson's dienophile (**4**) to give the Diels–Alder adduct (**5a**) as essentially the single diastereoisomeric product. This adduct was treated with two molar equivalents of (+)-(*R*)-α-methoxy-α-(trifluoromethyl)phenylacetyl (MTPA) chloride to give the di-MTPA ester (**6a**) and the NMR data for the compound were recorded. Now, a set of the complementary NMR data could have been obtained by the reaction of the opposite enantiomer of the *cis*-dihydrodiol (**8**) with Cookson's dienophile and then with (+)-(*R*)-MTPA chloride, which would have given the diastereoisomer **9**. Although the antipode **8** was unavailable, the *enantiomer* **7a** of the desired diastereoisomer **9** could be obtained by reaction of the Diels–Alder adduct **5a** with (−)-(*S*)-MTPA chloride. In an achiral environment, such as an NMR spectrometer, the two enantiomers **7a** and **9** would, of course, be identical.

The NMR spectra of the diastereoisomeric pairs **6** and **7** showed two distinctive and tentatively diagnostic features over a range of monocyclic *cis*-dihydrodiols: one of the methoxy peaks in the ^{1}H spectrum and one of the trifluoromethyl peaks in the ^{19}F spectrum. This means that future assignments should be possible by synthesis of only one diastereoisomer. Interestingly, this work brought to light an example of incomplete stereospecificity: the *cis*-dihydrodiol of fluorobenzene (**3c**) was found to have only *ca* 60% enantiomeric excess, possibly as a result of the similar steric bulk of the fluorine and hydrogen atoms. All other monosubstituted, monocyclic *cis*-dihydrodiols which have been investigated by this or other methods have been found to be a single enantiomer.

The second general method for absolute stereochemical assignment, from Stephenson *et al.*,[38] uses the circular dichroism (CD) spectra obtained from tricarbonyliron complexes of the *cis*-dihydrodiols. A series of optically pure *cis*-dihydrodiols (**3a,d**), their methyl ethers (**10**) and their acetate esters (**11**) were treated with diiron nonacarbonyl (**12**) to give a range of 1-substituted tricarbonyliron (η^4-cyclohexadiene)iron(0) complexes (**13**) (Scheme 4). The CD spectra obtained from these complexes were remarkably similar and gave confidence that this method could be used in a diagnostic fashion to assign absolute stereochemistry.

The drawback of both of these methods is that chemical transformations must be carried out first. This always carries a risk of losing small amounts of the opposite enantiomer through handling. Indeed, it has been found[18] that

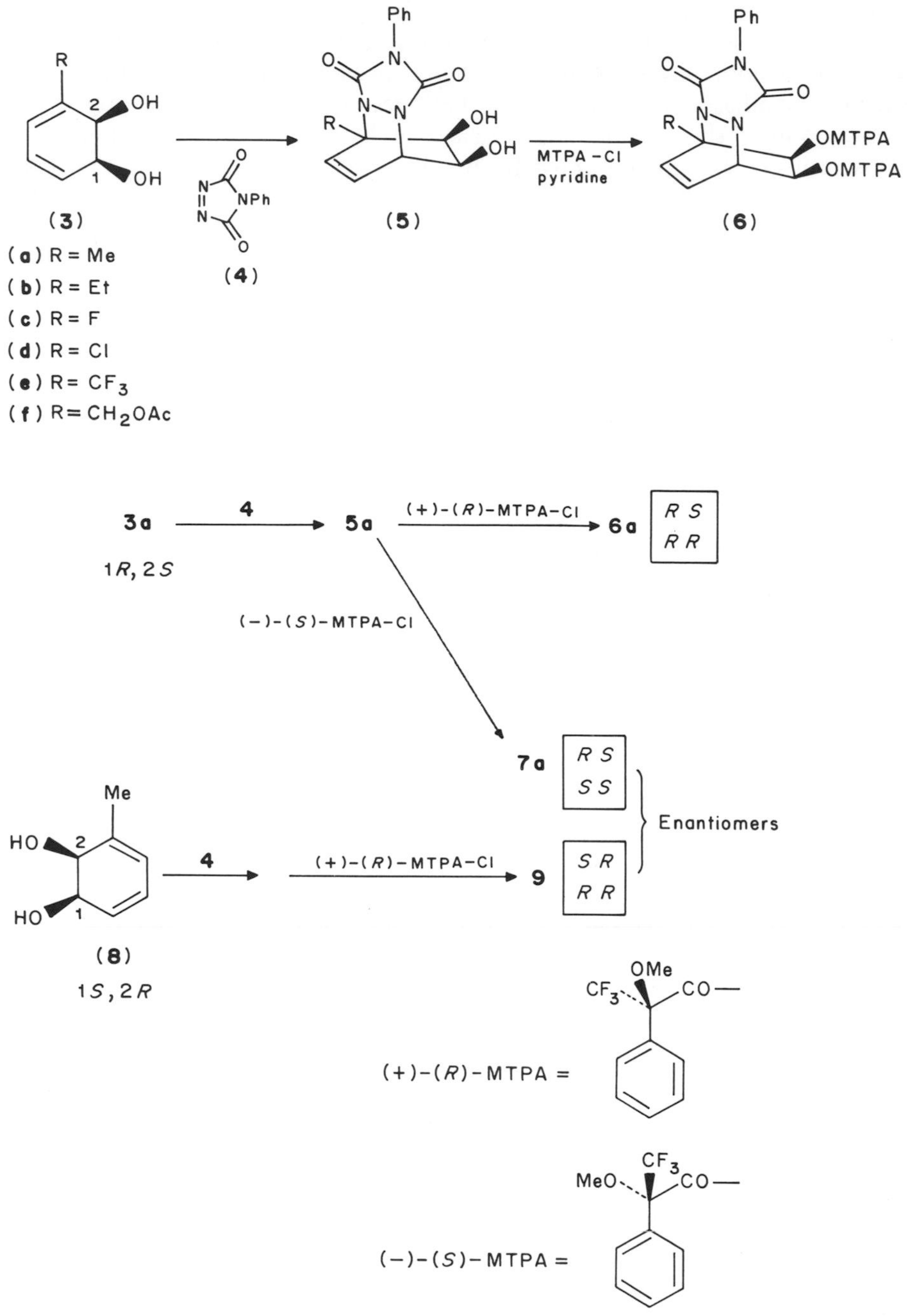

(3)
(a) R = Me
(b) R = Et
(c) R = F
(d) R = Cl
(e) R = CF_3
(f) R = CH_2OAc

(4)

(5)

MTPA-Cl
pyridine

(6)

3a 1*R*, 2*S* —4→ 5a —(+)-(*R*)-MTPA-Cl→ 6a *RS* *RR*

5a —(−)-(*S*)-MTPA-Cl→ 7a *RS* *SS*

(8) 1*S*, 2*R* —4→ —(+)-(*R*)-MTPA-Cl→ 9 *SR* *RR*

7a and 9: Enantiomers

(+)-(*R*)-MTPA =

(−)-(*S*)-MTPA =

Scheme 3

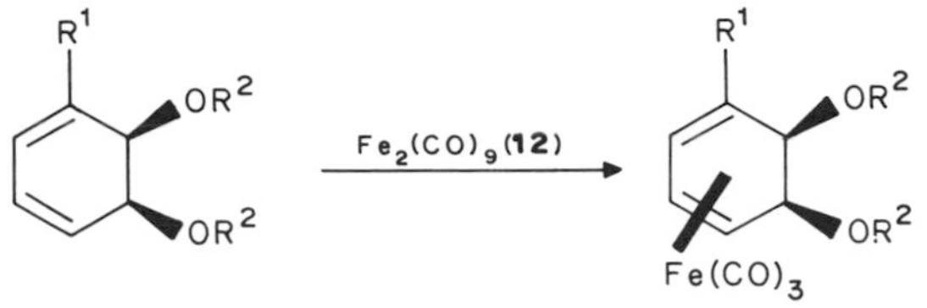

$Fe_2(CO)_9$ (**12**)

(**3a**) R^1 = Me, R^2 = H
(**3d**) R^1 = Cl, R^2 = H
(**10a**) R^1 = Me, R^2 = Me
(**10b**) R^1 = Cl, R^2 = Me
(**11a**) R^1 = Me, R^2 = COMe
(**11b**) R^1 = Cl, R^2 = COMe

(**13**) $Fe(CO)_3$

Scheme 4

the optically impure fluorobenzene *cis*-dihydrodiol (**3c**) can be recrystallized to be essentially optically pure.

6.2 INDUSTRIAL BIOTRANSFORMATION

ICI was the first chemical company to show a commercial interest in this biotransformation[39] and to develop a robust process capable of producing multi-kilogram quantities of arene *cis*-dihydrodiols.[17,40,41]

A strain of *Pseudomonas putida* was selected which had a high tolerance to the aromatic substrates which are normally toxic to microorganisms. The strain was also selected to be constitutive for dioxygenase activity, which simplified the fermentation. Thus, both fermentation and biotransformation parts of the process were separated physically in order to reduce problems caused by the inhibitory effect of the *cis*-dihydrodiols on the growing cells and also to allow more flexible use of manufacturing plant for the biotransformation reaction. The general techniques involved in the identification and development of a suitable candidate strain are described in more detail in Chapter 5.

In a typical industrial biotransformation, the cells are no longer growing when the substrate is introduced into the system; they act simply as a biological carrier for the complex enzyme catalyst. A carbon source is added to the mixture to provide a regeneration source for the NADH cofactor consumed in the reaction (see Section 6.1.2). When the biotransformation is complete, the cells and high molecular weight debris are removed and the *cis*-dihydrodiol may then be isolated from the aqueous mixture by conventional chemical process techniques.

Since the original ICI involvement, several other companies have published versions of the *cis*-dihydroxylation process to solve a wide range of synthetic problems. Shell[23,26] have been active for many years and, more recently, so have Japan Synthetic Rubber,[42] Idemitsu Kosan,[43] Sanraku,[44] Enzymatix,[45] 3M[46] and General Electric.[47] Several groups, both industrial and academic, have also investigated various aspects of this biotransformation process. Studies on the effect of reactor design on the kinetics of the reaction,[48,49] continuous reaction,[50] continuous extraction.[51] genetic engineering[52] and enzyme isolation have all been published. One chemical solution[53] to the problem of isolating the highly water-soluble *cis*-dihydrodiols from the biotransformation mixture involves the formation of phenylboronate esters (**16**) under mild aqueous conditions (Scheme 5). Addition of phenylboronic acid (**14**) to the *cis*-dihydrodiol solution under basic conditions forms the anionic species **15**. Adjustment of the pH to neutrality precipitates the insoluble phenylboronate ester (**16**).

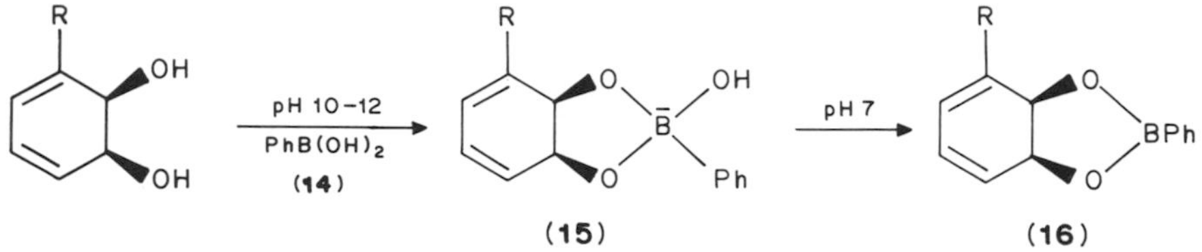

R OH OH — pH 10–12, $PhB(OH)_2$ (**14**) → R O O $\bar{B}$ OH Ph (**15**) — pH 7 → R O O BPh (**16**)

Scheme 5

6.3 COMMERCIAL PRODUCTS DERIVED FROM ARENE *CIS*-DIHYDRODIOLS

Perhaps because of the very novelty of retrosynthetic analysis of chiral compounds leading to an achiral aromatic precursor, there has not yet been much published industrial activity in this area. Given the long lead times required to bring an agrochemical or pharmaceutical product to the market place, and the relatively short time that arene *cis*-dihydrodiols have been available in commercial quantities, it may still be some time before an industrial-scale stereospecific synthesis based on a *cis*-dihydrodiol is seen.

One very recent example, and the first example of a patent which specifically claims uses for the chirality of *cis*-dihydrodiols, comes from Enzymatix.[45] This patent cites the cycloaddition reaction between toluene *cis*-dihydrodiol, protected as the acetonide (**17**), and *p*-toluenesulphonyl cyanide (**18**) to give the lactam **19**, which can be elaborated into antiviral compounds (Scheme 6).

Apart from this example, patents concerning the use of *cis*-dihydrodiols have been limited mainly to exploiting the novel aromatic substitution patterns which become available on rearomatization of the ring by either dehydration or dehydrogenation. These industrial processes are summarized in Scheme 7 and described in more detail by Brown.[2]

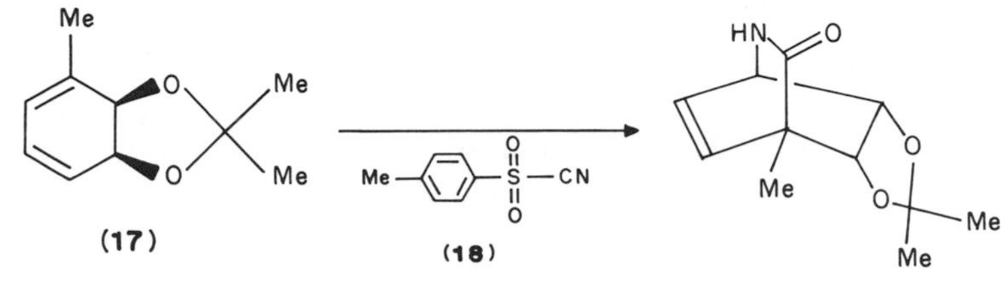
Me
O
O
Me
Me
(17)
Me
O
S
CN
O
(18)
HN
O
O
Me
O
Me
Me
(19)

Scheme 6

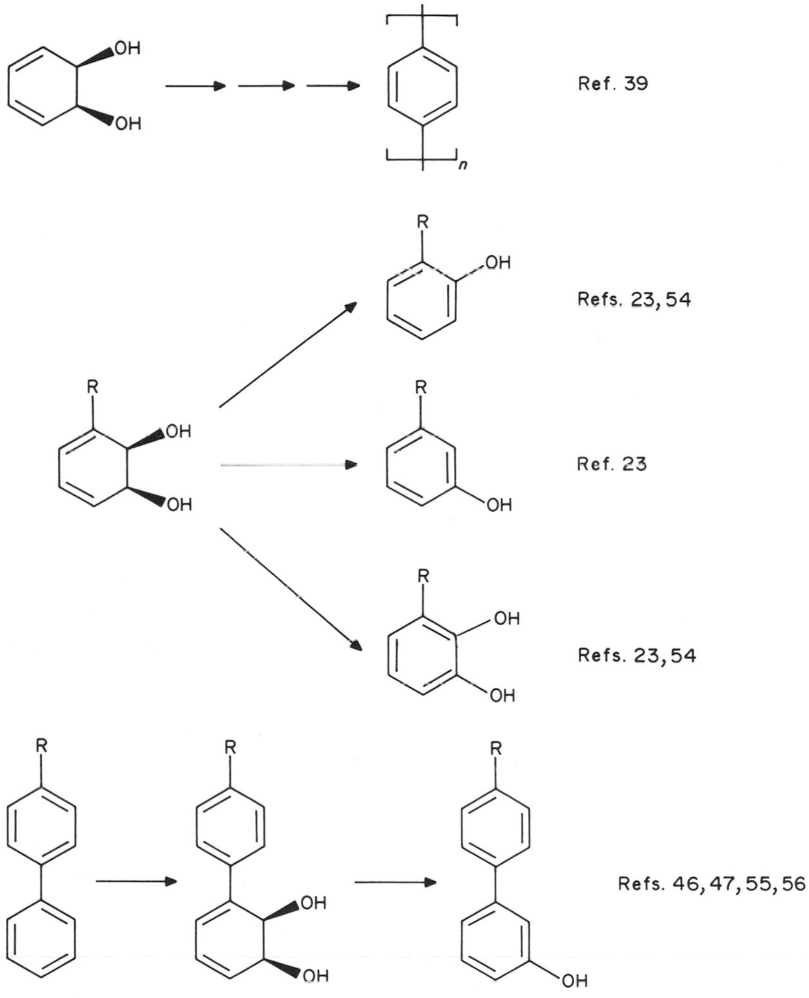
OH
OH
n
Ref. 39
R
OH
Refs. 23, 54
R
OH
OH
R
Ref. 23
OH
R
OH
Refs. 23, 54
OH
R
R
R
OH
OH
OH
Refs. 46, 47, 55, 56

Scheme 7

6.4 CHIRAL SYNTHESES USING ARENE *CIS*-DIHYDRODIOLS

While there may not yet be a large number of industrial synthetic uses which exploit the chiral centres of *cis*-dihydrodiols, there has been an exponential growth over the last 5 years in publications from academic groups which have used this strategy. The laboratories of Ley,[57] Hudlicky and Carless have been the most prolific in approaching natural product syntheses from *cis*-dihydrodiol precursors. While some of the chemical transformations described in this section are common to many of the syntheses, they dramatically illustrate the potential of *cis*-dihydrodiols for rapid and stereospecific access to biologically important chiral molecules.

6.4.1 PINITOLS

The naturally occurring cyclitol (+)-pinitol (**21**) has a diverse range of biological activities affecting growth[58] and feeding[59] of certain insect species as well as having anti-diabetic properties in mice.[60]

The first reported synthesis from a *cis*-dihydrodiol was by Ley *et al.*[61] in 1987 which gave racemic pinitol in six steps from benzene, including the protection and deprotection, with an overall yield of 49% (Scheme 8). The key steps were the epoxidation, which gave a 5:1 selectivity of attack at the desired face, and the osmium-catalysed dihydroxylation of the remaining double bond, which proceeded with complete facial selectivity.

This work was followed up in 1989[62] with a resolution of the intermediate alcohol (**20**) using menthoxyacetate esterification (Scheme 9). Having separated the diastereoisomers **22** and **23**, both antipodes of pinitol (**21**) were prepared separately.

Shortly after this resolution method, an elegant enantiodivergent synthesis of both (+)- and (−)-pinitol was published by Hudlicky *et al.*[19] (Scheme 10). In this approach, both enantiomers were prepared stereospecifically from the same enantiomer of bromobenzene-*cis*-dihydrodiol (**24**). The symmetry of the oxygenation pattern of the 'upper' and 'lower' halves of pinitol mean that they can be regarded as differing only by the transposition of the methyl from position 2 to position 3 (see Scheme 10). Thus, by carrying out the same chemical transformations but in a different order, Hudlicky *et al.* were able to prepare both enantiomers. The bromine atom which conferred the chirality to the starting material (**25**) is lost during the synthesis and so bromobenzene-*cis*-dihydrodiol is acting here as synthon for chiral benzene-*cis*-dihydrodiol.

A third approach to pinitol synthesis came from Carless *et al.*[63] Benzene-*cis*-dihydrodiol, protected as the bis-TBDMS ether (**26**), was subjected to low-temperature photosensitized oxidation by singlet oxygen to give a mixture of the endoperoxide **27** and the hydroperoxide **28**. Epoxidation of the remaining

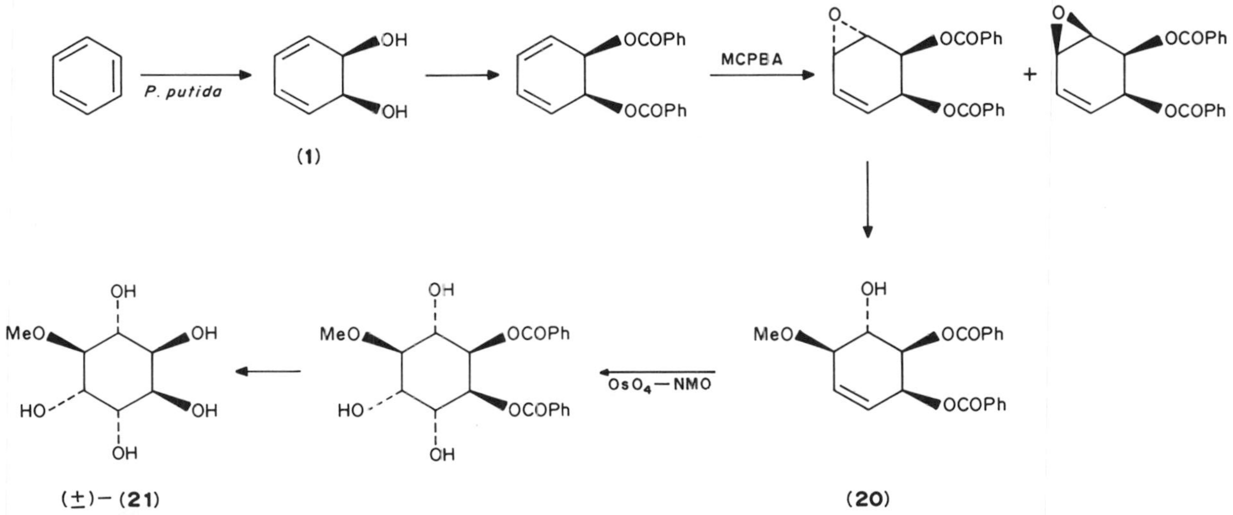

P. putida

(1)

MCPBA

(20)

OsO_4—NMO

(±)-(21)

MCPBA = 3-chloroperbenzoic acid

NMO = *N*-methylmorpholine-*N*-oxide

Scheme 8

(±)−(20)

(22) → → (+)−(21)

(23) → → (−)−(21)

Scheme 9

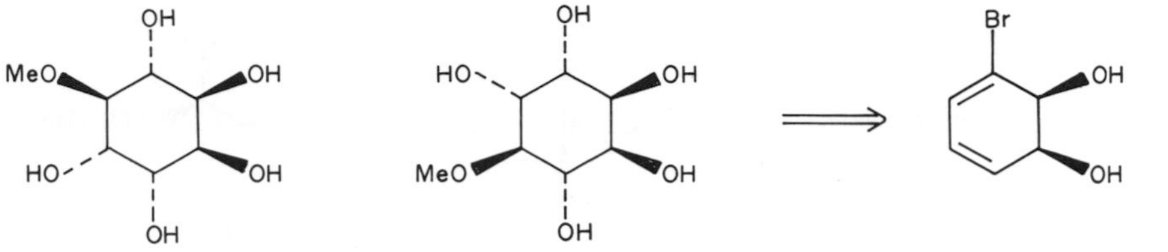

(+)-(**21**) (−)-(**21**) (**24**)

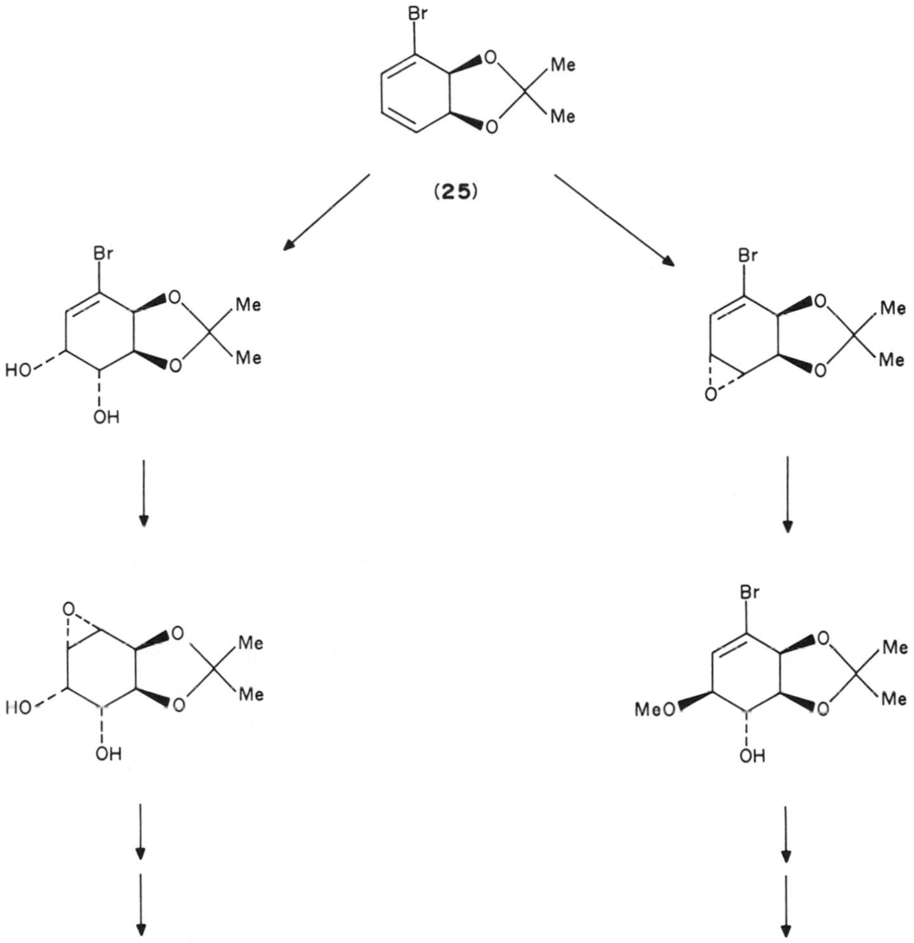

(**25**)

(+)-(**21**) (−)-(**21**)

Scheme 10

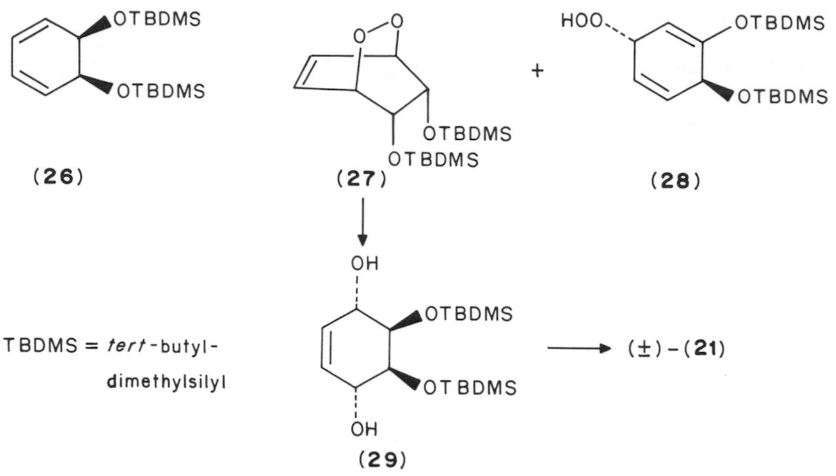

OTBDMS
OTBDMS
(26)
O
O
OTBDMS
OTBDMS
(27)
+
HOO
OTBDMS
OTBDMS
(28)
TBDMS = tert-butyl-
dimethylsilyl
OH
OTBDMS
OTBDMS
OH
(29)
(±)-(21)

Scheme 11

double bond in the tetrol **29** and stereocontrolled cleavage by acidic methanol completed the synthesis (Scheme 11).

6.4.2 INOSITOL PHOSPHATES AND ANALOGUES

The discovery of the physiological importance of *myo*-inositol-1,4,5-triphosphate (IP_3; **33**) and its analogues as secondary messengers in the stimulation of calcium release in a wide range of cell types and functions[64] has greatly increased the desirability of an efficient and flexible synthetic route to these compounds.[65] Ley and co-workers, again starting from benzene-*cis*-dihydrodiol (**1**), reported the synthesis of IP_3 and three analogues[66–68] (Scheme 12). As with pinitol (see Section 6.4.1), the initial work was on the racemic series.

The key stage in the synthesis was the stereo- and regiospecific opening of the epoxide **31** with an oxygen nucleophile, which proved to be a far from facile reaction. The problem was ultimately solved by the development of a new reagent, 5,5-dimethyl-1,3-dioxane-2-ethanol (**32**), which acts as a hydroxide equivalent. The three IP_3 analogues **36**, **37** and **38** (Scheme 13) resulted from the studies on the opening of the epoxide **31**. Further diversification of the same methodology gave access to inositol-1-phosphate, which is another metabolic product in the phosphatoinositol cycle.

The route was transformed into a synthesis of optically active IP_3 by a resolution but without an overall increase in the number of steps (Scheme 13). This time the first epoxide (**30**) was opened with the optically pure benzylic alcohol **34** to give two diastereoisomers (**35a** and **35b**), which were separated and converted respectively into the natural (−)-(*R*)- and the unnatural

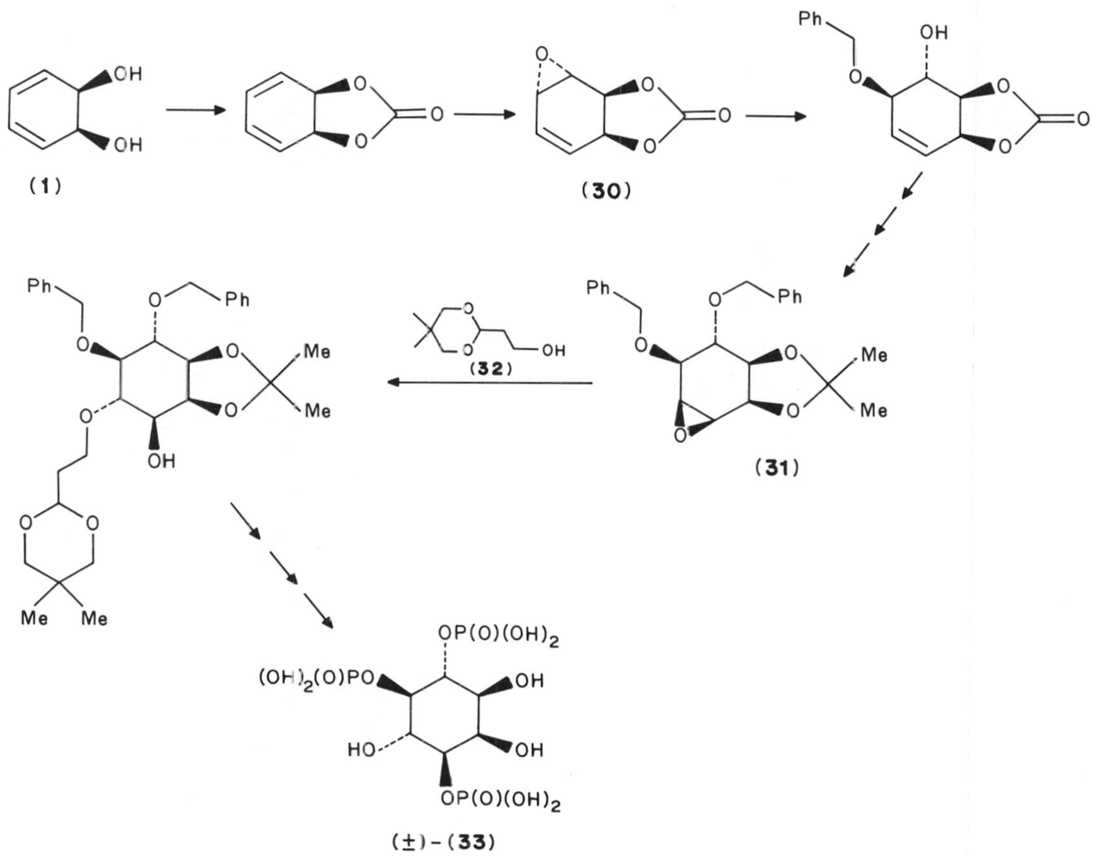

(1)

(30)

(31)

(32)

(±)-(33)

Scheme 12

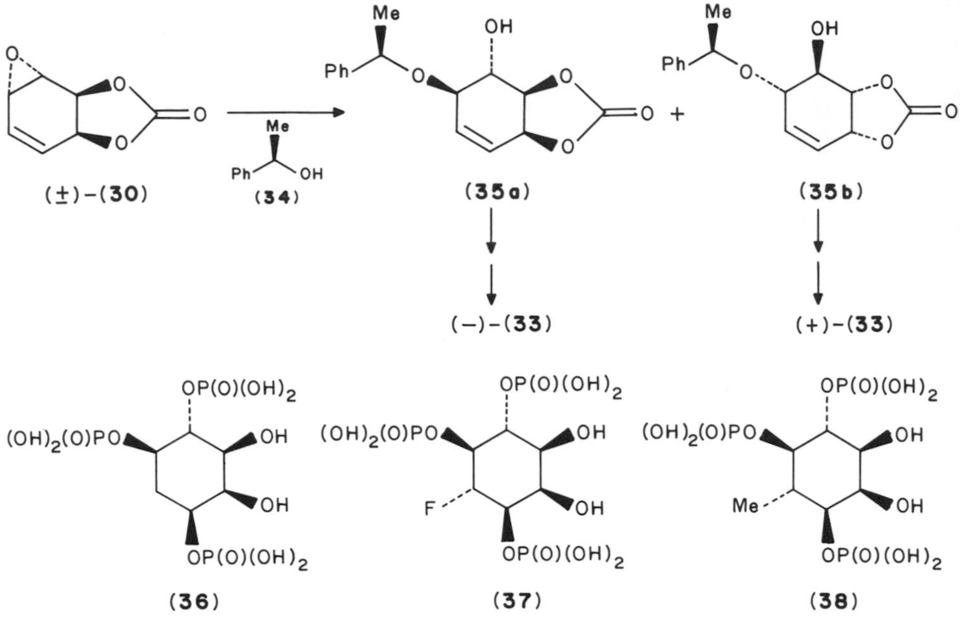

Scheme 13

(+)-(*S*)-isomers of IP_3 (**33**) using essentially the same chemistry as for the racemate.

6.4.3 CONDURITOLS AND DIHYDROCONDURITOLS

The conduritols (cyclohexenetetrols) and dihydroconduritols (1,2,3,4-cyclohexanetetrols) are subclasses of the cyclitol family which are widespread in nature in many of the possible isomeric forms. Interest in their synthesis has been prompted by their potential as glycosidase enzyme inhibitors.[69] Having fewer stereochemical centres and less functionality than the inositols, they are ideal targets for short stereospecific syntheses from arene *cis*-dihydrodiols.

The first workers to publish in this area were Carless and Oak,[70] who demonstrated syntheses of racemic conduritols A (**41**) and D (**42**) and their corresponding dehydro compounds (**43** and **44**) by the singlet oxygen strategy outlined earlier for the pinitol synthesis (Section 6.4.1) (Scheme 14).

The photosensitized oxidation of benzene-*cis*-dihydrodiol gave a mixture of two endoperoxides (**39** and **40**), which were reduced quantitatively with thiourea to give the racemic conduritol A (**41**) and D (**42**), respectively. The dehydroconduritols **43** and **44** were prepared simply by rearrangement of the endoperoxides **39** and **40** with organic base. All four compounds had thus been prepared in just three stages from benzene.

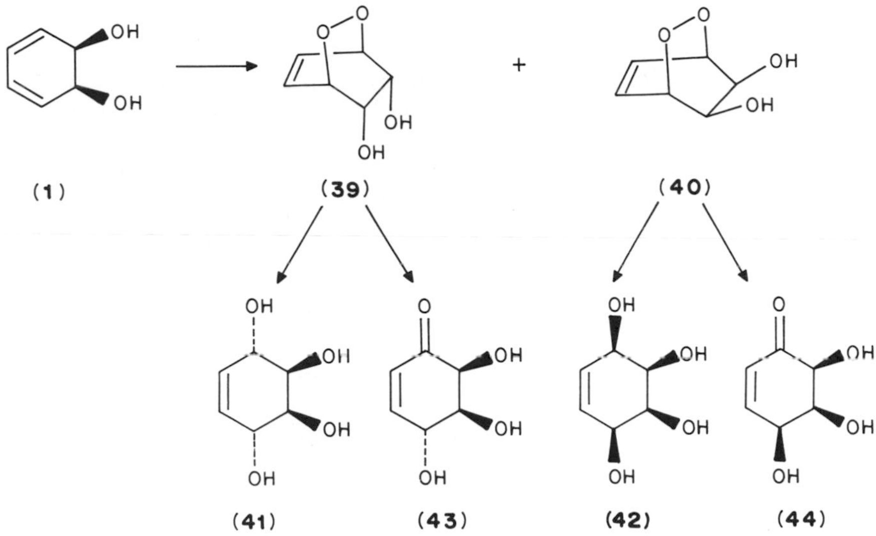

(1) (39) (40)

(41) (43) (42) (44)

Scheme 14

The same group followed up this work with an enantiospecific synthesis of (+)-conduritol C (**45**) from fluorobenzene-*cis*-dihydrodiol (**3c**) (Scheme 15)[71] using the fluorine atom as a sacrificial way of introducing chirality into the system.

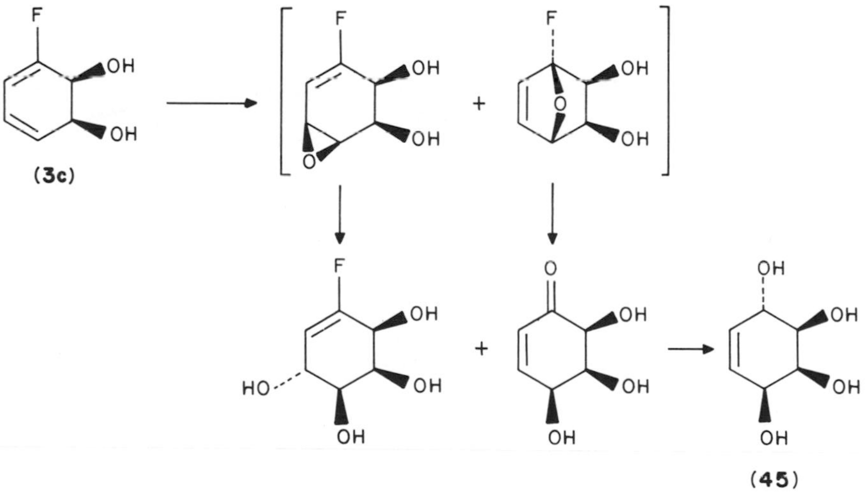

(3c)

(45)

Scheme 15

Ley and Redgrave[72] have also published a stereospecific synthesis, this time of conduritol F (**46**), using the chiral benzylic alcohol resolution method described earlier (Section 6.4.2) (Scheme 16). Sodium–liquid ammonia reduction of the diastereoisomers **35a** and **35b** gave the natural (+)- and unnatural (−)-isomers of conduritol F, respectively.

Hudlicky *et al.*[73] used the *cis*-dihydrodiol of chlorobenzene (**3d**) to effect the synthesis of (−)-dihydroconduritol C (**50**) (Scheme 17). Singlet oxygen oxidation

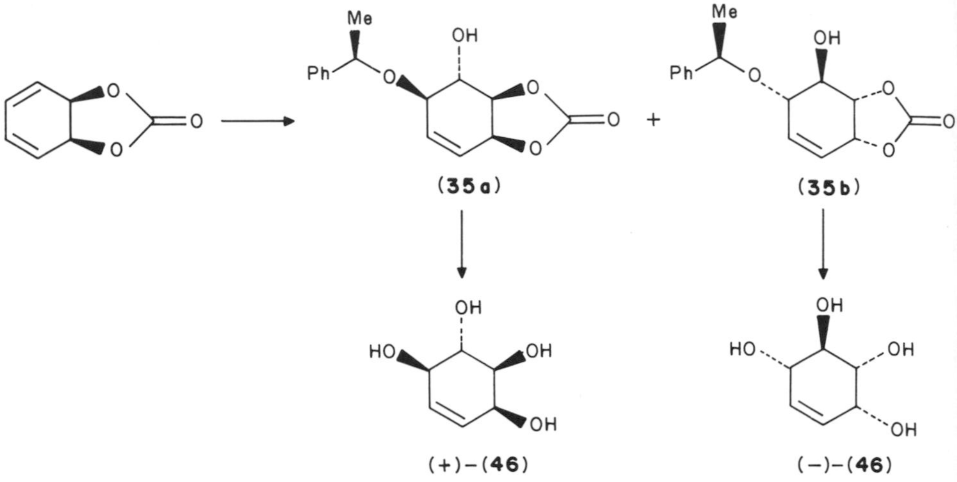

(35a) (35b) (+)-(46) (−)-(46)

Scheme 16

(47) $^1O_2$ (48) (−)-(50) (49)

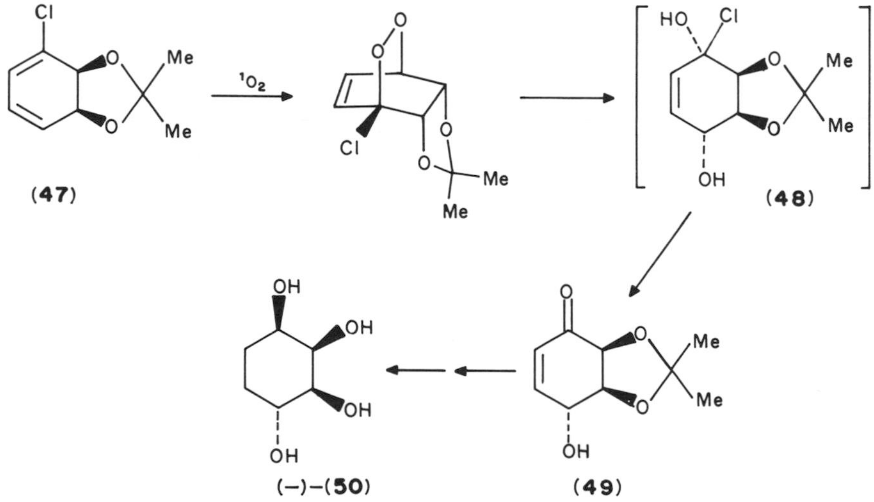

Scheme 17

of the protected *cis*-dihydrodiol **47** followed by reduction gave the enone **49** via HCl elimination from the *gem*-chlorohydrin **48**. After protection of the free hydroxy group, a sequence of catalytic reduction of the double bond, stereoselective ketone reduction and finally deprotection gave the desired (−)-enantiomer of dihydroconduritol C.

The approach of Johnson *et al.*[74] to the synthesis of both enantiomers of conduritol C (**54**) differs from that of the other groups in that a second enzyme-catalysed reaction is carried out to induce chirality from a prochiral substrate (Scheme 18). Starting with the protected benzene-*cis*-dihydrodiol **51**, singlet oxygen oxidation gave the prochiral *meso*-acetonide **52**, which was monoacetylated on the oxygen at the pro-*R* site using a lipase-catalysed transesterification. This optically active intermediate (**53**) could then be elaborated into either enantiomer of conduritol C by Mitsunobu inversion of the appropriate alcohol.

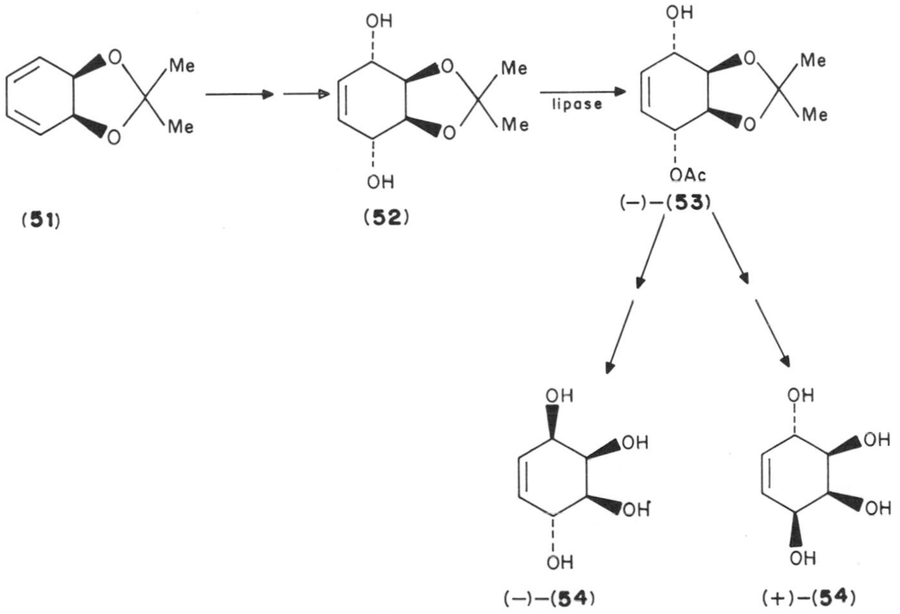

(**51**) (**52**) lipase (−)-(**53**) (−)-(**54**) (+)-(**54**)

Scheme 18

6.4.4 OTHER CYCLITOLS AND INTERMEDIATES

Carless *et al.*[63] have used their strategy of singlet oxygen oxidation of *cis*-dihydrodiols to prepare three cyclitol intermediates (**57**–**59**) (Scheme 19). Rather than reduction to the tetrol **29**, used in the pinitol synthesis (Section 6.4.1), this time the endoperoxide **27** was subjected to photolytic rearrangement to give a dynamic mixture of the β,γ-epoxy ketones **55** and **56**. These were converted into

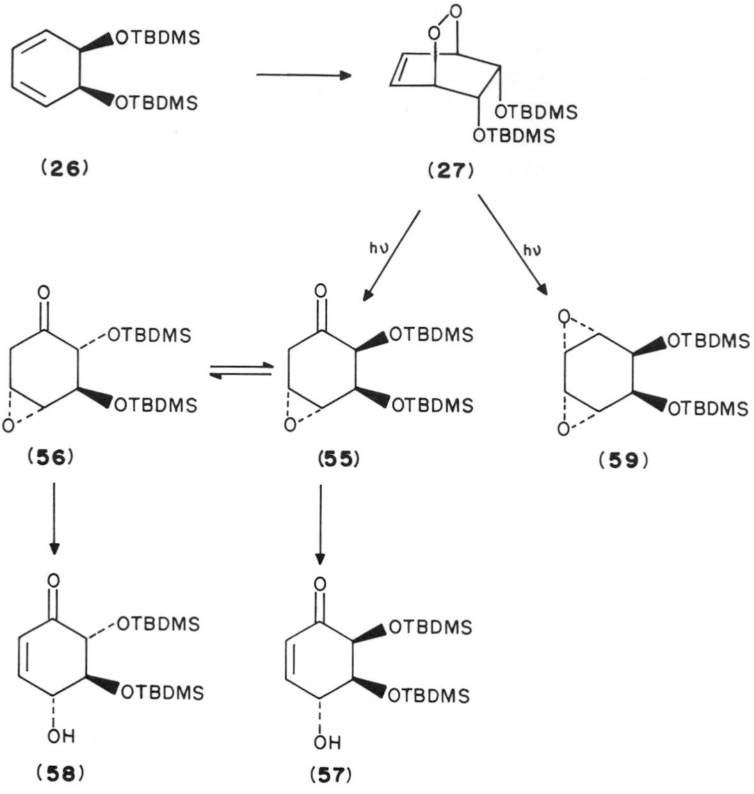

Scheme 19

the hydroxy enones **57** and **58** via base-catalysed tautomerism. The photolysis of **27** also yielded a small amount of the bisepoxide **59**, which could also be prepared in better yield by the use of a tetraporphorinylcobalt catalyst. All three intermediates (**57**–**59**) have obvious synthetic potential for further elaboration to cyclitol structures, including the inositols.

Using the Ley epoxidation strategy, Carless and Oak[75] have synthesized the *C*-methyl-*myo*-inositol (−)-laminitol (**64**), which is an algal metabolite (Scheme 20). Regio- and stereoselective epoxidation of the trisubstituted double bond of the protected toluene-*cis*-dihydrodiol **60** followed by acid-catalysed cleavage gave the tetrol **62**. A second epoxidation gave **63**, but trifluoroacetic acid-catalysed opening of this epoxide gave a 1:1 mixture of the desired (−)-laminitol (**64**) and the optically inactive diastereoisomer (**65**).

There have also been two reported approaches to aminocyclitols. In the first, Braun *et al.*[76] demonstrated the stereospecific synthesis of the two aminocyclitols **72** and **73** from **66**, the diacetate of benzene-*cis*-dihydrodiol (Scheme 21).

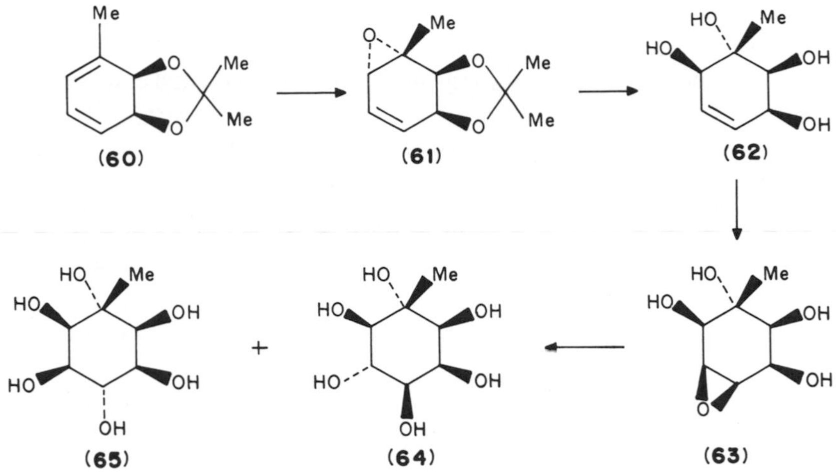

Scheme 20

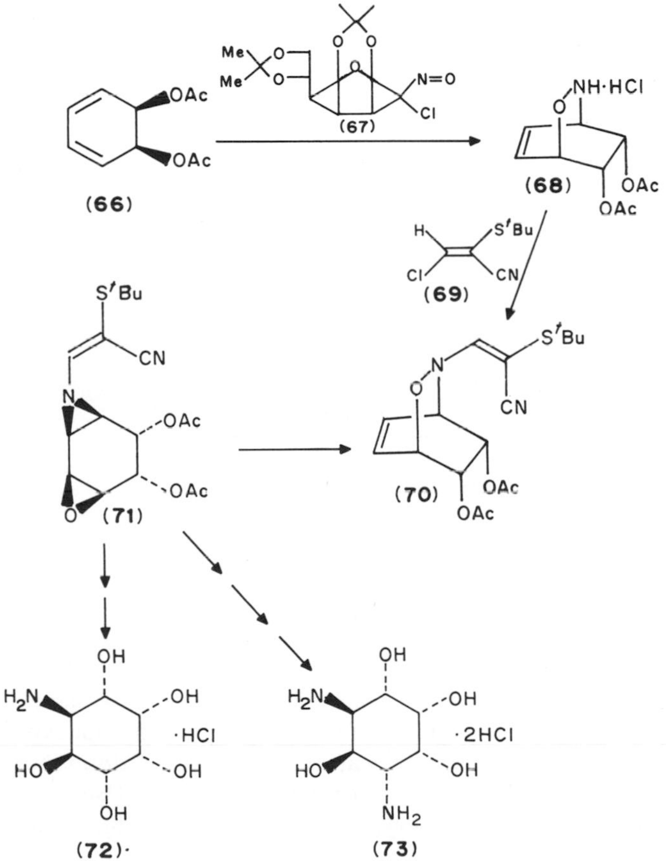

Scheme 21

Reaction of the *meso*-diene **66** with the optically active nitroso dienophile **67**, derived from D-mannose, followed by cleavage of the sugar residue gave the bicyclic oxazine **68**. *N*-Derivatization with the substituted vinyl chloride **69** gave **70**, which isomerized on heating to give the aziridine epoxide **71**. Selective ring-opening reactions, followed by deprotection, gave the desired aminocyclitols **72** and **73**.

In the second example, Hudlicky and Olio[77] used the reverse strategy of an achiral nitroso dienophile (**74**) with optically active dienes (**25** and **47**) in a regio- and stereoselective Diels–Alder reaction to give the oxazine **75**. Reduction of the N—O bond gave the protected conduramine A-1 (**76**) and further reduction of the double bond gave dihydroconduramine A-1 (**77**) (Scheme 22).

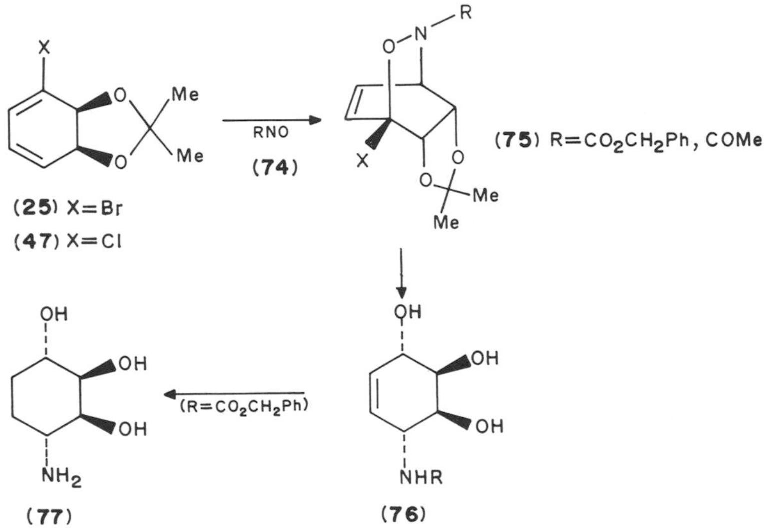

Scheme 22

6.4.5 ACYCLIC SUGARS

In addition to the cyclitol structures in the preceding sections, the possibility of accessing acyclic sugar molecules from *cis*-dihydrodiol precursors has been examined.

Hudlicky *et al.*[34] have shown that an enantiodivergent approach from chlorobenzene-*cis*-dihydrodiol (**3c**) can give both D- and L-erythrose derivatives (Scheme 23). Ozonolysis of the protected *cis*-dihydrodiol **47** gave the anomeric hemiacetal lactone **78** as a common precursor to both product enantiomers. In one case, the masked aldehyde was reduced with sodium borohydride to give the lactone **79**, which, on further reduction with DIBAL, gave the lactol **80**, a protected derivative of L-erythrose. The slightly longer conversion of the

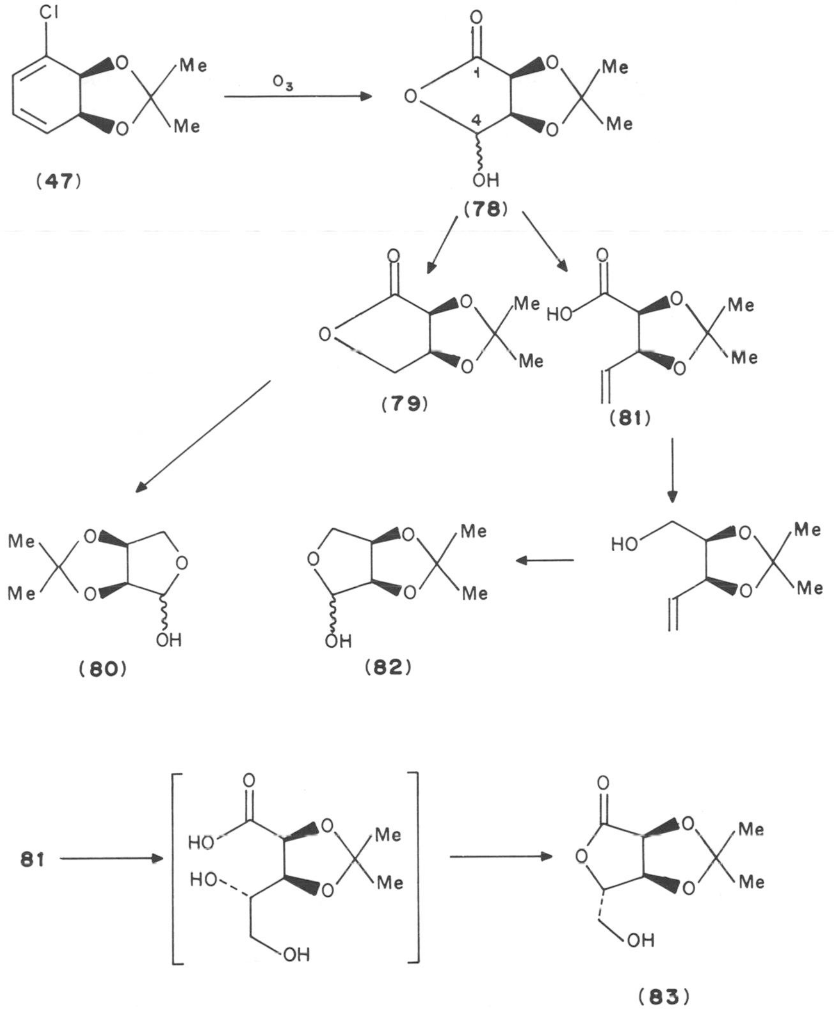

Scheme 23

lactone **78** to the antipode of **80** started with an olefination of the masked aldehyde, followed by reduction of the acid and a second ozonolysis to give the lactol **82**. Both routes avoided the situation in which carbons 1 and 4 in the lactone **78** gained identical substituents which would have destroyed the asymmetry of the molecule.

This work was extended[78] to the preparation of L-ribonic γ-lactone (**83**) (Scheme 23). The olefinated hydroxy lactone **81** was dihydroxylated *anti* to the isopropylidene group with osmium tetraoxide and spontaneous lactonization gave the L-ribonic γ-lactone derivative (**83**).

In 1991, Lehmann and Montz[79] published a racemic synthesis of the modified hexose *N*-acetylglucosamine **86** from benzene-*cis*-dihydrodiol (Scheme 24). The nitrogen was introduced into the molecule by sodium azide opening of the epoxide **84** and a series of selective protections and deprotections gave the monoacetate **85**. Conversion to an acyclic structure was effected by ozonolysis of **85**. A resolution of **86** was carried out by an enzyme catalysed D-galactosidation to give the pure enantiomers.

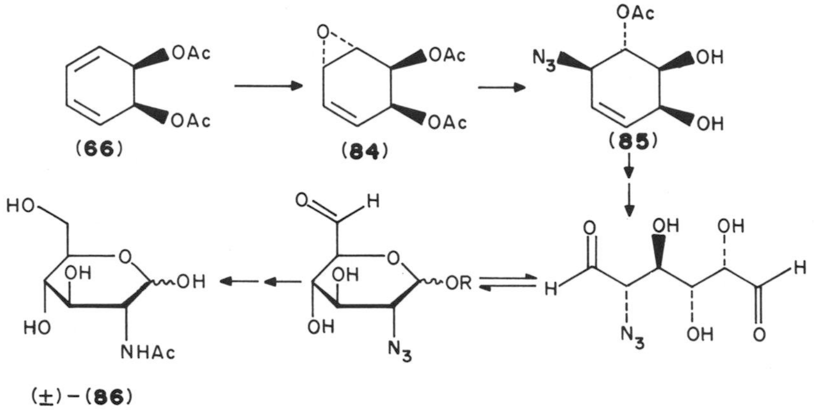

(66) (84) (85) (±)-(86)

Scheme 24

6.4.6 OTHER TOTAL SYNTHESES

A further three non-cyclitol syntheses based on *cis*-dihydrodiol starting materials have been reported by Hudlicky's group. In the first of these,[14] a prostaglandin intermediate (**89**) was prepared stereospecifically by controlled oxidation of the diene system of toluene-*cis*-dihydrodiol (**3a**) (Scheme 25). Careful ozonolysis of the protected *cis*-dihydrodiol **60** gave a diastereoisomeric mixture of the hemiacetals **87**, which retained the stereochemical integrity of the original diol. The same hemiacetals were also prepared by photosensitized oxidation of **60** followed by osmium tetroxide oxidation of the remaining double bond (the first example of the reaction of an arene *cis*-dihydrodiol with singlet oxygen). Dehydration of the hemiacetals **87** gave the dicarbonyl compound **88** and alumina-catalysed intramolecular aldol condensation gave the cyclopentenone **89**, which had previously been shown to be a key intermediate in the synthesis of prostaglandin $PGE_{2\alpha}$.

The natural product zeylana (**96**), derived from the roots of *Uvaria zeylanica* L. (Annonaceae), has a skeleton and functionality which are good candidates

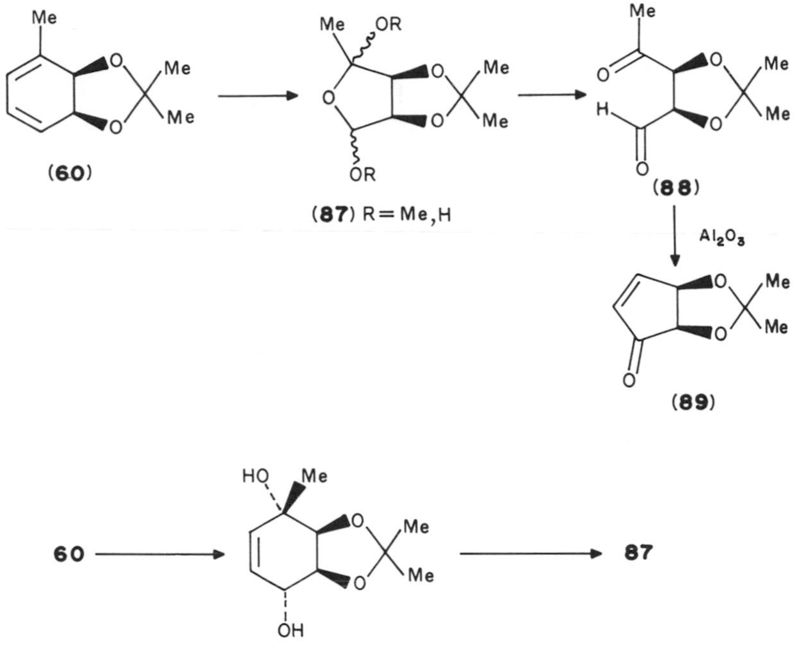

Scheme 25

for elaboration from *cis*-dihydrodiols. The strategy used by Hudlicky *et al.*[80] was based around an intramolecular Diels–Alder reaction of the styrene *cis*-dihydrodiol derivative **93** (Scheme 26). The first stage was to protect the reactive triene system of styrene *cis*-dihydrodiol as a reversible Diels–Alder adduct (**91**). Selective Mitsunobu inversion of the C-3 hydroxyl group allowed the formation of the *trans*-diester **92**. Regeneration of the endocyclic diene system by retro-Diels–Alder reaction set up the key intramolecular cyclization to give the tricyclic compound **94**. Ozonolysis of the exocyclic double and functional group manipulation gave zeylana acetate (**95**).

The final synthesis in this section is the only example to date of an alkaloid to be derived from an arene *cis*-dihydrodiol. Once again using the enantio-divergent strategy, Hudlicky *et al.*[35] have reported the synthesis of both enantiomers of the pyrrolizidine alkaloid trihydoxyheliotridane (**101**) (Scheme 27). The pair of enantiomeric erythroses **80** and **82**, derived from chlorobenzene-*cis*-dihydrodiol (see Section 6.4.5), were the key intermediates in this route. Extension of the aldehyde of **80** to the conjugated diene **97** and conversion of the alcohol to the azide **98** set up the functionality required for a [4 + 1]pyrroline annulation reaction. Ring expansion of the aziridine **99** to the pyrrolizidine system was achieved via an intramolecular 1,3-dipolar cyclization and reduction

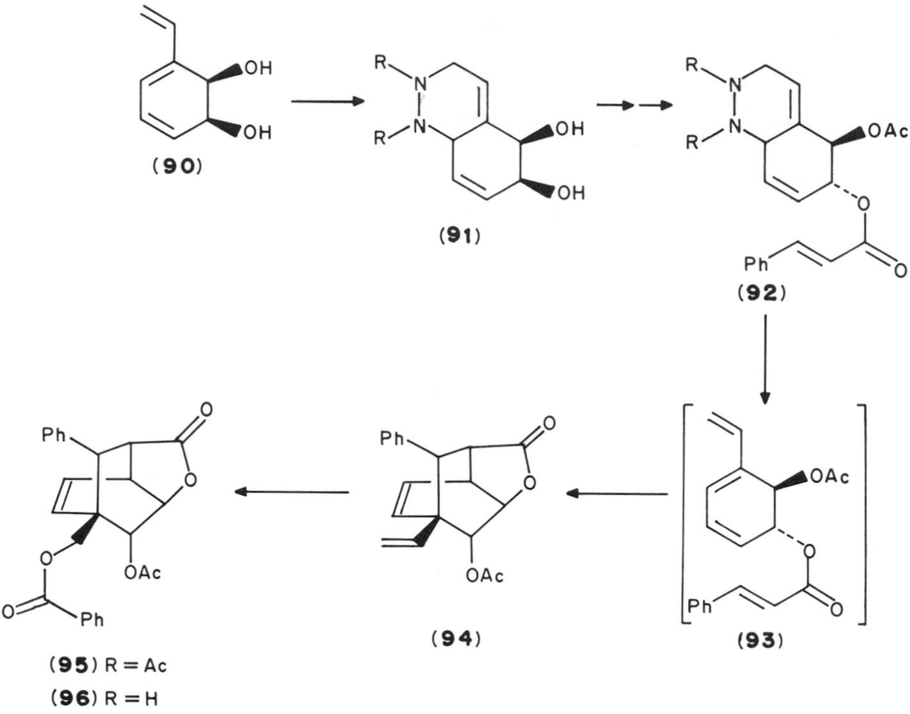

(90) (91) (92) (93) (94) (95) R = Ac (96) R = H

Scheme 26

of the ester **100** gave the isopropylidene derivative of (+)-trihydroxyheliotridane (**101**). Similar elaboration of the antipodal erythrose **82** gave the (−)-trihydroxyheliotridane acetal (**101**).

6.5 CYCLOADDITION REACTIONS

The strategy of [4 + 2]cycloaddition reactions of arene *cis*-dihydrodiols has already been encountered with the synthesis of zeylana (Section 6.4.6), the singlet oxygen reactions (Sections 6.4.1, 6.4.2 and 6.4.3), the aminocyclitol syntheses (Section 6.4.4), the determination of absolute stereochemistry (Section 6.1.2) and the synthesis of bicyclic lactams (Section 6.3). In general, the introduction of two new chiral centres with a diverse range of possible substituents makes this methodology potentially very powerful for chiral syntheses and cycloaddition reactions of arene *cis*-dihydrodiols have been the subject of study for a number of groups.

Diels–Alder reactions using the reactive endocyclic 1,3-diene system have

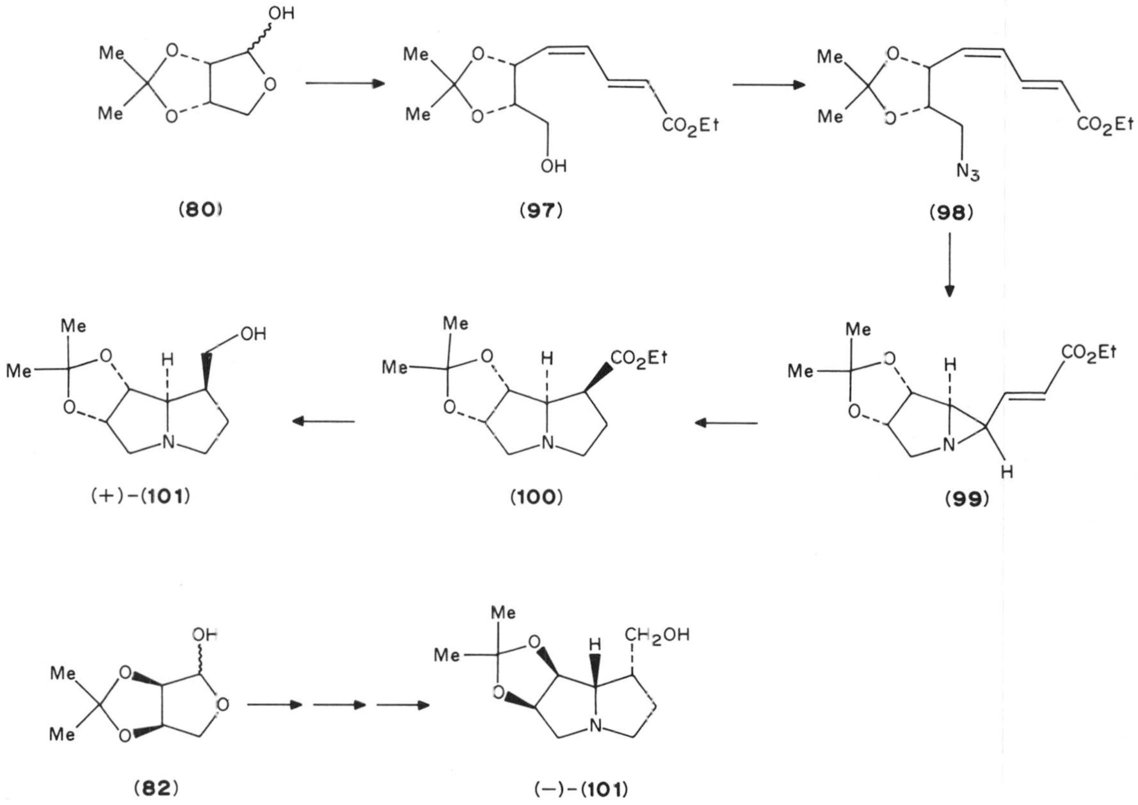

(80)

(97)

(98)

(99)

(100)

(+)-(101)

(82)

(−)-(101)

Scheme 27

been reported since the discovery of benzene-*cis*-dihydrodiol.[4,8,15] There has been evidence that the face of the diene *anti* to the diol oxygens is preferred by dienophiles,[4,8,15,61] but there has also been evidence in favour of *syn*-attack predominating.[18,82,83] An attempt[83] to rationalize the factors affecting selectivity, at least for one dienophile, *N*-phenylmaleimide (**103**), suggested that as the reactivity of the diene was increased by protecting the diol as a cyclic derivative, so the facial selectivity of the dienophile changed from *syn* (**105**) to *anti* (**104**) (Scheme 28). Whether this trend is true or not for other dienophiles and a wider range of *cis*-dihydrodiols has not been investigated rigorously.

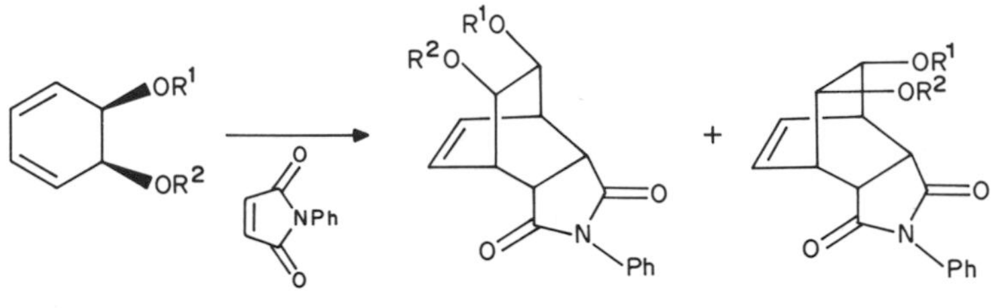

(**102**) (**103**) (**104**) (**105**)

$R^1 = R^2 = H, Ac, SiMe_3$
$R^1, R^2 = -SiMe_2-, -CMe_2-$

Scheme 28

The *cis*-dihydrodiol diene is electron-rich and its reactions with a wide range of electron-deficient dienophiles have been investigated. Double bonds in almost all permutations of carbon, nitrogen and oxygen are exemplified[18,82,84] as well as alkynes[82] and nitriles[44] (Scheme 29).

Apart from Diels–Alder-type [4 + 2]cycloadditions, other electrocyclic reactions of arene *cis*-dihydrodiols have been reported (Scheme 30). Scheme 29 showed an example of a *cis*-dihydrodiol acting as a dienophile for itself, but a photochemical [2 + 2] reaction has also been seen.[85] Another example[83] of a [2 + 2] reactions is the expected product (**107**) from the reaction of a diphenyl ketene (**106**) with the *cis*-dihydrodiol **51**, which was obtained along with the unusual [4 + 2] product (**108**). The only example[82] of a [6 + 4]cycloaddition in this series is that between the benzene-*cis*-dihydrodiol derivative **51** and tropone (**109**).

A final category of cycloaddition reactions are carbene additions. There have been a number of examples,[82,86] all of which have been regioselective for the more electron-rich double bond and the sterically less hindered face of the diene. There has not been much synthetic exploitation of these fused [6.3]-bicyclic compounds, but their potential is demonstrated by the formation[86] of the cyclic enol ether **113** from the ester **111** via reduction to the aldehyde **112**.

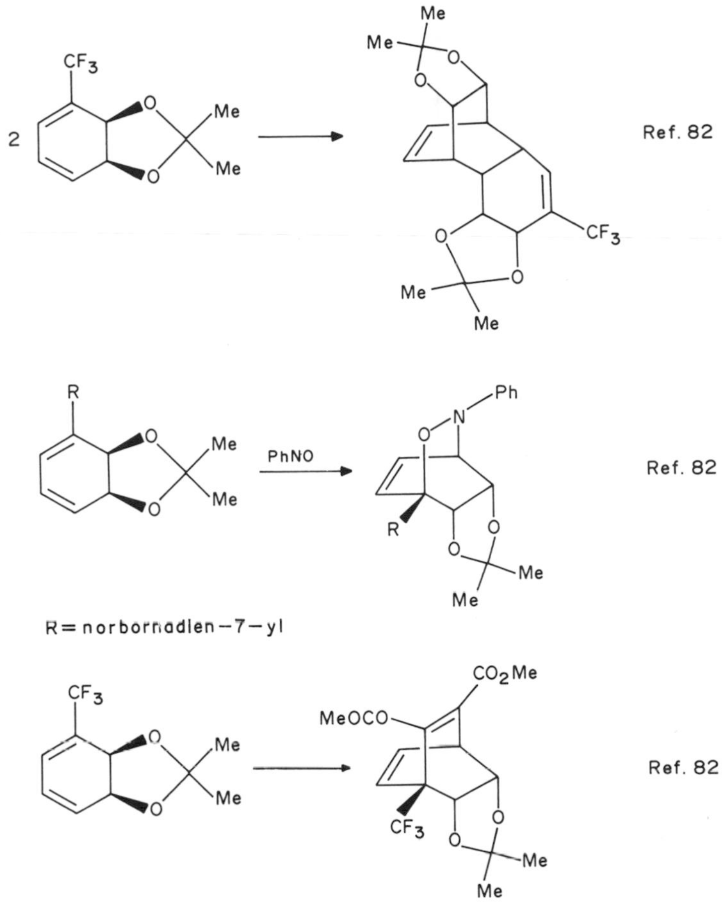

Ref. 82

PhNO

Ref. 82

R = norbornadien-7-yl

Ref. 82

Scheme 29

6.6 OTHER REACTIONS

A few other reactions of arene *cis*-dihydrodiols have been reported which have potential uses for chiral synthesis. One example is the coordination of the diene system to transition metals which was used (Section 6.1.2) in the CD method for absolute stereochemistry determination. Synthetic uses of these complexes have been limited so far, but do show promise[87] for stereospecific applications (Scheme 31).

Finally, at the end of 1991, Boyd *et al.*[20] published a method in which the halogen atom of bromo- (**24**) and iodobenzene-*cis*-dihydrodiol (**114**) can be replaced nucleophilically with other substituents (Scheme 32). A combination

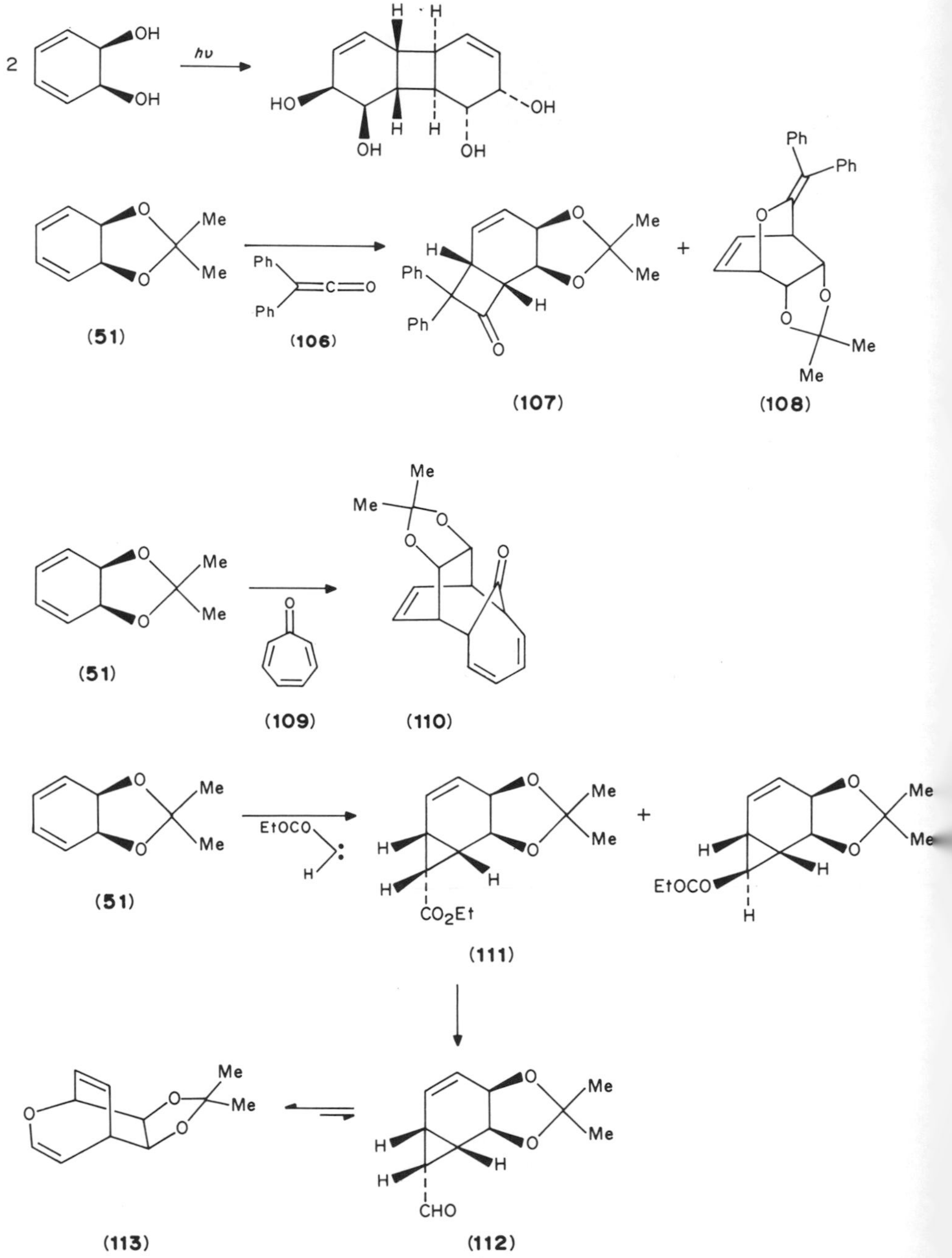

2 OH OH hν HO OH OH OH H H H H

(51) (106) (107) (108)

(51) (109) (110)

(51) (111)

(113) (112)

Scheme 30

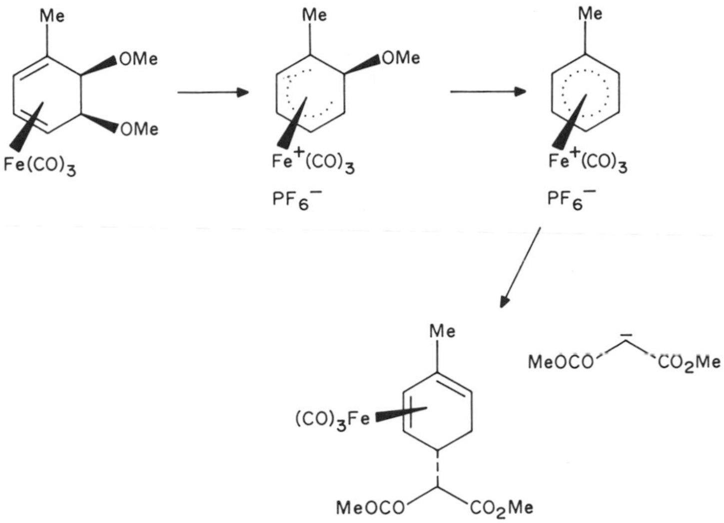

Scheme 31

(**24**) X=Br

(**114**) X=I

R = Me, Bu, CH=CH$_2$, CH$_2$CH=CH$_2$, C≡CSiMe$_3$, C≡CH, CN, SMe

Scheme 32

of biological and chemical synthesis should in this way be able to extend the effective substrate range of the biotransformation to include substituents which have formerly been either too labile or unacceptable to the organisms.

6.7 CONCLUSION

Arene *cis*-dihydrodiols have been shown, at least in the research laboratory, to be valuable synthetic building blocks for rapid and highly stereoselective syntheses of biologically active molecules containing multiple chiral centres. Although this strategy has not yet been translated into industrial-scale chiral synthesis, these compounds have established a niche. With the wide range of arene *cis*-dihydrodiols now available, either directly by biotransformation or by a combination of biological and chemical synthesis, it will only be a matter of time before this potential is realized commercially.

6.8 ACKNOWLEDGEMENTS

The author expresses his gratitude to Dr S. C. Taylor and Dr A. J. Blacker, ICI Biological Products, for their help in the preparation of this chapter and Dr R. A. McMordie, formerly of Queen's University, Belfast, whose PhD thesis was the source of much of the collected data in Table 1.

6.9 NOTES AND REFERENCES

1. Further information can be obtained from Dr M. J. Nicholds, ICI Specialties, Hexagon House, P.O. Box 42, Blackley, Manchester M9 3DA, UK.
2. Brown, S. M., in *Organic Synthesis; Theory and Applications* (ed. T. Hudlicky), JAI Press, in press.
3. Gibson, D. T., Koch, J. R., and Kallio, R. E., *Biochemistry*, **7**, 2653 (1968).
4. Gibson, D. T., Hensley, M., Yoshioka, H., and Mabry, T. J., *Biochemistry*, **9**, 1626 (1970).
5. Gibson, D. T., Cardini, G. E., Maseles, F. C., and Kallio, R. E., *Biochemistry*, **9**, 1631 (1970).
6. Axcell, B. C., and Geary, P. J., *Biochem. J.*, **146**, 173 (1975).
7. Hogn, T., and Jaenicke, L., *Eur. J. Biochem.*, **30**, 369 (1972).
8. Kobal, V. M., Gibson, D. T., Davies, R. E., and Garza, A., *J. Am. Chem. Soc.*, **95**, 4420 (1973).
9. Ziffer, H., Jerina, D. M., Gibson, D. T., and Kobal, V. M., *J. Am. Chem. Soc.*, **95**, 4048 (1973).
10. Jigani, Y., Onori, T., and Minoda, Y., *Agric. Biol. Chem.*, **39**, 1781 (1975).
11. Ziffer, H., Kabuto, K., Gibson, D. T., Kobal, V. M., and Jerina, D. M., *Tetrahedron*, **33**, 2491 (1977).
12. Catelani, D., Sorbini, C., and Treccani, V., *Experientia*, **27**, 1173 (1971).
13. Gibson, D. T., Roberts, R. L., Wells, M. C., and Kobal, V. M., *Biochem. Biophys. Res. Commun.*, **50**, 211 (1973).
14. Hudlicky, T., Luna, H., Barbieri, G., and Kwart, L. D., *J. Am. Chem. Soc.*, **110**, 4735 (1988).
15. Gibson, D. T., Gschwendt, B., Yeh, W. K., and Kobal, V. M., *Biochemistry*, **12**, 1520 (1973).
16. Whited, G. M., McCombie, W. R., Kwart, L. D., and Gibson, D. T., *J. Bacteriol.*, **166**, 1028 (1986).
17. Taylor, S. C., to ICI, *Eur. Pat.*, 76 606, 1983.
18. Boyd, D. R., Dorrity, M. R. J., Hand, M. V., Malone, J. F., Sharma, N. D., Dalton, H., Gray, D. J., and Sheldrake, G. N., *J. Am. Chem. Soc.*, **113**, 666 (1991).
19. Hudlicky, T., Price, J. D., Rulin, F., and Tsunoda, T., *J. Am. Chem. Soc.*, **112**, 9439 (1990).
20. Boyd, D. R., Hand, M. V., Sharma, N. D., Chima, J., Dalton, H., and Sheldrake, G. N., *J. Chem. Soc., Chem. Commun.*, 1630 (1991).

21. Frank, J. J., and Ribbons, D. W., *J. Bacteriol.*, **129**, 1356 (1977).
22. Reiner, A. M., and Hegeman, G. D., *Biochemistry*, **10**, 2530 (1971).
23. Ryback, G., to Shell Internationale, *Br. Pat. Appl.*, 2 203 150, (1988).
24. Taylor, S. J. C., Ribbons, D. W., Slawin, A. M. Z., Widdowson, D. A., and Williams, D. J., *Tetrahedron Lett.*, **28**, 6391 (1987).
25. Reinecke, W., Otting, W., and Knackmuss, H. J., *Tetrahedron*, **34**, 1707 (1978).
26. Schofield, J. A., to Shell Internationale, *Eur. Pat.*, 252 568, (1988).
27. Gibson, D. T., Mahadevan, V., and Davey, J. F., *J. Bacteriol.*, **119**, 930 (1974).
28. Omori, T., Matsubara, M., Matsuda, S., and Kodama, T., *Appl. Microbiol. Technol.*, **35**, 431 (1991).
29. Eaton, R. W., and Ribbons, D. W., *J. Bacteriol.*, **151**, 48 (1982).
30. Jeffrey, A. M., Yeh, H. J. C., Jerina, D. M., Patel, T. R., Davey, J. F., and Gibson, D. T., *Biochemistry*, **14**, 575 (1975).
31. Cerniglia, C. E., Gibson, D. T., and van Baalen, C., *J. Gen. Microbiol.*, **116**, 495 (1980).
32. Jerina, D. M., Selander, H., Yagi, H., Wells, M. C., Davey, J. F., Mahadevan, V., and Gibson, D. T., *J. Am. Chem. Soc.*, **98**, 5988 (1976).
33. Gibson, D. T., Mahadevan, V., Jerina, D. M., Yagi, H., and Yeh, H. J. C., *Science*, **189**, 295 (1975).
34. Hudlicky, T., Luna, H., Price, J. D., and Rulin, F., *Tetrahedron Lett.*, **30**, 4053 (1989).
35. Hudlicky, T., Luna, H., Price, J. D., and Rulin, F., *J. Org. Chem.*, **55**, 4683 (1990).
36. Nakajima, M., Tomida, I., and Takei, S., *Chem. Ber.*, **92**, 163 (1959).
37. Yang, N. C., Chen, M.-J., Chen, P., and Mak, K. T., *J. Am. Chem. Soc.*, **104**, 853 (1982).
38. Stephenson, G. R., Howard, P. W., and Taylor, S. C., *J. Chem. Soc., Chem. Commun.*, 127 (1991).
39. Ballard, D. G. H., Courtis, A., Shirley, I. M., Taylor, S. C., *J. Chem. Soc., Chem. Commun.*, 954 (1983); in *Proceedings of the First SPSJ International Polymer Conference*, 1984, p. 173; *Macromolecules*, **21**, 294 (1988).
40. Taylor, S. C., in *Enzymes in Organic Synthesis* (*Ciba Foundation Symposium III*), Pitman, London, 1985, p. 71; Taylor, S. C., *World Biotech. Rep.*, **1**, 5 (1986); Taylor, S. C., *Spec. Chem.*, **236**, 244 (1988).
41. Brown, S. M., and Taylor, S. C., *Perform. Chem.*, Nov., 18 (1986).
42. Japan Synthetic Rubber, *Eur Pat. Appl.*, 360 407, 1990.
43. Idemitsu Kosan *Jpn. Pat.*, 02 127 410, 1988.
44. Sanraku, *Jpn. Pat.*, 02 163 090, 1990.
45. Enzymatix *PCT Int. Appl.*, WO 90 12 798, 1990.
46. Minnesota Mining and Manufacturing, *Eur. Pat. Appl.*, 400 779, 1990; Williams, M. G., Olsen, P. E., Tautrydas, K. J., Bitner, R. M., Mader, R. A., and Wackett, L. P., *Appl. Microbiol. Technol.*, **34**, 316 (1990).
47. *General Electric*, *US Pat.*, 4 981 793, 1991.
48. Woodley, J. M., Brazier, A. J., and Lilly, M. D., *Biotechnol. Bioeng.*, **37**, 133 (1991).
49. Brazier, A. J., Lilly, M. D., and Herbert, A. B., *Enzyme Microb. Technol.*, *12*, 90 (1990).
50. Van den Tweel, W. J. J., Marsman, E. H., Vorage, M. J. A. W., Tramper, J., and De Bont, J. A. M., in *International Conference on Bioreactors and Biotransformations Gleneagles, Scotland, 9–12th Nov. 1987*, (ed. G. W. Moody and P. B. Baker). Elsevier Applied Science, Barking, 1987, p. 231.
51. Van den Tweel, W. J. J., De Bont, J. A. M., Vorage, M. J. A. W., Marsman, E. H., Tramper, J., and Koppejan, J., *Enzyme Microb. Technol.*, **10**, 134 (1988).
52. Kawakami, Y., Yarmoff, J. J., Yago, T., Maruo, H., and Nishimura, H., *Kenkyu Hokoku Asahi Garasu Kogyo Gi jutsu Shoreikai*, **51**, 199 (1987).
53. Herbert, A. B., Sheldrake, G. N., Somers, P. J., and Meredith, J. A., to ICI PLC, *Eur. Pat. Appl.*, 379 300, 1990.

54. Taylor, S. C., *Eur. Pat. Appl.*, to ICI PLC, 163 392, 1984; 243 065, 1984.
55. Arnold, F. E., in *Current and Future Chemistry of Aerospace Matrix Resins*, Proc. Am. Soc. Composites, 1987.
56. Hergerrother, P. M., *Polym. Prepr. Am. Chem. Soc. Div. Polym. Chem.*, **25**, 97 (1984).
57. For a summary of the Ley group strategy, see Ley, S. V., *Pure Appl. Chem.*, **62**, 2031 (1990).
58. Namata, A., Hokimoto, K., Shimada, A., Yamaguchi, H., and Takaishi, K., *Chem. Pharm. Bull.*, **27**, 602 (1979).
59. Reece, J. C., Chan, B. G., and Waiss, A. C., Jr, *J. Chem. Ecol.*, **8**, 1429 (1982).
60. Narayanan, C. R., Joshi, D. D., Miyumdar, A. M., and Dekhne, V. V., *Curr. Sci.*, **56**, 139 (1987).
61. Ley, S. V., Sternfield, F., and Taylor, S. C., *Tetrahedron Lett.*, **28**, 225 (1987).
62. Ley, S. V., and Sternfield, F., *Tetrahedron*, **45**, 3463 (1989).
63. Carless, H. A. J., Billinge, J. R., and Oak, O. Z., *Tetrahedron Lett.*, **30**, 3113 (1989).
64. For background, see Berridge, M. J., *Annu. Rev. Biochem.*, **56**, 159 (1987); Berridge, M. J., and Irvine, R. F., *Nature (London)* **341**, 197 (1989); **312**, 3115 (1984); Michell, R. H., *Biochim. Biophys. Acta*, **415**, 81 (1975).
65. Billington, D. C., *Chem. Soc. Rev.*, **18**, 83 (1989).
66. Ley, S. V., Parra, M., Redgrave, A. J., Sternfield, F., and Vidal, A., *Tetrahedron Lett.*, **30**, 3557 (1989).
67. Ley, S. V., and Sternfield, F., *Tetrahedron Lett.*, **29**, 5305 (1988).
68. Ley, S. V., Parra, M., Redgrave, A. J., and Sternfield, F., *Tetrahedron*, **46**, 4995 (1990).
69. Posternak, T., *The Cyclitols*, Hermann, Paris, 1962.
70. Carless, H. A. J., and Oak, O. Z., *Tetrahedron Lett.*, **30**, 1719 (1989).
71. Carless, H. A. J., and Oak, O. Z., *J. Chem. Soc., Chem. Commun.*, 61 (1991).
72. Ley, S. V., and Redgrave, A. J., *Synlett.*, 393 (1990).
73. Hudlicky, T., Price, J. D., Luna, H., and Anderson, C. M., *Synlett.*, 309 (1990).
74. Johnson, C. R., Ple, P. A., and Adams, J. P.; *J. Chem. Soc., Chem. Commun.*, 1006 (1991).
75. Carless, H. A. J., and Oak, O. Z., *Tetrahedron Lett.*, **32**, 1671 (1991).
76. Braun, H., Burger, W., Kresze, G., Schmidtchen, F. P., Vaerman, J. L., and Viehe, H. G., *Tetrahedron: Asymmetry*, **1**, 403 (1990).
77. Hudlicky, T., and Olio, H. F., *Tetrahedron Lett.*, **32**, 6077 (1991).
78. Hudlicky, T., and Price, J. D., *Synlett.*, 159 (1991).
79. Lehmann, J., and Moritz, A., *Liebigs Ann. Chem.*, 937 (1991).
80. Hudlicky, T., Seoane, G., and Pettus, T., *J. Org. Chem.*, **54**, 4239 (1989).
81. Cotterill, I. C., Roberts, S. M., and Williams, J. O., *J. Chem. Soc., Chem. Commun.*, 1628 (1988).
82. Pittol, C. A., Pryce, R. J., Roberts, S. M., Ryback, G., Sik, V., and Williams, J. O., *J. Chem. Soc., Perkin Trans. 1*, 1160 (1989).
83. Gilliard, J. R., and Burnell, D. J., *J. Chem. Soc., Chem. Commun.*, 1439 (1989).
84. Werbitsky, O., Klier, K., and Felber, H., *Liebigs Ann. Chem.*, 267 (1990).
85. Yang, N. C., Noh, T., Gan, H., Halfon, S., and Hrnjez, B. J., *J. Am. Chem. Soc.*, **110**, 5919 (1988).
86. Amon, C. M., Banwell, M. G., and Gravatt, G. L., *J. Org. Chem.*, **52**, 4851 (1987).
87. Howard, P. W., Stephenson, G. R., and Taylor, S. C., *J. Chem. Soc., Chem. Commun.*, 1603 (1988).

7 Biological Routes to Optically Active Epoxides

K. FURUHASHI
Nippon Mining Co., Ltd, Saitama, Japan

7.1 INTRODUCTION

Optically active epoxides are useful chiral synthons in organic synthesis. They have been used as intermediates in the synthesis of natural products and other biologically active substances[1] because they are easily derivatized in a regio-selective and stereoselective manner. Two methods which have been used to prepare optically active epoxides are synthesis from optically active precursors and asymmetric epoxidation. Most of the optically active epoxides employed in organic synthesis have been derived from optically active precursors, which were resolved chemically or enzymatically.[2] This approach has been applied to the synthesis of optically active epichlorohydrin and related compounds, and has been extensively investigated in recent years.[3]

Much effort has also been devoted to the development of asymmetric epoxidation reactions for prochiral olefinic compounds. Epoxidation reactions with optically active peroxide reagents [4] and with hydrogen peroxide in the presence

Chirality in Industry. Edited by A. N. Collins, G. N. Sheldrake and J. Crosby

of an optically active phase-transfer catalyst[5] and catalytic asymmetric epoxidation reactions employing optically active metal complexes such as iron porphyrins have been investigated[6]. Although the optical purity of the epoxides produced by these methods was low in most cases, a titanium tartrate catalytic system developed by Katsuki and Sharpless[7] produced excellent results in the epoxidation of allyl alcohol and related compounds. The Katsuki and Sharpless reaction has been the only practical method available for asymmetric epoxidation of olefinic compounds hitherto, although the substrates were found to be limited to allyl alcohol derivatives.

In addition to these synthetic approaches which utilize 'chemical' catalysts, the microbial epoxidation of olefinic compounds has been investigated during the last few decades, and was found in 1974[8] to be stereoselective. Since then, various microorganisms have been shown to produce optically active epoxides, especially from unfunctionalized olefins, which cannot be epoxidized in high optical purity by conventional synthetic methods. This chapter provides an introduction to the topic of epoxidation by microorganisms, and outlines the development of processes for various optically active epoxides which are performed on a commercial scale by Nippon Mining.

7.2 MICROBIAL EPOXIDATION OF OLEFINIC COMPOUNDS

Microbial epoxidation of olefinic compounds was first demonstrated by van der Linden in 1963.[9] Subsequently, May and Abbott[10] demonstrated that the ω-hydroxylation system of *Pseudomonas oleovorans* (Figure 1) catalyses the epoxidation of olefins in addition to the hydroxylation of alkanes and fatty acids thus showing that the epoxidation and hydroxylation are carried out by the same monooxygenase system. Since May and Schwartz[8] reported in 1974 that the enzymatic epoxidation of octa-1,7-diene to (R)-7,8-epoxy oct-1-ene by *P. oleovorans* proceeds with a high degree of stereoselectivity, many microorganisms have been isolated from soil and shown to produce epoxides from various olefinic compounds. All these microorganisms normally utilize aliphatic hydrocarbons as a source of energy, with only one exception (see below).

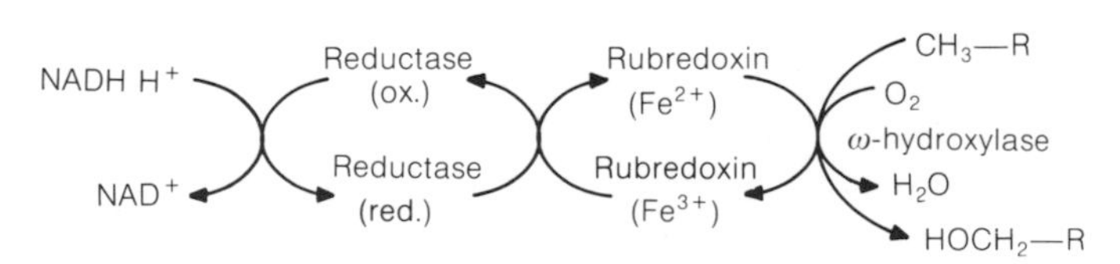

Figure 1. ω-Hydroxylation system of *Pseudomonas oleovorans.*

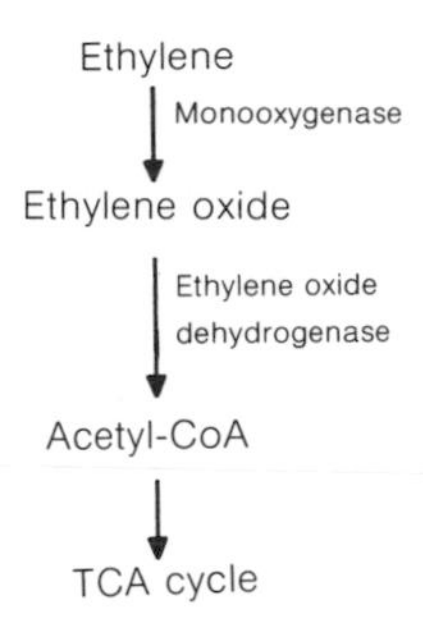

Figure 2. Proposed pathway for the utilization of ethylene in *Mycobacterium* E20

Epoxide-producing microorganisms can be divided into two groups according to the enzyme systems that take part in the epoxidation. In one group, which includes methylotrophs, as well as gaseous alkane-, long-chain alkane- and olefin-assimilating microorganisms, the epoxidation is carried out by hydroxylating enzyme systems. Methylotrophic bacteria, for example, are known to oxidize various hydrocarbons, including olefinic compounds, fortuitously.[11] The methane–monooxygenase systems of these bacteria epoxidize carbon–carbon double bonds in the same way as the ω-hydroxylation enzyme system of *P. oleovorans.* A cell-free extract of *Brevibacterium* sp. CRL57 grown on propane produces epoxides from gaseous olefins and styrene,[12] and results from inhibition studies indicate that a propane–monooxygenase system catalyses both epoxidation and hydroxylation reactions. *Nitrosomonas europaea*, which is the one exception mentioned above, produces epoxides with an ammonia–mono-oxygenase system.[13] This organism may also be classified in this first group.

The other group includes microorganisms which grow on short-chain olefinic compounds such as ethylene, propylene and but-1-ene. In *Mycobacterium* E20 isolated from soil with ethylene as the carbon source,[14] ethylene is assumed to be degraded by the pathway shown in Figure 2, and the monooxygenase involved in the epoxidation of ethylene does not oxidize saturated hydrocarbons. In these microorganisms, epoxides produced from gaseous olefins are not the products of fortuitous oxidation, but rather intermediates in the degradative pathway of the substrate olefins.

7.3 MICROBIAL PRODUCTION OF OPTICALLY ACTIVE EPOXIDES

The substrate specificity of *P. oleovorans* in epoxidation, and the conditions for production of epoxides, have been extensively investigated. Alk-1-enes in the range C_6–C_{12},[15] α,ω-dienes from C_6 to C_{12},[16] allylbenzene[17] and allyl ethers[18]

Table 1. Optical purity of epoxides produced by *P. oleovorans*

Epoxide	%*ee*
(*R*)-1,2-Epoxyoctane	70
(*R*)-1,2-Epoxydecane	60
(*R*)-7,8-Epoxy oct-1-ene	80
(*R*)-Butyl glycidyl ether	85
(*R*)-Allyl glycidyl ether	81
(*R*)-Benzyl glycidyl ether	75
(*R*)-Phenyl glycidyl ether	92
(*R*)-*p*-Methoxyphenyl glycidyl ether	98
(*R*)-*p*-Fluorophenyl glycidyl ether	99
(*R*)-*p*-Methoxyethylphenyl glycidyl ether	100

were asymmetrically epoxidized, while propylene, but-1-ene, *cis*-dec-5-ene and cyclohexane were hydroxylated.[16] The optical purities of the epoxides produced by *P. oleovorans* are listed in Table 1. In the presence of cyclohexane, the production of epoxides from octa-1,7-diene was enhanced, and the concentration of mono- and diepoxides in the reaction mixture reached 7.5 $g l^{-1}$ in 3 days.[19] De Smet *et al.*[20] reported that 28 $g l^{-1}$ of 1,2-epoxyoctane was accumulated by repeatedly transferring the organic layer, which consisted of the starting oct-1-ene and the product 1,2-epoxyoctane, to fresh, growing cultures of *P. oleovorans*[2].

Ohta and Tetsukawa[21] reported the production of (*R*)-1,2-epoxyhexadecane with 100% enantiomeric excess (*ee*) by cells of *Corynebacterium equi* which were grown on octane in the presence of hexadec-1-ene. Until Weijers *et al.*[22] reported the optical purity of short-chain epoxyalkanes produced by various microorganisms, *P. oleovorans*, *C. equi* and *Nocardia corallina* B-276, which will be described later, had been the only microorganisms known to produce optically active epoxides. Weijers *et al.* used complexation gas chromatography to determine the configuration and optical purity of the epoxides. Most of the bacteria tested, other than methylotrophic bacteria, produced optically active epoxides from propylene, but-1-ene, allyl chloride and *trans*-but-2-ene, although five strains of methylotrophs belonging to different genera produced racemic epoxides from these olefins. They examined 16 strains belonging to *Mycobacterium*, *Nocardia* and *Xanthobacter*, which grow on ethylene, propane, propylene, butane, but-1-ene, buta-1,3-diene, isoprene, hex-1-ene and vinyl chloride. These bacteria produced (*R*)-propylene oxide, (*R*)-1,2-epoxybutane, (*S*)-epichlorohydrin and (*R,R*)-*trans*-2,3-epoxybutane in high optical purity—in some cases, the optical purity of the 1,2-epoxides reached 98% *ee*. The optical purity of the epoxides was inherent in the microorganisms used, and did not change with the carbon sources used for growth. These authors also found that the epoxides produced by *N. europaea* had the opposite configuration to those produced by gaseous hydrocarbon-assimilating microorganisms. This important result demonstrated that both enantiomers of epoxides are available from biological epoxidation processes.

Although many microorganisms have been shown to produce optically active epoxides, there are several obstacles that must be overcome before these microorganisms can be applied to the large-scale production of epoxides, as follows:

1. Most of the microorganisms must be cultivated with gaseous hydrocarbons as the carbon source to gain high activity,[22] but such gases are difficult to handle on a large scale. These energy sources, which regenerate the reduced form of the cofactor, are indispensable to continued epoxidation with a monooxygenase. They include, in addition to gaseous hydrocarbons,[23] volatile intermediates from the degradation pathway of gaseous hydrocarbons and related compounds such as methanol,[23] ethanol and ethyl acetate.[24]
2. The epoxide products, especially those of short chain length, are toxic to microorganisms.[25] This toxic effect must be reduced to allow epoxide production to continue.
3. The recovery of epoxides of short chain length that are present in the gas phase in very low concentrations, and the prevention of evaporative loss of epoxides with moderate chain length by aeration, constitute the third problem area.

In practice, the second of these problems is usually the most serious in epoxide production. In the following section, the case of *Nocardia corallina* B-276 is introduced to illustrate how these difficulties, especially that of product inhibition, may be overcome.

7.4 CHARACTERISTICS OF *NOCARDIA CORALLINA* B-276

Nocardia corallina B-276, which was isolated from soil with propylene as the carbon source, grows on various carbohydrates and hydrocarbons such as ethylene, but-1-ene, butadiene, propane, and long chain *n*-alkanes and alk-1-enes.[26] Propylene-grown cells of *N. corallina* B-276 produced epoxides in the presence of chloramphenicol (an inhibitor of protein synthesis) whereas glucose-grown cells harvested at their exponential growth phase did not (Table 2). This indicates that the epoxidation enzyme system can be induced by propylene, but is not activated when the cells are grown in the presence of glucose (catabolite repression). The fact that the saturated alkane tetradecane was not oxidized probably indicates that *N. corallina* belongs to the second class of epoxide-producing organisms discussed in Section 7.2, in which the epoxide products are intermediates of olefin catabolism. Further, a spontaneous mutant that lost its ability to assimilate propylene did not produce propylene oxide.

Table 2. Oxidation of tetradec-1-ene and tetradecane by resting cells of *N. corallina* B-276

				With chloramphenicol	
Carbon source for growth	Substrate (200 mg)	Remaining substrate (mg)	Produced epoxide (mg)	Remaining substrate (mg)	Produced epoxide (mg)
Glucose	Tetradec-1-ene	154	27	194	1
Propylene	Tetradec-1-ene	142	48	151	38
	Tetradecane	149		198	

The most useful feature of *N. corallina* is that cells which have been grown on glucose (a convenient substrate) can, in some circumstances, exhibit almost the same activity as cells grown on propylene. This occurs when the inhibitory effect of glucose weakens, as in the stationary phase of growth after 24 h of cultivation in a glucose medium. Under these conditions epoxidizing activity appears spontaneously and increases with time (Figure 3). Glucose and other carbohydrates can also be used as energy sources for regeneration of the reduced form of the cofactor in the epoxidation of gaseous olefins, as is shown later.

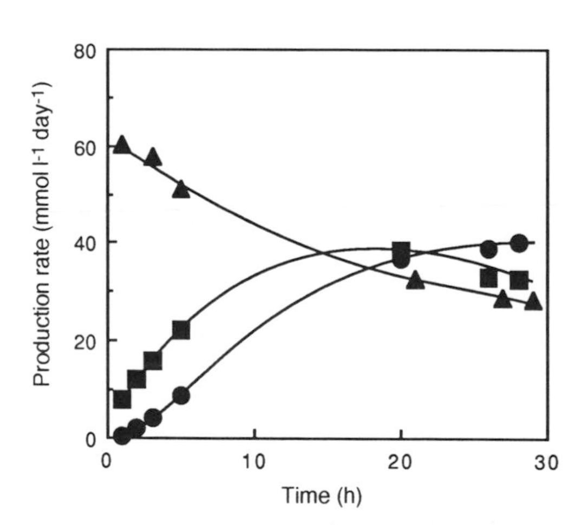

Figure 3. Effect of cultivation time on the epoxidation activity of *N. corallina* B-276. Reaction was carried out in 100 ml bubble tower reactors. Cell concentration, 3.8 mg ml^{-1}; initial glucose concentration, 28 g l^{-1}; reaction temperature, 30 °C; aeration rate, 6 vvm; propylene concentration, 10%. ●, 16 h cultivation; ■, 24 h cultivation; ▲, 48 h cultivation

7.5 SUBSTRATE SPECIFICITY OF *NOCARDIA CORALLINA* B-276

In most microorganisms, only short-chain olefins have been used as substrates for epoxidation. Figure 4 shows olefins which *N. corallina* B-276 can epoxidize, indicating the broad substrate specificity of the microorganism.[27] *N. corallina* B-276 produced epoxides from unfunctionalized aliphatic and aromatic olefins.

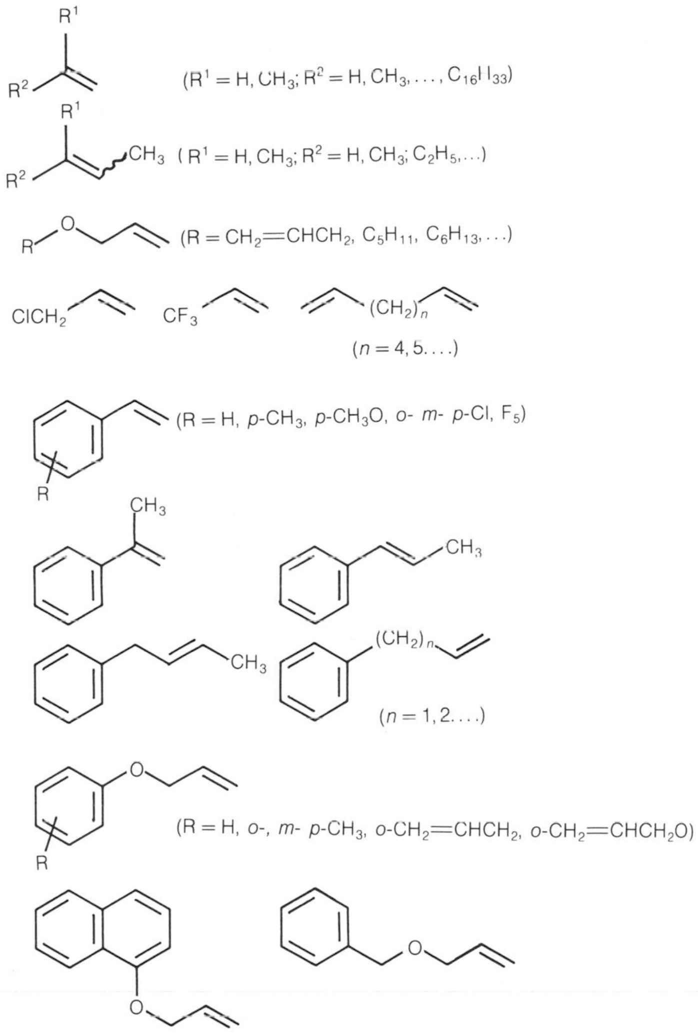

R^1, R^2 (R^1 = H, CH_3; R^2 = H, CH_3, . . . , $C_{16}H_{33}$)

R^1, R^2, CH_3 (R^1 = H, CH_3; R^2 = H, CH_3; C_2H_5,. . .)

R, O (R = CH_2=$CHCH_2$, C_5H_{11}, C_6H_{13}, . . .)

$ClCH_2$ CF_3 $(CH_2)_n$

(n = 4, 5. . . .)

(R = H, p-CH_3, p-CH_3O, o- m- p-Cl, F_5)

R

CH_3

CH_3

CH_3 $(CH_2)_n$

(n = 1, 2. . . .)

O

(R = H, o-, m- p-CH_3, o-CH_2=$CHCH_2$, o-CH_2=$CHCH_2O$)

R

O

O

Figure 4. Substrate specificity of *N. corallina* B-276 in the epoxidation of olefins

Table 3. Optically active epoxides produced by *N. corallina* B-276

Epoxides	$[\alpha]_D^{25}$ (neat)	Absolute configuration	Optical purity (% *ee*)
1,2-Epoxypropane	+11.6	R	83
1,2-Epoxyhexane	+12.1	R	66
1,2-Epoxyheptane	+15.1	R	94
1,2-Epoxyoctane	+14.4	R	91
1,2-Epoxynonane	+12.9	R	89
1,2-Epoxydecane	+11.6	R	86
1,2-Epoxyundecane	+10.9	R	86
1,2-Epoxydodecane	+10.2	R	87
1,2-Epoxytridecane	+10.1	R	92
1,2-Epoxytetradecane	+9.4	R	87
1,2-Epoxypentadecane	+8.2	R	81
1,2-Epoxyhexadecane	+7.7	R	82
1,2-Epoxyheptadecane	+7.9[a]	R	81
1,2-Epoxyoctadecane	+7.4[a]	R	84
cis-2,3-Epoxyoctane	+1.6		
trans-2,3-Epoxyoctane	+8.9		
2-Methyl-1,2-epoxypentane	−8.8	R	76
2-Methyl-1,2-epoxyhexane	−9.3	R	90
2-Methyl-1,2-epoxyheptane	−7.4	R	88
1,2-Epoxydec-9-ene	+12.7	R	
1,2,9,10-Diepoxydecane	+21.8	R,R	
Epichlorohydrin	+26.8	S	81
3,3,3-Trifluoro-1,2-epoxypropane	−9.2	S	75
Styrene oxide	+23.3	R	69
o-Chlorostyrene oxide	−22.3		86
m-Chlorostyrene oxide	+16.6		82
p-Chlorostyrene oxide	−4.3		72
Pentafluorostyrene oxide	+2.5	R	97
1,2-Epoxy-4-phenylbutane	+17.2		71
trans-2,3-Epoxy-1-phenylbutane	+16.2		
Pentyl glycidyl ether	+9.2	R	84
Hexyl glycidyl ether	+9.2	R	90
Heptyl glycidyl ether	+8.5	R	92
Octyl glycidyl ether	+7.7	R	92
Nonyl glycidyl ether	+6.9	R	86
Decyl glycidyl ether	+5.2	R	76
Undecyl glycidyl ether	+4.6	R	70
Dodecyl glycidyl ether	+4.8	R	88
Phenyl glycidyl ether	+18.0		67
o-Methylphenyl glycidyl ether	+18.7		73
m-Methylphenyl glycidyl ether	+0.3	S	5
p-Methylphenyl glycidyl ether	+16.2		79
o-Allylphenyl glycidyl ether	+13.1	S	53
o-Allyloxyphenyl glycidyl ether	+16.2	S	58
1-Naphthyl glycidyl ether	−78.2	R	71
Benzyl 2,3-epoxypropyl ether		S	15

[a] $c = 2.0$, hexane.

Alk-1-enes and alk-2-enes were good substrates for epoxidation, whereas only trace amounts of epoxides were produced from alk-3-enes, and alk-4-enes were not epoxidized. Mono- and diepoxides were produced from α,ω-dienes. Halogenated olefins such as allyl chloride and 3,3,3-trifluoropropene, allyl ethers, styrene and related olefins were also epoxidized.

Table 3 shows the specific rotation, configuration and optical purity of the isolated epoxides. The optical purity ranged from 66 to 94% *ee* in aliphatic epoxides and from 5 to 97% *ee* in aromatic examples. In aromatic epoxides, the optical purity and configuration changed depending on the position of substituents on the aromatic ring and on the ring type, but the optical purity of each product did not change when the reaction conditions were varied. Among these epoxides, (*R*)-1,2-epoxyalkanes (C_6–C_{18}), (*R*)-2-methyl-1,2-epoxyalkanes (C_6–C_{11}), (*R*)-alkyl glycidyl ethers (alkyl = C_5–C_{12}), (*R*)-1,2-epoxy dec-9-ene, (*R*,*R*)-1,2,9,10-diepoxydecane, pentafluorostyrene oxide and 3,3,3-trifluoro-1,2-epoxypropane are commercially available from Nippon Mining.

7.6 EFFECT OF SOLVENTS ON THE PRODUCTION OF EPOXIDES

The rate of epoxidation of alk-1-enes from C_6 to C_9 increased when hexadecane was added to cell suspensions of *N. corallina* B-276 grown on glucose, whereas these olefins alone were not good substrates for epoxidation (Figure 5).[28] In the epoxidation of oct-1-ene, an increase in the concentration of hexadecane at any oct-1-ene concentration enhanced the production of 1,2-epoxyoctane (Figure 6). In the epoxidation of styrene, however, the effects of styrene and hexadecane concentrations were different from those observed in the case of oct-1-ene (Figure 7). Although the production of styrene oxide was enhanced as the concentration of hexadecane was increased, it was reduced as the concentration of styrene increased. In addition, the rate of CO_2 evolution changed almost in parallel with the rate of styrene oxide production. Figure 8 shows that initial addition of the product (in this case 1,2-epoxyoctane) to the reaction mixture decreased the production of 1,2-epoxyoctane, indicating the presence of product inhibition. These results indicate that hexadecane reduced product inhibition in 1,2-epoxyoctane production, and both product and substrate inhibitions in styrene oxide production.

Other *n*-alkanes (C_{10}–C_{17}) produced almost the same effect as hexadecane on 1,2-epoxyoctane production, and long chain alk-1-enes, halogenated alkanes, alkylbenzenes and diesters of long-chain dibasic acids could also be used as solvents. When alk-1-enes were used as the solvent, the reaction was competitive and shorter alk-1-enes were preferentially epoxidized.

These two-phase, resting cell systems enabled Nippon Mining to produce many kinds of epoxides, including aliphatic epoxides of moderate chain length

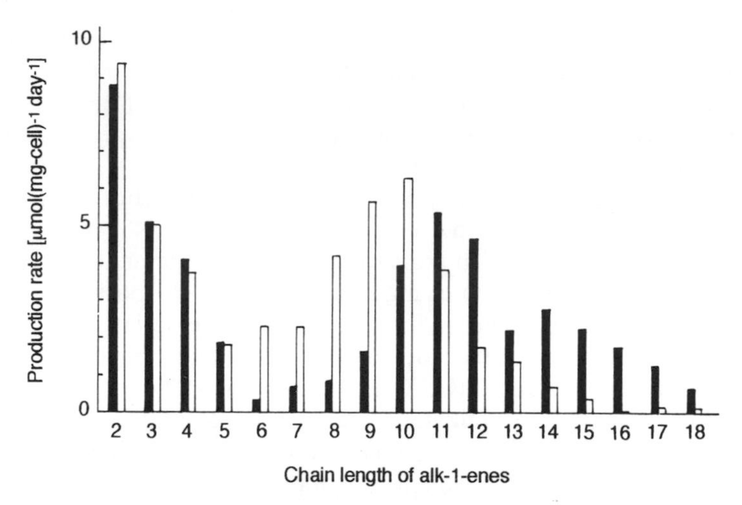

Figure 5. Rate of epoxidation of alk-1-enes by glucose-grown cells of *N. corallina* B-276. Reactions were carried out in stoppered 500 ml shake-flasks containing 20 ml of cell suspension (3.8 mg ml^{-1}) or in stoppered 40 ml Monod tubes containing 2 ml of cell suspension for 24 h at 30 °C under oscillation. The amounts of substrates and hexadecane employed in the reaction were as follows: ■, 1/0 [substrate (ml)/hexadecane (ml)]; □, 0.1/2. In the case of C_2–C_4 alk-1-enes, 4 ml of a gaseous substrate were injected into a stoppered Monod tube; 0.1 ml of hexadecane was employed when Monod tubes were used

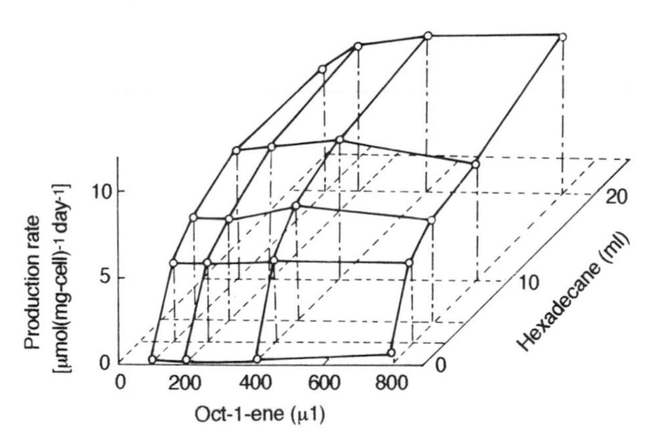

Figure 6. Effect of n-hexadecane on the production of 1,2-epoxyoctane by *N. corallina* B-276. Reactions were carried out under oscillation employing 20 ml of cell suspension (3.8 mg ml^{-1}) for 24 h at 30 °C

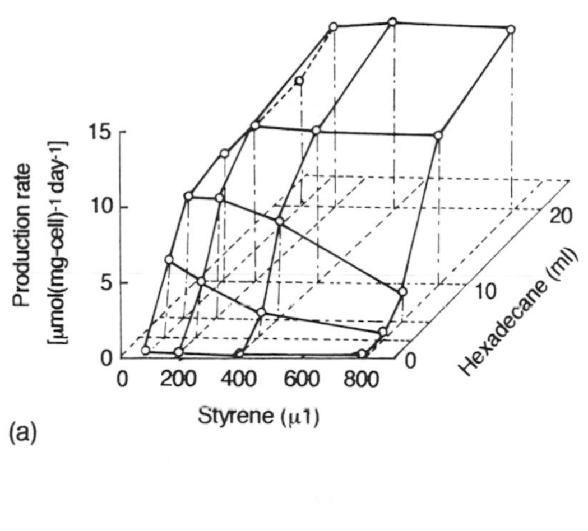

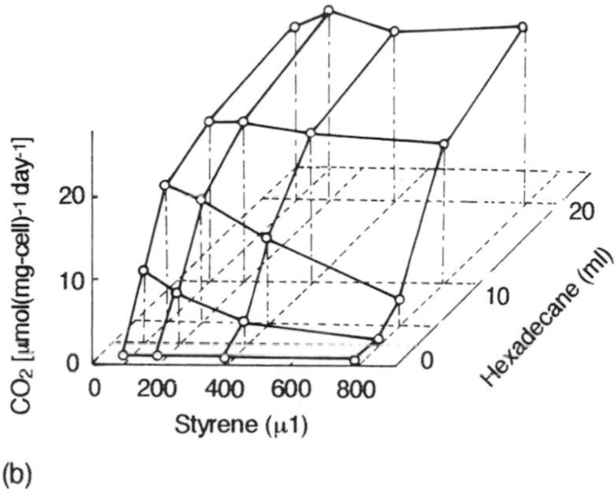

Figure 7. Effect of n-hexadecane on (a) the rate of production of styrene oxide and (b) the rate of CO_2 evolution by *N. corallina* B-276. Reactions were carried out under oscillation employing 20 ml of cell suspension (3.8 mg ml^{-1}) for 24 h at 30 °C

and most of the aromatic epoxides listed in Table 3. Table 4 shows the conversion and selectivity for the epoxidation reaction of alk-1-enes. In the production of 1,2-epoxides from C_{10}–C_{12} alk-1-enes, the selectivity for the desired transformation of alk-1-enes to 1,2-epoxides exceeded 90%, suggesting that hexadecane served as the energy source to regenerate the cofactor. However, the selectivity was lower in C_{13} and higher alk-1-enes where hexadecane was not added (in the epoxidation of these long-chain alk-1-enes, the addition of hexadecane decreased the rate of epoxidation).

As the presence of solvent lowers the vapour pressure of the substrates and products, and reduces their evaporative loss, the use of solvents made it practicable to epoxidize short-chain liquid olefins in aerated jar fermenters. Figure 9

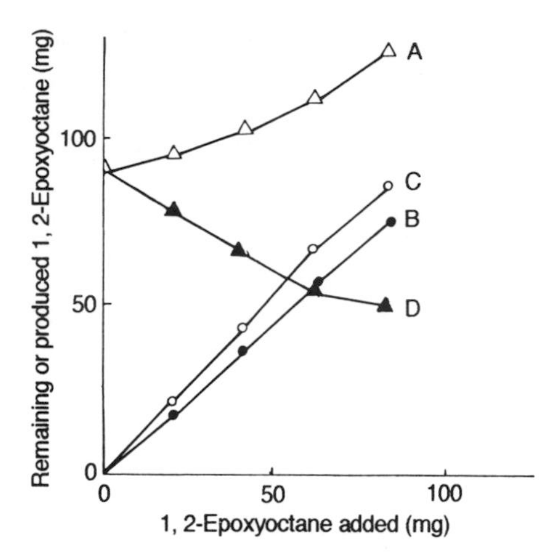

Figure 8. Product inhibition in the production of 1,2-epoxyoctane by *N. corallina* B-276. A, cell suspension + oct-1-ene + 1,2-epoxyoctane + n-hexadecane; B, cell suspension + 1,2-epoxyoctane + n-hexadecane; C, reaction medium + 1,2-epoxyoctane + n-hexadecane; D, A − B

Table 4. Epoxidation of alk-1-enes

Carbon No.[a]	Conversion (%)	Selectivity (%)
8	97.7	76.5
9	94.4	87.2
10	97.8	90.0
11	90.9	96.4
12	59.4	98.8
13	40.7	80.2
14	81.3	72.4
15	61.4	68.1
16	30.7	43.2
17	30.2	65.8
18	39.9	59.2

[a] C_8–C_{12} reaction conditions: cell suspension, 20 ml; substrate, 0.6 ml; hexadecane, 20 ml. C_{13}–C_{18} reaction conditions: cell suspension, 20 ml; substrate. 1 ml.

shows the time course of 1,2-epoxyoctane production by glucose-grown cells in a 4 m^3 fermenter after the reaction conditions had been optimized. The concentration of 1,2-epoxyoctane in the reaction mixture reached 30 $g l^{-1}$ after 120 h. The 1,2-epoxyoctane that accumulates in the solvent can be recovered and purified by distillation.

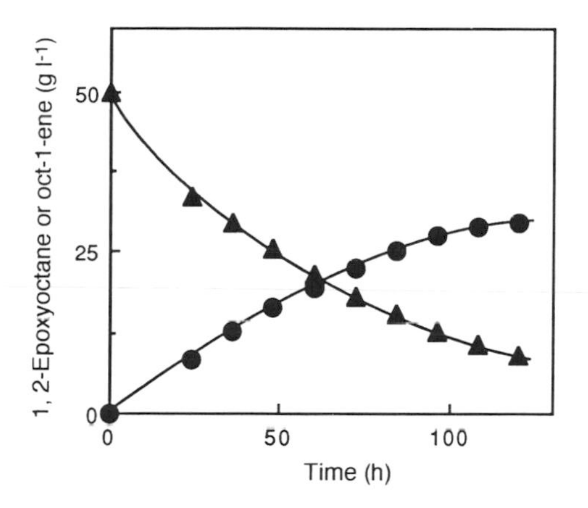

Figure 9. Time course of 1,2-epoxyoctane production by *N. corallina* B-276 in a $4\,m^3$ fermenter. ▲, Oct-1-ene; ●, 1,2-epoxyoctane

7.7. PRODUCTION OF LONG-CHAIN 1,2-EPOXYALKANES

N. corallina B-276 grows on C_{13}–C_{18} alk-1-enes and produces the corresponding epoxides.[29] 1,2-Epoxyalkanes with these chain lengths can be produced by a method which utilizes growing cells, because the toxicity of 1,2-epoxides in this

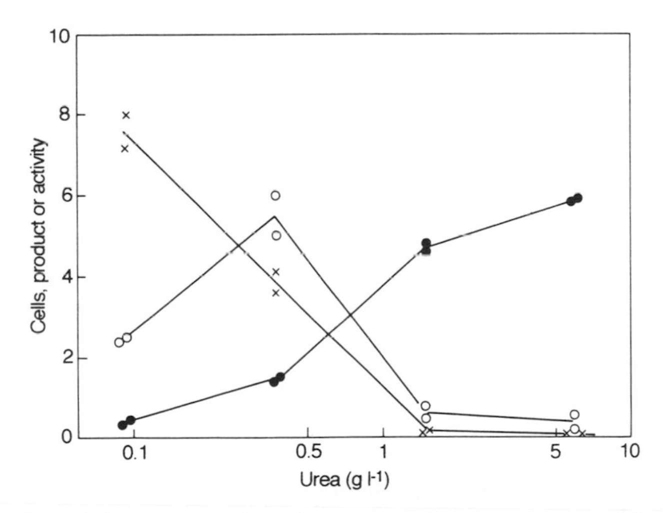

Figure 10. Effect of the concentration of urea as a nitrogen source on 1,2- epoxytetradecane production by *N. corallina* B-276. ○, 1,2-Epoxytetradecane (gl^{-1}); ●, cells (gl^{-1}); ×, activity (g epoxide per g cells per 96 h)

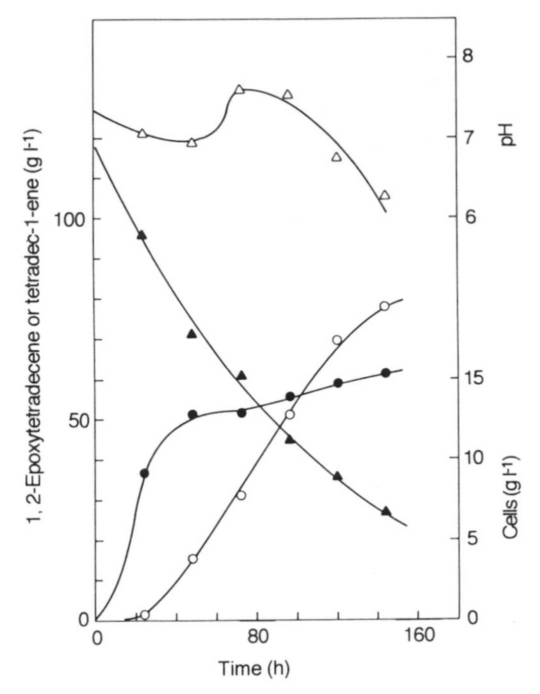

Figure 11. Time course of 1,2-epoxytetradecane production by *N. corallina* B-276 in a 5 l jar fermenter. ○, 1,2-Epoxytetradecane (gl^{-1}); ●, cells (gl^{-1}); ▲, remaining tetradec-1-ene; △, pH

range is less than that for 1,2-epoxides with shorter chain lengths. The production of 1,2-epoxides depends on both the specific activity of the cells and the cell concentration. The specific activity of the cells grown on alk-1-enes increased when their growth was suppressed by limiting components of the medium, such as the nitrogen source or potassium. Figure 10 shows that the specific activity to produce 1,2-epoxytetradecane increased as the concentration of urea decreased. Under these growth-limited conditions, it was found that the addition of yeast extract enhanced the growth without decreasing the specific activity of the cells. In a 5 l jar fermenter where the medium and cultivating conditions had been optimized, 80 $g l^{-1}$ of 1,2-epoxytetradecane accumulated in 6 days (Figure 11). The yield of 1,2-epoxytetradecane was 65% based on the tetradec-1-ene consumed.

7.8 EPOXIDATION OF GASEOUS OLEFINS

Glucose-grown cells of *N. corallina* B-276 have also been used in the epoxidation of gaseous olefins.[30] The severe toxicity of short-chain epoxides, regeneration

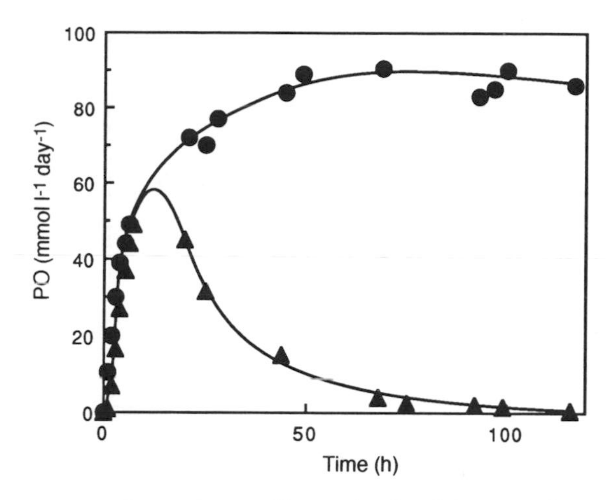

Figure 12. Time course of propylene oxide production by *N. corallina* B-276. Reaction was carried out in 100 ml bubble tower reactors. Cell concentration, 9.5 mg ml^{-1}; rate of addition of glucose, 10 gl^{-1}day^{-1}; reaction temperature, 30 °C; aeration rate, 3 vvm; propylene concentration, 20%. ▲, Without glucose; ●, with glucose

of the reduced form of the cofactor and recovery of the products were the three major difficulties encountered in this reaction.

Figure 12 shows the time course for propylene oxide production in 100 ml bubble tower reactors. In the presence of glucose, the rate of propylene oxide production was maintained at a high level for more than 100 h, whereas the system without glucose reached its maximum at about 10 h of reaction and then decreased rapidly. This indicates that glucose served as an energy source to regenerate the reduced form of the cofactor. Other carbohydrates such as sucrose and fructose, organic acids and long-chain aliphatic hydrocarbons are also possible energy sources for this reaction.

Product inhibition by propylene oxide was observed for *N. corallina* B-276[31] and for other microorganisms.[25,32] However, the concentration of propylene oxide in the reaction mixture varied with the aeration rate under steady-state conditions, owing to transfer of propylene oxide from the reaction mixture to the gas phase by aeration. By raising the aeration rate, the concentration of propylene oxide could be decreased with a corresponding increase in the rate of propylene oxide production (Figure 13). This reversible phenomenon was also observed in the production of 3,3,3-trifluoro-1,2-epoxypropane. In this way, a reduction in reversible product inhibition and irreversible enzyme inactivation could be achieved during the production of short chain epoxides.

Figure 14 shows the time course for propylene oxide production carried out at an aeration rate of 6 vvm (volume of gas per volume of medium per minute) with a mixture of air and propylene, after the reaction conditions had been

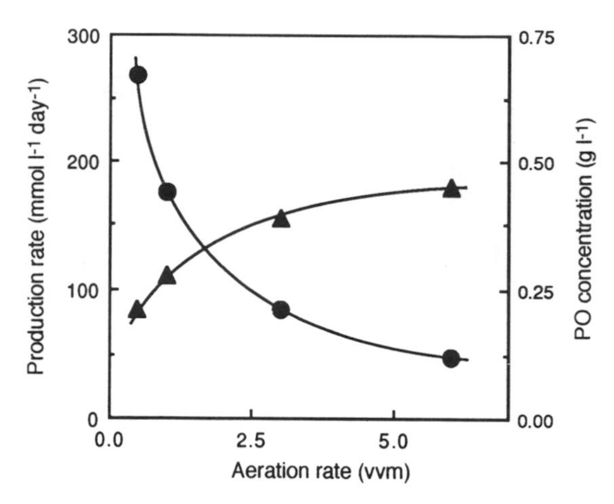

Figure 13. Effect of aeration rate on the production rate of propyene oxide. ▲, Production rate; ●, propylene oxide concentration

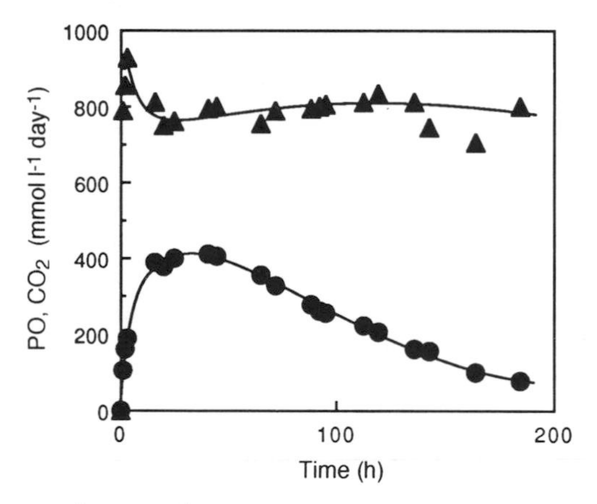

Figure 14. Time course for propylene oxide production by *N. corallina* B-276. in a 2.6 l jar fermenter. Cell concentration, 17.5 mg ml^{-1}; initial concentration of glucose, 20 g l^{-1}; rate of addition of glucose, 30 g l^{-1} day^{-1}; rate of addition of urea, 0.4 g l^{-1} day^{-1}; reaction temperature, 35 °C; aeration rate, 6 vvm; propylene concentration, 5%; ●, propylene oxide; ▲, carbon dioxide

optimized. The production rate reached 0.4 mol $l^{-1}day^{-1}$ after 20 h of reaction, and then decreased with a half-life of 4 days.

The recovery of the product which was present in very low concentrations in the gas phase was the last problem to be overcome in epoxide production from gaseous olefins. Although the products with low boiling points, such as propylene oxide and 3,3,3-trifluoro-1,2-epoxypropane, which appear in the exhaust gas can be trapped at dry-ice temperatures in laboratory-scale experi-

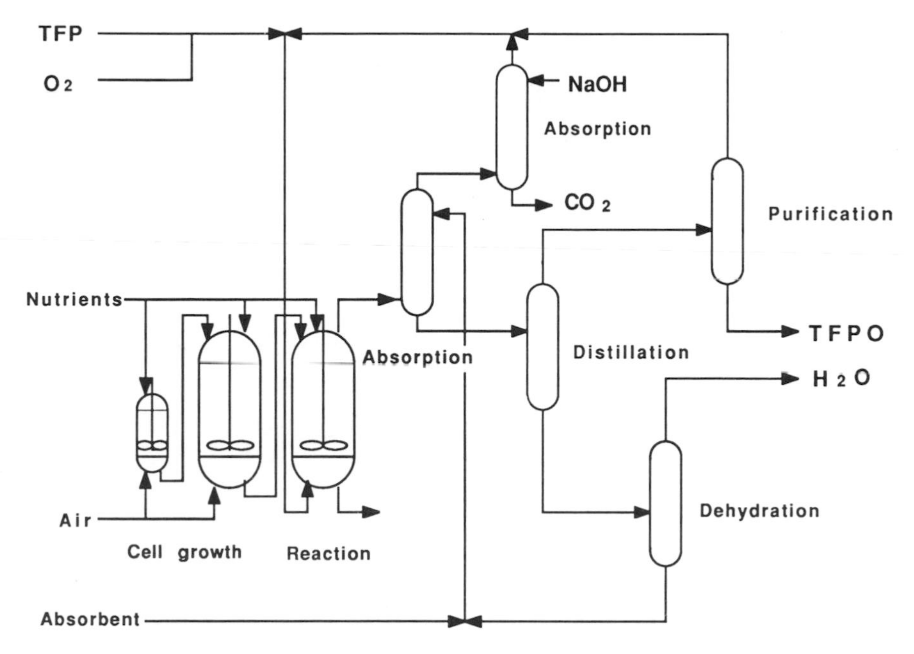

Figure 15. Process flow diagram for the production of 3,3,3-trifluoro-1,2-epoxypropane

ments, solvent absorption was also investigated, and, finally, adopted in large-scale production. Figure 15 shows the process flow diagram for the production of 3,3,3-trifluoro-1,2-epoxypropane. The recovery plant includes five towers for the absorption, distillation, purification, CO_2 absorption and dehydration steps, and the product in the gas phase is recovered quantitatively by the solvent absorption process.

7.9 USES OF OPTICALLY ACTIVE EPOXIDES

Among the many optically active epoxides produced by *N. corallina* B-276, the aliphatic epoxides have been used as starting materials in the synthesis of ferroelectric liquid crystals,[33] which are new optically active liquid crystals that have been extensively investigated in recent years. From these optically active epoxides, many kinds of optically active alcohols, carboxylic acids and halohydrins have been produced (Figure 16)[34] and used as key intermediates in the synthesis of liquid crystals. The liquid crystal displays that will employ these ferroelectric liquid crystals are expected to come onto the Japanese market in the near future.

Optically active 2-methyl-1,2-epoxyalkanes are useful precursors of optically active tertiary alcohols (Figure 16),[35] which can be used in the synthesis of

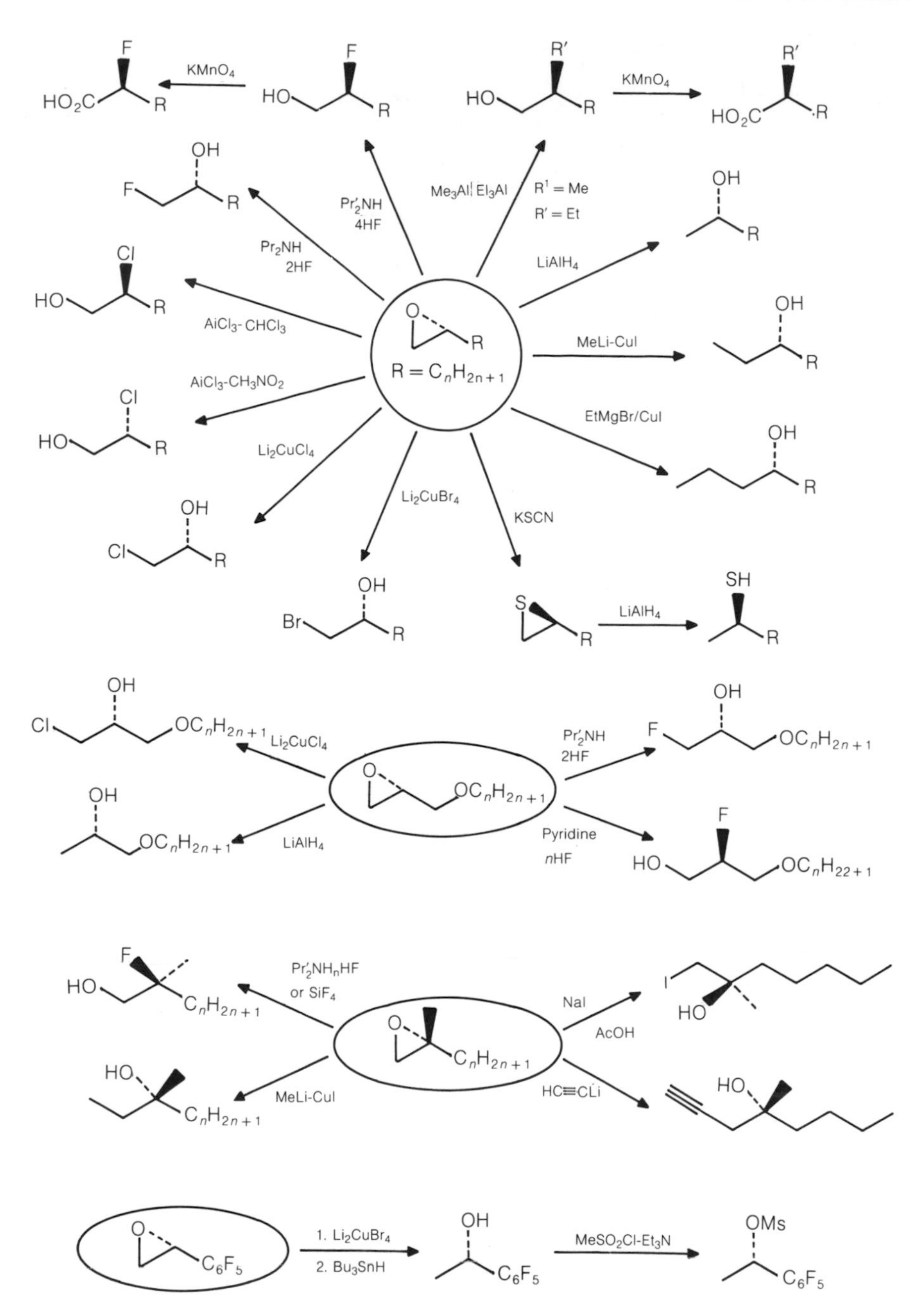

Figure 16. Chiral synthons derived from optically active epoxides

pharmaceuticals such as prostaglandins. Other optically active epoxides such as styrene oxide, phenyl glycidyl ethers and related compounds will also be important as intermediates in the synthesis of pharmaceuticals and agrochemicals. In the production of optically active epichlorohydrin and 3,3,3-trifluoro-1,2-epoxypropane, which have large potential markets, further improvements in the production processes will be necessary before they become practical.

7.10 CONCLUSION

Microbial epoxidation has proved to be a feasible, industrial method for the asymmetric epoxidation of olefinic compounds. Among the many microorganisms which catalyse asymmetric epoxidation, *Nocardia corallina* B-276 has been found to have the useful feature that cells grown on carbohydrates epoxidize various olefinic compounds efficiently. Substrates for the epoxidation by *N. corallina* B-276 are mainly unfunctionalized olefins including alk-1-enes, styrene and related compounds which are difficult to epoxidize in high optical purity by conventional synthetic methods. The productivity of epoxide synthesis by *N. corallina* B-276 has been improved in several ways, according to the chain length of the substrates, by altering the degree of product inhibition and by taking advantage of the physical properties of the products. A two-liquid-phase reaction system was applied to the epoxidation of olefins of moderate chain length and aromatic olefins by using a resting cell reaction with glucose-grown cells. The addition of organic solvents such as *n*-alkanes enhanced the production of epoxides by reducing the product inhibition. However, a growing cell reaction was applicable in the epoxidation of long-chain alk-1-enes, because of the weak product inhibition by epoxides with long chain length. In the epoxidation of gaseous olefins by glucose-grown cells, it was possible to raise the production rate and maintain it by transferring the product to the gas phase and keeping the concentration of the product in reaction mixture low by means of a high rate of aeration. Carbohydrates served as the energy source to regenerate the reduced form of the cofactor, and the product in the gas phase was recovered in good yield by a new recovery process employing absorption by solvents. Among the optically active epoxides produced, aliphatic epoxides are used as starting materials in the synthesis of ferroelectric liquid crystals, and other epoxides such as styrene oxide and related compounds will find their market as intermediates in the synthesis of pharmaceuticals and agrochemicals.

7.11 REFERENCES

1. Rossiter, B. E., in *Asymmetric Synthesis* (ed. J. D. Morrison), Academic Press, New York, 1986, Vol. 5, p. 193.

2. Crosby, J., *Tetrahedron*, **47**, 4789 (1991).
3. Kasai, N., Tsujimura, K., Unoura, K., and Suzuki, T., *Agric. Biol. Chem.*, **54**, 3185 (1990); Hamaguchi, S., Ohashi, T., and Watanabe, K., *Agric. Biol. Chem.*, **50**, 375 (1986); Ladner, W. E., and Whitesides, G. M., *J. Am. Chem. Soc.*, **106**, 7250 (1984).
4. Pirkle, W. H., Rinaldi, P. L., *J. Org. Chem.*, **42**, 2080 (1977); Kagan, H. B., Mimoun, H., Mark, C., and Schurig, V., *Angew Chem., Int. Ed. Engl.*, **18**, 485 (1979).
5. Helder, R., Hummelman, J. C., Laane, R. W. P. M., Wiering, J. S., and Wynberg, H., *Tetrahedron Lett.*, 1831 (1976).
6. Groves, J. T., and Myers, R. S., *J. Am. Chem. Soc.*, **105**, 5791 (1983); Naruta, Y., Tani, F., and Maruyama, K., *Chem. Lett.*, 1269 (1989).
7. Katsuki, T., and Sharpless, K. B., *J. Am. Chem. Soc.*, **102**, 5974 (1980).
8. May, S. W., and Schwartz, R. D., *J. Am. Chem. Soc.*, **96**, 4031 (1974).
9. van der Linden, A. C., *Biochim. Biophys. Acta*, **77**, 157 (1963).
10. May, S. W., and Abbott, B. J., *Biochem. Biophys. Res. Commun.*, **48**, 1230 (1972).
11. Higgins, I. J., Best, D. J., and Hammond, R. C., *Nature* (*London*), **286**, 561 (1980).
12. Hou, C. T., Patel, R., Laskin, A. I., Barnabe, N., and Barist, I., *Appl. Environ. Microbiol.*, **46**, 171 (1983).
13. Hyman, M. R., and Wood, P. M., *Arch. Microbiol.*, **137**, 155 (1984).
14. de Bont, J. A. M., and Harder, W., *FEMS Microbiol. Lett.*, **3**, 89 (1978).
15. Abbott, B. J., and Hou, C. T., *Appl. Microbiol.*, **26**, 86 (1973).
16. May, S. W., Schwartz, R. D., Abbott, B. J., and Zaborsky, O. R., *Biochim. Biophys. Acta*, **403**, 245 (1975).
17. de Smet, M.-J., Kingma, J., Wynberg, H., and Witholt, B., *Enzyme Microb. Technol.*, **5**, 352 (1983).
18. Fu, H., Shen, G.-J., and Wong, C.-H., *Recl. Trav. Chim. Pays-Bas*, **110**, 167 (1991).
19. Schwartz, R. D., and McCoy, C. J., *Appl. Environ. Microbiol.*, **34**, 47 (1977).
20. de Smet, M.-J., Wynberg, H., and Witholt, B., *Appl. Environ. Microbiol.*, **42**, 811 (1981).
21. Ohta, H., and Tetsukawa, H., *J. Chem. Soc., Chem. Commun.*, 849 (1978).
22. Weijers, C. A. G. M., van Ginkel, C. G., and de Bont, J. A.M., *Enzyme Microb. Technol.*, **10**, 214 (1988).
23. Suzuki, M., *Biosci. Ind.*, **49**, 492 (1991).
24. Habets-Crutzen, A. Q. H., and de Bont, J. A. M., *Appl. Microbiol. Biotechnol.*, **26**, 434 (1987).
25. Habets-Crutzen, A. Q. H., and de Bont, J. A. M., *Appl. Microbiol. Biotechnol.*, **22**, 428 (1985).
26. Furuhashi, K., Taoka, A., Uchida, S., Karube, I., and Suzuki, S., *Euro. J. Appl. Microbiol. Biotechnol.*, **12**, 39 (1981).
27. Furuhashi, K., *J. Synth. Org. Chem. Jpn.*, **45**, 162 (1987).
28. Furuhashi, K., Shintani, M., and Takagi, M., *Appl. Microbiol. Biotechnol.*, **23**, 218 (1986).
29. Furuhashi, K., and Takagi, M., *Appl. Microbiol. Biotechnol.*, **20**, 6 (1984).
30. Furuhashi, K., *Ferment. Ind.*, **39**, 1029 (1981).
31. Furuhashi, K., *Ferment. Ind.*, **45**, 468 (1987).
32. Hou, C. T., *Appl. Microbiol. Biotechnol.*, **19**, 1 (1984).
33. Yoshizawa, A., Nishiyama, I., Fukumasa, M., Hirai, T., and Yamane, M., *Jpn., J. Appl. Phys.*, **28**, L1269 (1989); Shiratori, N., Yoshizawa, A., Nishiyama, I., Fukumasa, M. Yokoyama, A., Hirai, T., and Yamane, M., *Mol. Cryst. Liq. Cryst.*, **199**, 129 (1991).
34. Takagi, M., Uemura, N., and Furuhashi, K., *Ann. N. Y. Acad. Sci.*, **613**, (Enzyme Eng. 10), 697 (1990).
35. Takahashi, O., Umezawa, J., Furuhashi, K., and Takagi, M., *Tetrahedron Lett.*, **30**, 1583 (1989).

8 The Production and Uses of Optically Pure Natural and Unnatural Amino Acids

J. KAMPHUIS, W. H. J. BOESTEN, B. KAPTEIN, H. F. M. HERMES, T. SONKE, Q. B. BROXTERMAN, W. J. J. VAN DEN TWEEL AND H. E. SCHOEMAKER
DSM Research, Geleen, The Netherlands

8.1 INTRODUCTION

Amino acids and their derivatives are becoming increasingly important as intermediates for pharmaceuticals and agrochemicals.[1] The growing requirement

Chirality in Industry. Edited by A. N. Collins, G. N. Sheldrake and J. Crosby

for natural and unnatural amino acids was anticipated in the 1970s by DSM, which went on to develop a generally applicable process based on the enzymic resolution of racemic amino acid amides with an aminopeptidase.

This chapter provides a description of the DSM process for α-H-amino acids which has been in commercial operation since 1987, and outlines particular applications of these materials as intermediates. Recent developments, including a process for α-alkylamino acids[2,3] and one step processes which use an L- or D-aminopeptidase in combination with an amino acid amide racemase to accomplish complete conversion are also described.

8.2 APPLICATIONS OF OPTICALLY PURE α-H-AMINO ACIDS

D-Phenylglycine is a raw material for the semi-synthetic antibiotic ampicillin, and D-*p*-hydroxyphenylglycine is used in the production of amoxicillin.

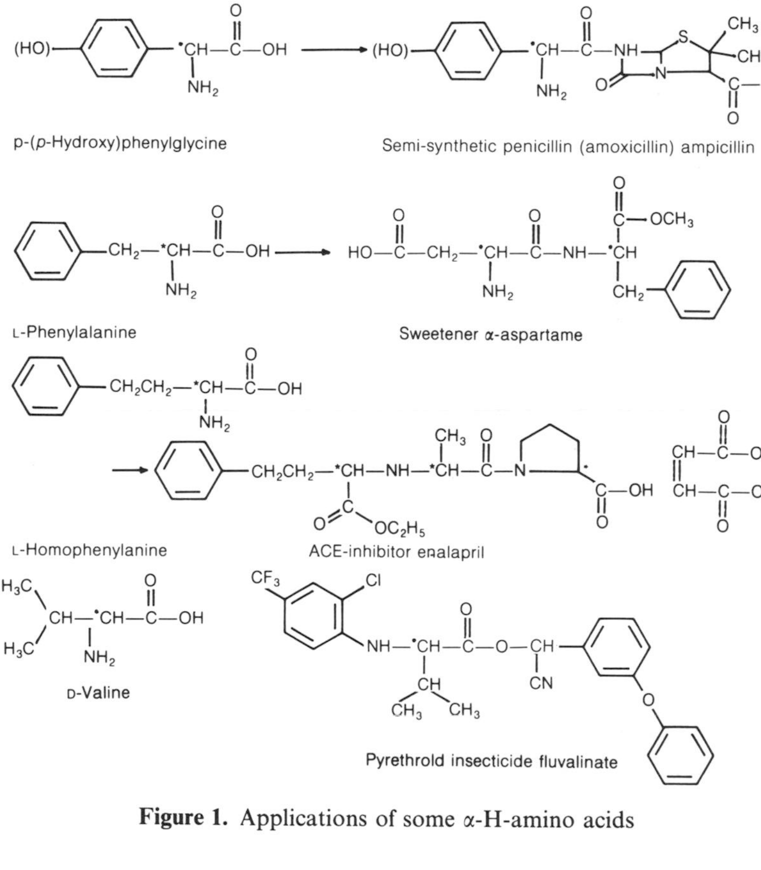

p-(*p*-Hydroxy)phenylglycine

Semi-synthetic penicillin (amoxicillin) ampicillin

L-Phenylalanine

Sweetener α-aspartame

L-Homophenylanine

ACE-inhibitor enalapril

D-Valine

Pyrethroid insecticide fluvalinate

Figure 1. Applications of some α-H-amino acids

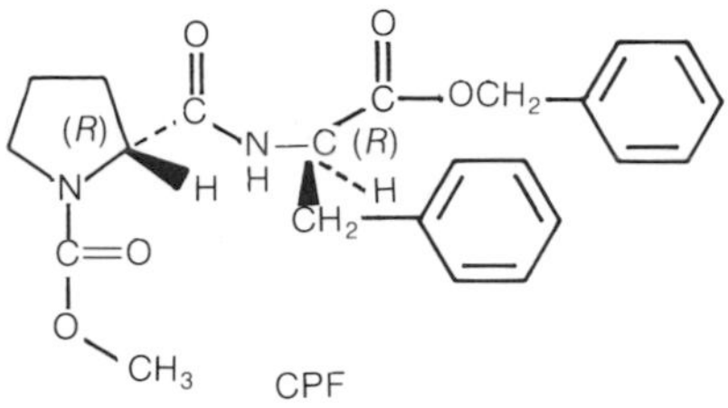

Figure 2. A dipeptide consisting of two, unnatural amino acids which blocks HIV infection

World-wide, the sales of these two antibiotics amount to $1500 and $1700 million per annum. DSM is a major producer of D-phenylglycine and D-*p*-hydroxy-phenylglycine, both of which are made by 'classical' resolution processes on a scale of more than 1000 tonnes per annum using camphorsulphonic acid and bromocamphorsulphonic acid, respectively, as resolving agents.[2,4,5]

L-Homophenylalanine is a versatile chiral building block for several ACE inhibitors (e.g. enalapril,[6] Figure 1), D-valine is used commercially by Zoëcon in the synthesis of fluvalinate,[7] and Monsanto uses L-phenylalanine in the production of the dipeptide sweetener aspartame (see Chapter 11).[7] At present, the major method for phenylalanine production is still fermentation,[8] although several other industrial methods are now operated (see Chapters 9, 19 and 20).

Biochemie (a subsidiary of Sandoz) have recently commercialized the production of cyclosporin A, an immunosuppressant which is used in the treatment of transplant rejection and auto-immune diseases.[9] L-Valine is an essential component of the nutrient medium which is used to make cyclosporin A by a fermentation process.

Many applications of D-amino acids are found in the peptide area. D-Serine, D-leucine, D-tryptophan, D-phenylalanine, D-proline and others find applications as intermediates for promising anti-cancer and anti-AIDS drugs (see Figure 2), and most are commercially available, or are about to be commercialized.[10]

8.3 DEVELOPMENT OF A GENERALLY APPLICABLE ROUTE TO CHIRAL AMINO ACIDS

In the mid-1970s, a research programme directed towards the development of a general route to chiral amino acids was begun, with D-phenylglycine as the first target molecule of interest. For the key resolution step, the use of enzymes appeared attractive. Some technical and economic restrictions were identified as follows:

i. a stable and cheap starting substrate was required (preferably an immediate precursor of the amino acid);
ii. the resolution step should be broadly applicable to a wide variety of substrates using the same biocatalyst;
iii. the separation and work-up procedure should be efficient and convenient.

From the literature, leucine aminopeptidase, known for its stereoselective hydrolytic action on peptides, was identified as a potential biocatalyst. In principle, the substrates, amino acid amides, are available from the corresponding aldehydes. Calf eye lenses were extracted in order to obtain crude leucine aminopeptidase,[11] and this preparation was found to retain a high activity for the hydrolysis of leucine amides, with L-leucine being obtained in high enantiomeric purity.

A broad screening programme was set up in order to find microorganisms with aminopepidase activity. In fact, the identification of the strain with highest activity and enantioselectivity towards amino acid amides owed much to serendipity, as the organism had been intended for a different screening programme. This strain was identified as belonging to the species *Pseudomonas putida*,[12] and in collaboration with Novo-Nordisk Industries, fermentation of the microorganism was optimized.

Meanwhile, a synthetic route to the racemic α-H-amino acid amides was being developed.[13] The first step, a Strecker reaction using hydrogen cyanide and ammonia, was already well known technology at DSM. Conversion of the aminonitrile products to the corresponding amides in water using a catalytic amount of acetone and base is a method applicable to most substituted aminonitriles, and so, in principle, a general route to both L- and D-amino acids was now at hand.

A solution to the problem of separating the D-amino acid amides from the L-α-H-amino acids was suggested by the observation that selective extraction of the amide from the aqueous hydrolysate was possible using an excess of benzaldehyde in an organic solvent. Without the benzaldehyde the extraction proved unsuccessful. It was postulated that a Schiff base of the amino acid amide was formed, and that this was being extracted.[14] This was shown to be the case by the addition of one equivalent of benzaldehyde (with respect to the amino acid amide) to the enzymic hydrolysate. Surprisingly, the Schiff base crystallized and could be isolated easily by filtration. Morever, these Schiff bases were prone to racemization in basic or even acidic media,[15,16] and therefore the complete conversion of racemic amino acid amides to L-α-H-amino acids could be achieved by racemizing and recycling the D-Schiff base of the amino acid amide. The process is outlined in Figure 3.

In testing the scope of this approach, it was found that the enzyme system is active with a broad range of substrates (see Tables 1 and 2) but that the presence

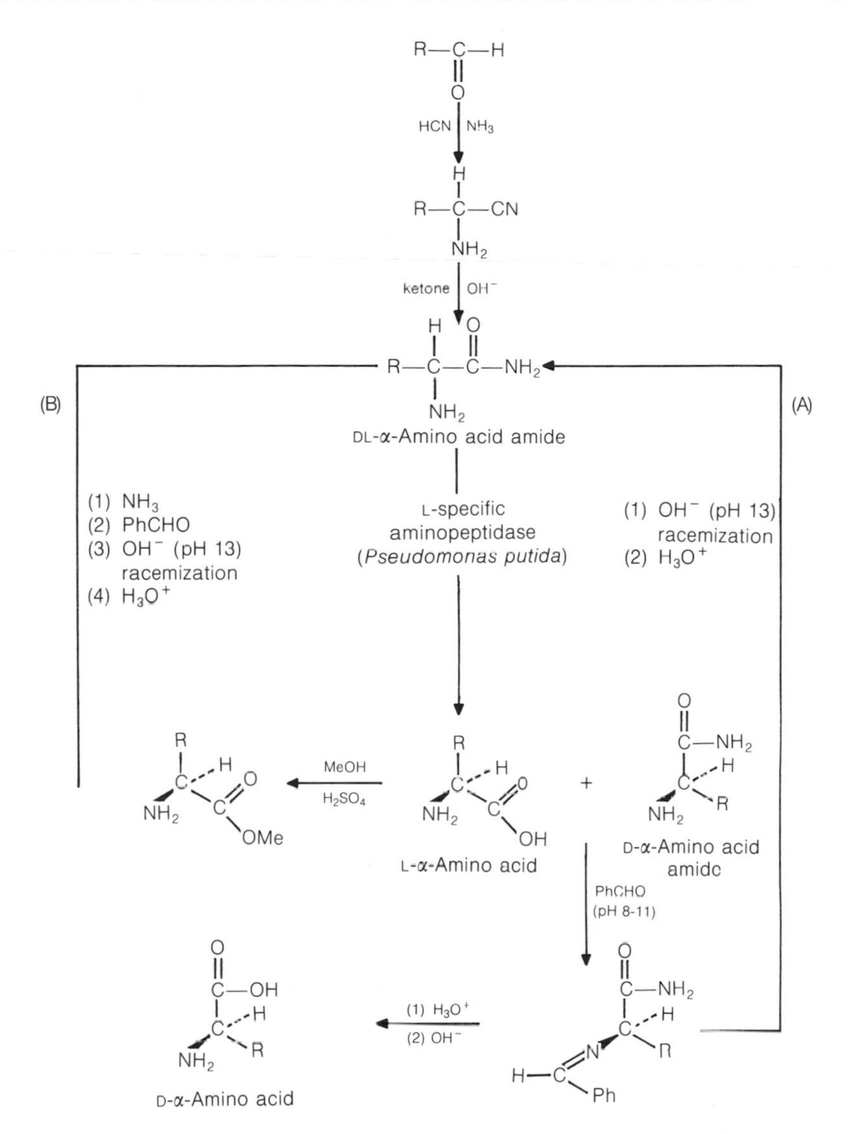

Figure 3. The DSM process for the production of L-and D-α-H-amino acids. From refs 3 and 13b

of an α-hydrogen on the α-carbon is a prerequisite for activity. The stereoselectivity of the process approaches 100% and no side products are observed. Permeabilized whole cells of *Pseudomonas putida* may be used and the enzyme is active over a broad pH range (8.0–9.5). Moreover, there is no substrate

Table 1. Substrate specificity of the biocatalyst from *P. putida* ATCC 12633 (selected examples). From ref. 43

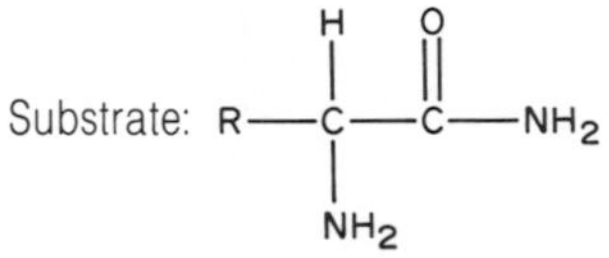

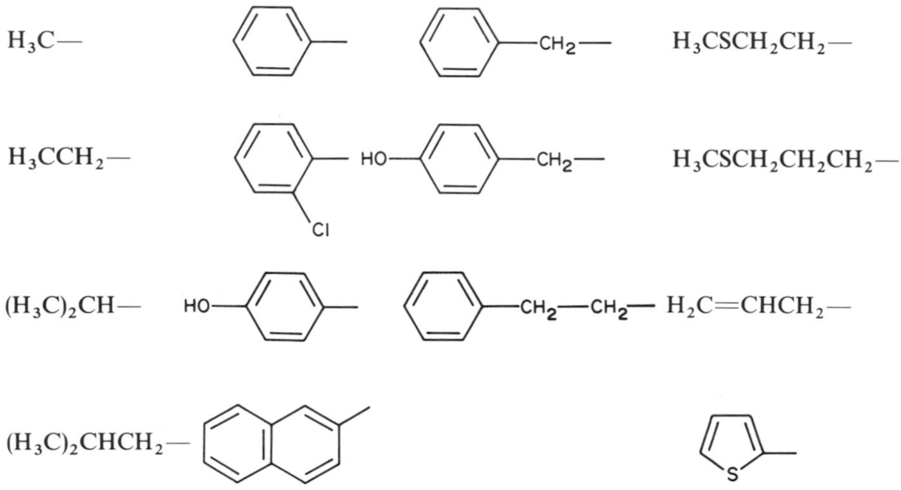

Substrate: R—C(H)(NH$_2$)—C(=O)—NH$_2$

H_3C—		—CH_2—	$H_3CSCH_2CH_2$—
H_3CCH_2—	Cl	HO— —CH_2—	$H_3CSCH_2CH_2CH_2$—
$(H_3C)_2CH$—	HO—	—CH_2—CH_2—	$H_2C{=}CHCH_2$—
$(H_3C)_2CHCH_2$—			S

inhibition and substrate concentrations of up to 20% may be used. The activity of the biocatalyst may be conveniently determined by a combination of automated sample preparation and high-performance liquid chromatography.[17]

The process was successfully commercialized in 1987 and several amino acids are now prepared on a multi-tonne scale.

8.4 ENZYME PURIFICATION

The wide substrate specificity led the authors to expect that more than one aminopeptidase was present in the whole-cell preparation. When the whole-cell extract was subjected to the purification scheme outlined in Figure 4, a homogeneous, pure enzyme fraction [as shown by sodium dodecyl sulphate polyacrylamide gel electrophoresis (SDS-PAGE) and gel filtration (Figure 5)], was isolated. The characteristics of this purified enzyme are given in Figure 6 and Table 3.[18–21]

Table 2. Further examples of amino acids resolved with *P. putida*

e.g. R^2 = cyclohexyl
R^1 = H

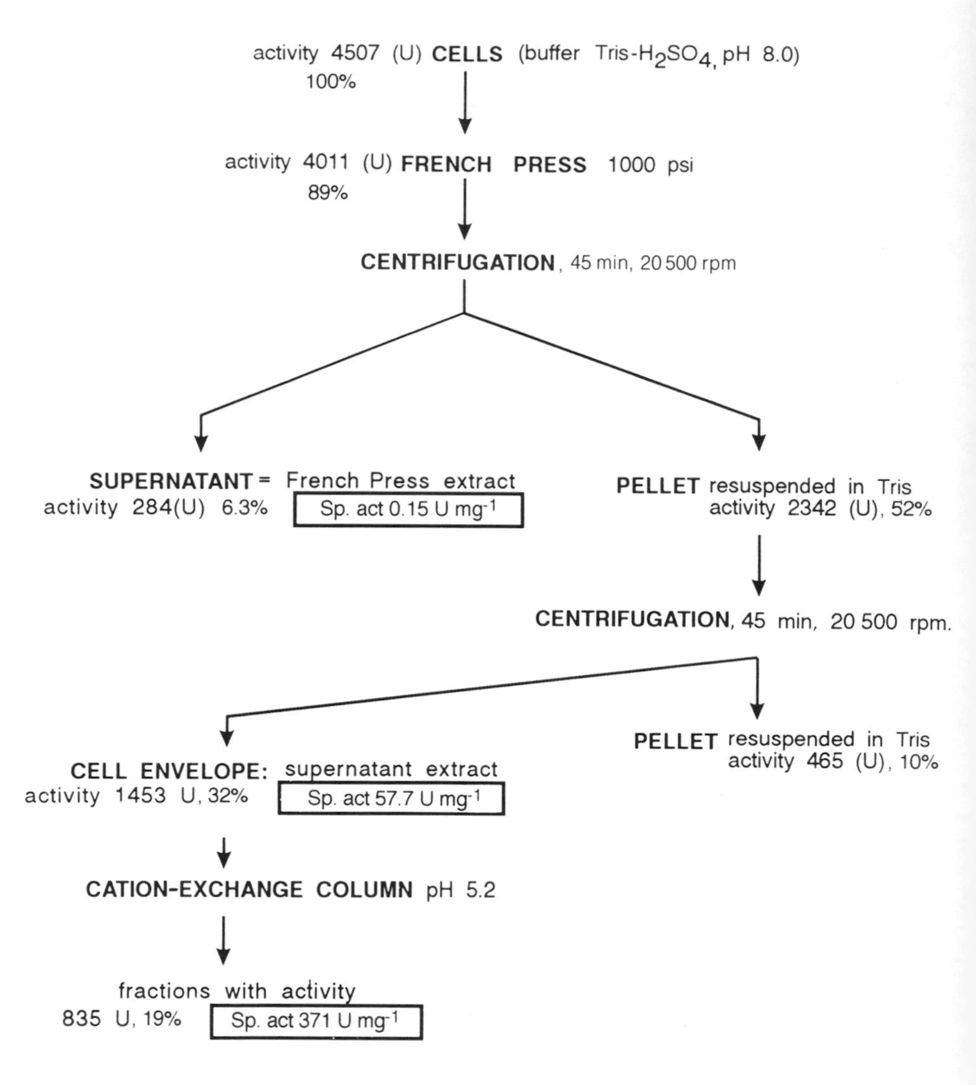

Figure 4. Purification scheme: L-aminopeptidase from *Pseudomonas putida* ATCC 12633

From the purification studies, it became clear that the whole cells contained a number of aminopeptidase activities. After disruption of the cells a cell-free extract was obtained which, when submitted to Mono S cation-exchange chromatography, was shown to contain four different fractions, all showing activity. However, only one of these exhibited a strictly Mn^{2+}-dependent enzyme

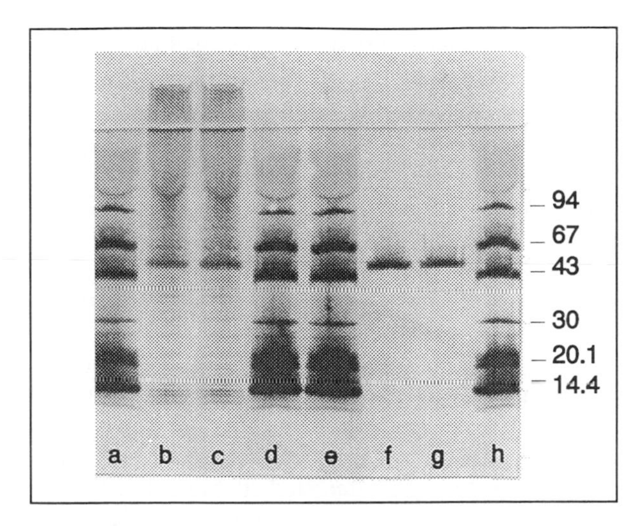

Figure 5. Purification of an L-aminopeptidase from *Pseudomonas putida* ATCC 12633. SDS-PAGE (8–25% gradient gel) of the enzyme fractions, Lanes: a, d, e, h = molecular weight standards; band c = 100 ng of protein from the cell envelope extract; f and g = 50 ng of protein after Mono S column. From ref. 21

Isoelectric point = 10.5
MW = *ca* 400 000 (gel permeation chromatography)
= 52 000–53 000 (SDS-PAGE)
Octamer
Optimum pH = 9.0–9.5
Optimum temperature = 40 °C
Stability: low in absence of substrate
60 min, 30 °C, 65% activity remains
Stability in the presence of substrate dramatically enhanced:
60 min, 30 °C, 100% activity remains

Figure 6. Characteristics of the purified L-aminopeptidase of *P. putida*

Table 3. Purified L-aminopeptidase from *P. putida*: substrate specificity

Substrate	Activity	Substrate	Activity
Butyramide	+/−	L-Phe-L-Leu	+ + +
Phenylglycinamide	+ +	L-Phe-L-Phe	+ + +
Phenylalaninamide	+	L-Leu-L-Phe	+ + +
Prolinamide	−	L-Leu-L-Leu	+ + +
Leucinamide	+ + +		
Alaninamide	+/−		
Methioninamide	+		
Valinamide	+/−		
Tryptophanamide	+		
Aliphatic amides	−		

activity.[21] From the fact that the different fractions showed different relative activities towards two representative substrates (L-valine and L-phenylglycine amides), the presence of different L-aminopeptidases in the cell-free extract of *Pseudomonas putida* ATCC 12633 with overlapping substrate specificity could be inferred. Notably, the Mn^{2+}-dependent enzyme was capable of stereoselectively hydrolysing a large number of different substrates (see Table 3).[22]

8.5 APPLICATIONS OF α-ALKYL-SUBSTITUTED AMINO ACIDS

A well known example of the use of α-alkyl-substituted amino acids in the pharmaceutical industry is L-α-methyl-3,4-dihydroxyphenylalanine, which is used in the treatment of high blood pressure[23] (see Figure 7). More recently, medicinal chemists have become interested in bio-active peptides containing α-substituted amino acids, as they tend to freeze specific conformations and dramatically slow enzymic processes. The rational design of conformationally constrained compounds which mimic or block the biological effects of physiologically important peptides represents a major goal of modern medicinal chemistry.

L-α-Methyl-*p*-tyrosine (R = H)
L-α-Methyl-DOPA (R = OH) 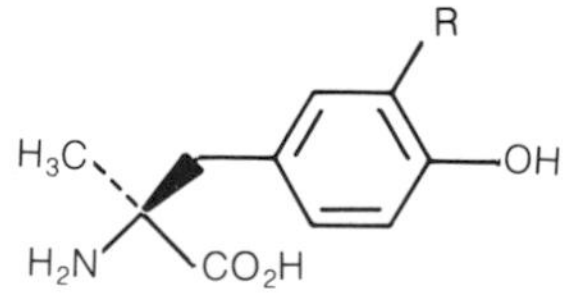
antihypertensives

D-α-Methyltryptophan
(for dipeptoid analogues of the neuropeptide cholecystekinin-B)

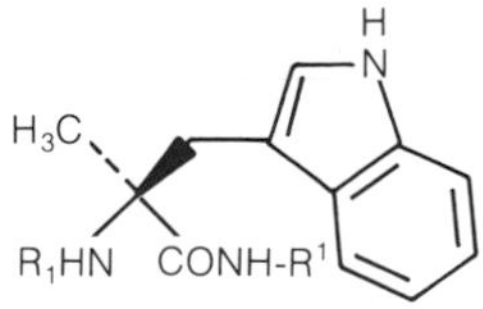

D-α-Methylvaline
Arsenal (herbicide)

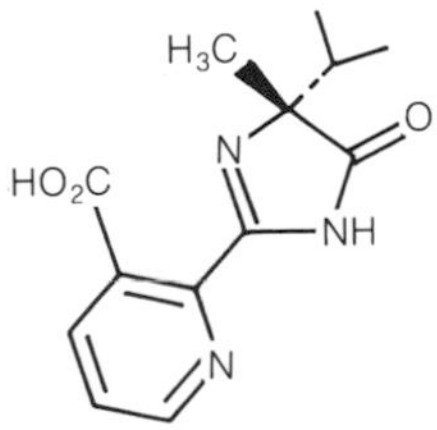

Figure 7. Applications of α,α-disubstituted amino acids

These conformational analogues should not only retain the pharmacological properties of the parent peptide, but should also exhibit improved bioavailability and pharmacokinetics.[24] In connection with this research, and following a request from Zoëcon to prepare α-methylvaline amide for the insecticide Arsenal. DSM developed a route to L- and D-α-alkyl-substituted amino acids.[25,26]

8.6 DEVELOPMENT OF A GENERALLY APPLICABLE PROCESS FOR THE SYNTHESIS OF α-ALKYL-SUBSTITUTED AMINO ACIDS

Although the biocatalyst from *Pseudomonas putida* cannot be used to resolve α-alkyl-substituted amino acid amides, it was possible to select a new biocatalyst from *Mycobacterium neoaurum* which was capable of stereoselectively hydrolysing a range of α-alkyl-substituted amino acid amides.[27,28]

The basis of this process is essentially the same as that for α-H-amino acids (see Figure 8), and has been reviewed elsewhere.[1,2] This process allows the production of both D- and L-α-alkyl-substituted amino acids, and a remarkably wide substrate specificity is observed in addition to a very high stereoselectivity (see Table 4). The reaction is carried out in water using either permeabilized whole cells of *Mycobacterium neoaurum* ATCC 25795 or crude enzyme preparations.

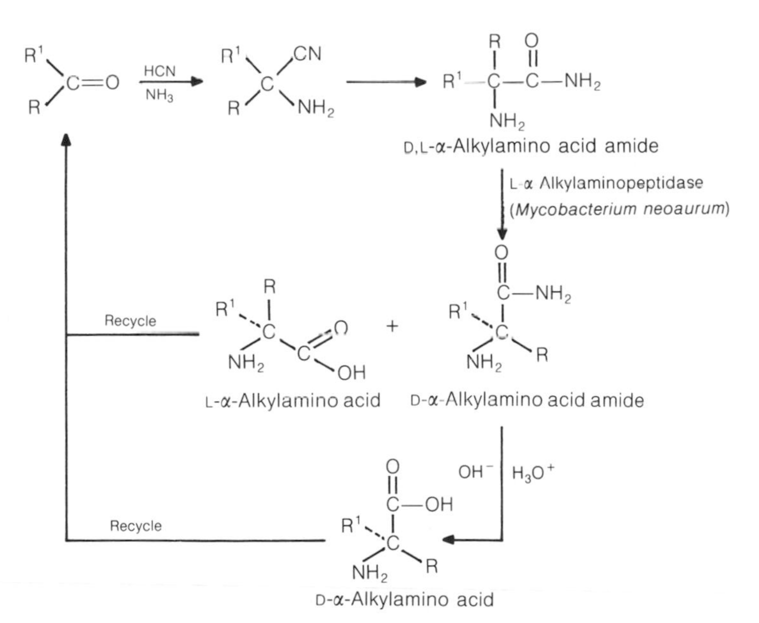

Figure 8. The DSM process for the production of L- and D-α-alkyl-substituted amino acids

Table 4. Substrate specificity and other characteristics of the enantioselective hydrolysis of α-alkyl-substituted amides with *Mycobacterium neoaurum*

$R(R^1)C(NH_2)CONH_2 \xrightarrow[\text{pH 8.5, 37 °C}]{\text{amino amidase}} R(R^1)C(NH_2)CO_2H + R^1(R)C(NH_2)CONH_2$

L-acid D-amide

R	R^1	Time	Conv.	*ee* acid (%)	*ee* amide (%)	*E*
CH_2CH_3	CH_3	5 h	0.47	80	72	19
$CH_2CH{=}CH_2$	CH_3	7 h	0.41	76	54	14
$CH(CH_3)_2$	CH_3	24 h	0.50	>96	>96	>200
$CH_2C_6H_5$	CH_3	24 h	0.50	>96	>96	>200
$CH_2C_6H_5$	CH_2CH_3	2 d	0.50	>96	>96	>200
$CH_2C_6H_5$	$CH_2CH{=}CH_2$	7 d	0.46	—	—	—
C_6H_5	CH_3	24 h	0.50	96	>96	>200
C_6H_5	CH_2CH_3	7 d	0.13	94	14	>50
C_6H_5	$CH_2CH{=}CH_2$	10 d	<0.05	—	—	—

E = enantiomer ratio (see Section 1.2.2.3, Chapter 1.)

The enzyme is active over a broad pH range (8.0–10.5) within which no uncatalysed hydrolysis of the amide is observed. The extent of the conversion is easily calculated by measuring the amount of ammonia produced.

Some restrictions with respect to the substrate specificity are indicated in Table 4. If R is large (e.g. phenyl), increasing the size of R^1 from methyl to allyl results in a progressively slower rate of stereoselective enzymic hydrolysis. If R^1 is small (e.g. methyl), decreasing the size of R from phenyl to allyl or ethyl results in a progressively faster rate of enzymic hydrolysis, accompanied, however, by decreasing stereoselectivity. Enantiomerically pure α-alkyl-substituted amino acids can be obtained in these cases by using two consecutive enzymic resolutions.

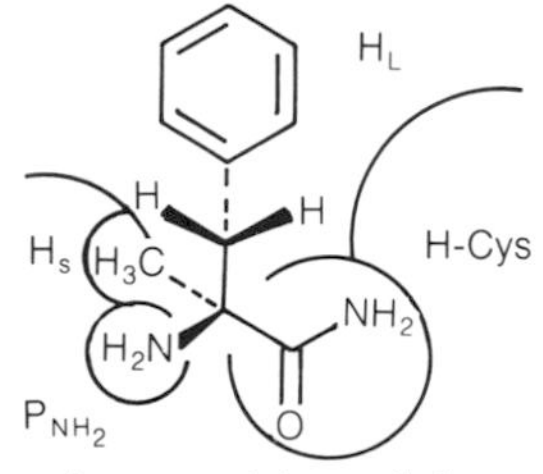

—The amidase has a cysteine activity
—H_S: maximum size isopropyl, allyl
—H_L: bulky substituents at the β-position retard the hydrolysis
—P_{NH_2} polar group necessary for stereospecific reaction

Figure 9. Active site model

On the basis of these results, an active site model was developed which is depicted in Figure 9.[28,29]

8.7 PURIFICATION OF THE L-AMIDASE FROM *M. NEOAURUM*

The L-amidase enzyme in *Mycobacterium neoaurum* ATCC 25795 was purified to homogeneity by disintegrating the whole cells, subjecting the cell-free extract to ammonium sulphate fractionation, gel filtration, and then ion-exchange chromatography. The purity of the enzyme was demonstrated by SDS-PAGE (see Figure 10).[30]

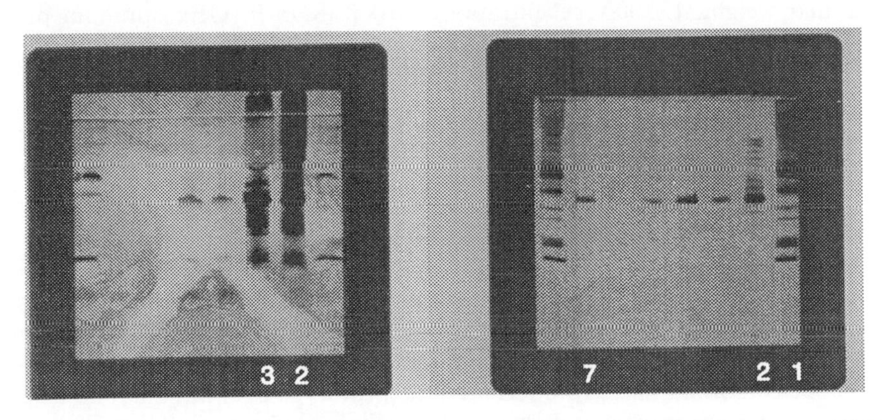

Figure 10. Purification of an L-aminoamidase from *Mycobacterium neoaurum* ATCC 25795. SDS-PAGE (8–25% gradient gel) of the enzyme fractions. Lanes: (right) 1 = molecular weight standards; 2 = protein after gel filtration; 7 = protein after anion-exchange column (Mono Q); (left) 2 = French Press extract; 3 = ammonium sulphate precipitation. From ref. 30

Table 5. Purified L-amidase from *Mycobacterium neoaurum*: substrate specificity

Substrate	activity	Substrate	Activity
Acetamide	+	α-Ethylphenylglycinamide	–
Propionamide	+	α-Allylphenylglycinamide	–
Butyramide	–	α-Propylphenylglycinamide	–
Phenylalaninamide	++	α-Isopropylphenylalaninamide	–
Valinamide	+++	*Peptides*:	–
Nicotinamide	–	L-Phe-L-Leu	–
Mandelic acid amide	–	L-Phe-*p*-nitroanilinamide	–
N-Acetylphenylalanine	–		

Table 6. K_m and V_{max} values for the purified L-amidase from *M. neoaurum*

Substrate	V_{max} (μmol min^{-1} mg^{-1})	K_m (mmol l^{-1})
D, L-α-Me-Val-NH_2	75	100
D, L-α-Me-Leu-NH_2	125	10
D, L-α-Me-Phe-NH_2	100	5
D, L-Iso-Val-NH_2	165	30
D, L-α-Allyl-Ala-NH_2	240	5
D, L-α-Me-homo-Phe-NH_2	5	15
D, L-α-Me-PG-NH_2	5	155
D, L-Pro-NH_2	175	40

The purified enzyme has the following characteristics: isoelectric point, 4.25; molecular weight, 136000 (gel filtration), 40000 (SDS-PAGE); optimum pH, 7.5–9.5; optimum temperature, 45 °C; high stability up to 60 °C. The substrate specificity and K_m and V_{max} values for the purified amidase are given in Tables 5 and 6.[31,32]

8.8 NEW APPLICATIONS OF CHIRAL α-ALKYL-SUBSTITUTED AMINO ACIDS

Recently it has been shown that α-methyl-α-alkylglycyl residues produce an interesting new type of conformational constraint in peptides. Thus, peptides rich in the achiral α,α-dimethylglycine (Aib) residue prefer regularly folded backbone conformations. More recently, the prototype of chiral α-amino acids of this family, α-methyl-α-ethylglycine, has been examined and its conformational behaviour found to be close to that of Aib, including an unexpected indifference to screw sense.[33]

In order to extend this work with chiral α-amino acids of this family, a variety of L-(α-Me)Val derivatives and peptides (up to the pentamer level) have been synthesized and characterized at the University of Padua.[34] The results indicate that (i) the (α-Me)Val residue is a strong type I/III β-turn and helix former and (ii) the relationship between (α-Me)Val chirality and helix screw sense is the same as that of C-α-monosubstituted protein amino acids.[35] Similar studies have been performed with L-(α-Me)Phe analogues of physiologically active peptides such as the formyl methionyl tripeptide chemoattractant[36] shown in Figure 11.

L-(α-Me)Phe is also incorporated in two classses of dipeptide sweeteners as a substitute for L-phenylalanine. The (α-Me)Phe analogue of aspartame (Figure 12, $R^1 = H$) was synthesized and was found to possess about the same sweetness as aspartame but showed superior stability at pH 4. In addition, the corresponding *N*-formylcarbamoyl-(α-Me)Phe analogue of aspartame exhibited

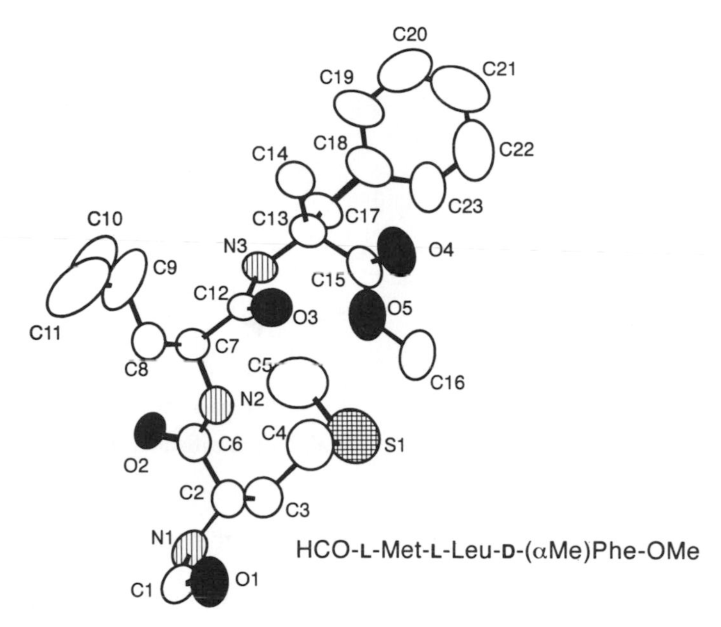

Figure 11. An application of D-(α-Me)Phe in a chemotactic peptide

Figure 12. L-(α-Me)Phe analogues of aspartame

a comparable sweet taste. *N*-Formylcarbamoylaspartame[37] is a recently discovered aspartame derivative which is very stable at high pH values (6–8). In contrast, neither the corresponding *N*-formyl ($R^1 = CHO$) nor the *N*-carbamoyl ($R^1 = CONH_2$) derivatives of both aspartame and the (α-Me)Phe analogue of aspartame were sweet.[38]

8.9. RECENT DEVELOPMENTS AT DSM TOWARDS THE GOAL OF 100% CONVERSION IN ONE STEP

8.9.1 USE OF L- OR D-AMINOPEPTIDASES IN COMBINATION WITH A RACEMASE

As with all resolution procedures, conversion of amino acid amides by aminopeptidases is limited to 50% for each pass. This limitation may be circumvented

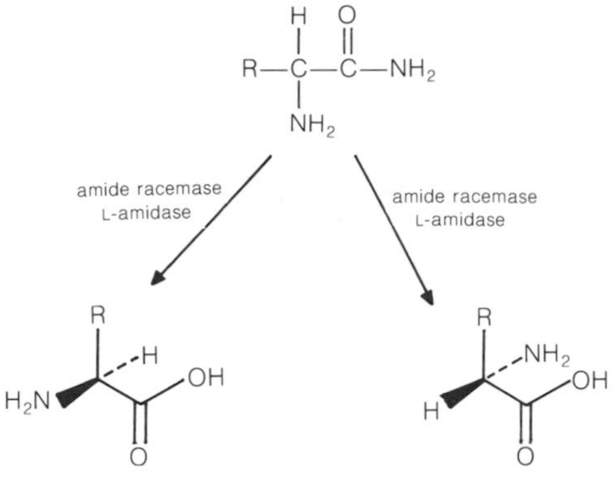

Figure 13. Use of L- or D-aminopeptidase in combination with a racemase. From Ref. 3

if an amino acid amide racemase is used in combination with an L- or a D-aminopeptidase (Figure 13).

Using *in vivo* protein engineering, the authors were able to select a mutant strain with, in addition to the original L-specific aminopeptidase activity, a novel D-aminopeptidase and amino acid amide racemase activity. The authors' research in this area is now directed towards the generation of L- and D-aminopeptidase-negative strains, and the specific activities of the new enzymes should, preferably, be raised. Racemization *in situ* with the aid of racemases is central to the development of economic processes for amino acids. This topic is surveyed in Chapter 20 (Section 20.4).

8.9.2 CRYSTALLIZATION-INDUCED ASYMMETRIC TRANSFORMATIONS OF RACEMIC AMINO ACID AMIDES

The Schiff bases of chiral amino acid amides racemize readily. The combination of diastereoisomeric salt formation with *in situ* racemization would, in theory, result in a one step 100% conversion of racemic amino acid amides into L- or D-amino acid amides, as shown in Figure 14.

In studies involving this approach it was demonstrated that D-phenylglycine could be obtained in this way, starting from benzaldehyde, hydrogen cyanide and ammonia. Quantitative conversion of the aminonitrile thus obtained to the amino acid amide is effected by using a catalytic amount of ketone and a small amount of base. Addition of one equivalent of benzaldehyde gives the Schiff base of the racemic amino acid amide. A simpler, innovative approach that also works well is a one-pot procedure whereby, directly from benzaldehyde, the Schiff base of the phenylglycine amide is obtained by sequential addition of hydrogen cyanide, ammonia and an extra equivalent of benzaldehyde with a

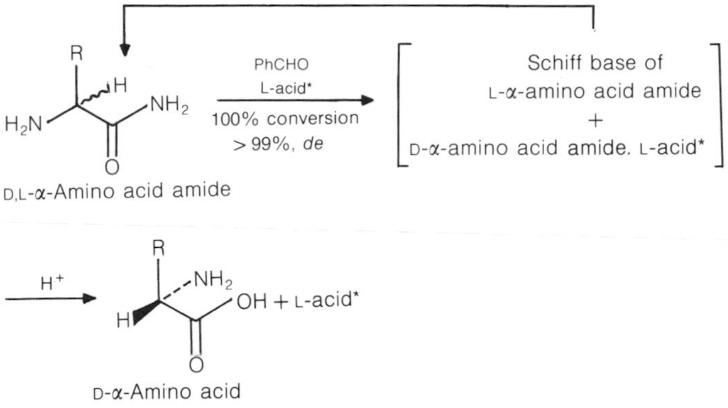

PhCHO
L-acid*
100% conversion
> 99%, *de*

D,L-α-Amino acid amide

Schiff base of
L-α-amino acid amide
+
D-α-amino acid amide. L-acid*

H^+

OH + L-acid*

D-α-Amino acid

Figure 14. Asymmetric transformation (resolution with *in situ* racemization)

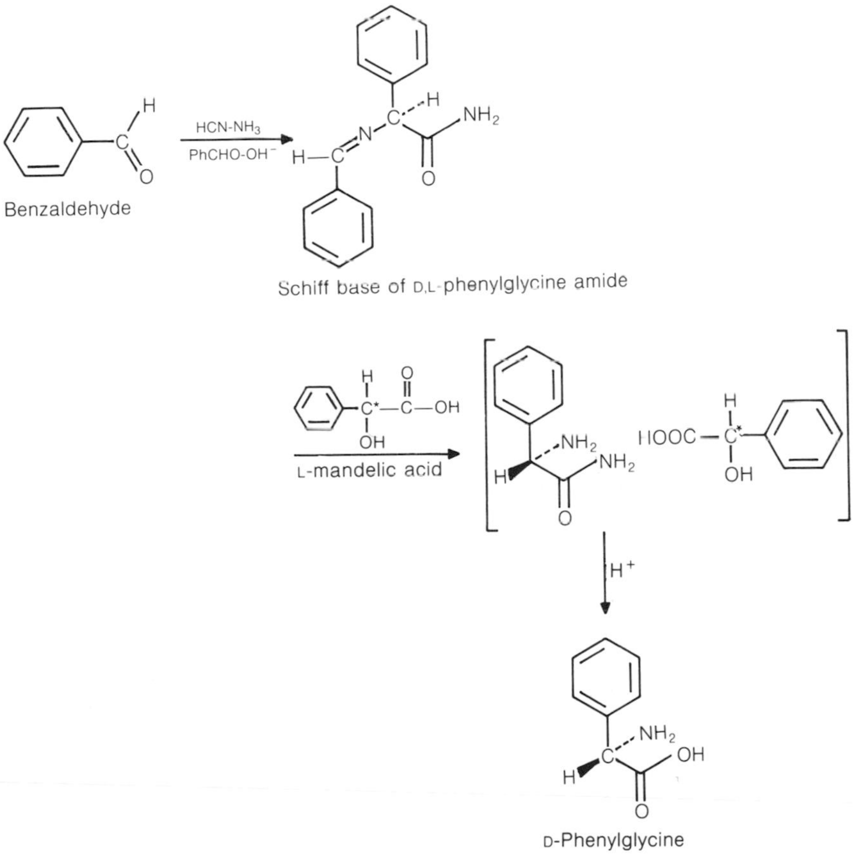

Benzaldehyde

$HCN\text{-}NH_3$
$PhCHO\text{-}OH^-$

Schiff base of D,L-phenylglycine amide

L-mandelic acid

H^+

D-Phenylglycine

Figure 15. Asymmetric transformation: D-phenylglycine

Table 7. Substrate specificity and relative activity of *Ochrobactrum anthropi*

Substrate	Specificity/ conversion	Relative activity (%)	Substrate	Specificity/ conversion	Relative activity (%)
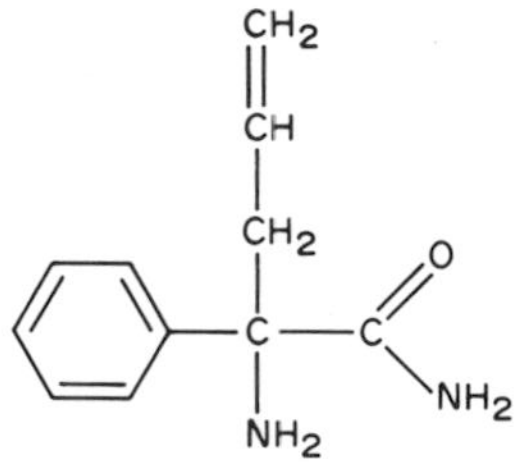 α-Allyl-PG-amide	DL ⟶ 50%	4	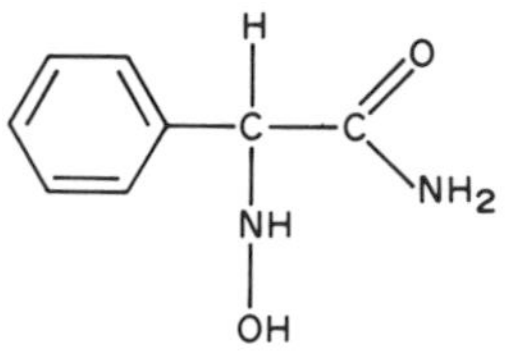 *N*-OH-PG-amide	DL ⟶ 50%	25
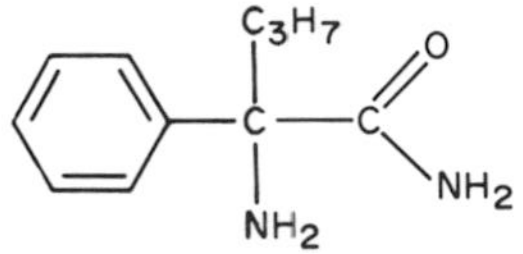 α-Propyl-PG-amide	DL ⟶ 50%	1	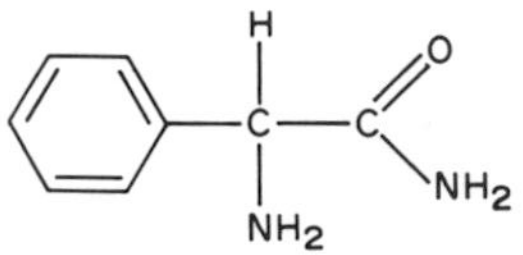 PG-amide	L ⟶ 100% DL ⟶ 50%	100
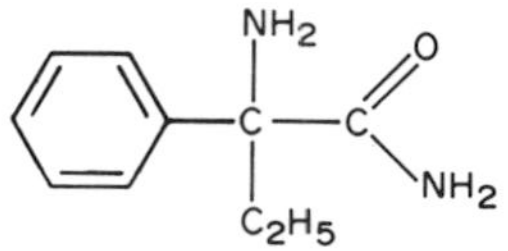 α-Ethyl-PG-amide	DL ⟶ 50%	4	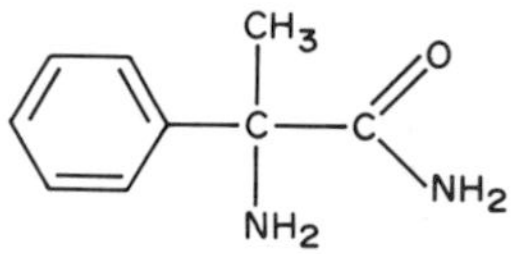 α-Methyl-PG-amide	DL ⟶ 50%	2

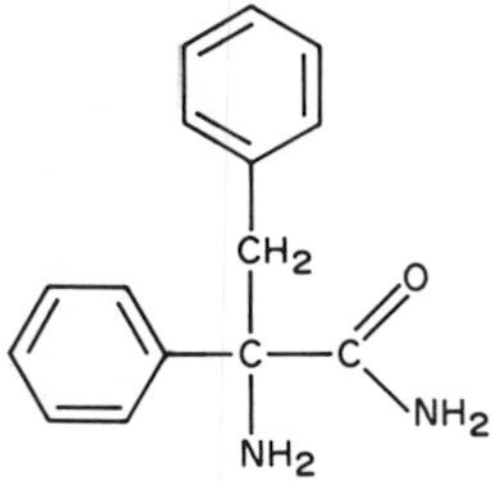 α-Benzyl-PG-amide	DL ⟶ 0%	
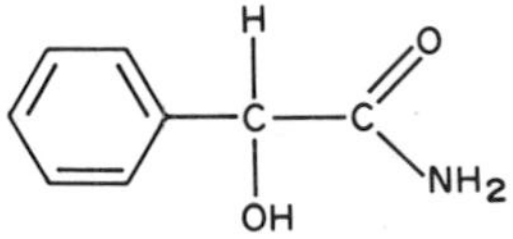 Mandelic acid amide	L ⟶ 100% DL ⟶ 50% D ⟶ 0% *E* value 300	5
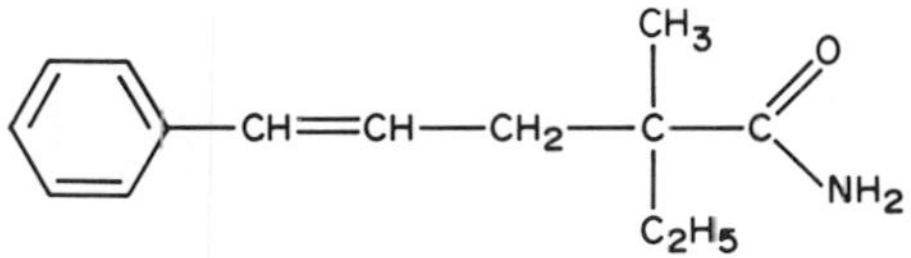 α-cinnamoylalaninamide	DL ⟶ 50%	16.5

catalytic quantity of base (Figure 15). A simple and very efficient chiral resolving acid is added, and crystallization of the D-phenylglycine amide L-acid salt begins immediately. Nearly quantitative conversion of the racemic amino acid amide is observed. Hydrolysis of the D-amide to the acid is accomplished under acidic conditions, while the L-acid may be quantitatively recycled by extraction.[39]

Research into the scope and limitations of this asymmetric transformation is now being undertaken at DSM.[40]

8.10 *OCHROBACTRUM ANTHROPI*: A NEW BIOCATALYST WITH AMIDASE ACTIVITY

Owing to the limitations of *Mycobacterium neoaurum* and *Pseudomonas putida*, and the interest of DSM in the resolution of very bulky α,α-disubstituted amino acids, *N*-hydroxyamino acids and α-hydroxy acid amides, another screening programme was begun. Surprisingly, a strain capable of hydrolysing these structurally very different substrates was isolated. This strain was identified as *Ochrobactrum anthropi*, and some of the preliminary results have been described.[41] The substrate specificity is remarkably wide, and ranges from α-H-, α-alkyl-, *N*-hydroxy acid amides to α-hydroxy acid amides (see Table 7).

Whole cells of *Ochrobactrum anthropi* are used and the optimum pH is 8.0. The biocatalyst is very thermostable (optimum temperature 50 °C), and high activity and enantioselectivity are combined with low substrate and product inhibition. More research in this area is in progress at DSM.[42,43]

8.11 CONCLUSIONS

Very efficient procedures have been developed for the production of optically pure natural and unnatural amino acids in which both enantiomers of α-H- and α-alkylamino acids can be obtained. The enzymes used have been purified to homogeneity and their substrate range studied. More recently, protein engineering techniques have been used to obtain new biocatalysts with a still broader scope.

New developments towards the goals of 100% conversion and absolute stereoselectivity are promising, and further research in this productive area is in progress at DSM.

8.12 REFERENCES

1. Wagner, I., and Musso. H., *Angew. Chem., Int. Ed. Engl.*, **22**, 816 (1983); Barrett, G. C., *Chemistry and Biochemistry of Amino Acids*, Chapman and Hall, London, 1985.

2. Sheldon, R. A., Schoemaker, H. E., Kamphuis, J., Boesten, W. H. J., and Meijer, E. M., in *Stereoselectivity of Pesticides* (ed. E. J. Ariens, J. J. S. van Rensen and W. Welling), Elsevier, Amsterdam, 1988, pp. 409–451.
3. Kamphuis, J., Boesten, W. H. J., Broxterman, Q. B., Hermes, H. F. M., van Balken, J. A. M., Meijer, E. M., and Schoemaker, H. E., *Adv. Biochem. Eng. Biotechnol.*, **42**, 134 (1990).
4. Williams, R. M. *Synthesis of Optically Active α-Amino Acids*, Pergamon Press, Oxford, 1989.
5. Sheldon, R. A., Hulshof, L. A., Bruggink, A., Leusen, F. J. J., van der Haest, A. D., and Wijnberg, H., *Chim. oggi*, **9**, 23 (1991).
6. Broxterman, Q. B., Boesten, W. H. J., Kamphuis, J., Kloosterman, M., Meijer, E. M., and Schoemaker, H. E., in *Opportunities in Biotransformations* (ed. L. G. Copping, R. E. Martin, J. A. Pickett, C. Bucke and A. W. Bunch), Elsevier Applied Science, Barking, New York, 1990, pp. 148–169.
7. Kamphuis, J., Boesten, W. H. J., Schoemaker, H. E., and Meijer, E. M., in *Conference Proceedings on Pharmaceutical Ingredients and Intermediates* (ed. B. G. Reuben), Expoconsult, Maarsen, 1991, pp. 28–38.
8. de Boer, L., and Dijkhuizen, L., *Adv. Biochem. Eng. Biotechnol.*, **41**, 1 (1991).
9. Ice, J., and Agathos, S. N., *Biotechnol. Lett.*, **11** (2), 77 (1989); Kobel, H., and Traber, R., *Eur. J. Appl. Microbiol. Biotechnol.*, **14**, 237 (1982).
10. Roth, H. J., and Fenner, H., *Arzneistoffe–Pharmazeutische Chemie III*, Georg Thieme, Stuttgart, 1988, pp. 202–204.
11. de Jong, W. W., and Bloemendal, H., *J. Biol. Chem.*, **257**, 7077 (1982); Allen, M. P., Yamada, A. H., and Carpenter, F. H., *Biochemistry*, **22**, 3778 (1983); Srin, S. H., and van Wart, H. E., *Biochemistry*, **21**, 5528 (1982); DSM/Stamicarbon, *US Pat.*, 3 971 000, 1976.
12. NOVO-DSM/Stamicarbon, *US Pat.*, 4 080 259, 1978.
13. (a) DSM/Stamicarbon, Br. Pat., 1 548 032, 1976; (b) Meijer, E. M., Boesten, W. H. J., Schoemaker, H. E., and van Balken, J. A. M., *Biocatalysts in Organic Synthesis* (ed. J. Tramper, H. C. van der Plas and P. Linko), Elsevier, Amsterdam, 1985, pp. 135–156.
14. DSM/Stamicarbon, *US Pat.*, 4 172 846, 1979.
15. DSM/Stamicarbon, *Dutch Pat. Appl.*, 8 501 093, 1985; *Eur. Pat. Appl.*, EP 1 816 75, 1985.
16. DSM/Stamicarbon, *Eur. Pat. Appl.*, EP 1 442 585, 1991.
17. Duchateau, A. L. L., Hillemans, M. G., Schepers, C. H. M., Ketelaar, P. E. F., Hermes, H. F. M., and Kamphuis, J., *J. Chromatogr.*, **566**, 493 (1991).
18. Shadid, B., van der Plas, H. C., Boesten, W. H. J., Kamphuis, J., Meijer E. M., and Schoemaker, H. E., *Tetrahedron*, **46**, 913 (1991).
19. Vriesema, B. K., ten Hoeve, W., Wijnberg, H., Kellogg, R. M., Boesten, W. H. J., Meijer, E. M., and Schoemaker, H. E., *Tetrahedron Lett.*, **26**, 2045 (1986).
20. Vanrobays, M., VanderHaeghe, H., Boesten, W. H. J., Schoemaker, H. E., Meijer, E. M., and Kamphuis, J., to be published; Mooiweer, H. H., *PhD Thesis*, University of Amsterdam, 1990.
21. Hermes, H. F. M., Sonke, T., Meijer, E. M., and Dijkhuizen, L., to be published.
22. DSM Research, to be published.
23. Pandey, R. C., Meng, H., Cook, J. C., and Rinehart, K. L., *J. Am. Chem. Soc.*, **99**, 5203, (1977).
24. Toniolo, C., and Benedetti, E., *ISI Atlas of Science: Biochemistry*, **1**, 1988 p. 255.
25. DSM/Stamicarbon, *Eur. Pat.* 0 150 854 (1984).
26. Kruizinga, W. H., Bolster, J., Kellogg, R. M., Kamphuis, J., Boesten, W. H. J., Meijer, E. M., Schoemaker, H. E., *J. Org. Chem.*, **53**, 1826, (1988).

27. Kamphuis, J., Hermes, H. F. M., van Balken, J. A. M., Schoemaker, H. E., Boesten, W. H. J., and Meijer, E. M., in *Amino Acids, Chemistry, Biology and Medicine* (ed. G. Lubec and G. A. Rosenthal), ESCOM Science Publ., Vienna, 1990, pp. 119–125.
28. Kaptein, B., Boesten, W. H. J., Broxterman, Q. B., Meijer, E. M., Polinelli, S., Schoemaker, H. E., and Kamphuis, J., in *Proceedings of the Second International Symposium on Chiral Discrimination, Rome*, 1991, (ed. D-Misiti), 1991, p. 58.
29. Duchateau, A. L. L. Hillemans, M. G., Kamphuis, J., Boesten, W. H. J., Schoemaker, H. E., and Meijer, E. M., *J. Chromatogr.*, **471**, 263 (1989).
30. DSM Research, unpublished results.
31. Hermes, F., PhD thesis, Rijksuniversiteit, Groningen, 1992.
32. Hermes, F., PhD thesis, Rijksuniversiteit, Groningen, 1992.
33. Valle, G., Crisma, M., Toniolo, C., Beisswenger, R., Rieker, A., and Jung, G., *J. Am. Chem. Soc.*, **111**, 6828 (1989).
34. Toniolo, C., Crisma, M., Bonara, G. M., Klajc, B., Lelj, F., Grimaldi, P., Rosa, A., Polinelli, S., Boesten, W. H. J., Meijer, E. M., Schoemaker, H. E., and Kamphuis, J., *Int. J. Pept. Protein Res.*, **38**, 242 (1991).
35. Valle, G., Crisma, M., Toniolo, C., Polinelli, S., Boesten, W. H. J., Schoemaker, H. E., Meijer, E. M., and Kamphuis, J., *Int. J. Pept. Protein Res.*, **37**, 521 (1991).
36. Toniolo, C., Crisma, M., Pegoraro, S., Valle, G., Bonara, G. M., Becker, E. L., Polinelli, S., Boesten, W. H. J., Schoemaker, H. E., Meijer, E. M., Kamphuis, J., and Freer, R., *Pept. Res.*, **4**, 66, (1991).
37. Boesten, W. H. J., Dassen, B. H. N., Kleinjans, J. C. S., van Agen, B., van der Wal, S., de Vries, N. K., Schoemaker, H. E., and Meijer, E. M., *J. Agric. Food Chem.*, **39**, 154 (1991).
38. Polinelli, S., Broxterman, Q. B., Schoemaker, H. E., Boesten, W. H. J., Crisma, M., Valle, G., Toniolo, C., and Kamphuis, J., *Bioorg. Med. Chem. Lett.*, **2**, 453 (1992).
39. DSM/Stamicarbon, *Eur. Pat. Appl.*, EP 442 584, 1991.
40. Kamphuis, J., Schoemaker, H. E., Kaptein, B., van den Tweel, W. J. J., and Boesten, W. H. J., *Ann. N. Y. Acad. Sci.*, in press.
41. DSM/Stamicarbon, *Dutch Pat. Appl.*, 9 100 038, 1991.
42. Kamphuis, J., Boesten, W. H. J., Hermes, H. F. M., van den Tweel, W. J. J., Van Balken, J. A. M., Kloosterman, M., Schoemaker, H. E., Broxterman, Q. B., and Meijer, E. M., *Biocatalysis, a versatile methodology in chemistry* in *Part I Proceedings of 3rd Netherlands Biotechnology Conferences* (ed. K. Ch. A. M. Luyben), Elsevier, Amsterdam, 1990, pp. 4–10.
43. Kamphuis, J., Meijer, E. M., Boesten, W. H. J., Broxterman, Q. B., Kaptein, B., Hermes, H. F. M., and Schoemaker, H. E., in *Production of Amino Acids and Derivatives* (ed. D. Rozzell and F. Wagner), Carl Hanser, Munich, 1992, in press.

9 The Use of Aminotransferases for the Production of Chiral Amino Acids and Amines

D. I. STIRLING
Celgene Corporation, Warren, NJ, USA

9.1 INTRODUCTION

Many commercial and development pharmaceuticals contain amino-substituted chiral centres, and these have been made by a number of methods. While some naturally occurring chiral amines, e.g. ephedrine, have been incorporated directly into target molecules, chiral amine products are usually obtained by chemical and biological resolutions. The most prevalent technology is chemical resolution of racemic amines with relatively inexpensive chiral acids such as D-tartaric acid. The resultant diasteroisomeric salts can usually be separated by chromatography or selective crystallization, but a disadvantage of this approach is the need, for reasons of economy, to racemize and recycle the unwanted enantiomer. In some selected cases the relatively inexpensive chiral amine α-methylbenzylamine can be used as a chiral auxiliary in the asymmetric synthesis of more elaborate chiral amine intermediates and products.[1]

Biological methods for the production of homochiral amines have, until recently, been directed at the production of α-amino acids. Although most of

Chirality in Industry. Edited by A. N. Collins, G. N. Sheldrake and J. Crosby

these are made by traditional fermentation technology, they can be synthesized directly or as racemic mixtures and resolved using enzyme-based systems. A number of enzyme types may be used for such processes, e.g. oxidases, dehydratases, dehydrogenases, ammonia lyases, *N*-acylases, esterases and aminotransferases (transaminases) (see also Chapters 8, 19 and 20).

Aminotransferases, which are ubiquitous in nature, are of critical importance to nitrogen metabolism (Scheme 1). Non-enzymatic transamination was first demonstrated in the early 1930s. Braunstein and co-workers discovered the enzymatic equivalent in the mid-1930s, and since then substantial progress has been made in the elucidation of the structure of the enzymes involved and their biological significance.[2–4]

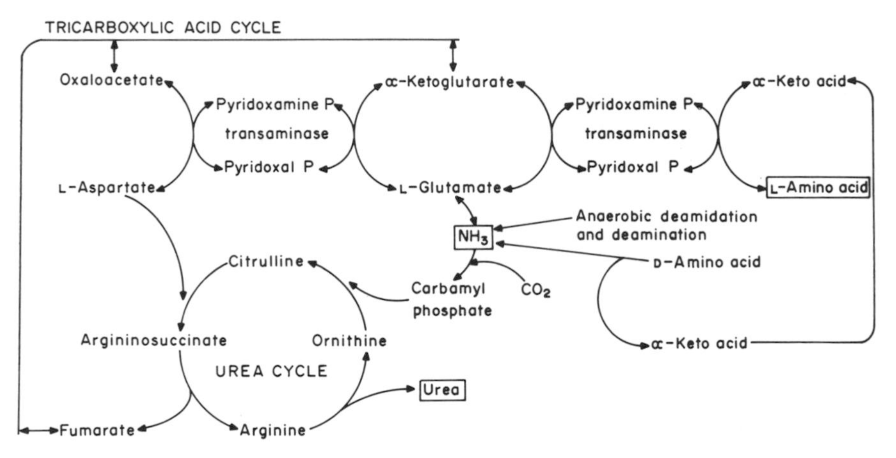

Scheme 1. The importance of aminotransferases to the tricarboxylic acid and urea cycles. Reproduced with permission from ref. 3

Aminotransferases catalyse the transfer of an amino group, a proton and a pair of electrons from a primary amine substrate to the carbonyl group of an acceptor molecule. Scheme 2 depicts the net reaction, which is reversible and may be considered, formally, as two half-reactions—the oxidative deamination of an amino donor, followed by the reductive amination of an amino acceptor. The biological significance of transfer of an amino group warranted the classification of aminotransferases as a special sub-group (E.C. 2.6.1) within the transferases.

$$R^1\text{—}\underset{\displaystyle H}{\overset{\displaystyle NH_2}{C}}\text{—}R^2 + R^3\text{—}\overset{\displaystyle O}{\overset{\|}{C}}\text{—}R^4 \underset{\text{aminotransferase}}{\rightleftharpoons} R^1\text{—}\overset{\displaystyle O}{\overset{\|}{C}}\text{—}R^2 + R^3\text{—}\underset{\displaystyle H}{\overset{\displaystyle NH_2}{C}}\text{—}R^4$$

Scheme 2. General aminotransferase reaction

Most aminotransferases have a requirement for the coenzyme pyridoxal-5-phosphate (PLP) (**1**). Although similar to benzaldehyde, the reactivity of the PLP aldehyde group is significantly modified by the presence of an adjacent hydroxyl substituent and the ring nitrogen in the *para* position. This reactivity, combined with the various ways in which the coenzyme can interact with the apoenzyme, leads to the diversity of reactions exhibited by PLP-dependent enzymes, e.g. transaminations, aldol reactions, racemizations and decarboxylations.

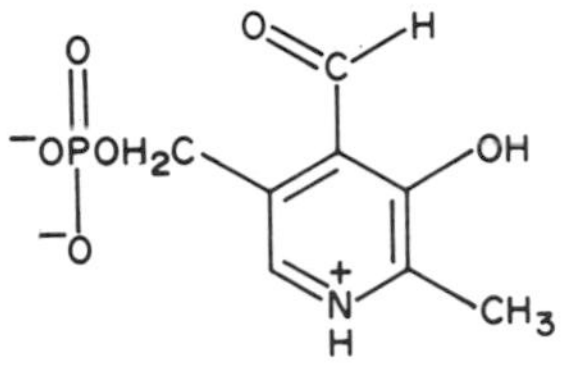

Pyridoxal phosphate (PLP)

(**1**)

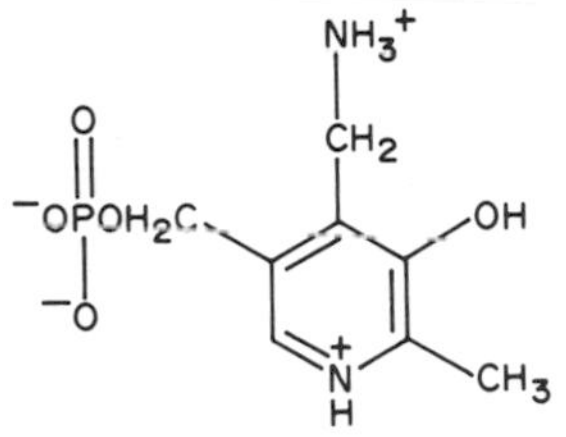

Pyridoxamine phosphate (PMP)

(**2**)

The initial step in the transamination reaction involves the interaction of the amino donor with the enzyme-bound PLP to form a Schiff base intermediate. Prototropic rearrangement followed by hydrolysis yields the pyridoxamine phosphate form (PMP) (**2**) of the enzyme plus the carbonyl-containing product. The second half-reaction involving the amino acceptor produces the corresponding amine product and regenerates the PLP form of the enzyme. A characteristic of reactions catalysed by aminotransferases is their ready reversibility.

The primary metabolic role of aminotransferases, illustrated in Scheme 1, is the transfer of an amino group from α-amino acids to 2-keto acids. Particularly important substrates are 2-ketoglutaric acid, oxaloacetic acid and pyruvic acid. For this reason, the commercial utilization of these enzymes, to date, has been focused on the production of α-amino acids. However, it has recently been shown that aminotransferases can be used to produce homochiral amine products that do not contain an α-carboxylic acid group.[5,6]

9.2 THE SCOPE OF AMINOTRANSFERASES AS INDUSTRIAL BIOCATALYSTS

A chiral amine product can be produced using an aminotransferase either by a kinetic resolution of the racemic amine or by asymmetric synthesis from the corresponding prochiral carbonyl compound. Thus, either racemic amines or prochiral ketones can be used as raw materials, depending on their relative cost and availability. In the case of kinetic resolutions, the unwanted enantiomer of the substrate may often be recycled in processes which make use of the same aminotransferase (see Section 9.4).

Collectively, aminotransferases exhibit very high reaction rates and stereoselectivity, features which make them attractive for commercial applications. Some microbial aminotransferases, in particular, possess a broad substrate specificity with respect to the amino donor.[7] A further advantage of these enzymes as catalysts for industrial processes is their relative stability. A potential drawback of aminotransferases is incomplete reaction is due to the position of the equilibrium for the reversible interconversion of products and raw materials (the equilibrium constant for the overall reaction is often about 1 for a typical α-amino acid–2-keto acid couple). This problem would be of particular significance when operating in the synthetic mode, and ways would have to be found to circumvent the thermodynamic constraint for a cost-effective synthetic process.

9.3 THE PRODUCTION OF AMINO ACIDS

The potential of aminotransferases for the production of amino acids has been well documented.[8–17] These reports describe the stereoselective syntheses of L-α-amino acids using a suitable amino donor. Enzyme systems, and whole-cell systems, both free and immobilized, have been developed, and the amino donor may be regenerated *in situ* by growing organisms.[10] The aforementioned problem of the reaction equilibrium has been cleverly overcome by coupling the transamination process to a second irreversible reaction. For example, when L-aspartic acid is used as the amino donor in reactions catalysed by aspartate

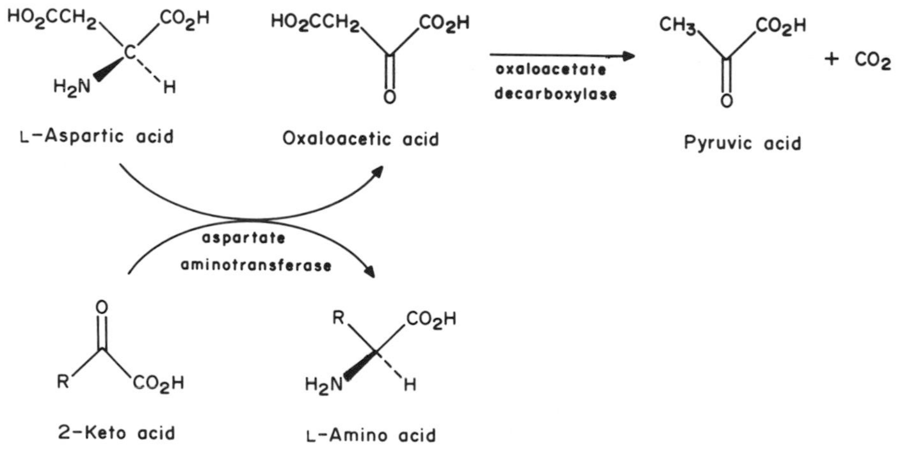

Scheme 3. Production of L-amino acids

aminotransferase, the products are the L-amino acid corresponding to the amino acceptor employed, plus oxaloacetic acid (Scheme 3). This 2-keto acid byproduct can be efficiently decarboxylated by oxaloacetate decarboxylase to give pyruvic acid and carbon dioxide. By coupling the reactions in this way, the conversion of the amino acceptor to the desired amino acid product can be increased to more than 90%. This has been demonstrated for L-phenylalanine in both whole cell[14,15] and enzyme-based systems.[13,17]

The continuous removal of oxaloacetic acid by decarboxylation allied with the immobilization of the aminotransferase and decarboxylase enzymes can result in an efficient, continuous process for the production of amino acids. An immobilized enzyme system based on the *E. coli* aspartate aminotransferase has been used to produce a variety of amino acids, including examples which do not occur in nature, e.g. L-2-aminoadipic acid and L-4-phenyl-2-aminobutanoic acid.[13,17]

In order to develop a similar process for the production of the branched-chain amino acids L-leucine, L-isoleucine and L-valine, the narrow substrate specificity of the aminotransferase from *E. coli* had to be overcome. The problem was solved by the introduction of a third enzyme, the branched-chain amino acid aminotransferase encoded by the *ilvE* gene of *E. coli.*[17] This enzyme will perform reductive aminations of all three α-keto acids (i.e. 2-ketoisocaproic acid, 2-keto-3-methylbutanoic acid and 2-ketoisovaleric acid) to the corresponding L-amino acids. The enzyme cannot, however, utilize L-aspartic acid as the amino donor, and so L-glutamic acid, the amino donor preferred by the enzyme, was generated *in situ* from 2-ketoglutarate with a second transaminase as outlined in Scheme 4. Thus, the production of L-leucine employs 2-ketoisopentanoic acid and L-aspartic acid as raw materials, with L-glutamate playing only a catalytic role.[17]

These co-immobilized enzyme systems can have half-lives of at least 3 months,

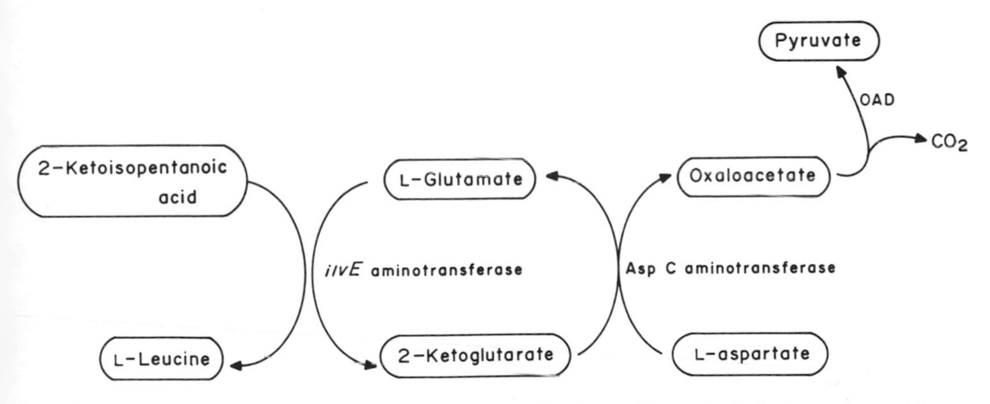

Scheme 4. Coupled enzyme systems for the production of branched-chain amino acids (OAD = oxaloacetate decarboxylase)

and give overall conversions of >90%. Volumetric productivities of around 500 g per kilogram of biocatalyst per hour have been achieved with product concentrations in excess of 0.5 molar.[17,18] Thus, by the early 1980s the commercial potential of aminotransferases for the production of α-amino acids had been realized in several industrial processes, although the utility of these enzymes in the manufacture of other chiral amines remained largely unexplored.

9.4 THE PRODUCTION OF CHIRAL AMINES

A number of aminotransferase enzymes which can utilize substrates where the amino group is not vicinal to a carboxylate group have been described. In these enzymes, which belong to the ω-aminotransferase class, the amino donor is generally restricted to ω-amino acids and α,ω-diamino acids (Scheme 5). However, one enzyme, the ω-amino acid/pyruvate aminotransferase from *Pseudomonas* sp. F-126, has been shown to utilize a variety of primary monoamines, such as ethylamine, butylamine, heptylamine and benzylamine, in addition to ω-amino acids.[19] Limited activity was also observed with the secondary amine substrate *sec*-butylamine.[19]

	Donor	*Acceptor*	*Hydrogen removed*
E.C. 2.6.1.8	α,ω-diamino acid	2-KG	
E.C. 2.6.1.13	L-Ornithine	2-KG	Pro-S
E.C. 2.6.1.18	β-Alanine (ω-APT)	PYR	Pro-R
E.C. 2.6.1.19	4-Aminobutyrate	2-KG	Pro-S
E.C. 2.6.1.29	α,ω-Diamine	2-KG	
E.C. 2.6.1.36	L-Lysine	2-KG	Pro-S
E.C. 2.6.1.46	2,4-Diaminobutyrate	PYR	
E.C. 2.6.1.55	Taurine	2-KG	

Scheme 5. ω-Amino transferases (2-KG = 2-ketoglutarate; PYR = pyruvate; ω-APT = ω-amino acid/pyruvate aminotransferase)

The amino acceptor for this enzyme is exclusively pyruvic acid. A narrow specificity with respect to amino acceptor is characteristic of this group of bacterial enzymes. After the formation of a Schiff base complex between the enzyme-bound pyridoxal phosphate and the amino donor, the aminotransferase reaction proceeds via proton abstraction. With the exception of glycine, the substrates of α-amino acid aminotransferases have only a single proton available for abstraction, whereas ω-aminotransferases have the choice of two chemically identical protons attached to the vicinal, prochiral carbon (Figure 1).

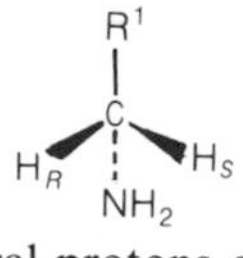

Figure 1. Prochiral protons of ω-amino acids

In view of the substrate specificity displayed by amino acid aminotransferases, one would expect the stereoselective removal of either the pro-(*S*) or the pro-(*R*) proton, and this has indeed been demonstrated for a number of ω-amino acid aminotransferases[20,21] (Scheme 5). For example, the ω-amino acid/pyruvate aminotransferase from *Pseudomonas* F-126 selectively removes the 4-pro-(*R*)-hydrogen from γ-aminobutyric acid[20] (Scheme 6).

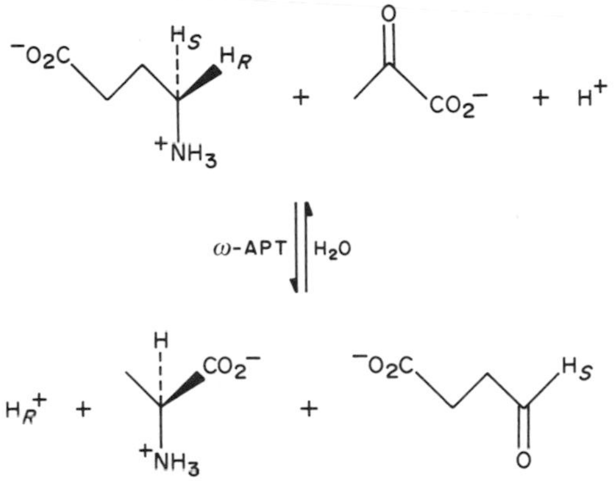

Scheme 6. Mode of action of ω-amino acid pyruvate aminotransferases

Based this information, the question posed by Celgene scientists was, "will these enzymes transaminate a substrate in which one of the protons is replaced by an alkyl group, and, if so, is the reaction still stereoselective?"

9.4.1 TRANSAMINATION OF SECONDARY AMINES

A variety of bacterial strains known to contain ω-amino acid aminotransferases were obtained from culture collections. In addition, new bacterial sources of this enzyme type were enriched in the ω-aminotransferase by including secondary amines in the culture medium as the sole nitrogen source. Once a range of suitable organisms was available, they were screened for the ability to deaminate 1-phenyl-3-aminobutane, an intermediate for the anti-hypertensive dilevalol. A sample of these results is given in Scheme 7. The *Bacillus megaterium* strain was a new isolate, whereas the *Pseudomonas aeruginosa* and *Pseudomonas putida* strains were from culture collections ATCC 15 692 and ATCC 39 213, respectively. Each of these strains apparently deaminated the amine, as evidenced by the appearance of benzylacetone in the culture medium. The low relative rate observed with the (*R*)-isomer of the amino donor alone suggested that the deamination reaction was stereoselective (entry 2), and the dependence on the presence of a keto acid (pyruvic acid) for activity suggested that the

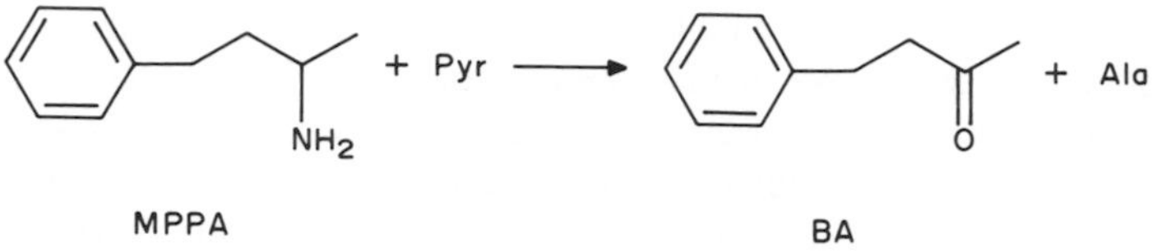

Entry	MPPA Isomers	Pyruvate	Inactivator	Relative rates of production of BA: *B. megaterium*	*P. aeruginosa*	*P. putida*
1	*RS*	+	–	100	100	100
2	*R*	+	–	15	4	3
3	*RS*	–	–	0	0	0
4	*RS*	+	Hydroxylamine	10	0	3
5	*RS*	+	Gabaculine	13	0	0

Scheme 7. Chiral amine aminotransferases (MPPA = 1-methyl-3-phenylpropylamine; BA = benzylacetone)

transamination was taking place (entry 3). The latter conclusion was supported by the strong inhibition of the reaction by the known aminotransferase inhibitors hydroxylamine and gabaculine (entries 4 and 5). Hence, the premise that secondary amines may be transaminated stereoselectively by the ω-aminotransferases was shown to be correct.

The enzyme from the *Bacillus megaterium* strain, which tolerated a broader range of substrates and exhibited superior reaction rates, was characterized in more detail. The purified protein was shown to be a dimer with a molecular weight of 110 000 and was confirmed to be a pyridoxal phosphate-dependent aminotransferase. The enzyme deaminates a wide variety of secondary amines in a stereoselective manner. In the natural host, the gene which encodes the aminotransferase can be induced by a wide variety of amines. This gene has been isolated, cloned and sequenced.[22]

Given that most, if not all, the known ω-amino acid aminotransferases require a specific α-keto acid to act as the amino acceptor, the range of amino acceptors for the *Bacillus* enzyme is unusually broad. This enzyme can utilize a number of α-keto acids, in addition to a variety of ketones and aliphatic or aromatic aldehydes. This surprising tolerance of a variety of amino acceptors would allow a wide range of relatively cheap carbonyl compounds to be used in this role in commercial processes.

Scheme 8 displays a representative sample of the types of amino donors deaminated by this aminotransferase and the attendant stereoselectivity. The enzyme will process a structurally diverse group of secondary amines, i.e. alkylamines, benzylic amines, arylalkylamines, heterocyclic and multicyclic amines and β-hydroxyamines. In most cases the enzyme selectively deaminates the (*S*)-enantiomer. Indeed, for some substrates, e.g. α-methylbenzylamine and some of its derivatives, the enzyme is entirely enantiospecific. Exceptions to the

Amine	*Relative Rate of Conversion*	
	(*R*)-*Enantiomer*	(*S*)-*Enantiomer*
NH_2	0	100
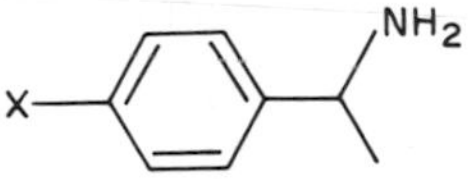 X, NH_2; X = H, Br, NO_2	0	100
NH_2	2	100
NH_2	11	100
NH_2	5	100
N, NH_2	0	100
NH_2, OH	100	0
NH_2, OH	100	0
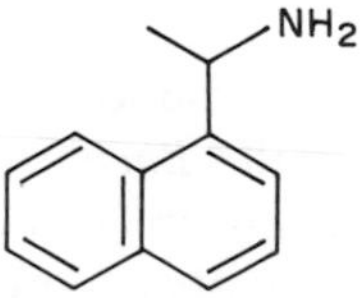 NH_2	0	100

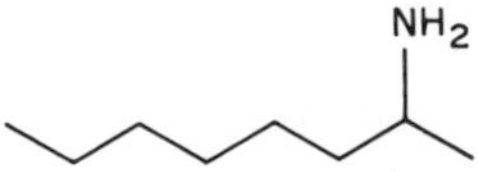

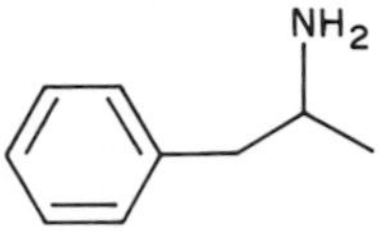

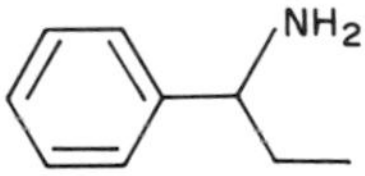

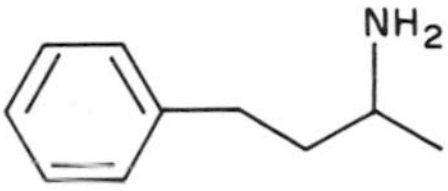

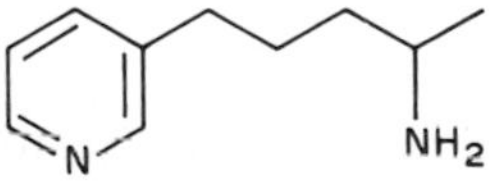

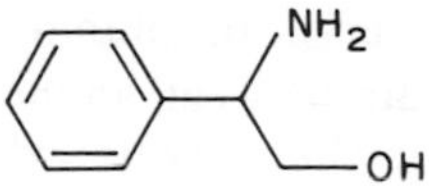

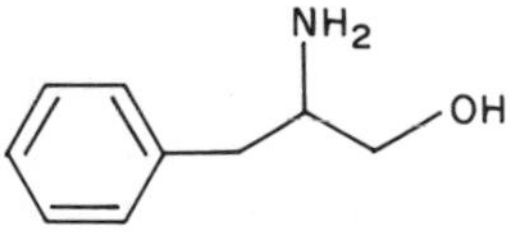

Scheme 8. Amino donors for *Bacillus megaterium* enzyme

general rule are the β-hydroxyamino compounds where, as a result of the operation of the sequence rules, the (*R*)-enantiomer is the preferred substrate.

As discussed earlier, it should be possible to profit from the reversible nature of aminotransferases and develop processes for both enantiomers of a given amine in high chiral purity using a single enzyme. For example, using the *Bacillus* enzyme, (*R*)-α-methylbenzylamine can be produced from the racemic mixture via a kinetic resolution [Scheme 9(i)]. (*S*)-α-Methylbenzylamine may

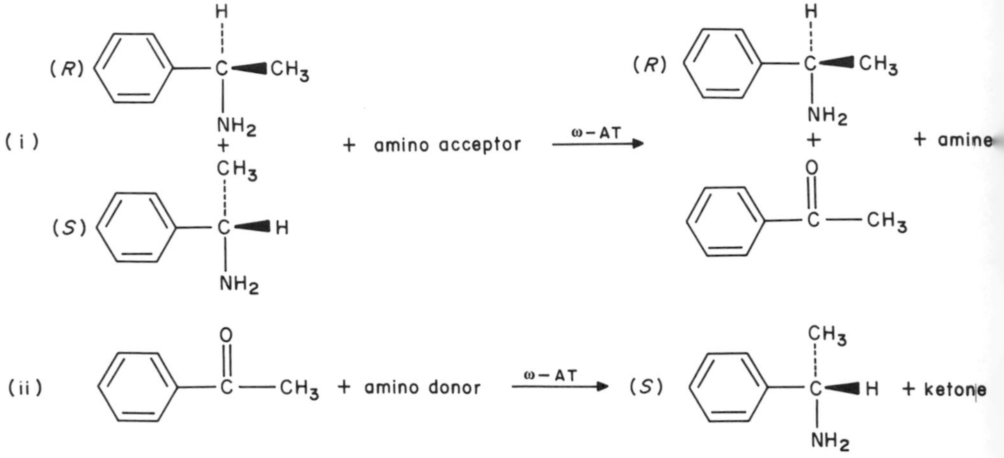

Scheme 9. Production of both enantiomers of α-methylbenzylamine in high *ee* using a single enzyme

be synthesized from the prochiral ketone using a suitable amino donor [Scheme 9(ii)]. In the cases where the enzyme is not specific enough for a particular enantiomer, a kinetic resolution can be used to obtain a high enantiomeric excess (*ee*). Figure 2 shows the relationship between amino donor conversion and *ee*. For this particular amino donor the enzyme exhibits a reaction rate ratio of 20:1 for the (*S*)-over the (*R*)-enantiomer. In the 'resolution mode' the *ee* of the (*R*)-enantiomer is dictated by the overall conversion, whereas in the 'synthesis mode,' the *ee* of the resulting (*S*)-enantiomer is fixed at about 92%. In such cases, where the desired chiral amine is the (*S*)-enantiomer, a complementary enzyme which favours the (*R*)-enantiomer would be required. Other microbial strains which contain such complementary aminotransferases have been isolated at Celgene.

Table 1 provides examples of chiral amines that can be produced either by resolution or synthesis by (*S*)-aminotransferases such as the *Bacillus* enzyme and by (*R*)-aminotransferases. Obviously, the preferred commercial route to a particular chiral amine will be dependent on the enantioselectivity of the enzymes available and the cost of the raw materials.

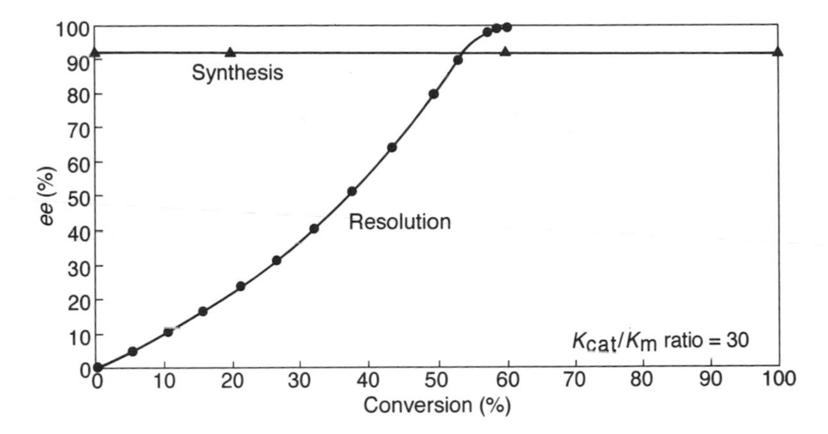

Figure 2. Kinetic resolution and chiral synthesis using *Bacillus megaterium* enzyme

The resolution of a racemic mixture involves the separation of the two enantiomers with, if possible, the unwanted enantiomer being racemized and recycled. Sometimes this racemization procedure can be difficult and expensive. In kinetic resolutions with aminotransferases, these problems with regard to recycling are minimized, as the prochiral ketone product can be re-used directly in the production of the racemic amine. For example, in the production of (*R*)-α-methylbenzylamine, the co-product acetophenone can be recycled to the racemic amine raw material (Scheme 10).

NH_3–H_2 reductive amination; ω-AT

(100) → NH_2 (100) Racemic amine → O (50) + NH_2 (50) (*R*)-MBA

Recycle

Scheme 10. Kinetic resolution of α-methylbenzylamine using an ω-aminotransferase

9.5 THE COMMERCIAL PRODUCTION OF CHIRAL AMINES

Celgene has developed processes for the commercial production of a wide variety of chiral amines. These processes are either whole cell or enzyme based, depending on the product and on the availability and cost of raw materials.

Table 1. Chiral amine resolution and synthesis using Celgene biocatalysts

Amino transferase	Product amine	By resolution: Conversion (%)	By resolution: % *ee*	By synthesis: % *ee*
(*S*)-AT	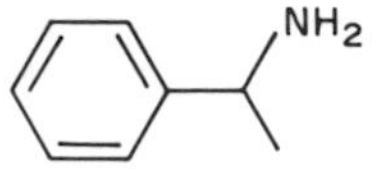 NH_2	50	>99 (*R*)	100 (*S*)
	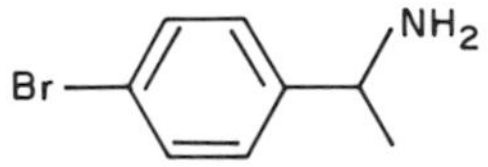 Br, NH_2	49	97.6 (*R*)	—
	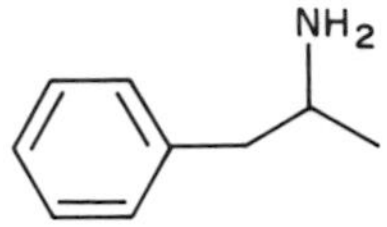 NH_2	49	98.6 (*R*)	96.4 (*S*)
	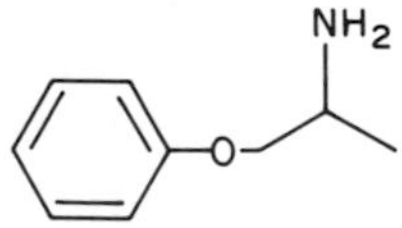 O, NH_2	52	>99 (*R*)	—
	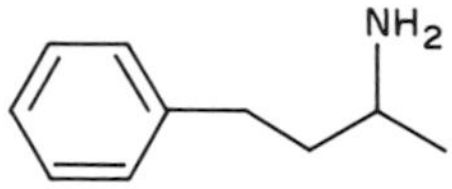 NH_2	58	>99 (*R*)	93 (*S*)
	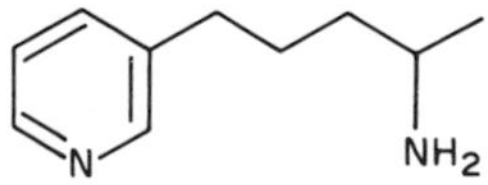 N, NH_2	49	>99 (*R*)	100 (*S*)
(*R*)-AT	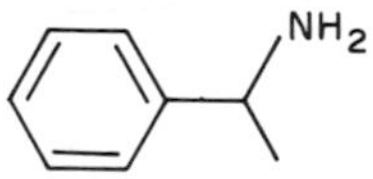 NH_2	50	>99 (*S*)	—
	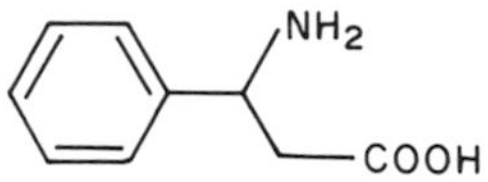 NH_2, COOH	50	>99 (*S*)	—
	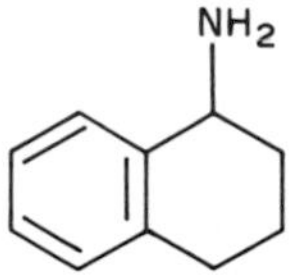 NH_2	57	>99 (*S*)	—

Currently a number of (*R*)- and (*S*)-amines are being produced commercially. Scheme 11 illustrates some of the structurally diverse chiral amines accessible by the Celgene technology: alkylamines, arylalkylamines, glycinols, aminopyrrolidines and even unnatural β-amino acids. The technology can be used for kilogram and tonne quantities. These chiral amines are valuable as

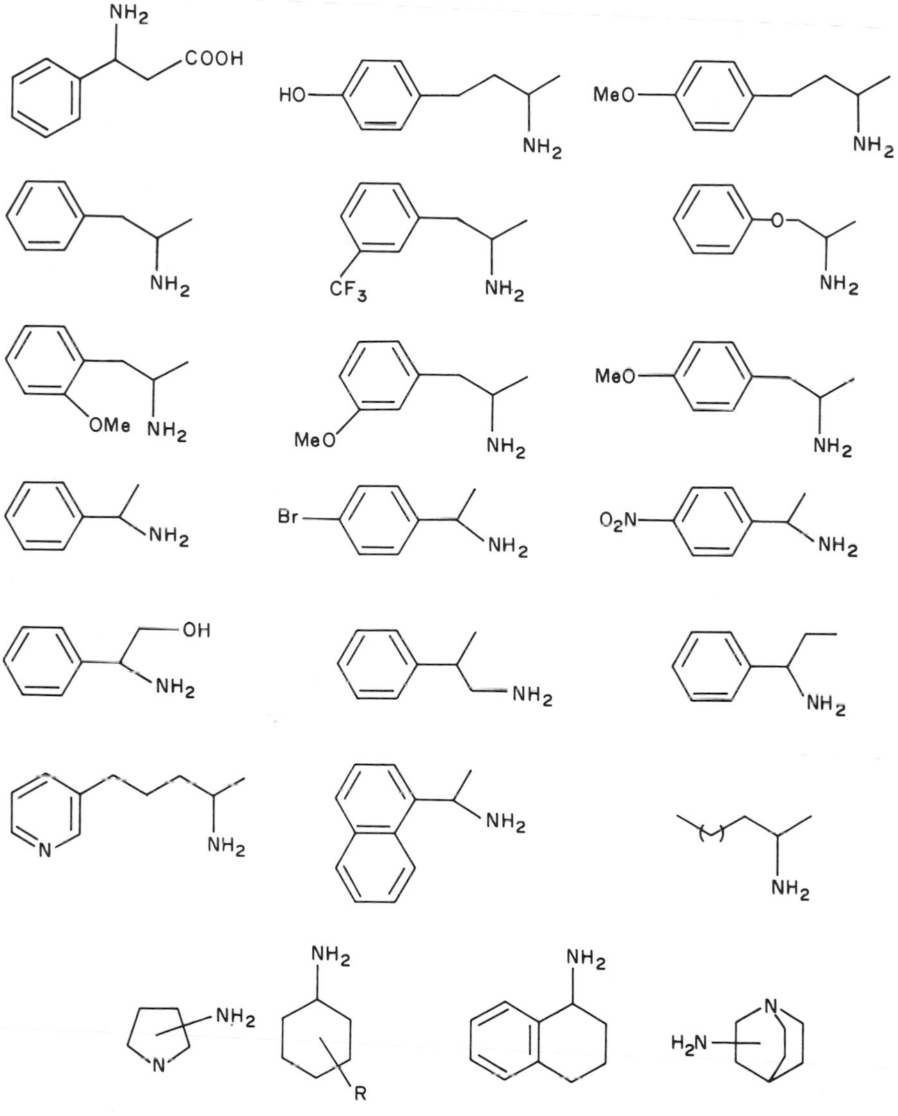

Scheme 11. Celgene chiral amine products

pharmaceutical intermediates and products, and they may also find large-scale applications as resolving agents and chiral auxiliaries.

9.6 ACKNOWLEDGEMENTS

The author thanks all the work and effort of the chiral amine group, in particular Mr A Zeitlin and Dr G Matcham. In addition, he thanks Ms P. Petras for her patient help in preparing this chapter.

9.7 REFERENCES

1. Jacewicz, V. W., *Br. Pat.*, 1 413 078, 1973.
2. Salerno, C., Giartosio, A., and Fasella, P., in *Vitamin B_6 Pyridoxal Phosphate, Part B* (ed. D. Dolphin, R. Poulson and O. Avramovic), Wiley, New York, 1986, pp. 117–168.
3. Cooper, A. J. L., and Meister, A., in *Transaminases* (ed. D. E. Metzler, and P. Christen), Wiley, New York, 1987, pp. 534–562.
4. Cooper, A. J. L., and Meister, A., *Biochemie*, **71**, 387 (1989).
5. Stirling, D. I., Zeitlin, A. L., and Matcham, G. W., *US Pat.*, 4 950 606, 1990.
6. Stirling, D. I., and Matcham, G. W., in *Proceedings of Chiral '90, Manchester, England*, 1990, pp. 111–113.
7. Yonaha, K., and Toyama, S., *Agric. Biol. Chem.*, **42**, 2363 (1978).
8. Ziehr, H., Hummel, W. I., Reichenbach, H., and Kula, M.-R., in *3rd European Congress on Biotechnology*, Verlag Chemie, Weinheim, 1984, Vol. 1, pp. 345–350.
9. Primrose, S. B., *Eur. Pat. Appl.*, 84 100 421.8, 1984.
10. Bulot, E., and Cooney, C. L., *Biotechnol. Lett.*, **7**, 93 (1985).
11. Fusee, M. C., *Ger. Pat. Appl.*, DE 3 427 495 A1, 1985.
12. Lawlis, B., Rastetter, W., and Snedecor, R., *Eur. Pat. Appl.*, 84 305 532.8, 1985.
13. Rozzell, D. J., *Eur. Pat. Appl.*, 84 110 407.8, 1985.
14. Wood, L. L., and Carlton, G. J., *Eur. Pat. Appl.*, 84 304 966.9, 1985.
15. Carlton, G. J., Wood, L. L., Updike, M. H., Lanty, L., and Hamman, J. P., Bio/Technology, **5**, 317 (1986).
16. Rozzell, J. D., *Methods Enzymol.*, **136**, 479 (1987).
17. Crump, S. P., Meier, J. S., and Rozzell, J. D., in *Biocatalysis* (ed. D. Abramowicz), 1990, pp. 115–133.
18. J. D. Rozzell, *US Pat.*, 4 876 766, 1989.
19. Burnett, G., Walsh, C., Yonaha, K., Toyama, S., and Soda, K., *J. Chem. Soc., Chem. Commun.*, 826 (1979).
20. Bouclier, M., Jung, M. J., and Lippert, B., *Eur. J. Biochem.*, **98**, 363 (1979).
21. Tanizawa, K., Yoshimara, T., Asada, Y., Sawoda, S., Misono, H., and Soda, K., *Biochemistry*, **21**, 1104 (1982).
22. Stirling, D. I., unpublished results.

10 Applications of Biocatalysts in the Synthesis of Phospholipids

A. E. WALTS, D. G. SCHENA, E. M. FOX,
J. T. DAVIS AND M. R. MISCHKE
Genzyme Corporation, Cambridge, MA, USA

10.1 INTRODUCTION

Phospholipids are of importance in a number of medical, diagnostic and research applications. They are the principal components of cellular membranes in virtually all living organisms and, as a result of their inherent biocompatibility, they are currently the topic of many areas of biomedical research. One application of phospholipids is as the basic component of liposomal and microemulsion drug-delivery systems.[1] A number of companies have been founded specifically to exploit the commercial promise of phospholipid-based drug-delivery systems and many of the major pharmaceutical companies support internal research and development in this area. A reliable, economical supply of medical-quality phospholipids in bulk is an important element in the successful commercialization of these phospholipid-based delivery systems. We describe here background information on phospholipids and liposomes, and discuss Genzyme's work

Chirality in Industry. Edited by A. N. Collins, G. N. Sheldrake and J. Crosby

involving applications of biocatalysts in the manufacture of high-purity, medical quality phospholipids.

10.2 PHOSPHOLIPIDS: STRUCTURE AND APPLICATIONS

The basic structure common to all phospholipids is shown in Figure 1. The molecule contains one chiral centre at C-2, and exists in nature as the L-enantiomer. Two distinct regions of polarity are present in the molecule: a hydrophobic fatty acyl tail region and a hydrophilic head region. The amphiphilicity imparted by these regions provides the physical properties of these molecules which render them useful in natural biological membranes and in synthetic liposome systems. When exposed to an aqueous environment under appropriate conditions, phospholipids assemble into bilayers wherein the hydrophilic head groups are exposed to the aqueous region and the hydrophobic tails are in contact with each other. Spheres formed from these bilayers are known as liposomes, and may contain a number of bilayers of phospholipids with an aqueous core. The ability to entrap water-soluble compounds within the aqueous core or lipid-soluble compounds within the fatty acyl tails accounts in many cases for the drug-delivery potential of liposomes. In certain instances phospholipids may form specific complexes with drugs, which may also provide useful species for drug delivery.[2]

Phospholipid-based systems have been utilized for a variety of applications besides liposomal drug delivery (Figure 2). For example, liposomes have proved useful in gene transfer protocols,[3] and diagnostic immunoassays[4] and as therapeutic agents based on their intrinsic physical properties. Phospholipids

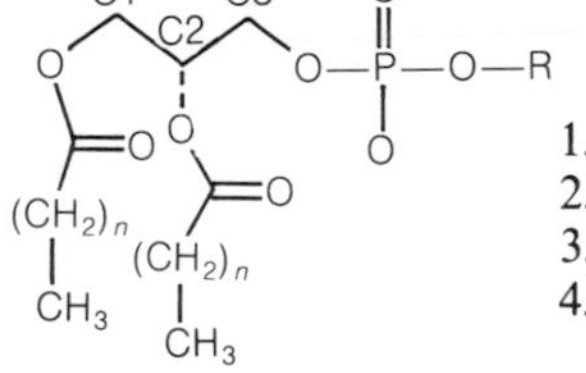

C1 C2 C3 O O—P—O—R O O =O =O $(CH_2)_n$ CH_3 $(CH_2)_n$ CH_3

1. Glycerol backbone
2. Hydrophobic acyl chains
3. Hydrophilic phosphate ester
4. Optically active—one chiral centre

Figure 1. Phospholipid structure

1. Emulsion formulations
2. Drug delivery—liposome encapsulation
3. Bioactive phospholipids
4. Diagnostic products

Figure 2. Phospholipid-based therapeutic products

are, for example, important components in lung surfactant preparations used to treat premature infants.[5]

The structural composition of phospholipids, i.e. fatty acyl chain length, homogeneity and structure and head group homogeneity and structure, is a determining factor in the physical and biological properties of derived liposomes. Heterogeneous phospholipids generally yield liposomes with poorly defined physical properties whereas homogeneous phospholipids yield liposomes with well defined and reproducible properties. In addition, by tailoring the structure of the fatty acyl or head group, specific release properties (e.g. temperature or pH release) and biological targeting may be achieved.[6] Structurally defined, pure phospholipids are therefore desirable for most biomedical applications.

10.3 METHODS FOR PHOSPHOLIPID SYNTHESIS

Examples of medically important phospholipids are shown in Figure 3, and include diacylphosphatidylcholine and diacylphosphatidylglycerol, which are commonly used to form liposomal and emulsion delivery systems. Diacylphosphatidylethanolamine is used routinely to prepare more complex molecules wherein the head group has been further modified via the terminal amine.[7]

Other molecules, such as diacylphosphatidylserine and diacylphosphatidylinositol, may be of interest with regard to their intrinsic biological properties. A number of strategies have been developed in efforts to obtain these compounds.[8]

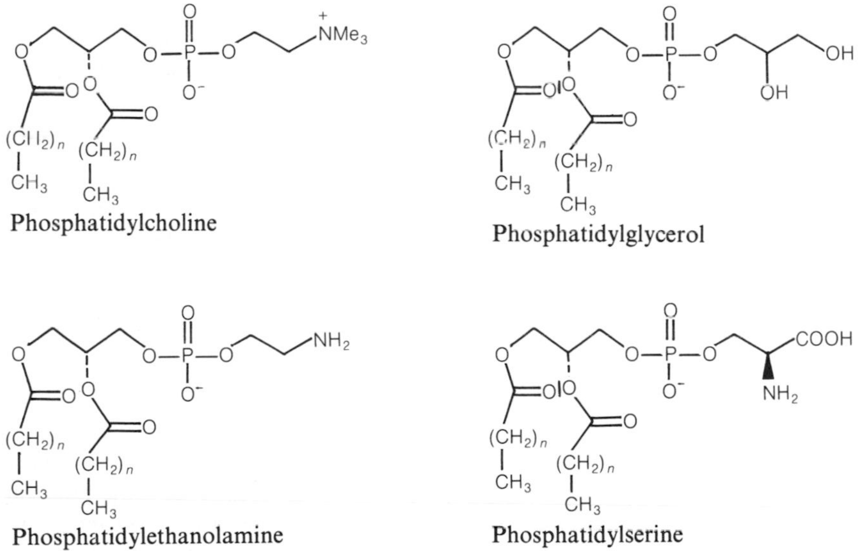

Phosphatidylcholine

Phosphatidylglycerol

Phosphatidylethanolamine

Phosphatidylserine

Figure 3. Medically important phospholipids

10.3.1 NATURAL PRODUCTS

Certain of the phospholipids of interest can be isolated directly from tissue sources. For example, diacylphosphatidylcholine is available in bulk quantities via extraction from soya or egg yolks, diacylphosphatidylserine from mammalian neural tissue and cardiolipin from mammalian cardiac tissue. Phospholipids extracted from tissue materials are used broadly in the food and cosmetic industry and have found some application in drug-delivery systems. Materials from natural sources are, however, heterogeneous with regard to the acyl chain structure and are often unstable owing to multiple unsaturation in the fatty acyl chains. Since these materials are naturally derived, the potential exists for contamination by trace amounts of proteins, nucleic acids and other lipids.

10.3.2 SEMI-SYNTHESIS

Using tissue-derived phosphatidylcholine as a starting material, various routes have been developed for partially degrading and then rebuilding the molecule. The acyl chains can be removed by basic chemical hydrolysis or by enzymatic cleavage with lipases.[8] The resulting glycerophosphocholine can then be chemically reacylated to provide a defined diacylphospholipid, e.g. dipalmitoylphosphatidylcholine.[9] Head groups other than choline are less easily accessed via this route. Phosphatidic acids, for example, require removal of the head group choline, while head groups such as glycerol require further manipulation. Perhaps most importantly, the semi-synthetic route relies on tissue-derived raw material which must be carefully purified to remove trace contaminants such as proteins, nucleic acids, lipopolysaccharides and non-phospho-type lipids.

10.3.3 TOTAL SYNTHESIS

Yet another method for preparing phospholipids is by total synthesis, wherein no naturally derived raw materials are required. A totally synthetic route poses certain challenges in devising technology suitable for large-scale economical production of phospholipids (Figure 4). These include the necessity for obtaining high enantiomeric purity, manipulation and purification of intermediates and products and a method to differentiate the hydroxyl groups at C-1 and C-2 for the preparation of complex lipids.

A common approach is to start with L-isopropylideneglycerol.[8,10] This route requires multiple steps to provide the desired phospholipids, and requires the availability of large quantities of the chiral starting material L-isopropylideneglycerol. An alternative route starting with (*S*)-glycidol has been reported.[11] This route requires blocking and deblocking steps, careful attention to the optical purity of the final products and a cheap and reliable source of the starting

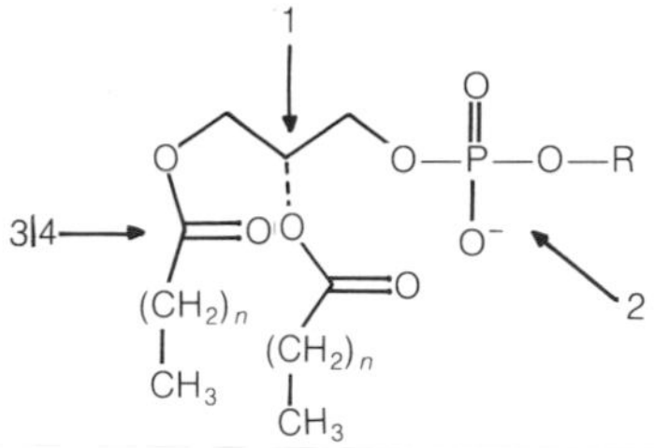

1. Optical activity
   —High enantiomeric excess desired
2. Phosphate moiety
   —Manipulation and purification
3. Purification of products
   —Amphiphilic properties
4. Differentiation of hydroxyl groups

Figure 4. Phospholipids: synthetic challenges

material. Yet another synthetic route to phospholipids utilizes *sn*-glycerol-3-phosphate (G-3-P) as a starting material.[12] This molecule contains the basic backbone of all phospholipids, and therefore provides a central starting material from which to elaborate virtually any phospholipid. This route requires the availability of bulk quantities of *sn*-glycerol-3-phosphate, which historically has been prohibitively expensive and available only in relatively small quantities. Ongoing work in Genzyme's research laboratories during the past several years has focused on developing methods for the large-scale production of G-3-P and for its conversion into medical-quality phospholipids.

10.4 *sn*-GLYCEROL-3-PHOSPHATE (G-3-P)

Crans and Whitesides[13] described a biocatalytic synthesis of G-3-P which is based on phosphorylation of glycerol by ATP, catalyzed by glycerol kinase. In order to be economically viable, a system to regenerate ATP from ADP during the synthesis is required. A number of regeneration systems have been developed[14] and Crans and Whitesides[13] have described the use of acetyl phosphate as phosphate donor, with acetate kinase as the coupling enzyme. In the light of the inherent simplicity and potential low cost of this procedure (i.e. glycerol as raw material), this route was chosen as part of the Genzyme strategy to prepare G-3-P.

Evaluation of this biocatalytic procedure for the commercial production of G-3-P suggested a number of challenges in the design of an economically and technically viable process. The following points were identified as critical in developing an economical procedure for G-3-P production:

cofactor regeneration procedure;
enzyme cost;
enzyme stability;
product isolation and purification.

Each of these factors can in principle be addressed by the use of an appropriate bioreactor for large-scale production. It was hoped that problems of enzyme instability could be addressed by a suitable immobilization system that might provide some level of enzyme stabilization. Appropriate enzyme stabilization would also allow for continuous production or repetitive discontinuous use, and thereby reduce the enzyme cost. Finally, an appropriate system would provide for cofactor regeneration and simplify the isolation of the product G-3-P.

Effective regeneration of ATP is critical to the success of a biocatalytic route to prepare G-3-P in ton quantities. A number of procedures to regenerate ATP have been developed and two methods for recycling ATP are illustrated in Figure 5. Each regeneration system poses specific challenges which must be addressed for large-scale production.[14] For example, a system based on acetyl phosphate–acetate kinase has the advantage of exceptionally low cost, but suffers from hydrolytic instability of acetyl phosphate ($t_{1/2} \approx 21$ hr at pH 7 and 25 °C). The phosphoenol pyruvate (PEP)–pyruvate kinase-based system displays exceptional stability ($t_{1/2}$ for PEP $\approx 10^3$ h at pH 7 and 25 °C), but is relatively expensive. Other considerations in selecting a regeneration system include free energy of hydrolysis of the phosphate donor, potential product inhibition by the phosphate donor product (i.e. acetate or pyruvate) and cost of the coupling enzyme.

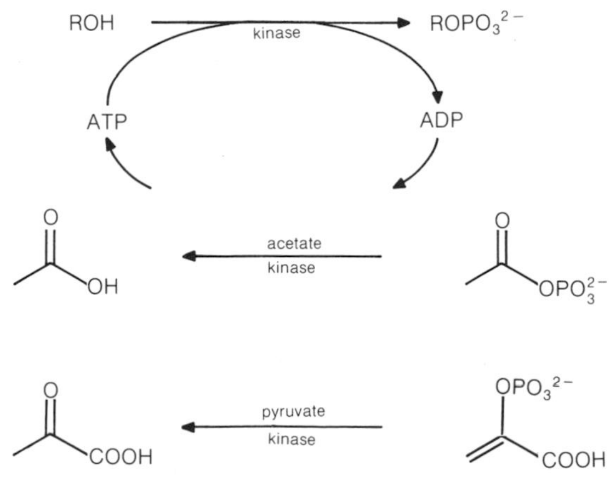

Figure 5. Nucleotide phosphate recycling technology

Genzyme have evaluated both the acetyl phosphate–acetate kinase and PEP–pyruvate kinase regeneration systems for large-scale application in the production of G-3-P. The PEP–pyruvate kinase system proved to be superior to the acetyl phosphate–acetate kinase system, and was chosen for further development. This choice was dictated primarily by the greater stability of PEP and compatibility of this system with the bioreactor used in production.

The development program culminated in the selection of a proprietary bioreactor which utilizes a solid matrix support. This system has been scaled up to provide routinely multi-kilogram batches of G-3-P, and additional scale-up work is ongoing. The biocatalytic synthesis of G-3-P and the operating parameters of this bioreactor system are summarized in Figure 6. This system features both the glycerol kinase and the pyruvate kinase immobilized together in one reactor. The process for G-3-P synthesis is illustrated schematically in Figure 7.

Of considerable note are the high concentrations of substrate and reagents which can be utilized. Glycerol is typically present at a concentration of 1 M and the phosphate donor is present at a similar concentration. ATP is used catalytically at *ca* 1–10 mol%, with turnover numbers of 10–100 achieved routinely. These concentration parameters provide for significant product output per unit reactor volume. At present, reactor space–time yields are defined by

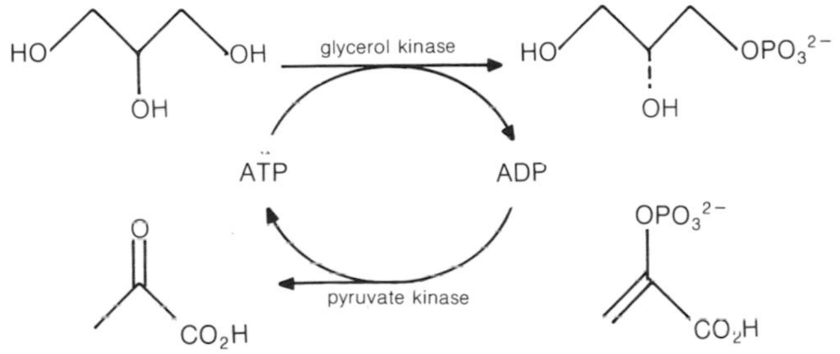

Bioreactor operating parameters

1. Temperature: ambient
2. pH: 7.5
3. Flow rate:
   Research scale: 50 ml min^{-1}
   Pilot scale: 300–500 ml min^{-1}
4. Solution components:
   Glycerol (1 M)
   PEP (1 M)
   ATP
   $MgCl_2$
   DTT—(dithiothreitol)
   NaN_3—(bacteriostat)

Figure 6. *sn*-Glycerol-3-phosphate

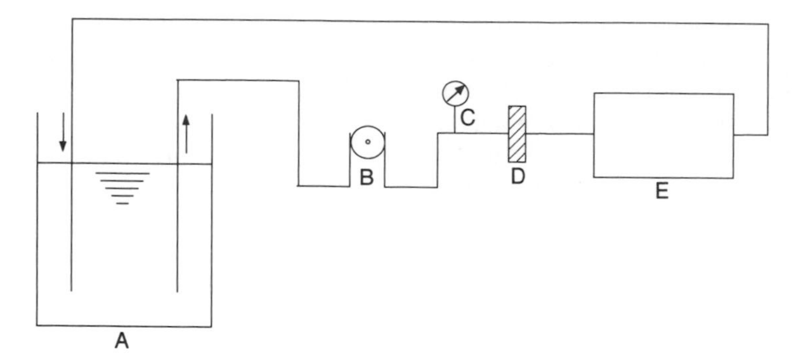

Figure 7. Biocatalytic synthesis of *sn*-glycerol-3-phosphate: schematic process. A = stirred reservoir of reaction solution; B = pump; C = pressure gauge; D = filter; E = immobilized enzyme reactor

the flux through the reactor which can be achieved, since the ATP is present in catalytic quantities. In the event, space–time yields of 0.01–0.02 mol l^{-1} h^{-1} are routinely achieved.

The reaction profile for an operational bioreactor in discontinuous use is shown in Figure 8. The data shown are for mole-scale reactions. Data for reaction 1 were obtained using a reactor which was 3.5 months old; those for reaction 2 are for the same reactor after 4.5 months. The reaction rates were found to be linear and constant for periods exceeding 4.5 months.

A significant stabilization of enzyme activity was noted when the enzymes were immobilized in the bioreactor. Reactor lifetimes of greater than 6 months

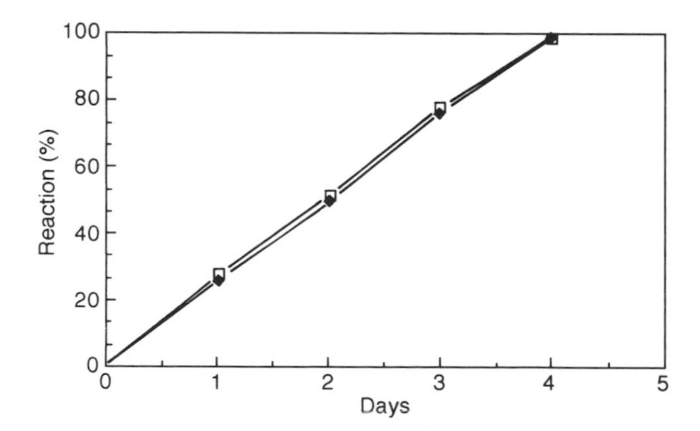

Figure 8. *sn*-Glycerol-3-phosphate reaction profile: □, run 1; ◆, run 2. Features: (1) linear reaction rate; (2) reproducible rate; (3) reaction rate directly dependent on cofactor concentration/flux. Parameters: mole scale; run 1, 3 months reactor age; run 2, 4.5 months reactor age

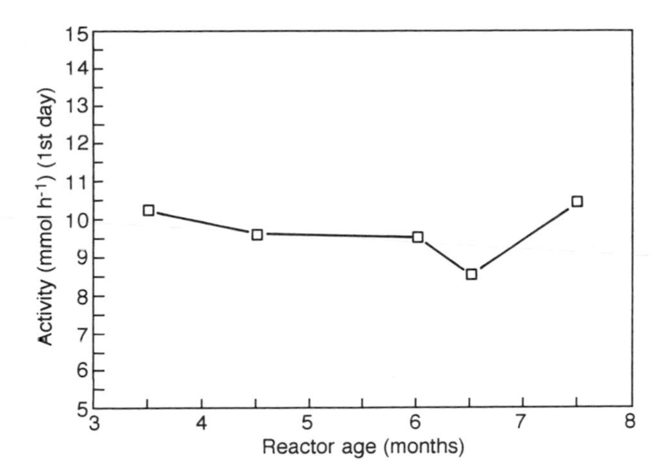

Figure 9. *sn*-Glycerol-3-phosphate bioreactor reproducibility. Features: initial reaction rate remains constant during multiple re-use; mole-scale reactions; > 7 months reactor age

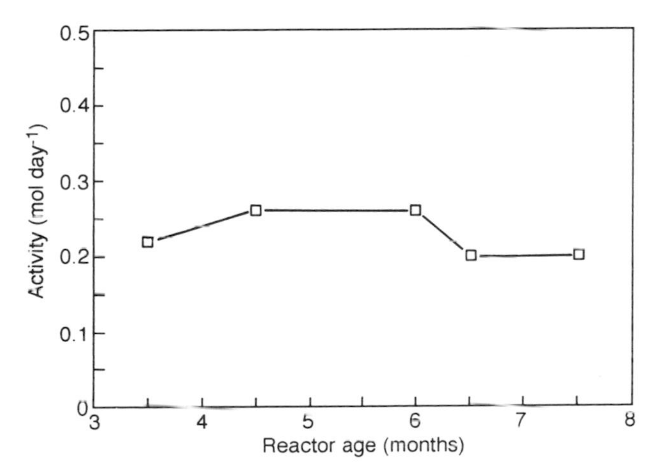

Figure 10. *sn*-Glycerol-3-phosphate bioreactor reproducibility. Average rate during 4-day reaction over > 7 months reactor age. Features: mole-scale reactions; multiple re-use

were typical, with the reactor in continuous use during this period. Figures 9 and 10 show data for reaction rates measured at various bioreactor ages. Figure 9 illustrates the reaction rate obtained during the first day of operation at various stages of reactor use, while Figure 10 shows the average daily reaction rate during a typical 4-day reaction. Figures 9 and 10 indicate that the reactor output remains constant over at least a 7.5-month period. This system did not require special stabilization techniques beyond standard protection from microbial growth with a bacteriostat.

The use of a bioreactor containing the immobilized enzymes is key to the large-scale production of G-3-P. It is believed that this bioreactor system is scalable to ton quantities of G-3-P, and provides an economical route to this important chiral molecule.

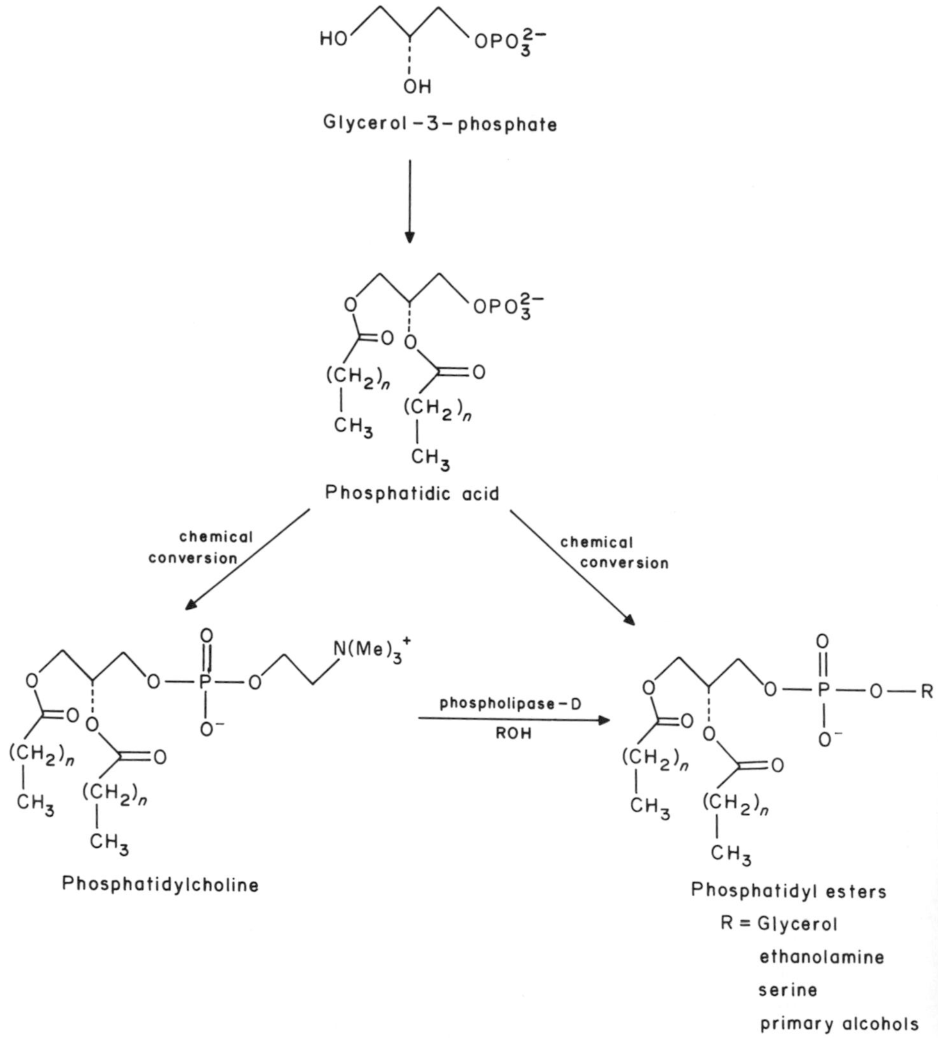

HO OPO$_3^{2-}$
OH
Glycerol-3-phosphate

O OPO$_3^{2-}$
O =O O =O
$(CH_2)_n$ $(CH_2)_n$
CH_3 CH_3
Phosphatidic acid

chemical conversion
chemical conversion

O O—P—O N(Me)$_3^+$
O O=O O$^-$
$(CH_2)_n$ $(CH_2)_n$
CH_3 CH_3
Phosphatidylcholine

phospholipase-D
ROH

O O—P—O—R
O O=O O$^-$
$(CH_2)_n$ $(CH_2)_n$
CH_3 CH_3
Phosphatidyl esters
R = Glycerol
ethanolamine
serine
primary alcohols

Scheme 1. Phospholipids—synthetic pathway

10.5 PHOSPHOLIPID SYNTHESIS

Following completion of the biosynthetic reaction, the G-3-P is isolated and purified as a proprietary salt developed at Genzyme.[15] The G-3-P is then used as a starting material to prepare a range of phospholipids (Scheme 1). In the first stage of the synthesis G-3-P is converted by a proprietary procedure into diacylphosphatidic acid (PA).[15] The availability of phosphatidic acids direclty from G-3-P provides a convenient, inexpensive method for commercial production of these important intermediates. The PA can be purified for use in therapeutic applications, or can be esterified to provide a range of lipid products (Figure 11). For instance, esterification with choline provides diacylphosphatidylcholine, while more complex syntheses render diacylphosphatidylglycerol or phosphatidylethanolamine available.

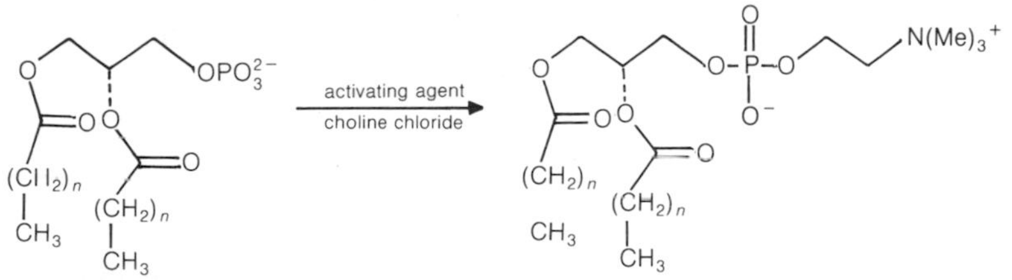

activating agent
choline chloride

1. High purity: ≥98%
2. High yields
3. Proprietary purification

Figure 11. Phosphate esters—general synthesis from phosphatidic acid

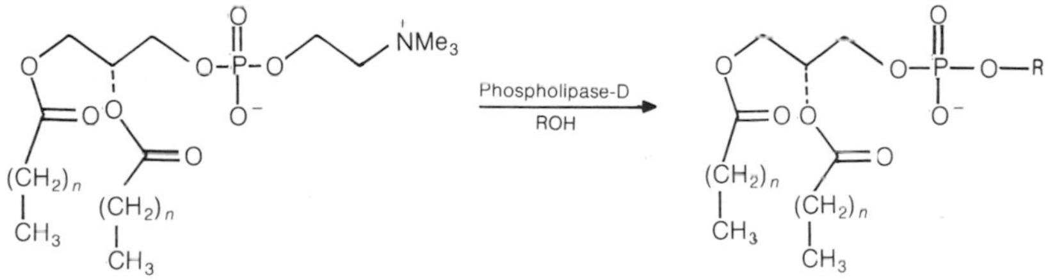

Phospholipase-D
ROH

Suitable alcohol acceptors

1. Glycerol
2. Ethanolamine
3. Serine
4. Carbohydrates
5. Novel therapeutic alcohols

Figure 12. Diacylphosphatidylcholine as a building block for more complex phospholipids

Diacylphosphatidylcholine can also serve as a building block for more complex phospholipids. This technology relies on the use of a phospholipase-D enzyme (PL-D) which catalyzes the transesterification of choline with a variety of primary alcohol species[16] (Figure 12). Glycerol, ethanolamine and serine are well known examples of alcohols which can be transesterified with phosphatidylcholine.

In recent work, a proprietary microbial source of phospholipase-D has been developed and is currently being used to prepare complex phospholipids. The use of microbial PL-D avoids the many problems associated with the use of a tissue-derived (e.g. cabbage) PL-D. In particular, the requirements for organic cosolvents are less stringent, higher reagent concentrations can be used and the buffer composition can be simplified. This technology facilitates, for example, the production of phosphatidyglycerol and phosphatidylethanolamine from phosphatidylcholine. A proprietary batch reaction system for production of phosphatidylglycerol has been developed, and currently operates at a multi-kilogram level. This technology allows the routine synthesis of products with chemical purities $>98\%$.

10.6 CONCLUSION

We have demonstrated that biocatalysis provides an effective and economical method for the manufacture of *sn*-glycerol-3-phosphate. The use of biocatalysis to prepare G-3-P is a central part of the program at Genzyme to prepare totally synthetic, medical-quality phospholipids, and it should play an important role in making these compounds available at low cost and in bulk quantities with high purity.

10.7 REFERENCES

1. Gregoriadis, G. (ed.), *Liposomes as Drug Carriers*, Wiley, New York, 1988.
2. Hume, D., and Nayar, R., *Lymphokine Res.* **8**, 415 (1989).
3. Mannino, R. J., and Gould-Fogerites, S., *Biotechniques* **6**, 682 (1988).
4. You, Y.-H., Lagocki, P. A., and Hu, R. C., *Eur. Pat. Appl.*, 0 301 333, 1989.
5. Tanaka, Y., *et al. J. Lipid Res*, **27**, 475 (1986).
6. Nye, J., and Synder, S., *PCT*, No. WO87/07150, 1987.
7. Heath, T. D., *Methods Enzymol.*, **149**, 111 (1987).
8. Eibl, H., *Chem. Phys. Lipids*, **26**, 405 (1980).
9. Patel, K. M., Morrisett, J. D., and Sparrow, J. T., *J. Lipid Res.*, **20**, 674 (1979).
10. Baer, E., and Kates, M., *J. Am. Chem. Soc.*, **72**, 942 (1950).
11. Burgos, C. E., Ayer, D. E., and Johnson, R. A., *J. Org. Chem.*, **52**, 4973 (1987).
12. Gupta, C. M., Radhakrishnan, R., and Khorana, H. G., *Proc. Natl. Acad. Sci. USA*, **74**, 4315 (1977).
13. Crans, D. C., and Whitesides, G. M., *J. Am. Chem. Soc.*, **107**, 7019 (1985).

14. Crans, D. C., Kazlauskas, R. J., Hirschbein, B. L. Wong, C.-H., Abril, O., and Whitesides, G. M., *Methods Enzymol.*, **136**, 263 (1987).
15. Schena, D. G., and Davis, J. T., to Genzyme Corp., *US Pat. Appl.*, USSN 577 351, 1989.
16. Juneja, L. P., Hibi, N., Yamane, T., and Shimizu, S., *Appl. Microbiol. Biotechnol.*, **27**, 146 (1987).

11 The Industrial Production of Aspartame

K. OYAMA
Tosoh Corporation, Kanagawa, Japan

11.1 INTRODUCTION

Aspartame (α-L-aspartyl-L-phenylalanine methyl ester, or APM) (**1**) is a synthetic sweetener which is about 200 times sweeter than sucrose. The compound contains the same two amino acids as the *C*-terminal dipeptide of the digestive hormone gastrin, and its sweetness was discovered by chance during the synthesis of this hormone by Mazur *et al.*[1] of G. D. Searle. The corresponding methyl ester had been synthesized earlier by Davey *et al.*[2] of ICI, but its sweetness was not recognized.

(**1**)
α-Aspartame (APM)

Chirality in Industry. Edited by A. N. Collins, G. N. Sheldrake and J. Crosby

The two essential amino acids which constitute aspartame (L-aspartic acid and L-phenylalanine) are commonly found in proteinaceous foods such as milk and meat. Digestion and further metabolism of the compound occur, therefore, in the same manner as for proteins containing these amino acids, and it was expected that it would be a safe food additive. Its physiological safety was confirmed by long-term toxicological studies known to be the most comprehensive in the history of approval procedures for food additives.[3] The use of aspartame as a sweetener was first approved in France in 1979, followed by most other countries within a few years. Owing to its pleasant sweetness without a bitter aftertaste, it was well received as a low-calorie sweetener for many foodstuffs including soft drinks, table-top sweeteners, dairy products, instant mixes, dressings, jams, confectionery and toppings and for pharmaceuticals.

Aspartame is used mostly in western countries, the largest market being the USA, followed by the European Community, Canada, etc. The original owners of the intellectual property rights, G. D. Searle, were acquired by Monsanto, and since then the aspartame division has been operated as Nutrasweet. The use patent as a sweetener has expired in most countries, although in the USA Nutrasweet have obtained an extension of the patent until the end of 1992.

11.2 SYNTHETIC METHODS

11.2.1 CHEMICAL METHODS

Many chemical syntheses have been reported, most of which involve conventional peptide synthesis. Research directed towards their development into industrial processes has been concerned largely with the problems of increasing the yield of a desired product by suppressing the formation of by-products, decreasing capital investments by simplifying the process and improving the purity of the APM. All of the chemical methods with industrial potential involve dehydration of the two carboxylate groups of aspartic acid to form acid anhydrides, which are then coupled with phenylalanine or its methyl ester, as shown in Scheme 1.

All of the above methods inevitably produce 20–40% of β-aspartame (**2**) together with the desired isomer α-aspartame (**1**). The β-isomer has a bitter taste and must, therefore, be removed from the product, which entails a difficult separation from the α-isomer. Further, since the amino acid raw materials are expensive, they must be recovered from the β-isomer by-product after hydrolysis.

In addition to the production of β-isomers, each method has its own particular problems. The reaction without protection (equation 1) is the simplest method, but it produces large amounts of oligopeptides.[4] This problem may be circumvented by protection of the aspartate amine group as the *N*-formyl compound

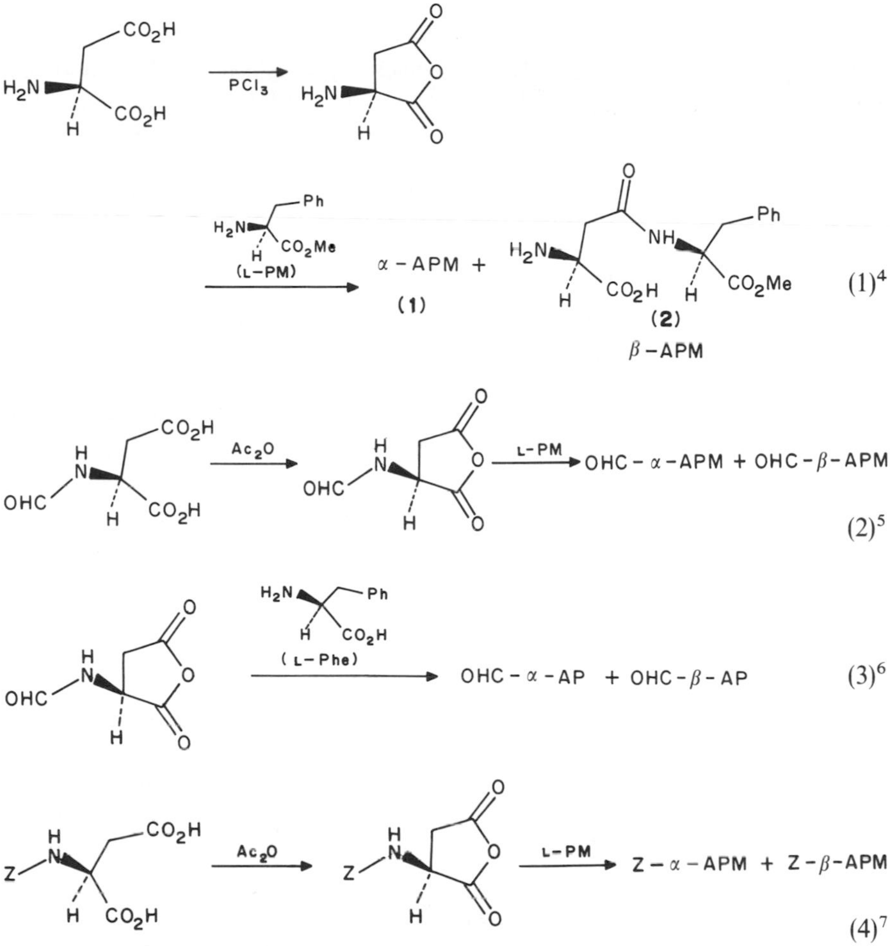

H_2N, CO_2H, CO_2H, H $\xrightarrow{PCl_3}$ H_2N, H, O, O, O

$\xrightarrow[\text{(L-PM)}]{H_2N,\ Ph,\ H,\ CO_2Me}$ α-APM (1) + H_2N, O, NH, Ph, H, CO_2H, H, CO_2Me (2) β-APM (1)[4]

OHC, H, N, CO_2H, CO_2H, H $\xrightarrow{Ac_2O}$ OHC, H, N, O, O, O, H $\xrightarrow{\text{L-PM}}$ OHC-α-APM + OHC-β-APM (2)[5]

OHC, H, N, O, O, O, H $\xrightarrow[\text{(L-Phe)}]{H_2N,\ Ph,\ H,\ CO_2H}$ OHC-α-AP + OHC-β-AP (3)[6]

Z, H, N, CO_2H, H, CO_2H $\xrightarrow{Ac_2O}$ Z, H, N, O, O, O, H $\xrightarrow{\text{L-PM}}$ Z-α-APM + Z-β-APM (4)[7]

PM = phenylalanine methyl ester
AP = L-aspartyl-L-phenylalanine
Z = benzyloxycarbonyl

Scheme 1

as in equation 2. However, in this case, the removal of the formyl group from the product is accompanied by cleavage of the ester bond.[5] The benzyloxycarbonyl protecting group, which is more easily removed, but less practical for large-scale use, has also been used in this sequence (equation 4).[7] Another strategy involves the coupling of the *N*-formylaspartic anhydride intermediate with unesterified phenylalanine (equation 3).[6] Here, esterification of the dipeptide, after deformylation, results in a mixture of the diester and monoesters.

Aside from these serious problems of selectivity, these chemical methods do not offer any stereoselectivity, and therefore only relatively expensive, pure L-amino acids may be used as raw materials. Moreover, the process conditions must be established very carefully to avoid racemization, a condition which applies equally to the recovery of amino acids from the β-isomer of the product.

11.2.2 ENZYMIC METHODS

The synthesis of peptides by reversing the proteinase-catalysed hydrolysis of a peptide bond has been known for some time. As early as 1909, Mohr and Strohschein[8] reported that under certain conditions some dipeptides can be synthesized in the presence of papain, the proteolytic enzyme isolated from papaiya. However, the method did not receive much attention as a tool for peptide synthesis until the 1970s, when the trend to utilize enzymes for the production of valuable substances inaccessible by other means became widespread. When the research described below was begun in the mid-1970s, the area was still in its infancy, and confined to rather basic studies.[9]

In 1976, the authors discovered a novel enzymic condensation reaction between phenylalanine methyl ester (PM) and aspartic acid whose amine group and side-chain carboxylic acid group were both protected, as shown in equation

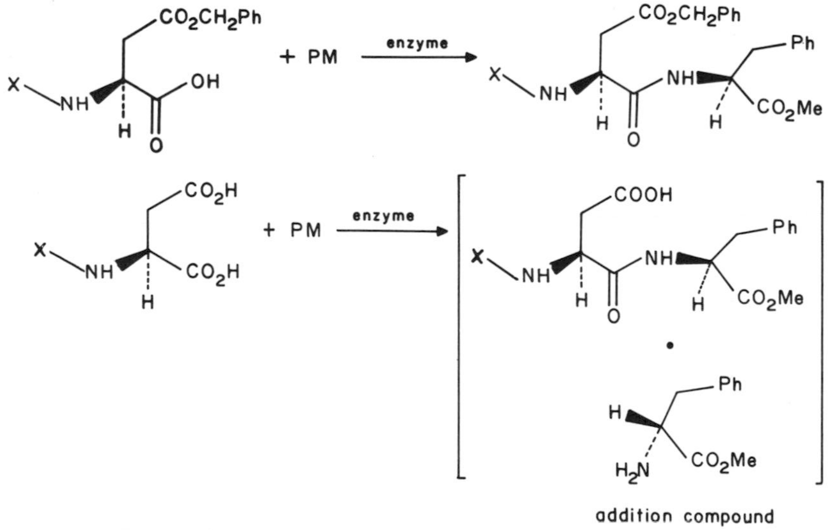

(5)

(6)

addition compound

PM = phenylalanine methyl ester
X = amine protecting group

Scheme 2

5 in Scheme 2.[10] Moreover, aspartic acid protected only at the amine group could also be made to react with PM (equation 6).[11]

In the latter reaction, it was found that the peptide bond formation took place exclusively at the α-carboxylic acid of aspartic acid even though the β-carboxylic acid was unprotected. Further, when inexpensive racemic amino acids were used, only L-substrates participated in the peptide bond formation to give the L–L-dipeptide, which was then deposited as an addition compound exclusively with the D-enantiomer of the excess of PM. The remarkable regio- and stereoselectivity of the enzyme-catalysed reaction compared with chemical methods encouraged Tosoh to develop it into an industrial process.

11.3 THE TOSOH PROCESS

11.3.1 THE ENZYME

An extensive programme of screening of commercially available proteolytic enzymes was conducted to find an enzyme suitable for the industrial process. Among the enzymes tested, thermolysin, a metalloproteinase from *Bacillus proteolyticus* Rokko and related enzymes (E.C. 3.2.24) was found to be the most suitable. The enzyme contains a zinc ion and four calcium ions,[12] the former playing a key role in the activity and the latter in the stability of the enzyme. The bacterial strain was found in the Rokko Hot Spring in central Japan.[13] Thermolysin possesses, therefore, an unusually high tolerance to heat, showing optimum activity at 50–60 °C, and even at 80 °C it retains 50% of its full activity after 1 h. Such robustness is particularly favourable from the viewpoint of industrial applications.

The family of enzymes are also known as 'neutral proteinases' because they exhibit optimum activity at around neutral pH.[12] This property, coupled with the fact that metalloproteinases do not have esterase activity, renders the enzymes very suitable for the present application, since the hydrolysis of the ester bonds in the substrate and the product dipeptide should be disfavoured. Moreover, of all the enzymes studied, the metalloproteinases showed the highest activity when operated in a synthetic mode to yield the addition compound.

11.3.2 KINETICS AND REACTION MECHANISM

The thermolysin-catalysed reaction between *Z*-Asp and PM was subjected to detailed kinetic study.[14] The reaction is first order with respect to the enzyme, and a bell-shaped pH–rate profile with the highest activity at pH 7.5 is observed. The kinetic observations with regard to substrate concentrations are shown in Figures 1–3. The reaction is first order in L-PM (Figure 1), while rate saturation is observed with respect to *Z*-L-Asp (Figure 2). A Lineweaver–Burk plot of

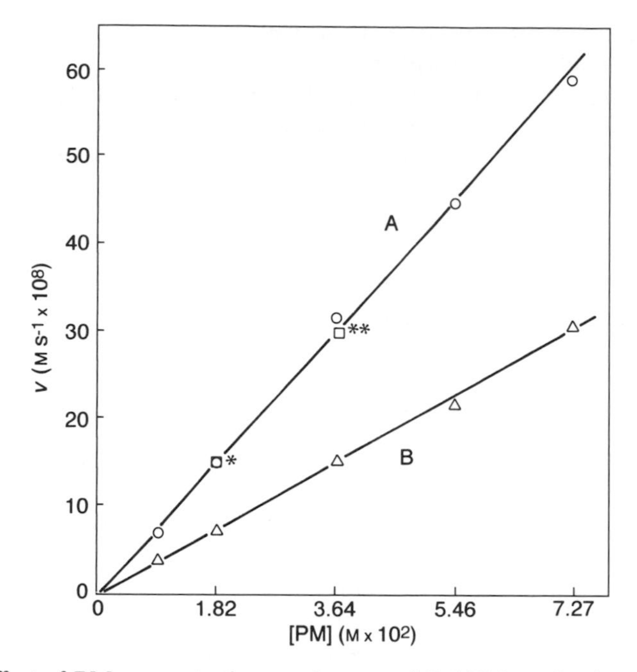

Figure 1. Effect of PM concentration on the rate of Z-APM production. Lines A and B represent plots of the rate versus [L-PM] and [D,L-PM], respectively, where □* denotes the reaction rate at [L-PM] = 1.82×10^{-2} M in the presence of [D-PM] = 9.09×10^{-3} M and □** denotes the reaction rate at [L-PM] = 3.64×10^{-2} M in the presence of [D-PM] = 1.82×10^{-2} M

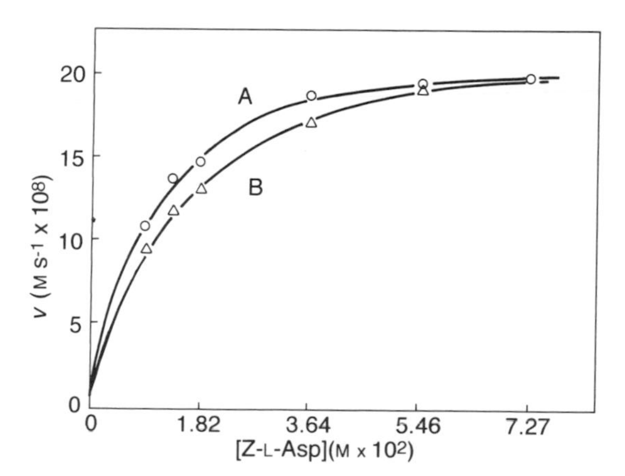

Figure 2. Effect of Z-L-Asp concentration on the rate of Z-APM production. A represents the reaction in the absence of Z-D-Asp and B that in the presence of Z-D-Asp (9.09×10^{-3} M)

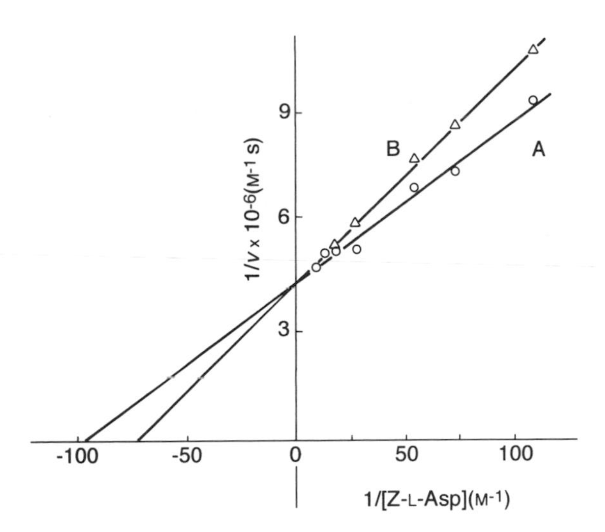

Figure 3. Lineweaver–Burk plots based on Figure 2. A, without Z-D-Asp; B, with Z-D-Asp

the latter yields a straight line (Figure 3), indicating that Z-L-Asp is bound first to the enzyme to form a Michaelis–Menten complex, followed by rate-determining attack of L-PM at the complex to give the dipeptide as shown in Scheme 3.

$$\text{Z-L-Asp} + \text{Enz} \underset{k_{-1}}{\xrightarrow{k_{+1}}} \text{Z-L-Asp}\cdot\text{Enz}]$$

$$[\text{Z-L-Asp}\cdot\text{Enz}] + \text{L-PM} \xrightarrow{k_2} \text{Z-APM} + \text{Enz}$$

Scheme 3

The rate law for the scheme can be expressed by the following equation:

$$v = \text{d}[Z\text{-APM}]/\text{d}t$$

$$= k_2[\text{Enz}]_0[\text{L-PM}]/\{1 + (k_{-1} + k_2[\text{L-PM}])/k_{+1}[Z - \text{L-Asp}]\} \quad (7)$$

If the second step is rate determining, i.e. if $k_{-1} \gg k_2[\text{L-PM}]$, equation 7 can be expressed by the equations

$$v = k_2[\text{Enz}]_0[Z\text{-L-Asp}][\text{L-PM}]/(K + [Z\text{-L-Asp}]) \quad (8)$$

$$1/v = \{K/(k_2[\text{Enz}]_0[\text{L-PM}])\} \times \{1/[Z\text{-L-Asp}]\} + 1/(k_2[\text{Enz}]_0[\text{L-PM}]) \quad (9)$$

where $K = k_{-1}/k_{+1}$. Equation 8 suggests that the reaction is first order with respect to [L-PM], while rate saturation would be seen with respect to [Z-L-Asp]. That the Lineweaver–Burk plot with respect to the latter would yield a straight line is suggested by equation 9. The kinetic observations (Figures 1–3) fit very well with equations 8 and 9, thus supporting the assumed reaction scheme. Further, it was found that D-PM is inert to the reaction, neither enhancing nor inhibiting the reaction rate. (Nonetheless, it is still useful, since it acts as a base to form the insoluble addition compound with the product dipeptide.) On the other hand, *Z*-D-Asp retards the reaction as a competitive inhibitor. This kinetic information has been invaluable not only for a clear understanding of the reaction, but also for optimization of the reaction conditions.

11.3.3 EQUILIBRIUM

The synthesis reaction is in equilibrium with the reverse (hydrolysis) reaction, with the equilibrium strongly favouring hydrolysis. There are several techniques which could be used to shift the equilibrium toward the synthesis. It is known, for example, that addition of a water-miscible solvent can greatly shift the equilibrium to favour the synthesis reaction.[15] In the case of the reaction illustrated by equation 6, it was found that addition of methanol increased the equilibrium constant threefold over that obtained in pure water.[16] A second approach is to add a water-immiscible organic solvent, resulting in a two-phase

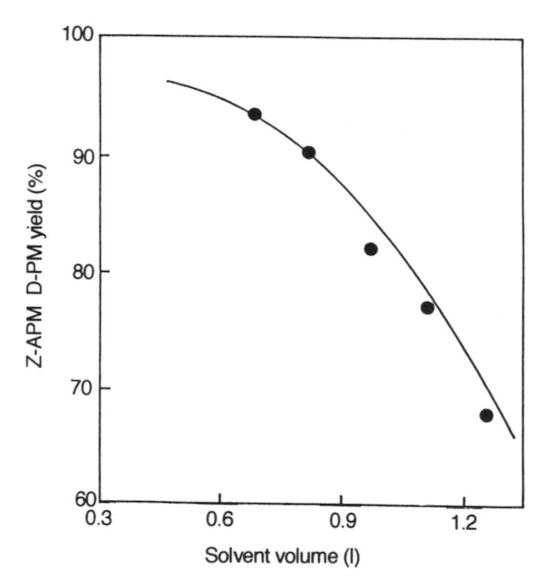

Figure 4. Dependence of the yield of the equilibrium *Z*-L-Asp + DL-PM ⇌ *Z*-APM·D-PM on the solvent volume ($K = 1.5\,l\,mol^{-1}$, $S = 0.005$ M, *Z*-L-Asp = 0.2 mol, DL-PM = 0.5 mol). The line was calculated by equation 10, and the points were obtained experimentally.

system. With an appropriate choice of solvent system, the product can be effectively extracted into an organic phase, while the reactants remain in the aqueous phase.[16–18]

A third approach, which was adopted successfully, is to remove the product by precipitation. It was found that the enzymic condensation represented by equation 6 yields the dipeptide which forms, via its free side-chain carboxylic acid, a sparingly soluble addition compound with excess PM. The synthesis reaction can, therefore, be driven by the precipitation of this addition compound, to an extent limited only by the solubility of this product in the solvent system. Thus, the maximum yield can be calculated according to the equation

$$K = S/\{[A(1-\alpha) - S][B(1-\alpha) - S]\} \qquad (10)$$

where K is the equilibrium constant, S is the solubility of Z-APM·PM complex, α is the conversion to the product and A and B are the initial concentrations of Z-L-Asp and L-PM, respectively.[19] The value of K was found to be $1.501\,\mathrm{mol}^{-1}$ and the solubility of the addition compound of Z-AMP with D-PM in water at 40 °C was 0.502 mg per 100 ml. Thus, from equation 10, the maximum theoretical yield of the reaction between Z-L-Asp and DL-PM may be calculated at various substrate concentrations. The calculated and measured yields are in excellent agreement (Figure 4), and at high substrate concentrations an almost quantitative yield can be expected.

11.4 CHARACTERISTICS OF THE TOSOH PROCESS

In the mid-1970s when this research was initiated, there were only a few successful examples of the application of enzymes to commercial processes. Although the past 20 years have witnessed a dramatic increase in the number of organic syntheses which utilize enzymes, many of these are based on simple reactions such as hydrolysis, isomerization and addition reactions (e.g. ammonia lyase) etc. Tosoh developed the enzymatic method for the synthesis of aspartame into an industrial process by a long and extensive programme of research, which led eventually to the current commercial plant operated by Holland Sweetener, a joint venture between Tosoh and DSM. Production commenced in 1988, and a large expansion of the plant capacity is expected in 1993 based on further improvements accomplished by joint research between the two parent companies. The Tosoh process is an excellent example of the industrial utilization of enzymes, since it has the following outstanding features:

i. While all of the chemical methods inevitably produce 20–40% of β-aspartame, this by-product is not present in the enzyme-catalysed process.
ii. The enzyme is completely stereoselective, which means that racemic or L-amino acids can be used, depending on price and availability.

iii. There is no racemization during the peptide synthesis. In the chemical methods, care must be taken to avoid racemization in several steps which include the recovery of the amino acid substrates.
iv. The condensation (which is a dehydration) takes place in an aqueous solution under very mild conditions.

The features of regio- and stereoselectivity have been major topics of research in the field of asymmetric synthesis with conventional chemical catalysts. In the present example, the enzyme can accomplish these tasks simultaneously in a single step.

11.5 CONCLUSION

The process described above is the first example of the large-scale production (hundreds to thousands of tonnes) of a peptide by an enzymic method; the only other industrial example being the enzymic conversion of porcine insulin to human insulin. The latter process cannot be compared with aspartame production, however, since the manufacture of the hormone takes place on not much more than a laboratory scale.

The successful commercialization of the above process for aspartame has firmly established enzyme-catalysed reactions as very effective methods for the synthesis of short peptides. Other viable methods include chemical synthesis (solution and solid phase) and those which involve gene manipulation. As each technique has advantages for particular products, the choice of the most appropriate method will depend on the target peptide to be synthesized.

11.6 NOTES AND REFERENCES

1. Mazur, R. H., Schlatter, J. M., and Goldkamp, A. H., *J. Am. Chem. Soc.*, **91**, 2684 (1969).
2. Davey, J. M., Laird, A. H., and Morley, J. S., *J. Chem. Soc. C*, 555 (1966).
3. There are many reports on the study of the safety of aspartame. The most comprehensive is *Authentication Review of Selected Materials Submitted to the Food and Drug Administration Relative to Application of Searle Laboratories to Market Aspartame*, Universities Associated for Research and Education in Pathology, 1978.
4. Ajinomoto, *US Pat.*, 3 798 206 (1974).
5. Stamicarbon, *US Pat.*, 3 879 372 (1975).
6. Nutrasweet, *US Pat.*, 4 946 988 (1990).
7. Ajinomoto, *US Pat.*, 3 7860 39 (1974).
8. Mohr, E., and Strohschein, F., *Chem. Ber.*, **42**, 2521 (1909).
9. Oyama, K., and Kihara, K., *Chemtec*, 100 (1984).
10. Isowa, Y., Oyama, K., and Ichikawa, T., *Abstracts of Papers, 35th Annual Autumn Meeting of the Chemical Society of Japan, Sapporo*, 1976, 4B23.

11. Isowa, Y., Ohmori, M., Ichikawa, T., Mori, K., Nonaka, Y., Kihara, K., Oyama, K., Satoh, H., and Nishimura, S., *Tetrahedron Lett.*, 2611 (1979).
12. For reviews of thermolysin, see Matsubara, H., *Methods Enzymol.*, **19**, 642 (1970); Matsubara, H., and Feder, J., in *The Enzymes* (ed. P. D. Boyer), Academic Press, New York, 1971, Vol. 3, p. 721.
13. Endo, S., *J. Ferment. Technol.*, **40**, 346 (1962).
14. Oyama, K., Kihara, K., and Nonaka, Y., *J. Chem. Soc., Perkin Trans. 2*, 356 (1981).
15. Homandberg, G. A., Matis, J. A., and Laskowski, M., Jr, *Biochemistry*, **17**, 5220 (1978).
16. Oyama, K., Irino, S., and Hagi, N., *Methods Enzymol.*, **136**, 503 (1987).
17. Oyama, K., in *Biocatalysis in Organic Media* (ed. C. Laane, J. Tramper and M. D. Lilly), Elsevier, Amsterdam, 1987, p. 209.
18. Oyama, K., Nishimura, S., Nonaka, Y., Kihara, K., and Hasimoto, T., *J. Org. Chem.*, **46**, 5241 (1981).
19. Oyama, K., Irino, S., Harada, T., and Hagi, N., *Ann. N. Y. Acad. Sci.*, **434**, 95 (1984).

12 New Preparative Methods for Optically Active β-Hydroxycarboxylic Acids

T. OHASHI and J. HASEGAWA
Kaneka Corporation, Osaka, Japan

This chapter describes novel work carried out by the authors on biological methods for β-hydroxycarboxylic acids, and serves as another illustration of the way in which microorganisms can be adapted to accomplish specific chemical transformations of industrial interest, in this case the enantiospecific β-hydroxylation of carboxylic acids.

12.1 BACKGROUND

Of a wide range of simple enantiomerically pure compounds which have been elaborated into important industrial products, the β-hydroxycarboxylic acids have proved to be of particular utility owing to the ready availability of the

Chirality in Industry. Edited by A. N. Collins, G. N. Sheldrake and J. Crosby

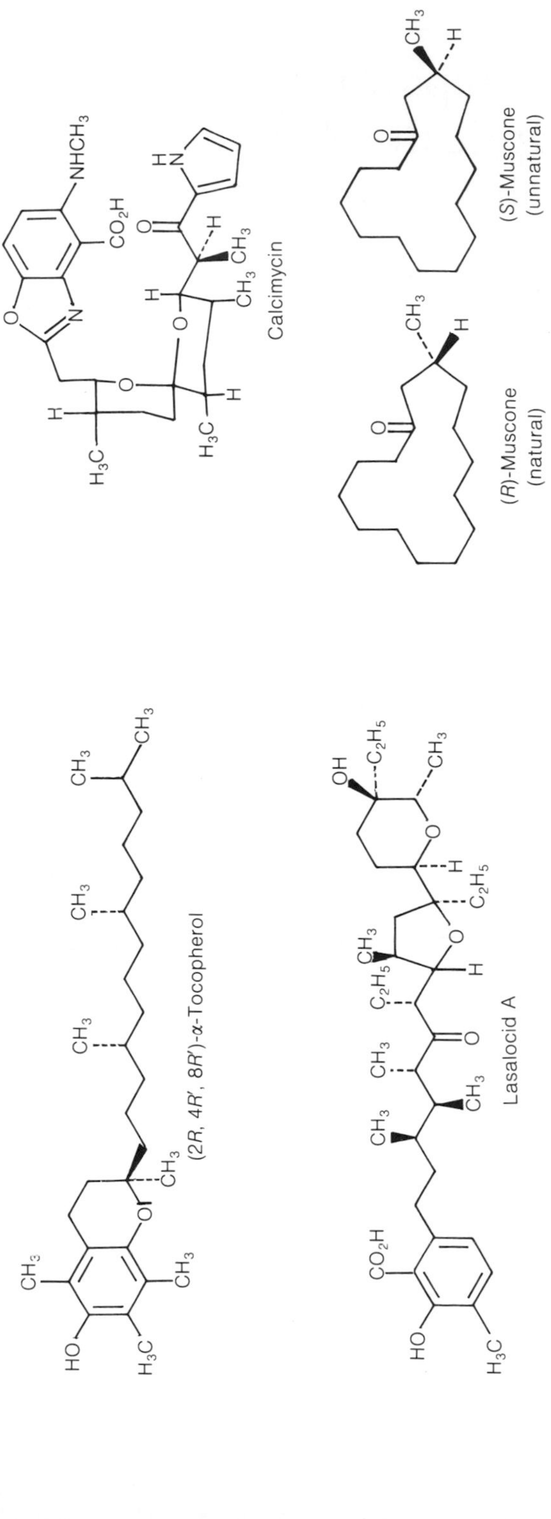

(2R, 4R', 8R')-α-Tocopherol

Lasalocid A

Calcimycin

(R)-Muscone (natural)

(S)-Muscone (unnatural)

Figure 1. Compounds derived from L-(+)-β-hydroxyisobutyric acid

L-(+)-form of the simpler members of this series and their useful dual functionality. Thus, optically active β-hydroxycarboxylic acids and their derivatives have been used as the starting materials for the syntheses of vitamins, flavour components, antibiotics and pheromones.[1–5] For example α-tocopherol,[6] lasalocid A,[7] calcimycin[8] and (*R*)- and (*S*)-muscone[9] have all been synthesized from L-(+)-β-hydroxyisobutyric acid (Figure 1), which is produced in a fermentation process. The less readily available D-enantiomer of this compound is a useful precursor of the anti-hypertensive agent captopril, which contains the D-(−)-3-mercapto-2-methylpropanoyl moiety[10–12] (see Chapter 1).

Several biological methods for the preparation of β-hydroxycarboxylic acids have been reported, and these may be classified as follows.

12.1.1 FERMENTATION PROCESSES

These have been used to prepare D-(−)-β-hydroxybutyric acid and poly[D-(−)-β-hydroxybutyric acid] from carbohydrates and alcohols utilising *Azotobacter*, *Bacillus* and *Alcaligenes* species.[13–15] Similarly, the yeast *Candida sorbosa* mediates the production of β-hydroxy-α-methylglutaric acid from glucose.[16]

12.1.2 OXIDATION OF ALIPHATIC GLYCOLS

This is another useful method for certain simple members. β-Hydroxypropionic acid and racemic β-hydroxybutyric acid have been prepared from propane-1,3-diol and DL-butane-1,3-diol respectively, using *Gluconobacter*,[17] *Hansenula*[18] and *Arthrobacter*.[19] D-(−)-β-Hydroxyisobutyric acid has been prepared from 2-methylpropane-1,3-diol using *Gluconobacter roseus*.[20]

12.1.3 β-HYDROXYLATION OF CARBOXYLIC ACIDS

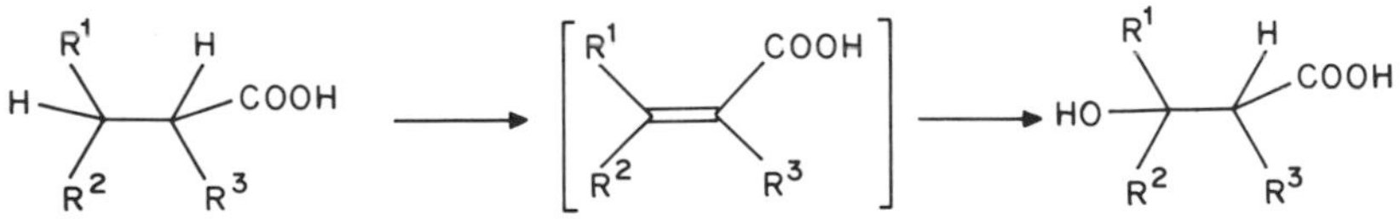

This transformation has been used widely to prepare optically active β-hydroxycarboxylic acids. The process takes place in two stages: an initial dehydrogenation to the α,β-unsaturated carboxylic acid and a subsequent hydration. These steps utilize the enzymes of the β-oxidation pathway of lipid catabolism, and so the β-hydroxy acids produced are generally of the 'natural' L-(+)-form. Both saturated carboxylic acids and their α,β-unsaturated counterparts have been used as raw materials. For example, β-hydroxypropionic acid has been prepared from acrylic acid in a process mediated by *Fusarium*,[21] and *Pseudomonas putida*[22] has been used to prepare L-(+)-β-hydroxyisobutyric acid from isobutyric acid. The preparation of C_6–C_{12} L-(+)-β-hydroxycarboxylic acids from the corresponding *trans*-α,β-unsaturated carboxylic acids by microbial hydration catalysed by resting cells of *Mucor* sp. is also known.[23]

12.1.4 HYDROGENATION OF β-KETOCARBOXYLIC ACIDS

$$R{-}CO{-}CH_2COOH \longrightarrow R{-}CH(OH){-}CH_2COOH$$

This is a common biochemical transformation, which has been harnessed to produce several useful β-hydroxy acids and esters. L-(+)-β-Hydroxybutyric acid, D-(−)-β-hydroxycaproic acid and D-(−)-β-hydroxycaprylic acid have been prepared from the corresponding β-ketocarboxylic acids by microbial hydrogenation using *Saccharomyces cerevisae*.[24] Ethyl L-(+)-β-hydroxybutyrate has been obtained from ethyl acetoacetate by the same method.[25]

The authors considered microbial β-hydroxylation to be particularly suitable for further refinement. The availability and low cost of the carboxylic acid precursors would favour the large-scale use of processes based on this method, but one major disadvantage was that where optically-active β-hydroxycarboxylic acids were obtained, only the L-(+)-form was observed. As noted earlier, the D-(−)-β-hydroxycarboxylic acids have potential as precursors of pharmaceuticals, and this provided one of the incentives for the development of a robust, general method for the D-form of these intermediates.

12.2 SELECTION OF MICROORGANISMS WHICH CONVERT ISOBUTYRIC ACID INTO β-HYDROXYISOBUTYRIC ACID[26]

The authors began the study with a search for organisms which could metabolize isobutyric acid to the simplest branched β-hydroxycarboxylic acid, β-hydroxyisobutyric acid. The latter is, in fact, an intermediate of valine catabolism in mammalian species[27,28] and microorganisms,[29,30] and one generally accepted pathway for its production is given in Figure 2. Previous studies on the stereochemistry of the production of β-hydroxyisobutyric acid from isobutyric acid and methacrylic acid by *P. putida* had demonstrated the production of the L-(+)-form.[31,32]

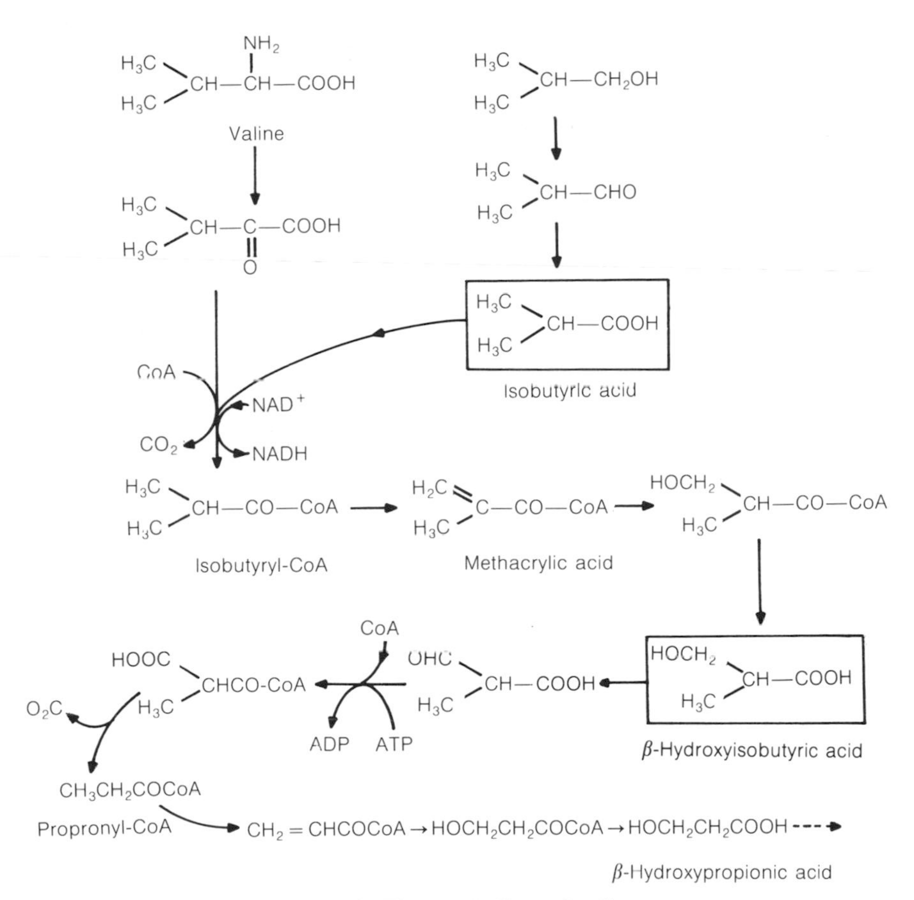

Figure 2. The catabolism of valine

The authors screened 725 strains of microorganisms for the ability to β-hydroxylate isobutyric acid. A selection of results displayed in Table 1 reveals that this capability is widely distributed in various moulds, yeasts and bacteria. The conversion under the screening conditions varied from 25 to 45% and, although the majority of organisms produced the expected L-(+)-β-hydroxyisobutyric acid, a minority, notably *C. rugosa*, yielded the D-(−)-form.

The enantiomeric purities of the methyl esters of D-(−)- and L-(+)-β-hydroxyisobutyric acid produced by *C. rugosa* IFO 0750 and IFO 1542 were estimated to be 99% and 97% respectively, by a chiral high-performance liquid chromatographic method. In comparison, the specific optical rotation reported for the methyl ester of D-(−)-β-hydroxyisobutyric acid from 2-methylpropane-1,3-diol by *G. roseus*[20] represents an optical purity of only 83% (−22.5° reported *vs* −26.3° in this work).

Table 1. Distribution of β-hydroxyisobutyric acid-producing microorganisms

Microorganism	β-HIBA[a]		Microorganism	β-HIBA[a]	
	OR[b]	PR[c]		OR[b]	PR[c]
Yeasts:			Moulds:		
Bullera alba	(+)	+++[d]	*Aspergillus niger*	(−)	+[d]
Candida parapsilosis IFO 0708	(−)	+	*Endomyces ovetensis*	(+)	+
IFO 1022	(+)	+	*Endomyces geotrichum*	(+)	++
Candida rugosa IFO 0591	(−)	+++	*Endomyces reessii* CBS 179.60	(+)	++++
IFO 0750	(−)	++++	*Endomyces tetrasperma*	(+)	+
IFO 1364	(+)	++	*Geotrichum amycelicum*	(+)	++
IFO 1542	(+)	+++	*Geotrichum fragrans*	(+)	++
Cryptococcus laurentii	(+)	+	*Geotrichum gracile*	(+)	++
Cryptococcus terresus	(+)	+++	*Geotrichum loubieri*	(+)	++++
Debaryomyces hansenii	(+)	++	*Geotrichum vanryiae* CBS 439 64	(+)	+
Dekkera intermedia	(+)	+	*Geotrichum klebahnii*	(+)	++
Hanseniaspora valbyensis	(+)	++++	*Helicostylum nigricans*	(+)	+
Hansenula anomala IFO 0120	(+)	++++	*Mucor alternans*	(+)	+
Hansenula henricii	(+)	+	*Fusarium merismoides*	(+)	+
Kloechera africana	(+)	++	*Choanephora circinans*	(−)	+
Pachysolen tannosphilus	(+)	++	*Wingea robertsii*	(−)	+
Pichia burtonii	(+)	++	*Zygorhynchus moelleri*	(−)	+

Microorganism			Microorganism		
Pichia membranaefaciens IAM 4904	(−)	+	*Moniliella tomentosa*	(+)	+
IAM 4258	(+)	+ + +			
Rhodosporidium toruloides	(+)	+	Bacteria:		
Saccharomyces cerevisiae	(+)	+ +	*Bacillus licheniformis*	(+)	+ +
Saccharomyces uvarum	(+)	+	*Bacillus megaterium*	(+)	+
Saccharomycopisis lipolytica IFO 1545	(+)	+ + +	*Brevibacterium ammoniagenes*	(+)	+ +
Schizosaccharomyces pombe	(+)	+	*Brevibacterium protophormiae*	(+)	+
Schwanniomyces persoonii	(+)	+ +	*Corynebacterium hydrocarbons*	(+)	+ +
Sporobolomyces salmonicolor	(+)	+	*Enterobacter cloacae*	(+)	+ + +
Sporidiobolus johnsonii	(+)	+ + +	*Micrococcus luteus*	(+)	+ + + +
Torulopsis gropengiesseri	(+)	+	*Micrococcus flavus*	(+)	+ + + +
Torulopsis candida	(−)	+ +	*Micrococcus lysodeikticus*	(+)	+ + + +
Torolopsis magnoliae	(+)	+	*Pseudomonas chlororaphis*	(+)	+ +
Trichosporon fermentans	(+)	+ + +	*Pseudomonas dacunhae*	(+)	+ + +
Trichosporon aculeatum ATCC 22310	(+)	+ + + +	*Pseudomonas riboflavina*	(+)	+
Trichosporon pullulans	(+)	+ +	*Sarrcina variabilis*	(+)	+ + +
Trichosporon fennicum	(+)	+ + + +	*Serratia plymuthica*	(+)	+
Rhodotorula rubra	(+)	+ +	*Nocardia corallina*	(+)	+
			Nocardia lyena	(+)	+
			Streptomyces griseus	(+)	+

[a] β-Hydroxyisobutyric acid.
[b] Optical rotation.
[c] Productivity.
[d] + + + +, $>5\ mg\ ml^{-1}$; + + +, $3–5\ mg\ ml^{-1}$; + +, $1–3\ mg\ ml^{-1}$; +, $<1\ mg\ ml^{-1}$.

Table 2. Production of β-hydroxyisobutyric acid from methacrylic acid

Microorganism[a]	Substrate	Cultivation period (h)	β-HIBA[b] produced ($mg\,ml^{-1}$)	Optical activity
Candida rugosa IFO 0750	IBA[c]	27	7.3	(−)
	MA[d]	44	4.1	(−)
Candida rugosa IFO 0591	IBA	27	8.1	(−)
	MA	44	3.6	(−)
Pichia membranaefaciens	IBA	44	3.2	(+)
IFO 4258	MA	44	1.5	(+)
Trichosporon aculeatum	IBA	20	5.5	(+)
ATCC 22310	MA	44	0.2	(+)
Endomyces reessii CBS 179.60	IBA	27	3.7	(+)
	MA	44	1.2	(+)

[a] Microorganisms were incubated in 500 ml Sakaguchi flasks at 30 °C with shaking. The medium (30 ml) contained 2% of isobutyric acid or methacrylic acid.
[b] β-Hydroxyisobutyric acid.
[c] Isobutyric acid.
[d] Methacrylic acid.

The authors also examined the ability of some organisms to effect the conversion of methacrylic acid into β-hydroxyisobutyric acid. The results, shown in Table 2, revealed that organisms with the ability to produce β-hydroxyisobutyric acid from isobutyric acid could also produce it from methacrylic acid, albeit with lower productivity.

12.3 PRODUCTION OF OTHER ALIPHATIC β-HYDROXYCARBOXYLIC ACIDS[33]

Having demonstrated the feasibility of stereoselective production of L-(+)- and D-(−)-β-hydroxyisobutyric acid from a 'natural' substrate, isobutyric acid, the authors proceeded to investigate the β-hydroxylation of various other aliphatic carboxylic acids by the same microorganisms. Accordingly, the organisms which displayed a high β-hydroxylating activity for isobutyric acid were incubated with various normal and iso carboxylic acids. However, of the normal carboxylic acids, only valeric acid gave rise to the β-hydroxy derivative in the broths of *C. rugosa*. Although propionic acid and butyric acid were readily metabolized by the microorganisms, the corresponding β-hydroxycarboxylic acids did not accumulate, and most normal carboxylic acids with longer chains were not metabolized at all. On the other hand, the iso carboxylic acids, isovaleric acid, isocaproic acid and α-methylbutyric acid, were converted into the corresponding β-hydroxy derivatives as shown in Table 3.

Table 3. β-Hydroxylation of carboxylic acids

Acid	*C. rugosa* IFO 0750		*C. rugosa* IFO 1542		*E. reessii* CBS 179.60		*T. fermentans* CBS 2529	
	Residual CA[a]	β-HA[b] formed	Residual CA[a]	β-HA[b] formed	Residual CA[a]	β-HA[b] formed	Residual CA[a]	β-HA[b] formed
Propionic acid	–[c]	–	–	–	–	–	–	–
Butyric acid	–	–	–	–	–	–	–	–
Valeric acid	–	+	–	+	++	–	++	–
Caproic acid	–	–	–	–	+++	–	+++	–
Heptanoic acid	–	–	++	–	+++	–	+++	–
Caprylic acid	+++	–	+++	–	+++	–	+++	–
Isobutyric acid	–	+++	–	+++	–	+++	–	+++
Isovaleric acid	++	–	–	+	+	–	+	+
Isocaproic acid	++	–	–	–	–	+	+++	–
α-Methylbutyric acid	+	+	–	±	–	±	–	+
α-Ethylbutyric acid	+++	–						

[a] Carboxylic acid.
[b] β-Hydroxycarboxylic acid.
[c] 'Observed' by thin-layer chromatography.

Although β-hydroxyvaleric acid and β-hydroxyisovaleric acid have been detected in the urine of patients with acidaemia[34,35] and β-hydroxyisocaproic acid has been extracted from tobacco leaves,[36] the present studies represent the first microbial preparations of these compounds in gram quantities.[37] As with the previous experiments with isobutyric acid as a starting material, *C. rugosa* IFO 0750 was unique in producing the D-(−)-form of the β-hydroxycarboxylic acids. The other organisms tested gave rise either to L-(+)-β-hydroxycarboxylic acids or, in the case of β-hydroxyisocaproic acid by *E. reessii* CBS 179.60, the racemic form.

12.4 PRODUCTION OF A MUTANT LACKING β-HYDROXYISOBUTYRATE DEHYDROGENASE[38]

The use of *C. rugosa* to produce D-(−)-β-hydroxyisobutyric acid from isobutyric acid, although novel, was not yet efficient enough to be adapted to an industrial scale. The conversion never exceeded 50% and the by-product, β-hydroxypropionic acid, was present at a concentration of about 10% of that of the D-(−)-β-hydroxyisobutyric acid. It is probable that β-hydroxypropionic acid is produced from D-(−)-β-hydroxyisobutyric acid via propionic acid by this microorganism, by the route illustrated in Figure 2. It seemed reasonable, therefore, that if a mutant unable to degrade D-(−)-β-hydroxyisobutyric acid could be derived from *C. rugosa* IFO 0750, a higher yield of this product would be expected. However, the selection of a mutant which did not degrade D-(−)-β-hydroxyisobutyric acid was far from trivial, because the organism in question did not grow well in media containing isobutyric acid or D-(−)-β-hydroxyisobutyric acid as a sole carbon source. However, *C. rugosa* assimilated both propionic acid and β-hydroxypropionic acid well, and so the selection of mutants unable to utilize propionic acid was comparatively easy. A group of mutants which were unable to assimilate propionic acid were derived from the parent strain and, of these, four mutants were found to produce D-(−)-β-hydroxyisobutyric acid in high yields without significant by-production of β-hydroxypropionic acid, as shown in Table 4.

The best strain, NPA-104, could utilize neither propionic acid nor β-hydroxypropionate as a sole carbon source, and the β-hydroxyisobutyrate dehydrogenase activity was reduced to 1.2% of that of the parent strain. Degradation of β-hydroxyisobutyrate by resting cells of strain NPA-104 was not observed at all.

Figure 3 depicts in graphical form the time dependence of the consumption of isobutyric acid and the production of D-(−)-β-hydroxyisobutyric acid by strain NPA-104 and the parent strain. The experiment was carried out with a medium containing 2% of isobutyric acid and 4% of glucose, with further isobutyric acid being added at intervals. Although the parent strain produced β-hydroxyisobutyric acid in reasonable amounts, the yield per unit of isobutyric acid consumed was poor. In the mutant strain, however, the β-hydroxyisobutyric

Table 4. D-(−)-β-Hydroxyisobutyric acid production by the mutants unable to assimilate propionic acid[a]

Mutant	IBA[b] consumed ($mg\,ml^{-1}$)	D-β-HIBA[c] produced ($mg\,ml^{-1}$)	Yield (%)	β-HPA[d]
NPA-101	23.9	24.4	87	−
NPA-102	19.9	22.7	96	−
NPA-103	25.9	22.7	74	±
NPA-104	32.3	34.5	91	−
NPA-105	29.6	29.3	84	−
Parent	33.7	13.2	33	+++

[a]Initial conditions: glucose, 4% (w/v); isobutyric acid, 2% (v/v). After 28 h of cultivation, 6% of glucose and 1% of isobutyric acid were added. After 51 h of cultivation, 1.5% of isobutyric acid was added. Cultivation was carried out for 68 h. The pH of the medium was adjusted to 7.0 twice a day.
[b]Isobutyric acid.
[c]D-β-Hydroxyisobutyric acid.
[d]β-Hydroxypropionic acid, observed on thin-layer chromatogram.

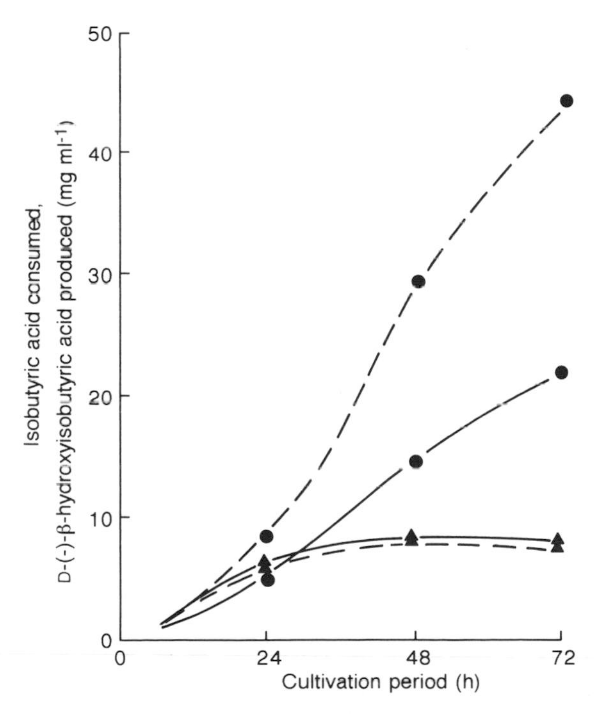

Figure 3. Production of D-(−)-β-hydroxyisobutyric acid by NPA-104. ●, Parent; ▲, NPA-104; dashed lines, IBA; solid lines, D-β-HIBA

acid production closely matched the consumption of the carboxylic acid precursor, but in this case activity ceased after about 24 h.

A possible explanation of these results may lie in the fact that the β-hydroxylation of isobutyric acid requires a net input of energy in the form of ATP. This is because ATP is required for the formation of isobutyryl-CoA from isobutyric acid in the sequence shown in Figure 2. In the parent strain, it appeared that about half of the β-hydroxyisobutyric acid produced was degraded further to provide an energy source for further β-hydroxylation. In strain NPA-104, however, the catabolic pathway for D-(−)-β-hydroxyisobutyric acid is blocked, and so this mutant requires an energy source for the β-hydroxylation of isobutyric acid.

In order to test this hypothesis, and to improve the system, the efficacy of glucose as an energy source for β-hydroxylation was investigated. Cultivation was carried out in media which contained initial concentrations of 4% of glucose and 2% of isobutyric acid, which were increased to 6% and 2.5% respectively, after 28 h. As shown in Figure 4, the production of β-hydroxyisobutyric acid by strain NPA-104 under these conditions increased markedly, whereas the parent strain actually performed less well than before.

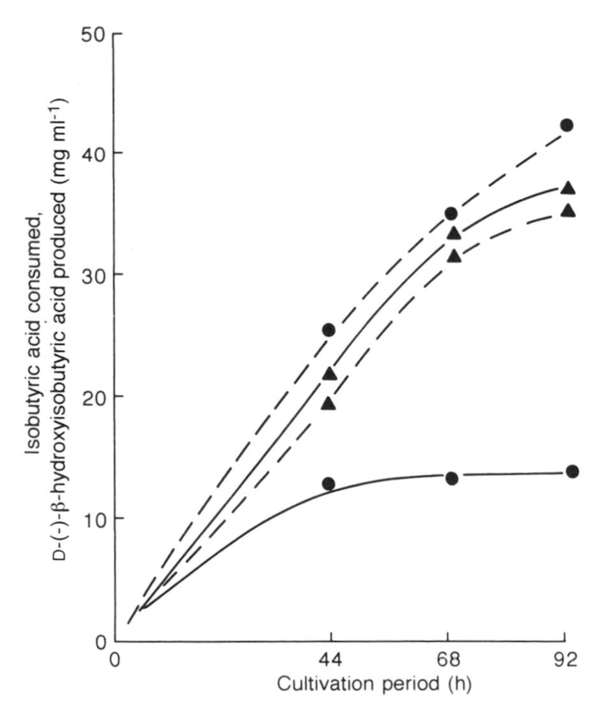

Figure 4. Effect of glucose on D-(−)-β-hydroxyisobutyric production by NPA-104. Symbols and lines as in Figure 3

Table 5. Effect of glucose on D-(−)-β-hydroxyisobutyric acid production by NPA-104[a]

	Glucose		44 h			68 h		
Test No.	Initial (%)	After 28 h (%)	IBA[b] consumed ($mg\,ml^{-1}$)	D-β-HIBA[c] produced ($mg\,ml^{-1}$)	Yield (%)	IBA[b] consumed ($mg\,ml^{-1}$)	D-β-HIBA[c] produced ($mg\,ml^{-1}$)	Yield (%)
1	4	—	10.9	10.7	83	10.7	10.8	86
2	6	—	12.0	12.1	83	12.1	13.1	92
3	8	—	12.1	12.4	87	13.2	14.9	96
4	4	2	21.1	21.7	87	25.1	24.9	84
5	4	4	20.0	22.1	94	29.7	30.2	86
6	4	6	20.3	22.1	92	36.0	34.5	87

[a]Initial concentration of isobutyric acid was 2% (w/v). In tests 1–3, 2% of isobutyric acid was added after 44 h of cultivation and in tests 4–6, 1% and 1.5% of isobutyric acid were added after 28 and 51 h of cultivation, respectively.
[b]Isobutyric acid.
[c]D-β-Hydroxyisobutyric acid.

A more detailed study of the effect of glucose concentration on the performance of strain NPA-104 produced further interesting results, as shown in Table 5. A high initial concentration of glucose, (entries 1–3) led to a good yield of D-(−)-β-hydroxyisobutyric acid per unit of isobutyric acid consumed, although the utilization of isobutyric acid was poor. In these cases, most of the glucose was probably utilized for cell growth during the early phase of development of the culture, and therefore there was little effect on the production of D-(−)-β-hydroxyisobutyric acid. If, however, extra glucose was added after 28 h to coincide with the phase of maximum D-(−)-β-hydroxyisobutyric acid production, the rate of conversion was much better (entries 4–6).

Hence, from mutants of *C. rugosa* which were unable to assimilate propionic acid, the authors succeeded in the selection of a strain which lacked β-hydroxyisobutyrate dehydrogenase. Under optimized conditions, and using glucose as an additional source of carbon, the molar yield of D-(−)-β-hydroxyisobutyric acid reached 96%.[39]

12.5 TOWARDS THE EFFICIENT PRODUCTION OF C_4–C_7 D-(−)-β-HYDROXYCARBOXYLIC ACIDS[40,41]

The authors' unsuccessful attempts to find culture conditions for the accumulation of normal β-hydroxycarboxylic acid from the corresponding carboxylic acids using *C. rugosa* were described in Section 12.3. Although normal carboxylic acids in the range C_1–C_7 were metabolized by this microorganism, the corresponding β-hydroxy acids could not be isolated, except in the case of D-(−)-β-hydroxyvaleric acid, which was obtained in low yield. This latter result suggested that a general method for the production of D-(−)-β-hydroxy acids by this organism might be possible, if the further catabolism of these products could be inhibited.

The next challenge, therefore, was to select a mutant which lacked the ability to oxidize normal β-hydroxy acids. A screening process was begun with a series of 64 mutants which were known to be unable to utilize butyric acid as a sole carbon source. It seemed likely that the inability of at least a proportion of these mutants to utilize this compound would be due to deficiencies in the

Table 6. Production of β-hydroxycarboxylic acids by the mutant unable to assimilate butyric acid[a]

Strain	pH	Growth	Residual BA[b] ($mg\,ml^{-1}$)	β-HBA[c] produced ($mg\,ml^{-1}$)	Yield (%)
NBA-1	9.3	+++	0.0	1.0	4
NBA-2	6.8	+	16.6	3.7	92
NBA-3	9.2	+++	0.0	4.0	17
NBA-4	9.4	+++	0.0	2.8	12
NBA-5	9.0	++	2.8	1.9	9
NBA-6	9.1	++	1.9	1.8	8
Parent	9.5	+++	0.0	0.0	0

[a]Cultivation was carried out in a test-tube (20 × 2.4 cm) containing 5 ml of GI medium (2% butyric acid added) at 30 °C for 3 days with shaking.
[b]Butyric acid.
[c]β-Hydroxybutyric acid.

Table 7. Growth responses of strain NBA-2 to glucose, carboxylic acids and β-hydroxycarboxylic acids[a]

Carbon source	Concentration (%, v/v)	Parent	NBA-2
Glucose	1.0	++++	++++
Acetic acid	1.0	+++	++
Propionic acid	1.0	+++	+++
Butyric acid	1.0	+++	−
L-β-Hydroxybutyric acid	1.0	+++	−
D-β-Hydroxybutyric acid	1.0	+++	−
Valeric acid	0.5	++	−
D-β-Hydroxyvaleric acid	0.5	+	−
Caproic acid	0.5	+	−
D-β-Hydroxycaproic acid	0.5	+	−
Heptanoic acid	0.5	+	−
D-β-Hydroxyheptanoic acid	0.5	+	−
Caprylic acid	0.5	±	−

[a]Tested in GII medium containing any one carbon source instead of glucose and agar. Cultivation was carried out in a test-tube containing 5 ml of medium at 30 °C for 3 days with shaking.

Table 8. Production of β-hydroxycarboxylic acids from carboxylic acids by strain NBA-2

				Residual carboxylic acid	β-Hydroxycarboxylic acid produced				
Carboxylic acid[a]	Concentration (%, v/v)	Strain	Cultivation period (h)	($mg\ ml^{-1}$)	Acid[b]	Concentration ($mg\ ml^{-1}$)	Yield (%)	Acid[b]	Concentration ($mg\ ml^{-1}$)
Butyric acid	1.0	Parent	23	0.0	β-HBA	0.0	(0)		—
	1.0	Mutant	72	0.0	β-HBA	10.7	(95)		—
Valeric acid	1.0	Parent	23	0.0	β-HVA	0.2	(2)		—
	1.0	Mutant	72	0.0	β-HVA	6.5	(60)		—
Caproic acid	1.0	Parent	48	0.0	β-HCA	0.0	(0)	β-HBA	0.0
	0.5	Mutant	72	0.0	β-HCA	4.4	(83)	β-HBA	0.9
	1.0	Mutant	72	4.5	β-HCA	5.2	(95)	β-HBA	0.5
Heptanoic acid	1.0[c]	Parent	48	0.0	β-HHA	0.0	(0)	β-HVA	0.0
	0.25	Mutant	48	0.0	β-HHA	1.3	(52)	β-HVA	0.3
	1.0[c]	Mutant	72	0.0	β-HHA	3.7	(37)	β-HVA	1.3

[a] Each substrate was added at the time of preparation of medium.
[b] β-HBA, β-hydroxybutyric acid; β-HVA, β-hydroxyvaleric acid; β-HCA, β-hydroxycaproic acid; β-HHA, β-hydroxyheptanoic acid.
[c] Initial concentration of heptanoic acid was 0.25%, and after 23 h of cultivation 0.75% of heptanoic acid was added.

enzymes which mediate the energy-releasing catabolism of the intermediate β-hydroxy acid, rather than an inability to hydroxylate the butyric acid itself.

In fact, six such strains with the ability to produce D-(−)-β-hydroxybutyric acid were isolated from the above screen (Table 6). The best of these strains, NBA-2, could produce β-hydroxybutyric acid in 92% yield from butyric acid, although the overall conversion was very poor. The growth response of this strain to various normal carboxylic acids and β-hydroxy acids was investigated. As shown in Table 7, the mutant grew well in a medium containing glucose, acetic acid or propionic acid as a carbon source, but hardly utilized C_4–C_7 β-hydroxycarboxylic acids.

The authors proceeded to evaluate the production of D-(−)-β-hydroxycarboxylic acids from saturated carboxylic acids by strain NBA-2. The organism was cultivated in a standard medium (with a source of carbon) to which various carboxylic acids were added. The results, shown in Table 8, were that, in addition to D-(−)-β-hydroxybutyric acid, D-(−)-β-hydroxyvaleric acid, D-(−)-β-hydroxycaproic acid and D-(−)-β-hydroxyheptanoic acid were all produced in good yield, although some degradation of these was observed when the substrates were completely consumed.

Table 9 lists the optical rotations of the D-(−)-β-hydroxy acids thus produced, which are in good agreement with the literature values, where these have been reported.[18,22,25]

As mentioned previously, the β-oxidation of lipids is known to proceed via the L-(+)-form of the β-hydroxyacyl-CoA intermediate. It is tempting to conclude from this study that *C. rugosa* IFO 0750 possesses a β-oxidation pathway which is unique among known organisms in the generation of a D-(−)-β-hydroxyacyl-CoA intermediate. This fact was capitalized upon in this study by the generation of *C. rugosa* mutants with the capacity to produce D-(−)-β-hydroxycarboxylic acids in good yields. This novel methodology has been developed by Kaneka for use on a large scale, and has furnished several D-(−)-β-hydroxycarboxylic acids in practical quantities for the first time. The

Table 9. Properties of D-(−)-β-hydroxycarboxylic acids produced by strain NBA-2

β-Hydroxycarboxylic acid	Form	B.p. (°C/mmHg)	$[\alpha]_D$ (*c*, solvent, °C)	n_D^{25}
D-β-Hydroxybutyric acid	Free acid	127/8	−23.1° (1, H_2O, 25)	1.442
	Methyl ester	68.5/17	−23.0° (neat, 20)	1.420
D-β-Hydroxyvaleric acid	Free acid	95/3	−26.9° (4, $CHCl_3$, 25)	1.446
	Methyl ester	71/10	−16.3° (neat, 20)	1.426
D-β-Hydroxycaproic acid	Free acid	114/2	−27.5° (2, $CHCl_3$, 22)	1.447
	Methyl ester	61/3	−10.3° (neat, 20)	1.428
D-β-Hydroxyheptanoic acid	Free acid	126/1	−25.8° (2, $CHCl_3$, 18)	1.448
	Methyl ester	64/2	−9.2° (neat, 20)	1.431

Table 10. β-Hydroxycarboxylic acids obtained in this study

$$\mathrm{H{-}\underset{R^2}{\overset{R^3}{C}}{-}\underset{R^1}{\overset{H}{C}}{-}COOH \xrightarrow{\text{microbial } \beta\text{-hydroxylation}} HO{-}\underset{R^2}{\overset{R^3}{C}}{-}\underset{R^1}{\overset{H}{C}}{-}COOH}$$

β-Hydroxy-carboxylic acid[a]	R^1	R^2	R^3	Microorganism[b]	Optical purity (% *ee*)
β-HPA	H	H	H	*C. rugosa* IFO 0750 M	—
D-β-HBA	H	H	CH_3	*C. rugosa* IFO 0750 M	95
L-β-HBA				*E. tetrasperma* CBS 765.70 M	88
D-β-HVA	H	H	CH_3CH_2	*C. rugosa* IFO 0750 M	93
L-β-HVA				*E. tetrasperma* CBS 765.70 M	82
D-β-HCA	H	H	$CH_3(CH_2)_2$	*C. rugosa* IFO 0750 M	96
D-β-HHA	H	H	$CH_3(CH_2)_3$	*C. rugosa* IFO 0750 M	95
D-β-HIBA	CH_3	H	H	*C. rugosa* IFO 0750 M	97
L-β-HIBA				*T. aculeatum* ATCC 22310	99
(−)-α-HMBA	CH_3CH_2	H	H	*C. rugosa* IFO 0750 M	98
(+)-α-HMBA				*T. fermentans* CBS 2529	95
β-HIVA	H	CH_3	CH_3	*E. reessii* CBS 179.60	—
DL-β-HICA	H	H	$(CH_3)_2CH$	*E. reessii* CBS 179.60	0

[a] β-HPA = β-hydroxypropionic acid; β-HBA = β-hydroxybutyric acid; β-HVA = β-hydroxyvaleric acid; β-HCA = β-hydroxycaproic acid; β-HHA = β-hydroxyheptanoic acid; β-HIBA = β-hydroxyisobutyric acid; α-HMBA = α-hydroxymethyl butyric acid β-HIVA = β-hydroxyisovaleric acid; β-HICA = β-hydroxyisocaproic acid.
[b] M, mutant.

range of β-hydroxycarboxylic acids which may be obtained readily by the method is given in Table 10, and Chapter 13 describes some applications of these intermediates as precursors to enantiomerically pure compounds of commercial importance.

12.6 NOTES AND REFERENCES

1. Fuzisawa, T., Sato, T., Kawara, T., Noda, N., Nishizawa, A., and Ohhikata, T., *Annu. Meet. Chem. Soc. Jpn.*, 1029 (1981).
2. Mori, K., *Tetrahedron*, **37**, 1341 (1981).
3. Mori, K., and Tanida, K., *Heterocycles*, **15**, 1171 (1981).
4. Fuzisawa, T., Sato, T., Kawara, K., and Ohashi, K., *Tetrahedron Lett.*, **22**, 4823 (1981).
5. Fuzisawa, T., Sato, T., Naruse, K., Mitsuyasu, K., and Kawara, T., *Annu. Meet. Chem. Soc. Jpn.*, 741 (1981).
6. Cohen, N., Eichel, W. F., Lopersti, R. J., Neukom, C., and Saucy, G., *J. Org. Chem.*, **41**, 3505 (1976).
7. Nakata, T., Kishi, Y., *Tetrahedron Lett.*, 2745 (1978).
8. Evans, D. A., Sacks, C. E., Kleschick, W. A., and Taber, T. R., *J. Am. Chem. Soc.*,**101**, 6789 (1979).
9. Branca, Q., and Fischli, A., *Helv. Chim. Acta*, **60**, 925 (1977).

10. Iwao, J., Iso, T., and Oya, M., *J. Synth. Org. Chem. Jpn.*, **38**, 1100 (1980).
11. Cushman, D. W., Chenung, H. S., Sabo, F. F., and Ondetti, M. A., *Biochemistry*, **16**, 5484 (1977).
12. Iso, T., Nishimura, K., Oya, M., and Iwao, J., *Eur. J. Pharmacol.*, **54**, 303 (1979).
13. Raffarty, M. L., *Jpn. Tokkyo Kokai*, JP 53-18794, 1978.
14. Raffarty, M. L., *Jpn. Tokkyo Kokai*, JP 53-20478, 1978.
15. Slepecky, R. A., and Low, J. H., *J. Bacteriol.*, **82**, 37 (1961).
16. Tabuchi, T., Nakahara, T., Kodama, K., Uchiyama, H., and Sakai, S., *Agric. Biol. Chem.*, **32**, 1641 (1981).
17. DeLay, J., and Kersters, K., *Bacteriol. Rev.*, **28**, 164 (1964).
18. Harada, T., and Hirabayashi, T., *Agric. Biol. Chem.*, **32**, 1175 (1968).
19. Yagi, O., *Hakko To Kogyo*, **37**, 1070 (1979).
20. Ohta, H., and Tetsukawa, H., *Chem. Lett.*, 1379 (1979).
21. Miyoshi, T., and Harada, T., *J. Ferment. Technol.*, **52**, 388 (1974).
22. Goodhue, C. T., and Schaeffer, J. R., *Biotechnol. Bioeng.*, **13**, 203 (1971).
23. Tahara, S., and Mizutani, J., *Agric. Biol. Chem.*, **42**, 879 (1978).
24. Lemieux, R. U., and Giguere, J., *Can. J. Chem.*, **29**, 678 (1951).
25. Deol, B. S., Ridley, D. D., and Simpson, G. W., *Aust. J. Chem.*, **29**, 2459 (1976).
26. Hasegawa, J., Ogura, M., Hamaguchi, S., Shimazaki, M., Kawaharada, H., and Watanabe, K., *J. Ferment. Technol.*, **59**, 203 (1981).
27. Robinson, W. G., Nagel, R., Bachhawat, B. K., Kupiecki, F. P., and Coon, M. J., *J. Biol. Chem.*, **224**, 1 (1956).
28. Baretz, B. H., and Tanaka, K., *J. Biol. Chem.* **253**, 4023 (1978).
29. Bannerjee, D., Sanners, L. E., and Sokatch, J. R., *J. Biol. Chem.*, **245**, 1828 (1970).
30. Nurmikko, V., Puukka, P., and Puukka, R., *Suom. Kemistil. B*, **45**, 195 (1972).
31. Aberhart, D. J., and Tann, C.-H., *J. Chem. Soc., Perkin Trans. 1*, 939, 1404 (1979).
32. Aberhart, D. J., *Bioorg. Chem.*, **6**, 191 (1977).
33. Hasegawa, J., Ogura, M., Kanama, H., Noda, N., Kawaharada, H., and Watanabe, K., *J. Ferment. Technol.*, **60**, 501 (1982).
34. Lawrence, S., Walter, W., William, L., *et al.*, *Biomed. Mass Spectrom.*, 198 (1978).
35. Landaas, S., *Clin. Chim. Acta*, **64**, 143 (1975).
36. Fukuzumi, T., *Okayama Tab. Shikensho Hokoku*, 103 (1971).
37. Two pathways for the catabolism of α-methylbutyric acid have been proposed, by Mamer *et al.* [*Biochem. J.*, **160**, 417 (1967)]. On the degradative pathway of isoleucine, α-methylbutyryl-CoA is converted into α-methyl-β-hydroxybutyryl-CoA via tigloyl-CoA, whereas in the degradation of alloisoleucine this CoA derivative is converted into β-hydroxymethylbutyryl-CoA via α-ethylacrylyl-CoA. In fact, both *C. rugosa* IFO 0750 and *T. fermentans* CBS 2529 converted α-methylbutyric acid into α-hydroxymethylbutyric acid, presumably via the alloisoleucine pathway.
38. Hasegawa, J., Hamaguchi, S., Ogura, M., and Watanabe, K., *J. Ferment. Technol.*, **59**, 257 (1981).
39. In the course of the studies on β-hydroxyisobutyric acid metabolism described above, the authors made another interesting observation. When racemic β-hydroxyisobutyric acid was incubated with the resting cells of some microorganisms, the residual β-hydroxyisobutyric acid isolated was the laevorotatory form. On further investigation, two types of cell-free extracts were found to catalyse the dehydrogenation of β-hydroxyisobutyric acids. One of these, isolated from *Micrococcus luteus*, had activity only for L-(+)-β-hydroxyisobutyric acid, whereas the other, from *C. rugosa* IFO 0750, was non-specific. The authors succeeded in crystallizing the latter enzyme, which was shown to catalyse the dehydrogenation of L-(+)-β-hydroxyisobutyric acid more rapidly than that of the D-(−)-form. A purified form of the former enzyme

did not degrade D-(−)-β-hydroxyisobutyric acid. Thus, it appears that at least two types of β-hydroxyisobutyric acid dehydrogenases exist in microorganisms, one of which is entirely stereoselective. This latter enzyme may, on further development, prove to have commercial potential for the kinetic resolution of novel β-hydroxy acid derivatives.

40. Hasegawa, J., Ogura, M., Kanema, H., Kawaharada H., and Watanabe, K., *J. Ferment. Technol.*, **60**, 591 (1982).
41. Hasegawa, J., Ogura, M., Kanema, H., Kawaharada, H., and Watanabe, K., *J. Ferment. Technol.*, **61**, 37 (1983).

13 D-(−)-β-Hydroxycarboxylic Acids as Raw Materials for Captopril and β-Lactams

T. OHASHI and J. HASEGAWA
Kaneka Corporation, Osaka, Japan

Chapter 12 details the development of a new and efficient process for the preparation of β-hydroxycarboxylic acids. In this chapter, the versatility of these intermediates as enantiomerically pure precursors to important commercial compounds is illustrated by reference to the authors' work on routes to captopril and β-lactams.

13.1 SYNTHESIS OF CAPTOPRIL FROM D-(−)-3-HYDROXYISOBUTYRIC ACID[1]

Captopril, (*S*)-1-(3-mercapto-2-methyl-1-oxopropyl)-L-proline (**6**), is an orally active, anti-hypertensive drug, which acts by inhibiting the angiotensin-converting enzyme (ACE).[2,3] The efficacy of captopril as an inhibitor of this enzyme depends critically on the configuration of the mercaptoalkanoyl side-chain; the compound with the D-configuration is about 100 times more active than the L-enantiomer.[4] Routes to captopril (see Method B in Figure 1) have involved coupling D-3-acetylthio- or D-3-benzoylthio-2-methylpropanoic acid (**9**) (or the acid chloride (**10**) of either compound), with L-proline, followed by a deacylation of the product. However, optically active 3-acylthio-2-methylpropanoic acids have hitherto been prepared only by chemical resolution of a racemic mixture using an optically active amine such as 1,2-diphenylethylamine or 2-amino

Chirality in Industry. Edited by A. N. Collins, G. N. Sheldrake and J. Crosby

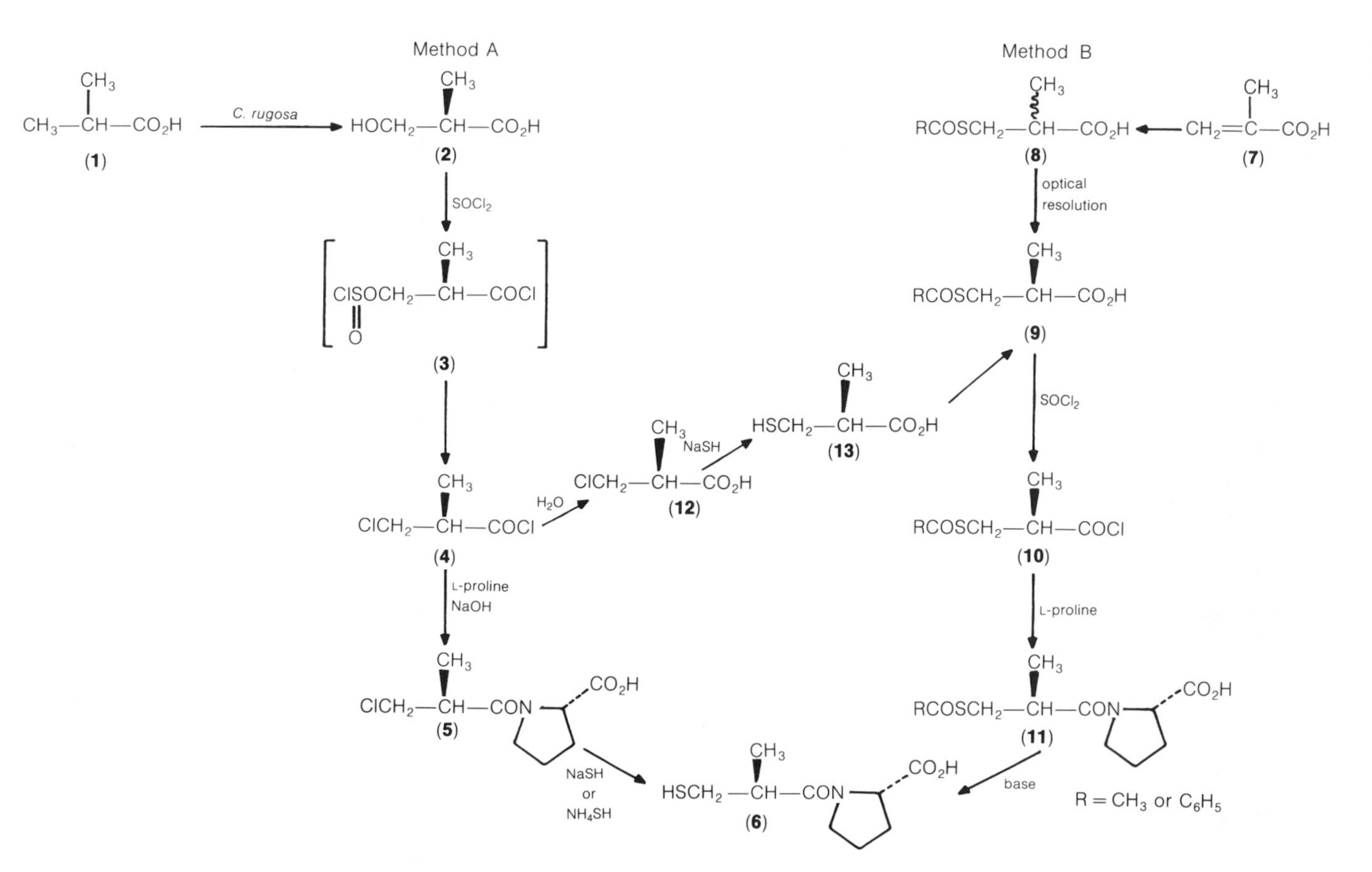

Figure 1. Syntheses of captopril from D-(−)-3-hydroxyisobutyric acid

butan-1-ol,[5–7] and there have been no reports of the synthesis of the enantiomerically pure 3-acylthio-2-methylpropanoic acids without chemical resolution.

Method A in Figure 1 illustrates a successful alternative synthesis of captopril from D-(–)-3-hydroxyisobutyric acid (**2**), which was developed by the authors. D-3-Chloro-2-methylpropanoyl chloride (**4**) was prepared by treating D-(–)-3-hydroxyisobutyric acid with thionyl chloride. This acid chloride was coupled with L-proline to afford the chloride **5**, which was directly converted into captopril (**6**) by reaction with a hydrosulphide or trithiocarbonate salt in hot water. This reaction proceeded with retention of stereochemistry. Alternatively, an intermediate from the original route, D-3-acetylthio-2-methylpropionic acid (**9**), could be derived from D-(–)-3-hydroxyisobutyric acid via the intermediate acids **12** and **13** (Figure 1).

Whereas the yield of captopril from methacrylic acid by Method B was estimated to be 19%, the corresponding yield from isobutyric acid by the new route (Method A) was 65%. Subsequent process development has confirmed the greater economy of Method A for the industrial production of captopril.

13.2 SYNTHESIS OF OPTICALLY PURE β-LACTAMS[8,9]

Since the discovery of the antibiotic thienamycin (**14**), compounds which contain the carbapenem and penem ring systems have attracted special interest. Studies of the relationship of structure to activity in this series have demonstrated the importance of the stereochemistry of the hydroxyethyl group, which must be in the (*R*)-configuration for optimum activity against microorganisms. Previous syntheses of carbapenem and penem compounds have often utilized the optically active β-lactam intermediate **15** which possesses a masked 3-(*R*)-hydroxyethyl group and an acetoxy group, which can behave as a leaving group, at the 4-position.[10–26]

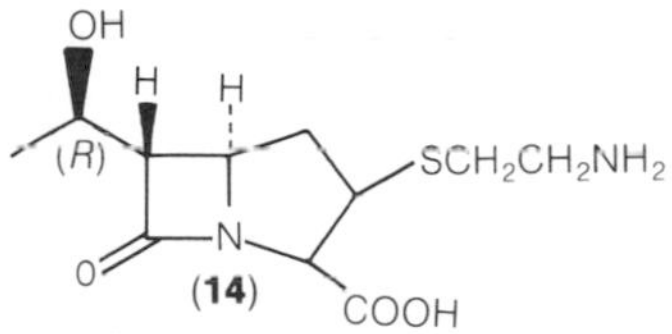

(**14**) Thienamycin

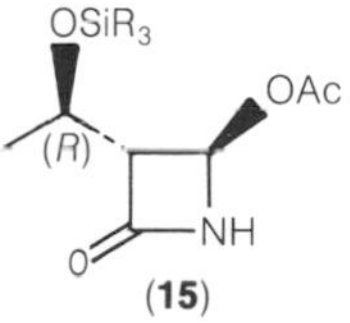

(**15**)

Some reference is made in Chapters 1 and 17 to the various synthetic strategies which have been adopted for the synthesis of this type of β-lactam intermediate. Most of these can be represented by the two disconnections illustrated in Figure 2. Each of these methods employs a diastereoselective cycloaddition step in which the precursor of the hydroxyethyl side-chain controls the relative stereochemistry of the chiral centers of the β-lactam product. The first approach, corresponding to disconnection A, employs a cycloaddition reaction between

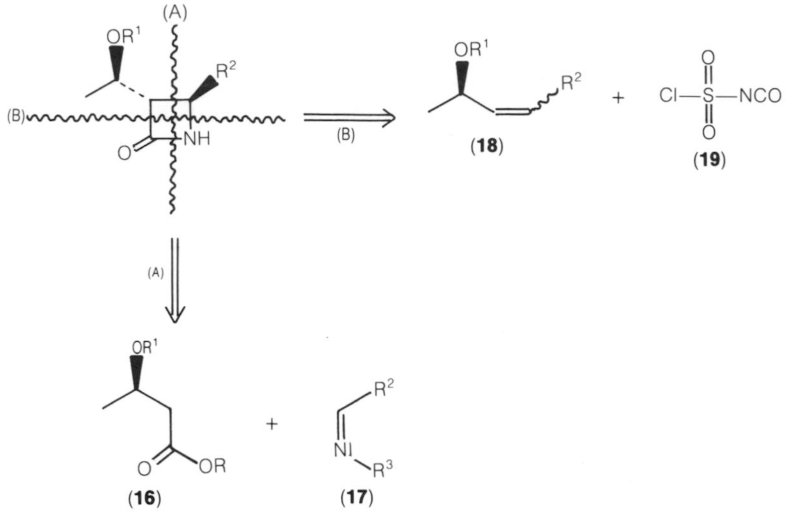

(A)
OR[1]
R[2]
(B)
NH
O
(B)
OR[1]
R[2]
(18)
+
Cl—S—NCO
(19)
(A)
OR[1]
O OR
(16)
+
R[2]
N
R[3]
(17)

Figure 2. Retrosynthetic analyses of masked 3-hydroxyethyl-β-lactams

the enolate of a 3-hydroxybutyric ester derivative (**16**) and various imines, and has been adopted by several groups[27–29].

The alternative disconnection B, which would involve the reaction of a silyl enol ether bearing an (*R*)-hydroxyethyl group (**18**) with an isocyanate derivative (**19**), appears also to offer an easy entry into this β-lactam series, but has been less widely used. Previous examples of the use of this strategy include the non-stereoselective cycloaddition reactions of chlorosulphonyl isocyanate with substituted imines shown in reactions 1 and 2 in Figure 3.[30] The reaction of an enantiomerically pure thio enol ether with chlorosulphonyl isocyanate has also been reported,[31] although in this case a disappointing 2:1 mixture of stereoisomers of the product was obtained (reaction 3).

This latter approach, corresponding to disconnection B, seemed worthy of further development, not least because of the attractive possibility of a short synthesis of either enantiomer of the required optically pure silyl enol ethers, such as compound **20**, from D- or L-3-hydroxybutyric acid.

The methodology developed by the authors for the latter transformation is outlined in Figure 4.

D-(−)-3-Hydroxybutyric acid was obtained by the microbial hydroxylation of butyric acid, as described previously. Conversion to the silyloxybutyric ester **21**, followed by reduction of the latter with diisobutylaluminium hydride or sodium bis(methoxyethoxy)aluminium hydride–amine complex, afforded the corresponding aldehyde **22** with negligible loss in optical purity. This aldehyde was converted to the silyl enol ether **23** by treatment with a second equivalent of trialkylsilyl chloride in the presence of triethylamine.

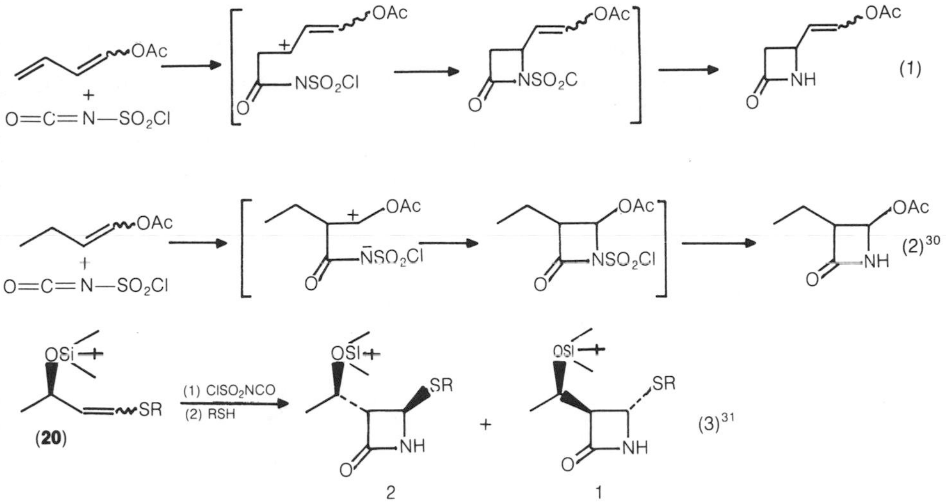

Figure 3. Previous β-lactam syntheses corresponding to disconnection (**B**)

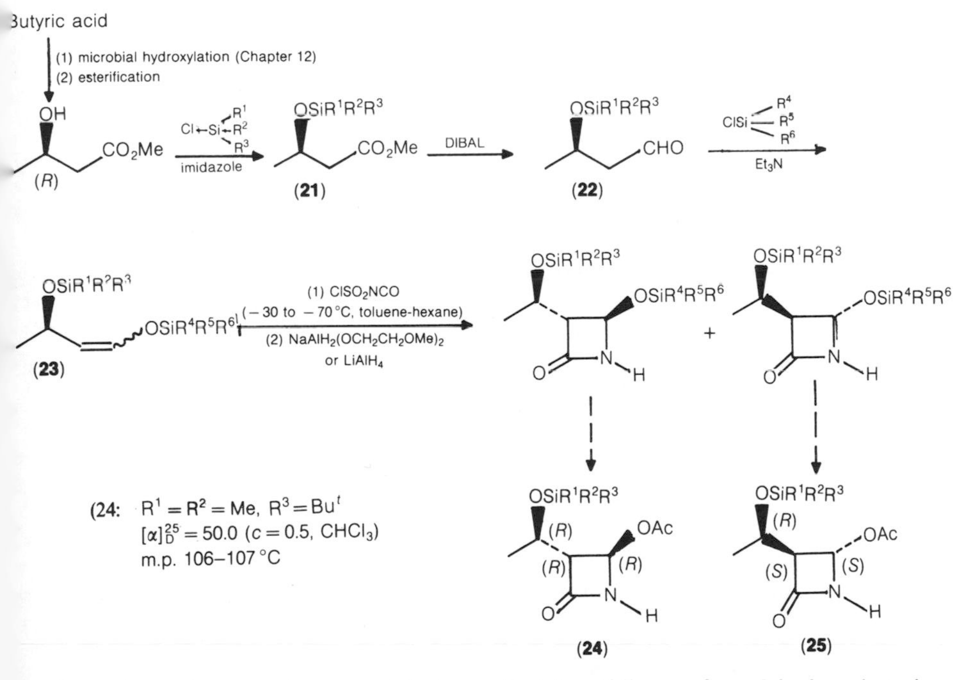

Figure 4. Syntheses of 4-acetoxy-3-hydroxyethyl-2-azetidinones from 3-hydroxybutyric acid

Various silyl enol ethers thus obtained were treated with chlorosulphonyl isocyanate in a hydrocarbon solvent at low temperatures, to effect the cycloaddition reaction. This was followed by an *in situ* cleavage of the chlorosulphonyl group by treatment with sodium bis(methoxyethoxy)aluminium hydride or lithium aluminium hydride. The stereochemistry of the β-lactam products was determined after conversion to the known compound 4-acetoxy-3-hydroxyethyl-2-azetidinone.

Table 1. Stereoselectivity of cycloadditions of silyl enol ethers

(i) $ClSO_2NCO$; (ii) Red. Al

(i) −60 °C in toluene

3*R*,4*R* (1) + 3*S*,4*S* (2)

R^1	R^2	*E*:*Z*	(1):(2)
$Si(CH_3)_2C(CH_3)_3$	$Si(CH_3)_3$	9:1	9:1
$Si(CH_3)_2C(CH_3)_3$	$Si(CH_3)_2CH_2CH(CH_3)_2$	5:1	5:1
$Si(CH_3)_2C(CH_3)_3$	$Si(CH_3)_2C(CH_3)_3$	1:1	3:2
$Si(CH_3)_2C(CH_3)_3$	$SiC(CH_3)_3(CH_3)Ph$	6:1	5:1
$Si(CH_3)_2C(CH_3)_3$	$SiC(CH_3)_3Ph_2$	1:0	5:1
$Si(CH_3)_2CH(CH_3)CH(CH_3)_2$	$Si(CH_3)_2C(CH_3)_3$	8:1	10:1
$Si(CH_3)_2C(CH_3)_2CH(CH_3)_2$	$Si(CH_3)_3$	8:1	8:1
$Si(CH_3)_2CH(CH_3)_2$	$Si(CH_3)_3$	7:1	8:1
$Si(CH_3)_2CH_2CH(CH_3)_2$	$Si(CH_3)_3$	4:1	4:1

The stereochemical outcome of the cycloaddition process was found to be influenced by the nature of the alkyl groups attached to both silicon atoms, as shown by the results in Table 1. The best combination of specificity for the 3(*R*)-enantiomer and for the *E*-regioisomer was obtained in the cases where R^2 in Table 1 was a trimethylsilyl group.

Table 2. Properties of novel 4-trialkylsilyloxy-2-azetidinones

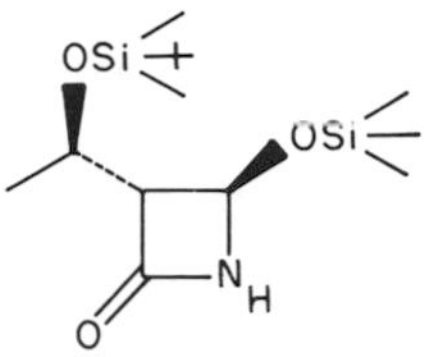

Structure	Properties
OSi, OSi, N H, O	$[\alpha]_D^{25} = -9.5°$ ($c = 1.0$, $CHCl_3$) m.p. 95–96 °C
OSi, OSi, N H, O	$[\alpha]_D^{25} = -8.3°$ ($c = 1.0$, CCl_4) m.p. 131–133 °C
OSi, OSi, N H, O	$[\alpha]_D^{25} = -33.1°$ ($c = 1.0$, CCl_4) m.p. 43–45 °C

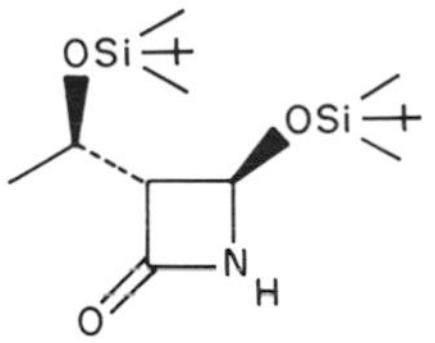

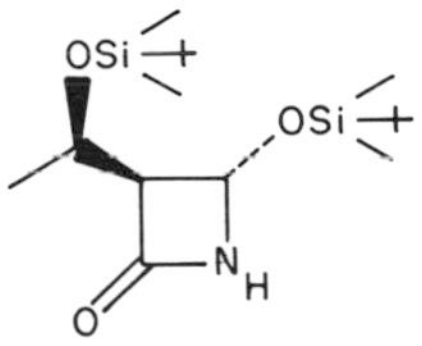

Table 3. The effect of reaction temperature on stereoselectivity

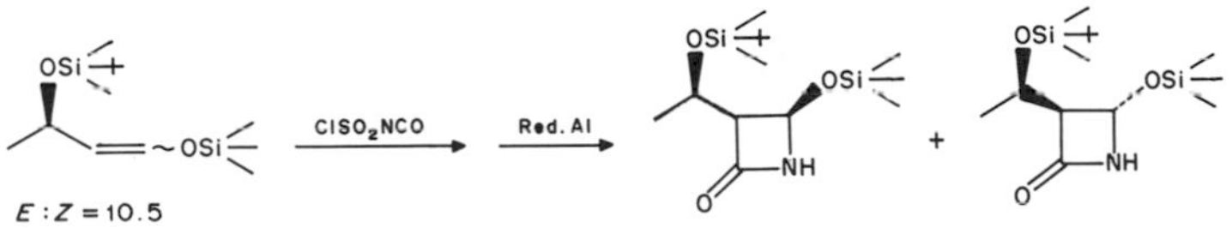

OSi, OSi; *E* : *Z* = 10.5 —$ClSO_2NCO$→ —Red. Al→ OSi, OSi, NH, O (3*R*,4*R*) + OSi, OSi, NH, O (3*S*,4*S*)

Reaction temperature (°C)	3*R*,4*R*:3*S*,4*S*
− 20	5.8:1.0
− 30	8.1:1.0
− 50	10.3:1.0
− 70	11.6:1.0

The properties of several novel optically active β-lactams prepared by this procedure are given in Table 2.

The diastereoselectivity of β-lactam formation could be increased further by lowering the reaction temperature as shown in Table 3. In addition, the reaction of the (*S*)-enantiomer of the silyl ether intermediate **26** with chlorosulphonyl isocyanate provided a ready route to the enantiomeric product, i.e. the (3*S*, 4*S*)-*trans*-β-lactam (Figure 5).

The effect of the *E*/*Z* geometry of the silyl enol ether on the stereochemistry of the β-lactam product was also investigated (Table 4). The ratio of (3*R*, 4*R*)-*trans* to (3*S*, 4*S*)-*trans* products did not appear to be affected by the proportions of (*E*)- and (*Z*)-isomers in the silyl enol ether starting material. However, the proportion of *cis*-β-lactam in the product increased in proportion to the amount of (*Z*)-isomer present in the silyl enol ether, suggesting that the *cis*-β-lactam is formed mainly from (*Z*)-silyl enol ether.

These novel β-lactams with a trimethylsilyl group at the 4-position were easily converted into the desired acetoxyazetidinones by acetylation with a combination of acetic anhydride and dimethylaminopyridine, or pyridine and acetic anhydride in the presence of an organic acid (Figure 6).

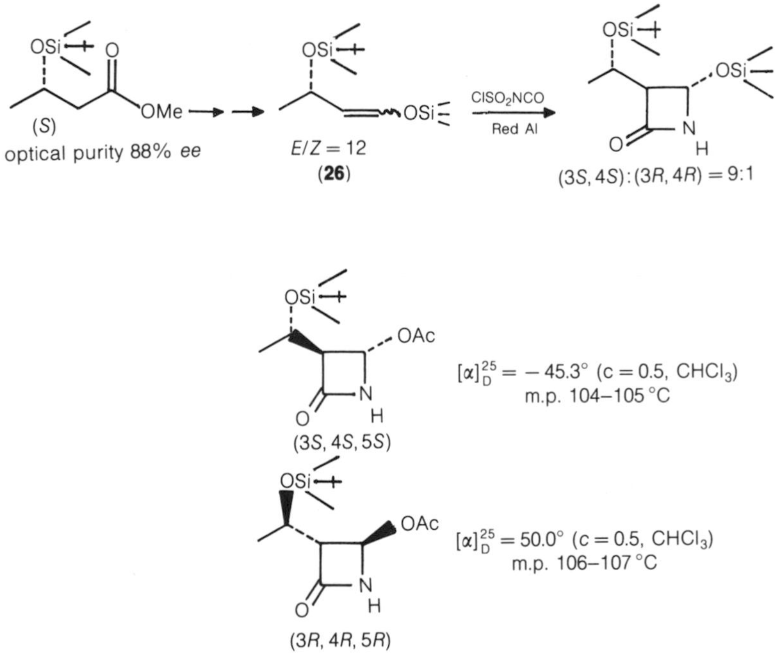

(*S*)
optical purity 88% *ee*

$E/Z = 12$
(**26**)

$ClSO_2NCO$
Red Al

(3*S*, 4*S*):(3*R*, 4*R*) = 9:1

(3*S*, 4*S*, 5*S*)

$[\alpha]_D^{25} = -45.3°$ ($c = 0.5$, $CHCl_3$)
m.p. 104–105 °C

(3*R*, 4*R*, 5*R*)

$[\alpha]_D^{25} = 50.0°$ ($c = 0.5$, $CHCl_3$)
m.p. 106–107 °C

Figure 5. Synthesis and properties of the enantiomeric *trans*-2-azetidinones

Table 4. The effect of silyl enol ether geometry on stereoselectivity

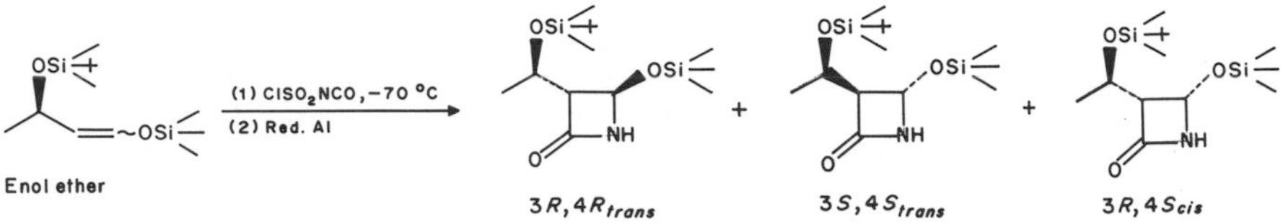

(1) $ClSO_2NCO$, −70 °C
(2) Red. Al

Enol ether

$3R,4R_{trans}$ + $3S,4S_{trans}$ + $3R,4S_{cis}$

Enol ether E:Z	3R,4R:3S,4S:cis
99:1	11:1:0.4
94:6	10:1:0.8
80:20	10:1:2.9

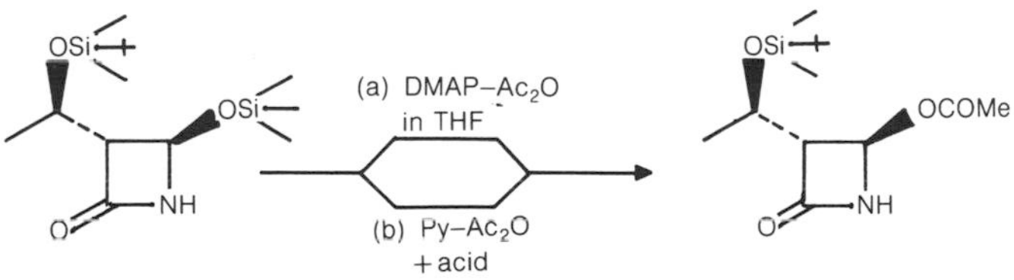

(a) DMAP–Ac_2O in THF

(b) Py–Ac_2O + acid

acid = *p*-toluenesulphonic acid, trichloroacetic acid or trifluoroacetic acid

Figure 6. Conversion of 4-trimethylsilyloxy azetidinones to 4-acetoxy-2-azetidinones

In summary, a stereoselective cycloaddition process for the desired (*R, R, R*)-monocyclic β-lactam has been established starting from D-(−)-3-hydroxybutyric acid. The (*R*)-stereochemistry at the β-position was induced by the (*R*)-configuration of the silyloxyethyl group, with the *trans* geometry of the β-lactam product being conferred by the *E*-stereochemistry of the silyl enol ether.

13.3 REFERENCES

1. Shimazaki, M., Hasegawa, J., Kan, K., Nomura, K., Nose, Y., Kondo, K., Ohashi, T., and Watanabe, K., *Chem. Pharm. Bull.*, **30**, 3139 (1982).
2. Ondetti, M. A., Rubin, B., and Cushman, D. W., *Science*, **196**, 441 (1977).
3. Ondetti, M. A., and Cushman, D. W., *Jpn. Kokai Tokkyo*, JP 52-116 457 (1977).
4. Cushman, D. W., Chenung, H. S., Sabo, F. F., and Ondetti, M. A., *Biochemistry*, **16**, 5484 (1977).
5. Iwao, J., Oya. M., Kato, E., and Watanabe, T., *Jpn. Kokai Tokkyo*, JP 54-151912 (1979).
6. Houbiers, J. P. M., *Jpn. Kokai Tokkyo*, JP 55-38 386 (1980).
7. Ohashi, N., Nagata, S., and Katsube, S., *Jpn. Kokai Tokkyo*, JP 56-7756 (1981).
8. Ohashi, T., Kan, K., Sada, I., Miyama, A., and Watanabe, K., *Jpn. Kokai Tokkyo*, JP 61-18 791 (1986).

9. Ohashi, T., Kan, K., Sada, I., Miyama, A., and Watanabe, K., *Jpn. Kokai Tokkyo*, JP 61-18 758 (1986).
10. Fujimoto, K., Iwano, Y., Hirai, K., and Sugawara, S., *Chem. Pharm. Bull.*, **34**, 999 (1986).
11. Alpegiani, M., Bedeschi, A., Giudichi, F., Perrone, E., and Franceschi, G, *J. Am. Chem. Soc.*, **107**, 6398 (1985).
12. Girijavallabham, V. M., Ganguly, A. K., McCombie, S. W., Pinto, P., and Rizvi, R., *Tetrahedron Lett.*, **22**, 3485 (1981).
13. Daniels, N., Johnson, G., and Ross, B. C., *J. Chem. Soc., Chem. Commun.*, 1006 (1983).
14. Laing, M., Schneider, P., Tosch, W., Scartazzini, R., and Zak, U., *Abstracts of the 25th Interscience Conference on Antimicrobial Agents and Chemotherapy, 1985*, p. 159, No. 376.
15. Leanza, W. J., Dininno, F., Muthard, D. A., Wilkening, R. R., Wildonger, K. J., Ratcliffe, R. W., and Christensen, B. G., *Tetrahedron*, **39**, 2505 (1983).
16. Reider, P. J., and Grabowski, E. J. J., *Tetrahedron Lett.*, **23**, 2293 (1982).
17. Tajima, Y., Yoshida, A., Takeda, N., and Oida, S., *Tetrahedron Lett.*, **26**, 673 (1985).
18. Nagao, Y., Kumagai, T., Tamai, S., Abe, T., Kuramoto, Y., Taga, T., Aoyagi, S., Nagase, Y., Ochiai, M., Inoue, Y., and Fujita, E., *J. Am. Chem. Soc.*, **108**, 4673 (1986).
19. Fuentes, M., Shinkai, I., and Salzmann, T. N., *J. Am. Chem. Soc.*, **108**, 4675 (1986).
20. Shibata, T., Iino, K., Tanaka, T., Hashimoto, T., Kameyama, Y., and Sugimura, Y., *Tetrahedron Lett.*, **26**, 4739 (1985).
21. Shiozaki, M., Ishida, N., Maruyama, H., and Hiraoka, T., *Tetrahedron*, **39**, 2399 (1983).
22. Ito, Y., and Terashima, S., *Tetrahedron Lett.*, **28**, 6625 (1987).
23. Tschaen, D. M., Fuentes, L. M., Lynch, J. E., Laswell, W. L., Volante, R. P., and Shinkai, I., *Tetrahedron Lett.*, **29**, 2779 (1988).
24. Georg, G. I., Kant, J., and Gill, H. S., *J. Am. Chem. Soc.*, **109**, 1129 (1987).
25. Sunagawa, J., Nozaki, Y., Sasaki, A., and Matsumura, H., *Jpn. Kokai Tokkyo*, JP 63-165 349 (1988).
26. Noyori, I., Ikeda, T., Ohkuma, T., Widhalm, M., Kitamura, K., Takaya, H., Akutagawa, S., Sayo, N., Saito, T., Taketomi, T., and Kumobayashi, H., *J. Am. Chem. Soc.*, **111**, 9134 (1989).
27. Ha, D.-C., Hart, D. J., and Yang, T.-K., *J. Am. Chem. Soc.*, **106**, 4819 (1984).
28. Chiba, T., and Nakai, T., *Chem. Lett.*, 651 (1985).
29. Iimori, T., and Shibazaki, M., *Tetrahedron Lett.*, **26**, 1523 (1985).
30. Kametani, T., Honda, T., Nakayama, A., Sasaki, Y., Mochizuki, T., and Fukumoto, K., *J. Chem. Soc. Perkin Trans. 1*, 2228 (1981).
31. Ishiguro, M., Iwata, T., Nakatsuka, T., and Kouzu, M., *Jpn. Kokai Tokkyo*, JP 61-207 373 (1986).

14 Chiral Cyanohydrins—Their Manufacture and Utility as Chiral Building Blocks

C. G. Kruse
Solvay Duphar Research Laboratories, Weesp, The Netherlands

14.1 INTRODUCTION

Owing to their broad synthetic potential, cyanohydrins have attracted attention as starting materials for the preparation of several important classes of

Chirality in Industry. Edited by A. N. Collins, G. N. Sheldrake and J. Crosby

compounds. These include α-hydroxy acids and esters, acyloins, α-hydroxyaldehydes, vicinal diols, β-amino alcohols and β-hydroxy-α-amino acids.[1] These structural moieties are present in a large number of industrial products, examples of which are pharmaceuticals, veterinary products, crop-protecting agents, vitamins and food additives. As enantiomeric purity has become an increasingly important criterion in these application areas, so research into new syntheses and applications of optically active cyanohydrines has increased markedly in the last decade.

It is probable that the use of enantiomerically pure cyanohydrins as building blocks for the production of chiral industrial chemicals will continue to grow, as this avoids the problems associated with the optical resolution or asymmetric synthesis of certain products. New routes to homochiral cyanohydrins represent, therefore, an opportunity to enlarge the pool of chiral starting materials which are available to the fine chemicals industry. Several criteria must be realized fully before optically pure cyanohydrins can be adopted as raw materials for industrial processes. These are:

(i) the availability of a range of methods for the manufacture of cyanohydrins with a high enantiomeric excess (*ee*) in an economically feasible way;
(ii) the preservation of optical purity during subsequent chemical transformations;
(iii) the possibility of chirality transfer by diastereoselective reactions at either the cyano group or the main organic residue.

This chapter highlights the progress that has been made in these areas up to mid-1991.

14.2 AN OVERVIEW OF PRODUCTION METHODS

Most chiral building blocks that are currently on the market have their origin in cheap natural products, such as amino acids, carbohydrates, terpenes and hydroxy acids. In the case of chiral cyanohydrins other methods must be followed, and both resolution processes and asymmetric synthesis have been

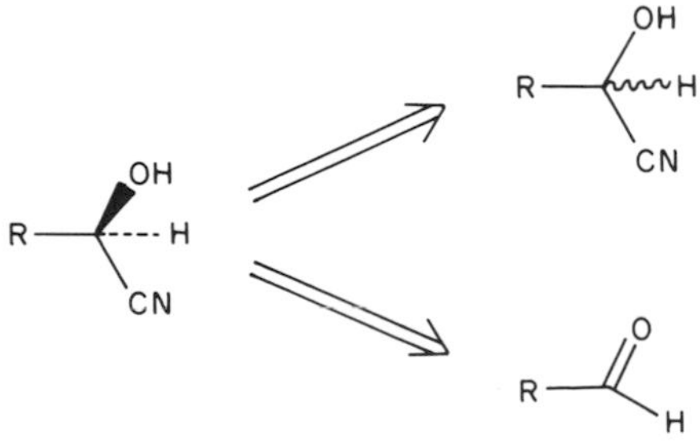

OH
R—C---H
CN

OH
R—C~~H
CN

O
R—C
H

Scheme 1

investigated extensively. As illustrated in Scheme 1 for aldehyde cyanohydrins, the precursor is usually either the cyanohydrin racemate or the prochiral aldehyde.

In the former case, the racemate must be resolved, either by the formation of a diastereoisomeric derivative or by a kinetic resolution process. The inherent disadvantage of these approaches is the need to separate the desired product from the by-product derived from the other enantiomer. In the case of prochiral aldehydes an enantioselective addition of (a precursor of) HCN may be achieved with the assistance of a chiral chemo- or biocatalyst.

14.2.1 RESOLUTION OF RACEMATES BY DIASTEREOISOMER FORMATION

Only a few methods have been published on the use of chiral auxiliaries for the separation of the optical isomers of cyanohydrins. Elliot *et al.*[2] reported the application of optically pure pentane-2,4-diol for this purpose. Cyanohydrin ethers (**1**) (Scheme 2) were obtained from a diastereoselective (*de* >90%) ring-opening reaction of the corresponding aldehyde acetals with cyanotrimethylsilane. The two-step procedure which had to be used for removal of the auxiliary group makes this method unattractive for industrial use, howcvcr.

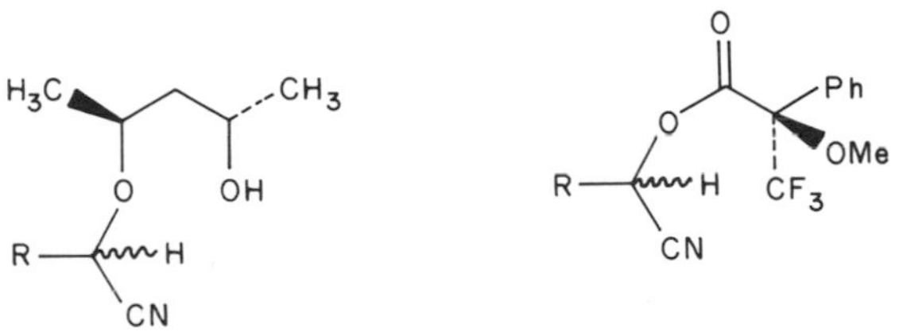

(**1**) (*de* >90%) (**2**)

Scheme 2

The conversion of cyanohydrin racemates into chiral norbornyl derivatives by Noe[3] serves only analytical purposes. The same holds for the conversion into MTPA esters (**2**) by treatment with Mosher's reagent.[4] The latter technique is of great value, as it permits an accurate determination of the optical purities of cyanohydrins by gas chromatography.

14.2.2 RESOLUTION OF RACEMATES BY KINETIC RESOLUTION

Many authors have claimed the application of biocatalysts for the production of optically active cyanohydrin esters, as illustrated in Scheme 3.

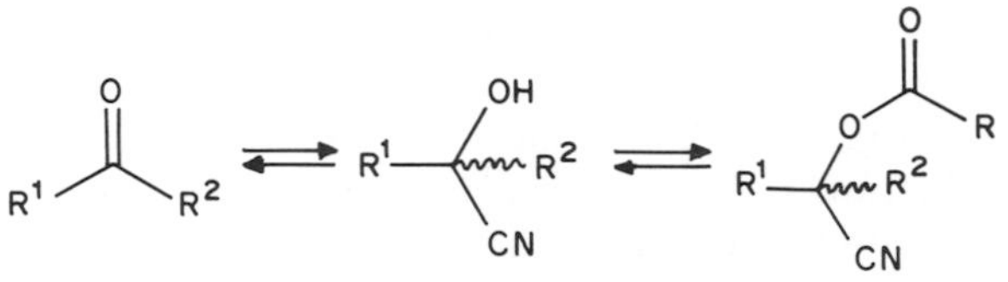

(3)(R/S) (4)(R/S)

Scheme 3

The principle of all these methods is the stereoselective recognition of esterification or hydrolysis transition states by enzymes. Processes may be carried out in two different modes. Esterification of the ketone cyanohydrin racemates **3** (R^1 and R^2 = alkyl) by bacterial preparations[5] leads to a mixture of the unconverted cyanohydrin **3** and the corresponding ester **4**, both enriched in one enantiomer. The reverse reaction, hydrolysis of the racemic ester **4** (R^1 and R^2 = alkyl) with similar microbial catalysts, as reported by Ohta and co-workers,[6] and of the ester **4** (R^2 = H) with different types of lipases,[7–9] leads to the optically enriched cyanohydrin **3** (R^2 = H) and the unconverted ester **4**. Since in both cases the cyanohydrin **3** can be reconverted to its carbonyl precursor, these enzymic reactions allow, in principle, a choice of efficient processes for the production of optically active cyanohydrin derivatives from the carbonyl raw materials.

14.2.3 ENANTIOSELECTIVE SYNTHESIS BY CHEMOCATALYSIS

The use of chiral basic catalysts for the stereospecific addition of hydrogen cyanide to aldehydes (Scheme 4) was first conceived in 1912 by Bredig and Fiske,[10] who reported the application of quinine for this purpose. Polymers with chiral amine groups were used by Tsuboyama[11] in the 1960s and copolymers of acrylonitrile and cinchona alkaloids by the Sumimoto group[12] in 1986.

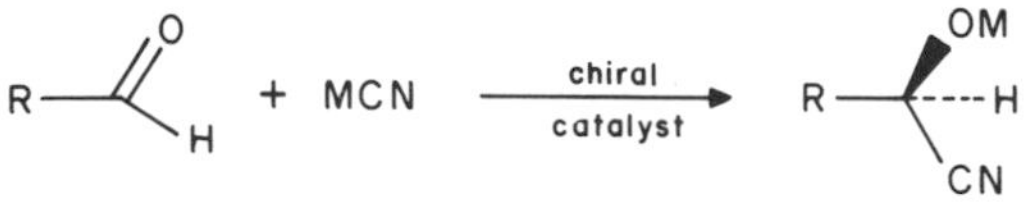

Scheme 4

In the last decade, a large number of academic and industrial laboratories have been involved with this topic. Modern catalysts for this reaction include hydantoins[13] and β-cyclodextrins.[14] All of these methods result in poor *ee* values (less than 50%), however. Apparently, it is difficult to obtain the rigid transition-state geometry which is necessary for good enantioselectivity.

Better results have been published with chiral Lewis acids based on complexes of aluminium,[15] titanium,[16] titanium–lithium[17] and boron[18] complexes which contain conformationally rigid ligands. The direct addition of cyanotrimethylsilane to carbonyl compounds can be achieved with these catalysts. The most promising development, however, was initiated by Inoue and co-workers[19] and developed later by Jackson and co-workers.[20] These authors investigated chiral basic dipeptides (Scheme 5), both cyclic as in compound **5**,[21] and acyclic, as in structure **6** (in the latter case as a titanium complex[22]). These compounds catalyse the addition of HCN to a wide variety of aldehydes in organic solvents.

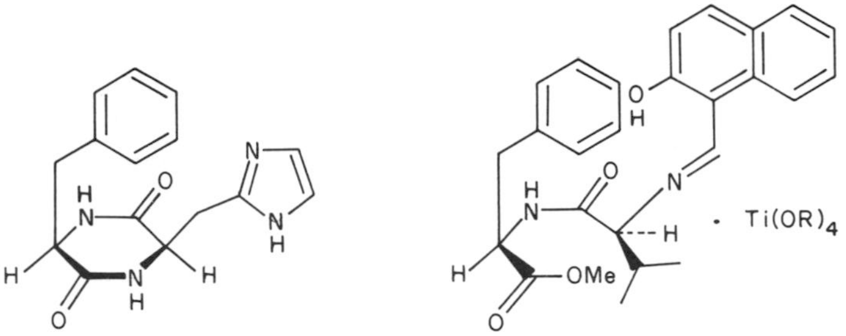

(**5**) (**6**)

Scheme 5

Although a number of examples of *ee* values >90% were reported, these methods still suffer from substantial drawbacks, such as critical experimental conditions with low temperatures (< −40 °C), dilute solutions and vulnerable catalysts, all of which hamper their industrial application.

14.2.4 ENANTIOSELECTIVE SYNTHESIS BY BIOCATALYSIS

The pioneering work of Becker and Pfeil[23] in the 1960s demonstrated for the first time that the hydroxynitrile lyase (oxynitrilase) enzyme isolated from almonds (E.C. 4.1.2.19) can be employed successfully for the production of (*R*)-cyanohydrins from aldehydes and HCN on a kilogram scale (Scheme 4, M = H). After a long lapse this work was continued in the late 1980s by the groups of Effenberger,[24] Brussee and Van der Gen[25] and Kula.[26] These workers have shown that the chiral cyanohydrins can be obtained in an optical purity of more than 90%, by carrying out the reaction either in aqueous solutions at relatively low pH[25,26] or in organic solvents.[24a,b] The substrate specificity of the enzyme appeared not to be very restricted; both saturated and α,β-unsaturated aliphatic and (hetero)aromatic aldehydes were converted. Moreover, by working in organic media, even aliphatic ketones were transformed into their corresponding (*R*)-cyanohydrins with high *ee* values.[24c] Studies with a related enzyme isolated from sorghum (E.C. 4.1.2.11) revealed the possibility of a general route to aldehyde (*S*)-cyanohydrins,[24,26] although only aromatic substrates were tolerated.[27]

On the basis of these results, comprehensive research efforts have been undertaken to find industrially useful procedures. This work is discussed in the next section.

14.3 THE OXYNITRILASES

14.3.1 ACCESSIBILITY FROM NATURAL SOURCES

Enzymes of the oxynitrilase family occur widely in higher plants. Their natural function is to catalyse the final step in the biodegradation pathway of cyanogenic glycosides (**7**) into the elements of glucose, a carbonyl compound and HCN (Scheme 6).

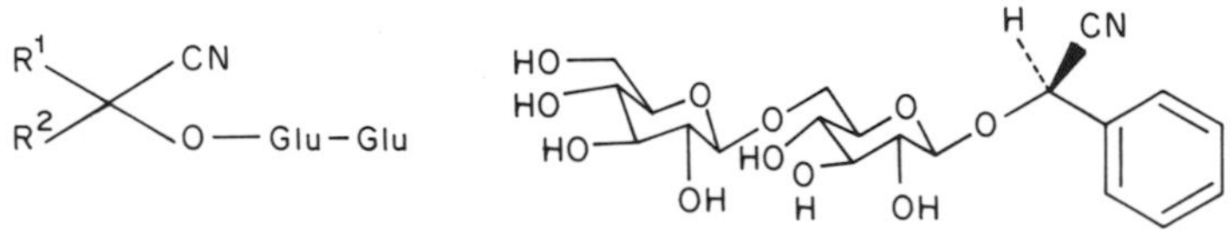

(**7**)

(**8**) Amygdalin

Scheme 6

A variety of secondary metabolites of type **7** have been isolated.[28] In almonds, for example, amygdalin (**8**) is present, and is decomposed by a mixture of β-glucosidases and (*R*)-mandelonitrile lyase. Rosenthaler,[29] in 1908, was the first to isolate these enzymes from an aqueous extract of defatted almonds termed emulsin. His attempts to use the unpurified oxynitrilase as a biocatalyst in the synthesis of (*R*)-mandelonitrile from benzaldehyde and excess HCN met with little success, however.

In 1963, Becker and Pfeil obtained a highly purified preparation of this (*R*)-oxynitrilase from bitter almonds.[30] The enzyme turned out to be a rather stable, yellow glycoprotein which contained flavin adenine dinucleotide (FAD), and had a molecular weight of 60 000. Since then, a large number of similar FAD-containing enzymes have been isolated from the seeds of higher plants from the Rosaceae subfamilies Prunoideae and Maloideae.[31] All these enzymes catalyse the *Si*-face addition of HCN to a large number of aliphatic and aromatic aldehydes (Figure 1).

Alternatively, a group of non-FAD-dependent oxynitrilases isolated from cyanogenic plants such as cassava, flax, clover, sorghum and the Olacaceae family catalyse the formation of chiral (*S*)-cyanohydrins by a *Re*-face addition

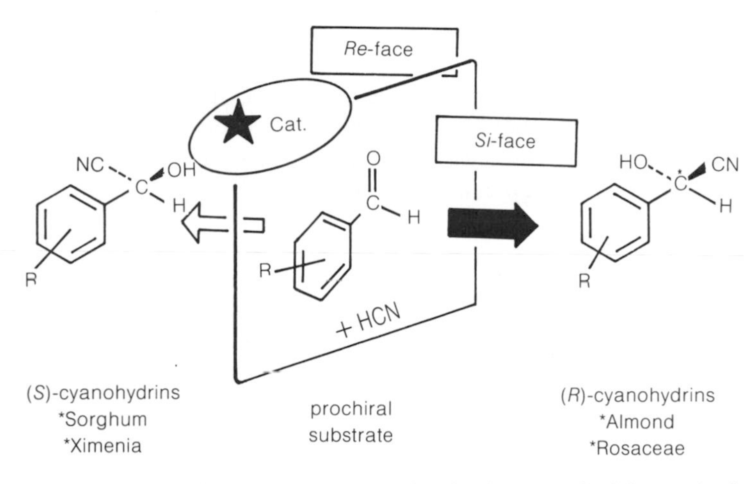

Figure 1. Oxynitrilase-catalysed enantioselective cyanohydrin synthesis

of HCN (Figure 1). These enzymes have molecular weights of approximately 100, 000, and are less stable than their (*R*)-oxynitrilase counterparts.[27]

Although both (*R*)- and (*S*)-oxynitrilases have been shown to be highly efficient biocatalysts,[27,31] only in the case of almonds (and to a lesser extent sorghum) is the amount of enzyme in the natural source sufficiently high to justify an economical large-scale isolation. However, Brussee and co-workers[25] have shown that substantial purification is not necessary, as crude aqueous extracts from almond meal can be employed successfully.

The cloning and overexpression of the genomes for oxynitrilases in prokaryotic or eukaryotic organisms will greatly expand the potential industrial applications of these enzymes in the near future.

14.3.2 HOMOGENEOUS AQUEOUS SYSTEMS

In many cases biocatalytic steps can only be carried out successfully if a technical solution has been found for the problem of bringing together a water-soluble catalyst with a lipophilic substrate. In the case of (*R*)-oxynitrilase from almonds, the addition of water-miscible organic solvents such as the lower alcohols (up to 30%, v/v) to an aqueous buffer of pH 5.5 has been investigated.[23,25] Only in the case of a limited number of substrates (derivatives of the natural substrate benzyaldehyde) could products with sufficiently high *ee* values be obtained.[24a,25] The poor enantioselectivity was attributed to a competing non-enzymic addition of HCN to the aldehyde.

By lowering the pH it was possible to suppress the chemical reaction, thus causing the biocatalytic process to predominate. This was demonstrated both for (*R*)- and (*S*)-oxynitrilase (from sorghum), although low substrate concentrations were used.[26,32] Other water-miscible co-solvents such as dimethylformamide did not give substantial improvements.[33]

In order to develop an industrially useful process, high throughputs must be obtained, and the enzyme should be used in the most cost-effective way. Technologies which have been developed to achieve conditions under which the substrate is in contact with high concentrations of the biocatalyst are discussed in the next section.

14.3.3 IMMOBILIZATION METHODS

The concept of catalyst immobilization on a solid support has been applied extensively in industrial processes, (see Chapter 19). Being relatively robust, the (*R*)-oxynitrilase enzyme lends itself well to the production of optically active cyanohydrins on a large scale. This opportunity has been recognised by Becker and Pfeil,[34] who reported the use of the purified enzyme in a continuous-flow reactor by immobilization on silica. This approach was futher improved by the group of Mattiason,[35] and Westman[36] reported the immobilization of (*R*)-oxynitrilase on an ethylene–maleic anhydride copolymer. The aldehyde substrates, together with a buffer containing HCN (2–3 equivalents) in a solvent mixture of water and methanol (1:3, v/v) were fed to the reactor, and the flow adjusted in such a way that complete conversion was achieved.

These immobilized enzyme preparations possess a remarkable stability. A significant problem associated with these aqueous systems, however, remains the competition from the chemical addition reaction, which may result in dramatic decreases of *ee* values of the isolated products. A major advance by Effenberger *et al.*[37] in 1987 was the demonstration that the production of many cyanohydrins in high optical purity could also be carried out in apolar organic solvents saturated with an aqueous buffer solution of an appropriate pH. The purified (*R*)-oxynitrilase enzyme was applied in an immobilized preparation by absorption or precipitation on chemically modified cellulose. An alternative immobilization method which involves an encapsulation of the enzyme in lyotropic liquid crystals embedded in a porous support has been reported by Kula *et al.*[38] An attractive alternative was published by Zandbergen *et al.*,[39] who simply used the crude, defatted almond meal as a natural support.

The main advantage of the organic reaction medium is the almost complete suppression of the chemical reaction. By carrying out the process in a flow reactor, high conversions and high *ee* values can be achieved, at least in principle, with all substrates for the enzyme. The application of selectively permeable membranes can also be envisaged.

In fact, the Solvay Duphar research group[33] found that appreciable problems remained with the use of immobilized oxynitrilases. First, a large number of apolar hydrocarbons and chlorinated solvents adversely affect the stability of the (*R*)-oxynitrilase enzyme (Effenberger *et al.* reported only ethyl acetate[24a] and diisopropyl ether.[24b]) Second, the activity of the biocatalyst is drastically reduced in the immobilized preparations. Further, the residual catalytic activity was found to depend critically on the moisture content of the solvent, a disadvantage in both batch and continuous operations.

14.3.4 TWO-PHASE SYSTEMS

The application of heterogeneous systems of two liquid phases to industrial biocatalytic processes is an intriguing possibility.[40] The principle is to combine optimum enzyme performance in the buffered aqueous phase with a high productivity by employing the organic phase as a reservoir of both substrate and product. The efficiency of this method is due, in part, to the absence of inhibition of the enzyme by the cyanohydrin product. An attractive feature is the ease with which the catalyst may be separated from the product.

The first report on the use of this procedure for the preparation of optically active cyanohydrins with catalysis by (*R*)-oxynitrilase was recently published by Ognyanov *et al.*[41] The possibility of a competing chemical reaction in the aqueous phase was prevented elegantly by the use of a *trans*-hydrocyanation process, with acetone cyanohydrin as a donor of HCN (Scheme 7).

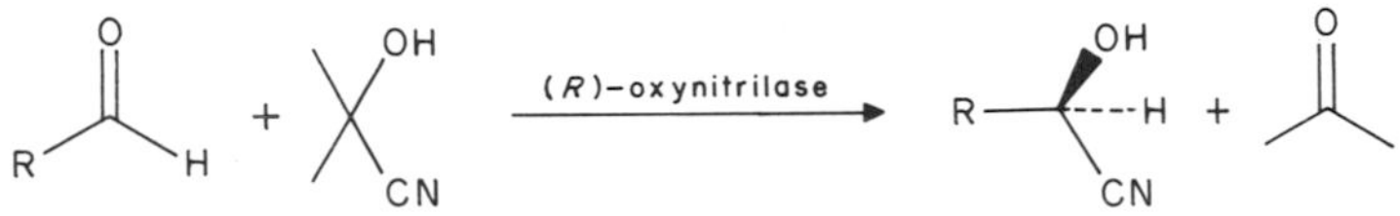

O
R H + OH CN (*R*)-oxynitrilase R OH H CN + O

Scheme 7

Several aldehyde substrates were transformed into the corresponding chiral cyanohydrins with reasonable yields and in high optical purity by this method. However, a close examination of the reaction conditions again reveals a low enzyme efficiency, resulting in long reaction times, and a large volume of organic solvent (diethyl ether in this case). These drawbacks were completely eliminated in a process developed by the Duphar group, which is the subject of a recent patent application.[42] The method combines a high enzyme activity (approximately equal to the theoretical value) with a high productivity by reducing the volume of the organic solvent. Further, it was shown that the enzyme possesses excellent stability under the applied conditions, so that re-use of the aqueous layer is possible. Moreover, the (*S*)-oxynitrilase from sorghum worked equally well.

With a variety of technological possibilites for two-phase processes under development, it is likely that highly efficient, cost-effective production methods for a wide variety of optically pure cyanohydrins on a multi-kilogram scale will soon be within reach.

14.4 THE USE OF PROTECTING GROUPS

The direct application of optically active cyanohydrins as chiral building blocks for industrial fine chemicals is prevented by their instability. In most cases, purification by means of distillation, crystallization or chromatography is accompanied by extensive degradation and racemization. Futhermore, chemical transformations of racemic cyanohydrins with basic, reducing or organometallic reagents often result in low yields of the desired products.[1,43] If unprotected chiral cyanohydrins are unsuitable candidates for the chiral pool, studies aimed at finding the most appropriate protecting groups may well hold the key to a broad industrial application of these compounds.[44]

For obvious reasons, only those groups which can be attached to the hydroxyl function under neutral or acidic conditions can be considered as protecting groups. Protection of optically active cyanohydrins (**9**) as silyl ethers has been accomplished with complete retention of enantiomeric purity and in high yields, by treatment with trialkyl- or triarylsilyl chlorides in the presence of imidazole or dimethylaminopyridine (Scheme 8). The resulting chiral, silylated cyanohydrins (**10**) are stable compounds which are easy to purify.[25,44] When more basic amines were used, decomposition and racemization were observed. Weak organic bases such as pyridine were successful only in the case of **10a**, as was shown by Effenberger *et al.*[45] The preservation of *ee* in the silyl ether **10** was established unambiguously by treatment of this compound with hydrogen fluoride in acetonitrile to give the cyanohydrin **9** with an unchanged optical purity.[25b]

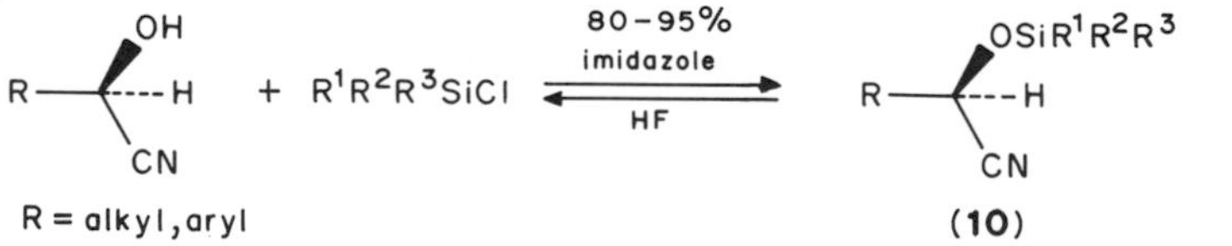

R—C(OH)(H)—CN + $R^1R^2R^3SiCl$ ⇌ (80–95%, imidazole; reverse: HF) R—C($OSiR^1R^2R^3$)(H)—CN

R = alkyl, aryl
(**9**) (**10**)

	R^1	R^2	R^3
(**a**)	Me	Me	Me
(**b**)	Bu^t	Me	Me
(**c**)	thexyl	Me	Me
(**d**)	Bu^t	Ph	Ph

thexyl = $(CH_3)_2CH—C(CH_3)_2—$

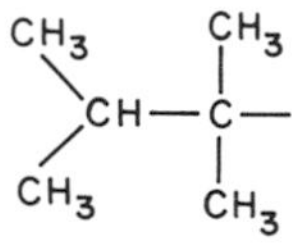

Scheme 8

Adopting the technology for the biocatalytic step as described in Section 14.3, and utilizing *tert*-butyldimethylsilyl chloride (TBDMSCl) as a protecting group, Solvay Duphar's subsidiary, Peboc, have introduced the first optically pure (*ee*

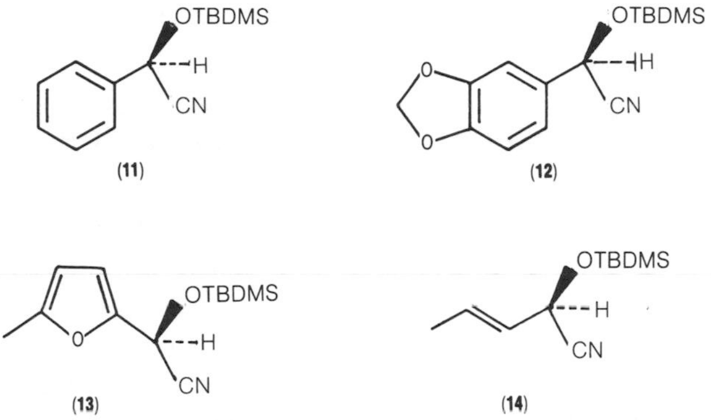

(11) (12)

(13) (14)

Figure 2. TBDMS-protected cyanohydrin enantiomers (*ee* >95%) available on the market (Peboc)

>95%), TBDMS-protected cyanohydrins (11–14) on to the market (Figure 2). Other protecting groups have also been coupled successfully to the cyanohydrin enantiomers in high yield and without racemization (Scheme 9).

The reaction of cyanohydrins with acid chlorides in pyridine gives access to cyanohydrin esters (**15**), whereas acid-catalysed reactions with vinyl ethers produce acetals, such as methoxyisopropyl (MIP) ethers (**16**), cyclic THF ethers (**17**) and THP ethers (**18**).

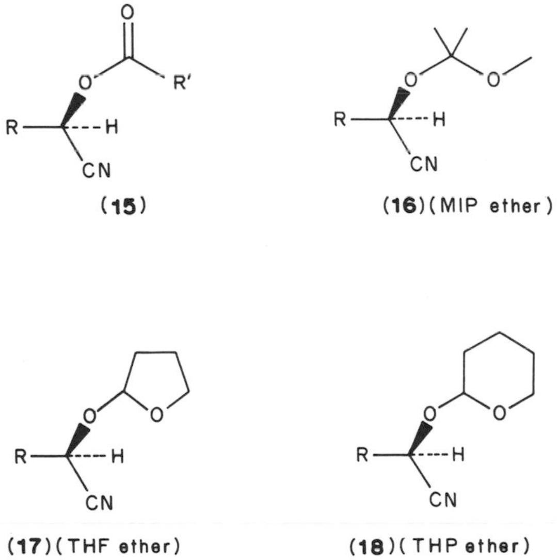

(**15**) (**16**)(MIP ether)

(**17**)(THF ether) (**18**)(THP ether)

Scheme 9

14.5 CHEMICAL TRANSFORMATIONS INTO CHIRAL PRODUCTS

14.5.1 α-HYDROXYCARBOXYLIC ESTERS AND ACIDS

One of the few chemical reactions on the nitrile group of chiral cyanohydrins that can be carried out without the need for protection is acid-catalysed hydrolysis. Owing to the instability of most cyanohydrins at higher temperatures, the preferred method for hydrolysis is that of Corson *et al.*:[46] first stirring with concentrated hydrochloric acid at ambient temperature until the formation of the amide **19** is complete, followed by heating to give the optically-active α-hydroxy acid **20** (Scheme 10), without racemization.[24b] The alternative procedure of Noe[3] starts from 4-methoxymandelonitrile (**9**) (R = 4-MeOPh; *ee* 99%).

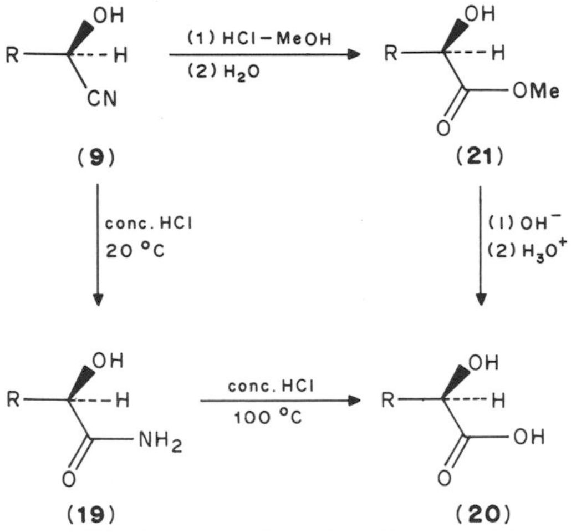

(1) HCl–MeOH
(2) H_2O
conc. HCl
20 °C
(1) OH^-
(2) H_3O^+
conc. HCl
100 °C
(**9**) (**21**) (**19**) (**20**)

Scheme 10

Treatment with hydrogen chloride and two equivalents of methanol in diethyl ether leads to the nearly quantitative formation of the imide hydrochloride salt, which, on hydrolysis, leads to the optically pure α-hydroxy methyl ester **21** (R = 4-OMePh) in high yield (95%). Subsequent hydrolysis proceeds smoothly under mild conditions with one equivalent of sodium hydroxide to give optically pure 4-methoxymandelic acid (**20**) (R = 4-MeOPh).[25b]

14.5.2 α-HYDROXYALDEHYDES AND KETONES

Racemic α-hydroxyaldehydes are known to be rather unstable.[47] Partial reduction of racemic *protected* cyanohydrins with sodium di(2-methoxyethoxy)-

aluminium hydride[47] or diisobutylaluminium hydride (DIBAL)[48] has been described, although the corresponding protected α-hydroxyaldehydes were isolated in only moderate yields. Similar poor results were obtained with DIBAL for the reduction of optically active compounds protected as silyl esters,[49] and Effenberger *et al.*[50] indicated that partial racemization of the product occurred during their multi-step procedure.

Brussee[51] studied the DIBAL reduction of the TBDMS ether **11** under a wide variety of conditions. With exactly one equivalent of DIBAL, the conversion of **11** was incomplete, presumably owing to over-reduction of the initially formed imine–aluminium complex. The best results were obtained at room temperature in THF with a rapid addition of 1.5 equivalents of DIBAL (Scheme 11). Under

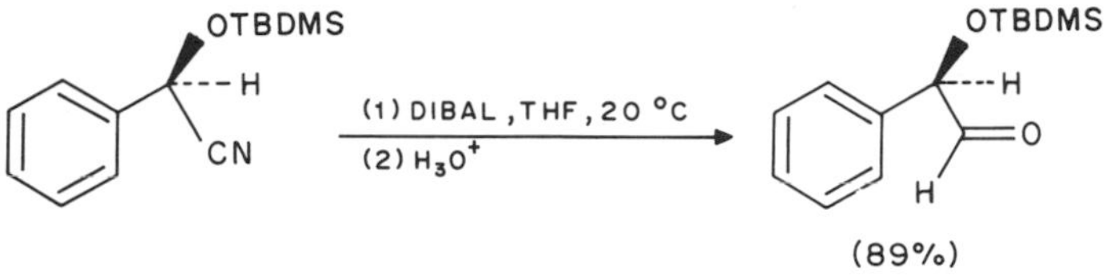

OTBDMS H CN (11) (1) DIBAL, THF, 20 °C (2) H_3O^+ OTBDMS H O H (89%) (22)

Scheme 11

OP H CN R (11,23–25) CH_3MgI OP H NMgI CH_3 R (26) P=TMS H_3O^+ P=TBDMS H_3O^+ OH H O CH_3 R (27a,b) OTBDMS H O CH_3 R (28a,b)

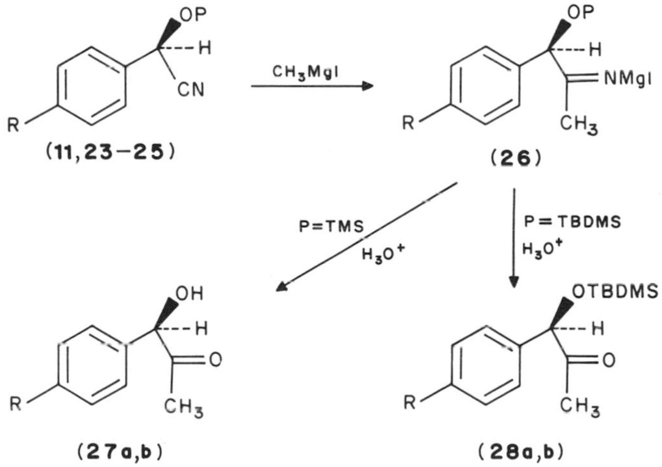

	R	P	Product	Yield (%)	*ee* (%)
11	H	TBDMS	**28a**	80	92
23	OMe	TBDMS	**28b**	80	99
24	H	TMS	**27a**	71	95
25	OMe	TMS	**27b**	78	96

Scheme 12

these conditions, protected, optically active α-hydroxyphenylacetaldehyde (**22**) was obtained in 80% yield.

The nitrile to ketone functional group transformation requires a Grignard-type addition reaction, followed by hydrolysis. Protection of the hydroxyl function is known to be necessary to obtain a reasonable conversion in the case of racemic cyanohydrins.[43b] Brussee *et al.*[25a,52] have investigated the formation of acyloins by reaction of protected mandelonitriles (**11** and **23**–**25**) with the Grignard reagent derived from methyl iodide. The initially formed imine adducts (**26**) were hydrolysed to unprotected acyloins (**27**) in the case of the TMS ethers, and to their protected analogues (**28**) in the case of the TBDMS ethers (Scheme 12).

High yields of acyloins with virtually unaffected optical purities were obtained, indicating that conservation of chirality had been achieved in spite of the presence of the relatively acidic benzylic α-proton in compounds **11** and **23**–**25**.

14.5.3 β-HYDROXYAMINES

Direct reduction of the nitrile group to the amino group in unprotected, enantiomerically pure cyanohydrins may give reasonable results in certain cases.[49,53] The best results were obtained starting with TBDMS-protected cyanohydrins.[52] Interestingly, **11** and **23** could be converted into optically active, deprotected 2-aryl-2-ethanolamines (**29**) in a single step by treatment with lithium aluminium hydride in THF, with complete retention of *ee* (Scheme 13).

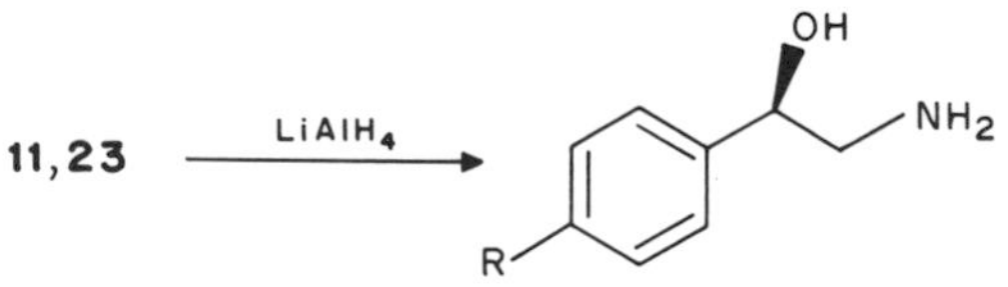

11,23 $\xrightarrow{LiAlH_4}$ [4-R-C6H4-CH(OH)-CH2-NH2]

	R	Yield (%)	*ee* (%)
(**29a**)	H	67	95
(**29b**)	OMe	68	99

Scheme 13

Although TBS ethers are assumed to be stable towards reducing agents, the presence of a second heteroatom at close range may facilitate an intramolecular reductive cleavage. A similar observation in the racemic 1-ethoxyethyl-protected series was observed by Schlosser and Birch.[47]

The preparation of chiral α-substituted β-ethanolamines by the sequence of a Grignard reaction followed by hydride reduction of TMS-protected cyanohydrins in either racemic[43b] or optically active[45,49] form is known to proceed with reasonable stereoselectivity. Jackson *et al.*[49] reported better selectivity for the *erythro* isomer with the 1-ethoxyethyl protecting group. The same phenomenon was observed by Brussee *et al.*[52] with the TBDMS-protected cyanohydrins **11** and **23**.[52] Moreover, the stability of the TBDMS group to acid allowed the isolation of the protected *erythro* products in a pure form as their hydrochloride salts (**32**) [after a sodium tetrahydroborate reduction of the intermediate Grignard adducts (**30**)]. Both aliphatic and aromatic Grignard reagents worked equally well (Scheme 14).

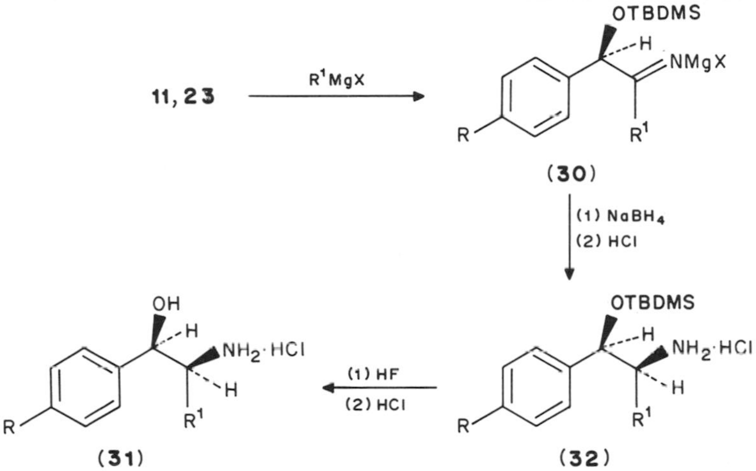

R	R^1	Yield (%) 31	Yield (%) 32
H	Me	61	73
H	Et	70	83
H	Ph	83	92
Me	Me	71	94
Me	Et	61	95
Me	Ph	78	87

Scheme 14

Removal of the TBDMS group was most conveniently achieved by treatment with aqueous hydrogen fluoride in acetonitrile, resulting in products (**31**) which were isolated as their hydrochlorides with diastereomeric and enantiomeric purities of more than 95%. These results demonstrate the feasibility of chirality transfer from the benzylic carbon, and extend further the possible uses of protected cyanohydrins as chiral building blocks.

In the pursuit of optically pure *N*-substituted derivatives of compounds **31** and **32**, reductive amination reactions with TBDMS-acyloin (**28a**) were first investigated.[54] Under standard conditions for this reaction, *erythro–threo* product mixtures were obtained in a ratio of about 80:20. Absolute stereocontrol could be achieved by first adding a magnesium salt to a solution of TBDMS-acyloin (**28a**) in acetonitrile. On addition of a primary amine, imine formation was rapid and quantitative at room temperature. The resulting imine complex (**33**) apparently forms a conformationally rigid bidentate structure, since its reduction with sodium tetrahydroborate at $-20\,^{\circ}C$ was completely diastereoselective, as shown in Scheme 15. The TBDMS-protected β-ethanolamines **34** were isolated in almost quantitative yields as enantiomerically pure products.

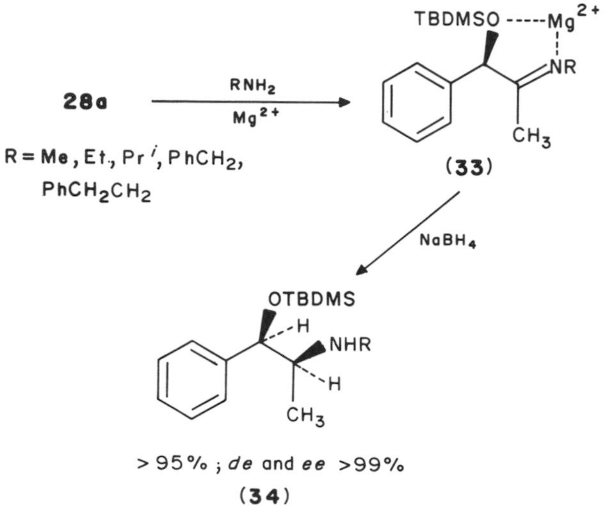

Scheme 15

The structural similarities of the magnesium complex **33** and the Grignard addition product **26** (P = TBDMS, R = H) from the cyanohydrin **11** (see Scheme 12) suggested the possibility of *transimination* reactions with this system. Thus, in a one-pot process, **11** was converted into TBDMS-protected ephedrine (**34**) (R = Me) by sequential treatments with methylmagnesium iodide, methanol, methylamine (to achieve transimination with liberation of ammonia) and sodium tetrahydroborate, in an overall yield of 92%.[55] Pure (1*R*, 2*S*)-ephedrine was obtained after deprotection with lithium aluminium hydride (Scheme 16).

14.5.4 AMINO ACIDS

The conversion of optically active cyanohydrins to optically pure α-amino acids is another interesting industrial opportunity. The first steps towards this goal

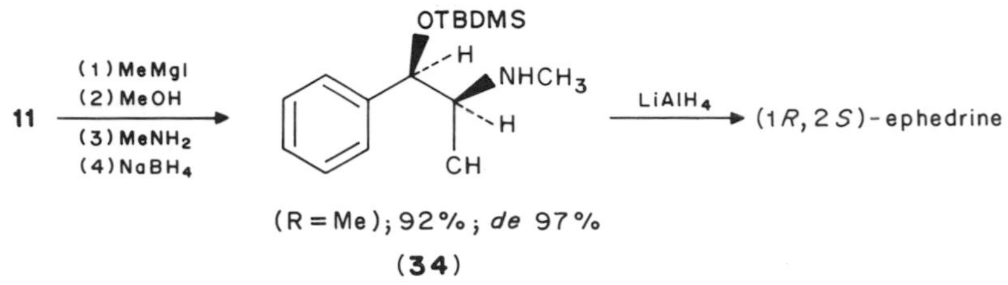

11 → (1) MeMgI (2) MeOH (3) MeNH$_2$ (4) NaBH$_4$ → (34) (R = Me); 92%; *de* 97% → LiAlH$_4$ → (1*R*, 2*S*)-ephedrine

Scheme 16

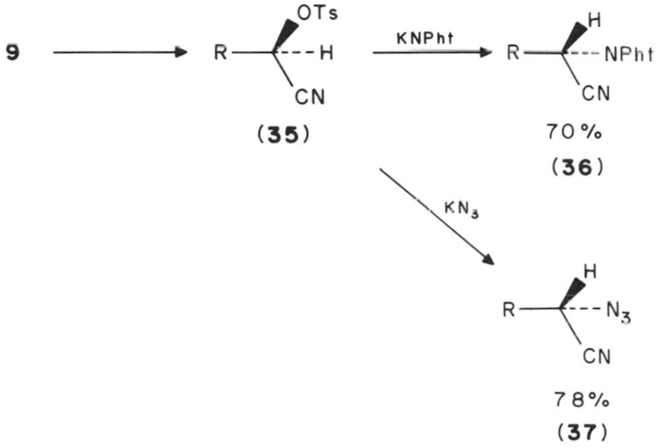

9 → (35) → KNPht → (36) 70%; (35) → KN$_3$ → (37) 78%

Scheme 17

were published recently by Effenberger and Stelzer.[56] The optically active aliphatic cyanohydrins **9** were converted into the corresponding tosylates **35**, which were treated with potassium phthalimide or azide in DMF (Scheme 17).

The resulting products, **36** and **37**, respectively, were formed in good yields with virtually complete inversion of configuration, indicating an S_N2 process. Unfortunately, aromatic substrates gave less favourable results. Another line of research was initiated by Brussee.[51] Treatment of the free imines of salts **30** and **33** with potassium cyanide in methanol resulted in the formation of β-hydroxy-α-aminonitriles. Further transformations of these Strecker-type products should lead to α-amino acids and β-hydroxy-α-amino acids, respectively.

14.6 INDUSTRIAL APPLICATIONS

The results outlined in the previous sections indicate that major progress has been made in applications of enantiomerically pure cyanohydrins as building blocks for chiral products. Currently, the most promising method for industrial manufacture is the biocatalytic process using the (*R*)-oxynitrilase enzyme, which has an abundant natural source in almonds. Clearly, this method is restricted to those substrates which fit into the active site of the biocatalyst, and several workers[25,27,41] have begun to probe the geometric constraints of the transition-state complex. Fortunately, the enzyme accepts a large number of low molecular weight aliphatic and aromatic aldehydes in addition to methyl ketones.[24c] The

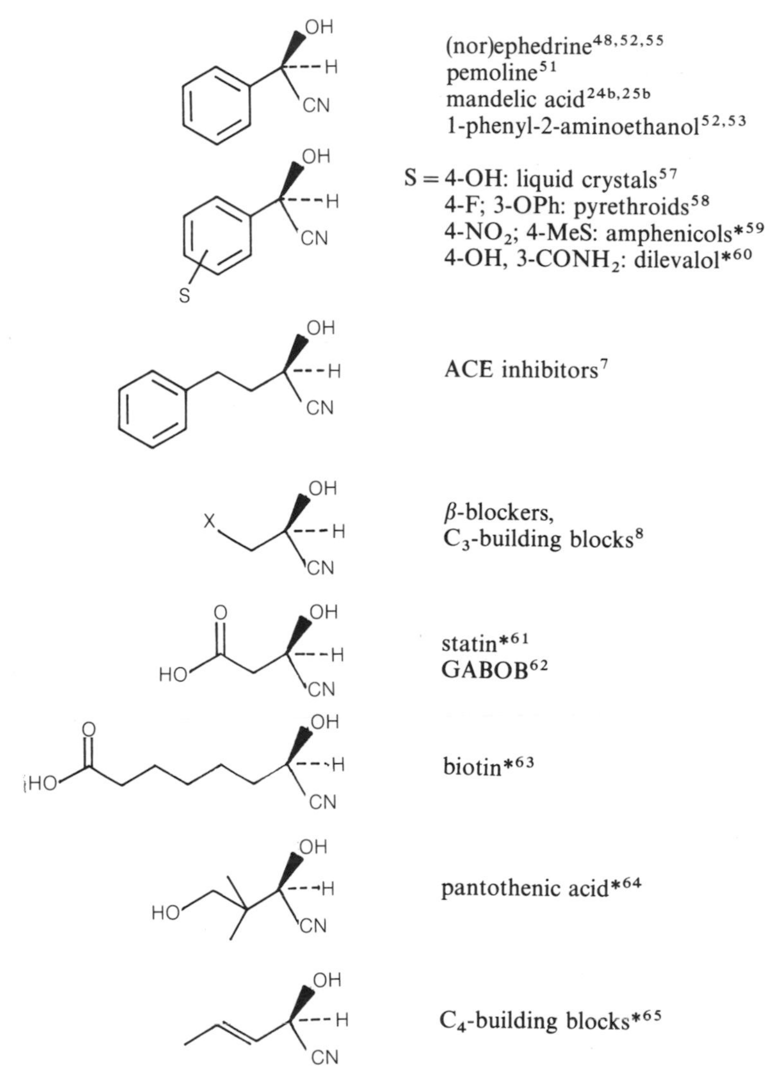

Figure 3. Chiral cyanohydrins as a potential source of commercially interesting optically pure products. *Denotes a different route.

less easily accessible (*S*)-oxynitrilase from sorghum shows a more restricted behaviour towards aldehyde substrates, only aromatic aldehydes being accepted. However, the recently published conversion of aliphatic (*R*)-cyanohydrins into their (*S*)-enantiomeric esters[56] offers a possible entry into this series also.

There remains a need for alternative industrially useful methods for the synthesis of optically pure cyanohydrins. Promising research topics include the development of new chemical catalysts, the cloning of different types of (*S*)-oxynitrilases and the use of lipases. In the further elaboration of cyanohydrins to useful products, the majority of successful transformations reported so far involve reactions of the cyano group.

Preservation of chirality and chirality transfer, especially of the protected compounds **10**–**18**, has given access to a large number of interesting products. The reactivity in the pendant organic group R (in compounds of type **9** and **10**) has yet to be fully explored in many readily available cyanohydrins, and the commercial product **14** is especially interesting in this respect. A large number of commercially interesting chiral intermediates and products can be produced already from chiral cyanohydrins. The main opportunities are listed in Figure 3.

In addition to the provision of useful intermediates for fine chemicals, applications of cyanohydrin derivatives as resolving agents (especially the α-hydroxy acids and β-ethanolamines) are likely. Practical syntheses of unnatural amino acids are expected to be within reach shortly. By virtue of their versatility, protected chiral cyanohydrins hold great promise for future applications in industry.

14.7 ACKNOWLEDGEMENTS

The author acknowledges especially his colleagues at Leiden University (A. van der Gen and J. Brussee) and Solvay Duphar (H. W. Geluk and G. J. M. van Scharrenburg) and their co-workers (W. T. Loos and J. B. Sloothaak). Without their numerous innovative contributions and stimulating discussions, this chapter could not have been written.

14.8 REFERENCES

1. Fuhrhop, J., and Penzlin, G., *Organic Synthesis, Concepts, Methods, Starting Materials*, Verlag Chemie, Weinheim, 1983, p. 47.
2. Elliot, J. D., Choi, V. M. F., and Johnson, W. S., *J. Org. Chem.*, **48**, 2294 (1983).
3. Noe, C. R., *Chem. Ber.*, **115**, 1591 (1982).
4. Dale, J. A., Dull, D. L., and Mosher, H. S., *J. Org. Chem.*, **34**, 2543 (1969).
5. Yuki Gosei Yakuhin, *Jpn. Pat.*, 3 219 325, 1987.
6. (a) Ohta, H., Kimura, Y., and Sugano, Y., *Tetrahedron Lett.*, **29**, 6957 (1988); (b) Ohta, H., Kimura, Y., Sugano, Y., and Sugai, T., *Tetrahedron*, **45**, 5469 (1989).

7. Sugai, T., and Ohta, H., *Agric. Biol. Chem.*, **55**, 293 (1991).
8. Matsuo, N., and Ohno, N., *Tetrahedron Lett.*, **26**, 5533 (1985).
9. Van Almsick, A., Buddrus, J., Hönicke-Schmidt, P., Laumen, K., and Schneider, M. P., *J. Chem. Soc., Chem. Commun.*, 1391 (1989).
10. Bredig, G., and Fiske, P. S., *Biochem. Z.*, **46**, 7 (1912).
11. (a) Chiral poly(ethyleneimines): Tsuboyama, S., *Bull. Chem. Soc. Jpn.*, **35**, 1004 (1962); (b) crosslinked analogues: Tsuboyama, S., *Bull. Chem. Soc. Jpn.*, **38**, 354 (1965).
12. Sumimoto, *Jpn. Pat.*, 3 165 355, 1986.
13. Sumimoto, *Jpn. Pat.*, 3 165 354, 1986.
14. Gountzos, H., Jackson, W. R., and Harrington, K. J., *Aust. J. Chem.*, **39**, 1135 (1986).
15. Mori, A., Ohno, H., Nitta, H., Tanaka, K., and Inoue, S., *Synlett*, 563 (1991).
16. Minamikawa, H., Mayakawa, S., Yamada, Y., Iwasawa, N., and Narasaka, K., *Bull. Chem. Soc. Jpn.*, **61**, 4379 (1988).
17. Reetz, M. T., Kyung, S. H., Bolm, C., and Zierke, T., *Chem. Ind, (London)*, 824 (1986).
18. Reetz, M. T., Kunish, F., and Heitmann, P., *Tetrahedron Lett.*, **27**, 4721 (1986).
19. (a) Oku, J., and Inoue, S., *J. Chem. Soc., Chem. Commun.*, 229 (1981); (b) Asada, S., Kobayashi, Y., and Inoue, S., *Makromol. Chem.*, **6**, 1755 (1985).
20. (a) Jackson, W. R., to ICI, *Br. Pat.*, 2 143 823, 1985; (b) Matthews, B. R., Jackson, W. R., Jayatilake, G. S., Wilshire, C., and Jacobs, H. A., *Aust. J. Chem.*, **41**, 1697 (1988).
21. (a) Kobayashi, Y., Asada, S., Hayashi, H., Motoo, Y., and Inoue, S., *Bull. Chem. Soc. Jpn.*, **59**, 893 (1986); (b) Tanaka, K., Mori, A., and Inoue, S., *J. Org. Chem.*, **55**, 181 (1990).
22. Mori, A., Nitta, H., Kudo, M., and Inoue, S., *Tetrahedron Lett.*, **32**, 4333 (1991).
23. Becker, W., and Pfeil, E., *J. Am. Chem. Soc.*, **88**, 4299 (1966).
24. (a) Effenberger, F., *Angew. Chem.*, **99**, 491 (1987); (b) Effenberger, F., Hörsch, B., Förster, S., and Ziegler, T., *Tetrahedron Lett.*, **31**, 1249 (1990); (c) Effenberger, F., Hörsch, B., Weingart, F., Ziegler, T., and Kühner, S., *Tetrahedron Lett.*, **32**, 2605 (1990).
25. (a) Brussee, J., Roos, E. C., and Van der Gen, A., *Tetrahedron Lett.*, **29**, 4485 (1988); (b) Brussee, J., Loos, W. T., Kruse, C. G., and Van der Gen, A., *Tetrahedron*, **46**, 979 (1990).
26. Niedermeyer, U., and Kula, M. R., *Angew. Chem.*, **102**, 423 (1990).
27. Smitskamp-Wilms, E., Brussee, J., Van der Gen, A., Van Scharrenburg, G. J. M., and Sloothaak, J. B., *Recl. Trav. Chim. Pays-Bas*, **110**, 209 (1991).
28. For a review, see Conn, E. E., *Biochem. Soc. Symp.*, **38**, 277 (1973).
29. Rosenthaler, L., *Biochem. Z.*, **14**, 238 (1908).
30. Becker, W., and Pfeil, E., *Biochem. Z.*, **337**, 156 (1963); for an improved method, see Hochuli, E., *Helv. Chim. Acta*, **66**, 489 (1983).
31. (a) Jorns, M. S., *Biochim. Biophys. Acta*, **613**, 203 (1980); (b) Gerstner, E., and Pfeil, E., *Hoppe-Seyler's Z. Physiol. Chem.*, **353**, 271 (1972).
32. (a) Niedermeyer, U., Kula, M. R., Wandrey, C., Makryaleus, K., and Dranz, S., to Degussa, *Eur. Pat.*, 326 063, 1989; (b) Kula, M. R., Niedermeyer, U., and Stürtz, I. M., to Degussa, *Eur. Pat.*, 350 908, 1989.
33. Geluk, H. W., Loos, W. T., Sloothaak, J. B., Van Scharrenburg, G. J. M., and Kruse, C. G., unpublished results.
34. Becker, W., and Pfeil, E., to Boehringer, *Ger. Pat.*, 1 593 260, 1966.
35. Wehtje, E., Adlercreutz, P., and Mattiason, B., *Appl. Microbiol. Biotechnol.*, **29**, 419 (1988).
36. Westman, T. L., to Monsanto, *US Pat.*, 3 649 457, 1968.
37. Effenberger, F., Ziegler, T., and Förster, S., to Degussa, *Eur. Pat.*, 276 375, 1987.
38. Kula, M. R., Stürtz, I. M., Wandrey, C., and Kragl, U., to Forschungszentrum Jülich, *Eur. Pat.*, 446 826, 1991.

39. Zandbergen, P., Van der Linden, J., Brussee, J., and Van der Gen, A., *Synth. Commun.*, **21**, 1387 (1991).
40. For excellent reviews, see: (a) Carrea, G., *Trends Biotechnol.*, **2**, 102 (1984); (b) Woodley, J. M., and Lilly, M. D., *Chem. Eng. Sci.*, **45**, 2391 (1990).
41. Ognyanov, V. I., Datcheva, V. K., and Kyler, K. S., *J. Am. Chem. Soc.*, **113**, 6992 (1991).
42. Geluk, H. W., Loos, W. T., Sloothaak, J. B., Van Scharrenburg, G. J. M., and Kruse, C. G., to Solvay Duphar, *Eur. Pat. Appl.*, 91 203 241.4, 1991.
43. (a) Amouroux, R., and Axiotis, G. P., *Synthesis*, 270 (1981); (b) Krepski, L. R., Heilmann, S. M., and Rasmussen, J. K., *Tetrahedron Lett.*, **24**, 4075 (1983).
44. Brussee, J., and Van der Gen, A., to Duphar, *Eur. Pat.*, 322 973, 1988.
45. Effenberger, F., Gutterer, B., and Ziegler, T., *Liebigs Ann. Chem.*, 269 (1991).
46. Corson, B. B., Dodge, R. A., Harris, S. A., and Yeaw, J. S., *Org. Synth., Coll. Vol.*, **I**, sccond cdition, 1961, p. 336.
47. Schlosser, M., and Birch, Z., *Helv. Chim. Acta*, **61**, 1903 (1978).
48. Cainell, G., Mezzina, E., and Panunzio, M., *Tetrahedron Lett.*, **31**, 3841 (1990).
49. Jackson, W. R., Jacobs, H. A., Jayatilake, G. S., Matthews, B. R., and Watson, K. G., *Aust. J. Chem.*, **43**, 2045 (1990).
50. Effenberger, F., Hopf, M., Ziegler, T., and Hudelmayer, J., *Chem. Ber.*, **124**, 1651 (1991).
51. Brussee, J., *Thesis*, University of Leiden, 1992. The enantiomeric purity of **21** could not be determined accurately by NMR.
52. Brussee, J., Dofferhoff, F., Kruse, C. G., and Van der Gen, A., *Tetrahedron*, **46**, 1653 (1990).
53. (a) Becker, W., Freund, H., and Pfeil, E., *Angew. Chem.*, **77**, 1139 (1965); (b) Ziegler, T., Hörsch, B., and Effenberger, F., *Synthesis*, 575 (1990).
54. Brussee, J., Van Benthem, R. A. T. M., Kruse, C. G., and Van der Gen, A., *Tetrahedron: Asymm.*, **1**, 163 (1990).
55. Brussee, J., and Van der Gen, A., *Recl. Trav. Chim. Pays-Bas*, **110**, 25 (1991).
56. Effenberger, F., and Stelzer, U., *Angew. Chem.*, **103**, 866 (1991).
57. Kusumoto, T., Hanamoto, T., Miyama, T., Takehara, S., Shoji, T., Osawa, M., Kuriyama, T., Nakamura, K., and Fujisawa, T., *Chem. Lett.*, 1615 (1990).
58. Sattar, A. K., Arbale A. A., and Kulkarni, G. H., *Synth. Commun.*, **20**, 2217 (1990).
59. (a) Chênevert, R., and Thiboutot, S., *Synthesis*, 444 (1989); (b) Giordano, C., Caviochioli, S., Levi, S., and Villa, M., *Tetrahedron Lett.*, **29**, 5561 (1988).
60. Clifton, J. E., *J. Med. Chem.*, **25**, 670 (1982).
61. Saiah, M., Bessoder, M., and Antonakis, K., *Tetrahedron: Asymm.*, **2**, 111 (1991).
62. Lu, Y., Miel, C., Kunesch, N., and Poisson, J., *Tetrahedron: Asymm.*, **1**, 707 (1990).
63. Volkmann, R. A., Davis, J. T., and Meltz, C. N., *J. Am. Chem. Soc.*, **105**, 5946 (1983).
64. Stiller, E. T., Harris, S. A., Finkelstein, J., Keresztesy, J. C., and Folkers, K., *J. Am. Chem. Soc.*, **62**, 1785 (1940).
65. Adam, G., and Seebach, D., *Synthesis*, 373 (1988).

ASYMMETRIC SYNTHESIS BY CHEMICAL METHODS

15 Naproxen: Industrial Asymmetric Synthesis[1]

C. GIORDANO M. VILLA and S. PANOSSIAN PANOSSIAN
Zambon Group SpA, Milan, Italy

15.1 THE DRUG

Naproxen[2–4] is the international non-proprietary name for (*S*)-2-(6-methoxy-2-naphthyl)propanoic acid (Figure 1). Naproxen, described for the first time[5] in 1967 by Syntex, was introduced on to the ethical pharmaceutical market in 1972[6] as an anti-inflammatory, analgesic and antipyretic drug in the form of the free acid, and later as the sodium salt (naproxen sodium). Thus, naproxen belongs to the family of non-steroidal anti-inflammatory drugs (NSAID),[7] most of which are carboxylic acids such as salicylic acid derivatives (aspirin, diflunisal, etc.),[2b,7] acetic acid derivatives (indomethacin, diclofenac, sulindaç etc.),[2b,7] 2-arylpropanoic acid derivatives (ibuprofen, suprofen, flurbiprofen, ketoprofen,

Chirality in Industry. Edited by A. N. Collins, G. N. Sheldrake and J. Crosby

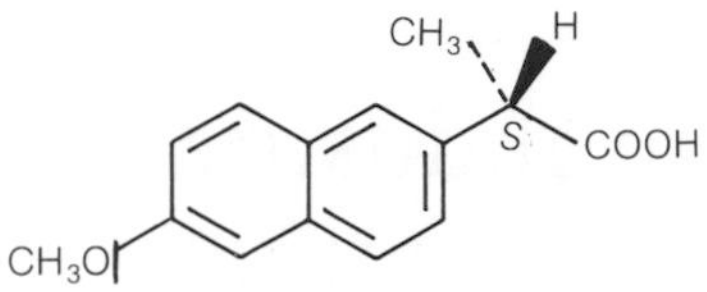

CH3 H S COOH CH3O

Figure 1. Naproxen

naproxen, etc.)[2b,7] and fenamate derivatives (mefenamic acid, flufenamic acid, etc.).[2b,7]

Since C-2 in 2-arylpropanoic acids is a stereogenic centre, these derivatives exist in two enantiomeric forms of which the (*S*)-enantiomer is the biologically more active. Naproxen, which is the only NSAI drug currently on the market in an enantiomerically pure form,[8] is 28 times more active as an anti-inflammatory agent than its (*R*) enantiomer.[4]

It is likely that in the near future other NSAI drugs such as ibuprofen, currently used as racemic mixtures, will be marketed as pure (*S*)-enantiomers.[9]

15.2 THE MARKET

The expiry of the patent protection on naproxen as a chemical entity in 1988 allowed the development of a generic market in many countries, the main exception being the USA where the patent protection will expire in December 1993. An O.T.C. (over-the-counter) form of naproxen is expected to be launched in a joint venture between Syntex and Procter & Gamble.[10]

While the authors estimate the current production of naproxen and naproxen sodium at *ca* 1000 tonnes per annum, a growth is forecast in view of the future generic and O.T.C markets in the USA. The bulk price in the free market in 1990 was about US $140–150 per kg.[11]

15.3 PRACTICAL SYNTHETIC APPROACHES TO NAPROXEN[12]

In view of the economic importance of the drug, both industrial and academic laboratories have devoted many efforts to devising new competitive strategies for the synthesis of naproxen. This chapter will not present an exhaustive analysis of the hundreds of papers and patents that have appeared, but will focus rather on the approaches which appear to be most practical and promising.

Virtually all the synthetic methodologies for preparing enantiomerically pure compounds have been applied to the synthesis of naproxen, i.e. optical resolution, incorporation of chiral building blocks and stereoselective reactions involving an enantiomerically pure reagent (stoichiometric or catalytic). The industrial production of naproxen, with the exception of the Zambon process,[13] involves the synthesis of racemic 2-(6-methoxy-2-naphthyl)propanoic acid and its optical resolution.

15.3.1 OPTICAL RESOLUTION

15.3.1.1 Synthesis of Racemic 2-(6-Methoxy-2-naphthyl)propanoic Acid and Its Derivatives

Although many syntheses of racemic materials have been reported,[14] to the authors' knowledge only three of them have been used industrially:

(a) Grignard coupling. Reaction of 6-methoxy-2-naphthylmagnesium bromide with a bromomagnesium or sodium salt of α-bromopropanoic acid[15] provides racemic 2-(6-methoxy-2-naphthyl)propanoic acid (Scheme 1). It is likely that Syntex still produces the racemic acid according to a Grignard-based technology.[16]

Scheme 1

(b) Darzens reaction. Base-catalysed condensation of 2-acetyl-6-methoxynaphthalene or of its 5-bromo analogue with an alkyl chloroacetate affords an α-alkyl naphthylglycidate, which is converted into 2-(6-methoxy-2-naphthyl)-propanoic acid[17] (Scheme 2).

Scheme 2

(c) Rearrangement of alkyl-arylketals. In 1980 a general method for the synthesis of racemic 2-arylpropanoic acids, involving a 1,2-aryl shift in ketals of α-haloalkyl aryl ketones,[14] was reported. In the case of racemic 2-(6-methoxy-2-naphthyl)propanoic acid, the method involves a 1,2-aryl shift in ketals of 2-halo-1-(6-methoxy-2-naphthyl)propan-1-one catalysed by Lewis acids[18] or by protic solvents[19] (Scheme 3).

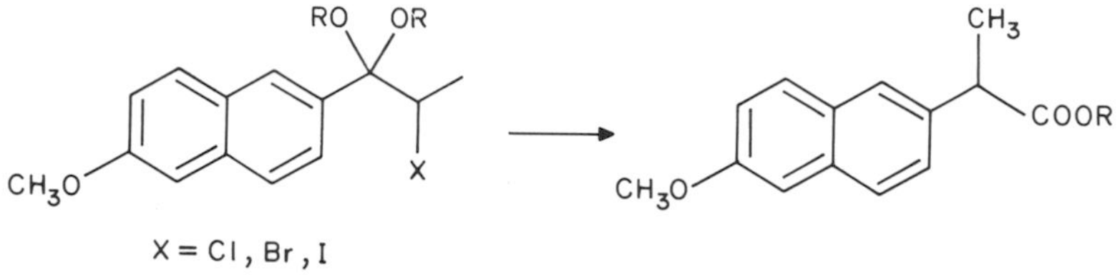

X = Cl, Br, I

Scheme 3

It is likely that Blaschim (Italy) industrialized[14a] the Lewis acid technology for the production of 2-(4-isobutylphenyl)propanoic acid (ibuprofen) and of racemic 2-(6-methoxy-2-naphthyl)propanoic acid. An alternative version employing ketals of 2-sulphonyloxy-1-(6-methoxy-2-naphthyl)propan-1-one has been described.[20]

15.3.1.2 Resolution Methods

The starting material for resolution is racemic 2-(6-methoxy-2-naphthyl)propanoic acid or its amide or ester derivatives. The resolution of the racemic acid is based on the formation of diastereomeric salts with optically active amines such as cinchonidine,[4,5,21] (−)-α-phenylethylamine[22] and *N*-alkyl-D-glucamine.[23] Of these, *N*-methyl-D-glucamine[23a] is the most practical and convenient. Racemization of undesired (*R*)-enantiomer, by heating with alkali, allows it to be recycled.

Among the resolutions recently described, a noteworthy process involves a second-order asymmetric transformation of amides derived from enantiomerically pure amino alcohols.[24]

Recently, the preparation of (*S*)-(+)-α-arylalkanoic acids, including naproxen, by reacting a racemic ester with an esterase capable of selectively hydrolysing the *S*-(+)-form, has been reported[25] (See also Chapter 5, Section 5.3).

The combination of the existing simple methods for the production of the racemic acid with the very efficient optical separations and the ease of recycling the undesired (*R*)-acid makes it difficult for asymmetric synthesis to be competitive.

15.3.2 INCORPORATION OF CHIRAL BUILDING BLOCKS

Natural ethyl (*S*)-lactate has been used as a chiral raw material for the preparation of dimethyl ketals of (*S*)-1-(6-methoxy-2-naphthyl)-2-sulphonyloxypropan-

1-one. The enantiomerically pure ketal was then converted[26] into naproxen according to the method[20] described for the synthesis of racemic 2-(6-methoxy-2-naphthyl)propanoic acid.

Analogously, the synthesis of naproxen has been accomplished[27] starting from the dimethyl ketal of (*S*)-1-(6-methoxy-2-naphthyl)-2-chloropropan-1-one, which was prepared from 6-methoxy-2-naphthylmagesium bromide and (*S*)-2-chloropropionyl chloride, derived from (*S*)-alanine, followed by ketalization.

15.3.3 STEREOSELECTIVE REACTIONS INVOLVING AN ENANTIOMERICALLY PURE REAGENT

15.3.3.1 Stoichiometric Approach

Practical approaches to enantiomerically enriched 2-arylalkanoic acids, consisting of highly diastereoselective protonation of 2-arylmethylketenes, are known. The protonation is carried out by using a stoichiometric amount of enantiomerically pure alcohol selected from sugar derivatives[28] or from α-hydroxy esters.[29] According to the latter method, protonation of 2-(6-methoxy-2-naphthyl) methylketene by ethyl-(*S*)-lactate in toluene, at low temperature, provides the ethyl-(*S*)-lactate ester of naproxen with 80% *de*[29b] (Scheme 4).

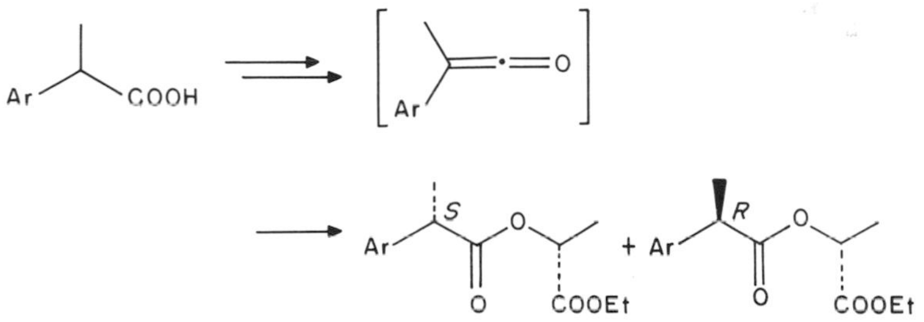

Ar COOH
Ar O
Ar S O O COOEt + Ar R O O COOEt

Scheme 4

15.3.3.2 Catalytic Approach

Conceptually, among the most attractive routes to optically active carboxylic acids are asymmetric hydroformylation of prochiral olefins and asymmetric hydrogenation of α,β-unsaturated carboxylic acids. Parrinello and Stille[30] reported that the hydroformylation of 2-vinyl-6-methoxynaphthalene, catalysed by platinum complexes containing chiral phosphine ligands, affords (*S*)-2-(6-methoxy-2-naphthyl)propanal with 81% *e.e.*, which is then oxidized to naproxen (Scheme 5).

In 1987, Noyori and co-workers[31] reported the homogeneous hydrogenation of 2-arylacrylic acids, including the 2-(6-methoxy-2-naphthyl) derivative, in the presence of a catalytic amount of enantiomerically pure ruthenium [2,2′-bis-

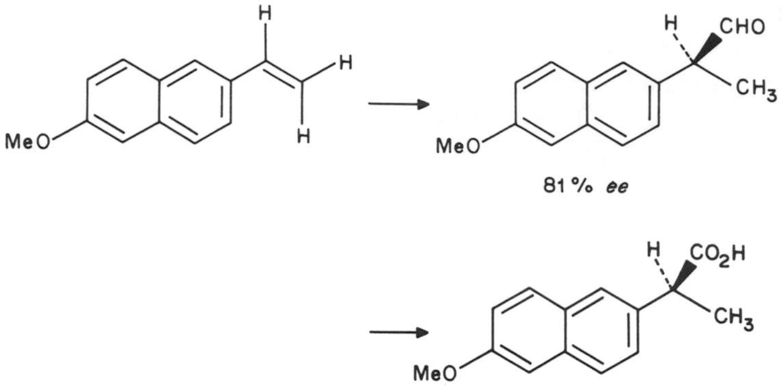

81 % *ee*

Scheme 5

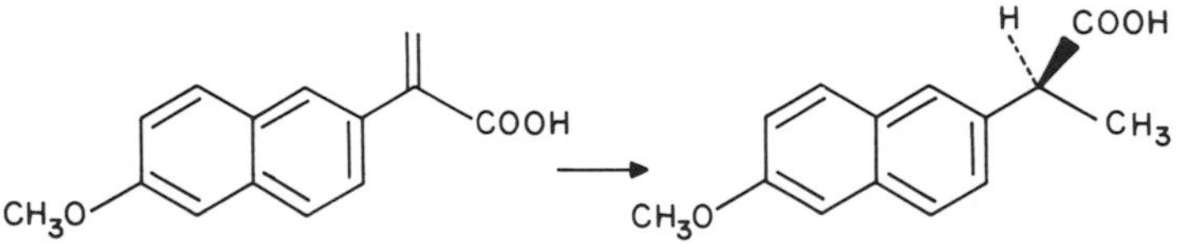

Scheme 6

(diarylphosphino)-1,1′-binaphthyl] diacetate under hydrogen pressure (135 atm). The method provides the acid with 97% *e.e.* (Scheme 6). A similar process has been claimed by Monsanto.[32]

Shell have described a stereoselective microbiological oxidation of 2-(6-methoxy-2-naphthyl)propane to naproxen.[33]

15.4 THE ZAMBON PROCESS

The Zambon process[34] differs from all the other industrial processes in that it affords naproxen as a single enantiomer, thereby avoiding optical resolution and recycling. The process represents one of the very few examples of non-enzymatic and non-microbiological industrial asymmetric synthesis.

The process consists of five steps, starting from 1-(6-methoxy-2-naphthyl)-propan-1-one (**1**) (Scheme 7).

STEP 1. Acid-catalysed ketalization of **1** with alkyl esters of (2*R*, 3*R*)-tartaric acid with azeotropic removal of water provides the ketal **2** in almost quantitative yield.

STEP 2. The highly diastereoselective bromination[35] of the homochiral diester ketal **2** with bromine gives rise to the formation of a mixture of diastereo-

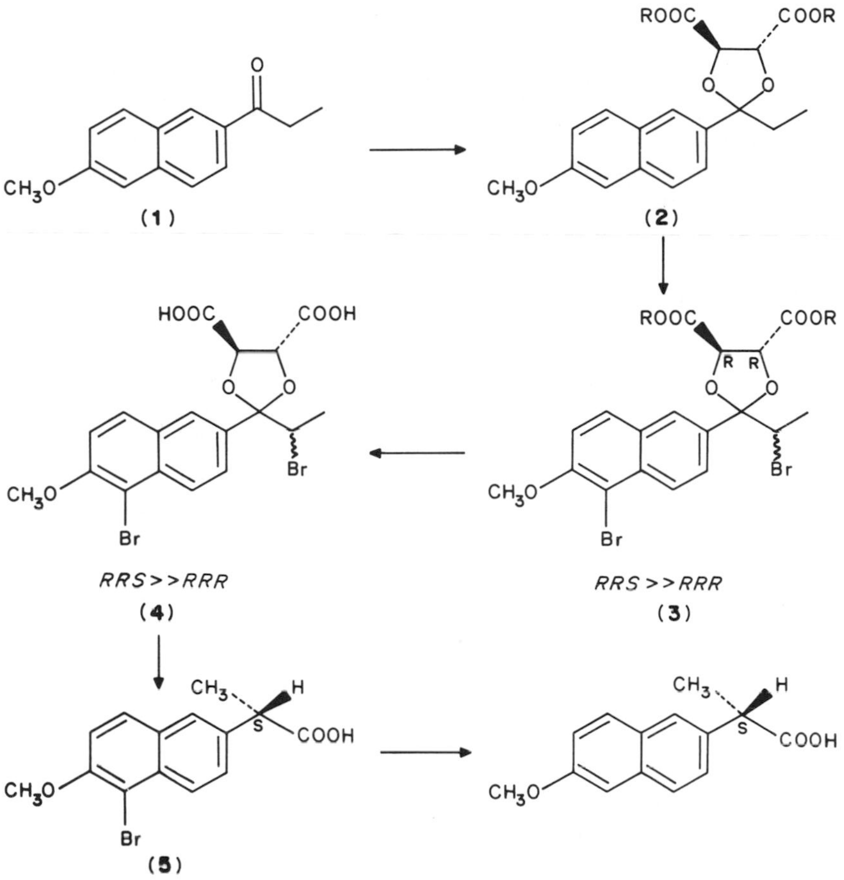

Scheme 7

meric bromoketals (**3**) wherein the desired (*S*)-epimer at the carbon bearing bromine dominates (*RRS*:*RRR* = 94:6). The bromination requires 2 mol of bromine per mole of substrate owing to a second bromination on the aromatic ring.

STEP 3. The 94:6 mixture of diastereomeric ketal diesters (**3**) is hydrolysed, under conventional conditions, to the corresponding α-bromoketal diacids (**4**) of the same *d.e.*[36]

STEP 4. The bromoketal diacids **4** provide (*S*)-2-(5-bromo-6-methoxy-2-naphthyl)propanoic acid (**5**) of 99% *ee* under aqueous acidic conditions (100 °C). The reaction occurs via 1,2-aryl shift with complete inversion of configuration at the migration terminus.[36]

STEP 5. The conversion of the bromo acid **5** into naproxen is accomplished, without loss of the enantiomeric purity, by catalytic reductive dehalogenation according to a known method.

The chiral auxiliary, a dialkyl ester of (2*R*, 3*R*)-tartaric acid, which is introduced in the first step and recovered as free (2*R*, 3*R*)-tartaric acid in the fourth step, has a twofold effect: it determines the high diastereoselectivity of the bromination (second step) and enhances the *ee* of the propanoic acid **5** with respect to that expected on the basis of the *de* of the ketal precursor **4**. Conceptually, this last feature is the equivalent of a kinetic resolution. For example, a 1:1 mixture of the epimeric α-bromoketals **4** provides an $86:14 = S:R$ ratio of 2-(5-bromo-6-methoxy-2-naphthyl)propanoic acid.[36]

Accordingly, the enantiomeric purity of naproxen depends on:

1. The enantiomeric purity of ketal **2**, which corresponds to the enantiomeric purity of the starting dialkyl ester of (2*R*, 3*R*)-tartaric acid (100% ee).
2. The diastereoselectivity of the bromination, which is 92%. The bromination represents the first example of diastereoselective halogenation of homochiral ketals. The diastereoselectivity of the bromination is the result of the interaction between the enol ether(s), generated *in situ* from the acetals, and bromine.[35]
3. The stereoselectivity of the rearrangement (100% inversion of configuration at the migration terminus).
4. The efficacy of the 'kinetic resolution' in the rearrangement step.

The industrial process is carried out in standard steel and glass-lined reactors and does not require any special apparatus. The productivity of each step is high and the overall yield from the ketone **1** is $> 75\%$.

Compared with other processes involving a 1,2-aryl shift but producing racemic 2-(6-methoxy-2-naphthyl)propanoic acid, the Zambon process provides enantiomerically pure naproxen in the same number of steps. The same technology has been shown to be useful for the large-scale preparation of other enantiomerically pure 2-arylalkanoic acids such as (*S*)-2-(4-isobutylphenyl)-propanoic acid [(*S*)-ibuprofen][9] and (*S*)-2-(4-chlorophenyl)-3-methylbutanoic acid, an intermediate for the insecticide esphenvalerate.[37]

15.5 NOTES AND REFERENCES

1. The chapter is dedicated to Dr Alberto Zambon, whose encouragement and support made the asymmetric synthesis of naproxen an industrial reality.
2. (a) *The Merck Index*, 11th edn, Merck, Rahway, NJ, 1989, No. 6337, p. 1014; (b) Elks, J., and Ganellin, C. R., *Dictionary of Drugs*, Chapman and Hall, London, 1990.
3. Riegl, J., Maddox, M. L., and Harrison, I. T., *J. Med. Chem.*, **17**, 377 (1974).
4. (a) Harrison, I. T., Lewis, B., Nelson, P., Rooks, W., Roszkowski, A., Tomolonis, A., and Fried, J. H., *J. Med. Chem.*, **13**, 203 (1970); (b) Hutt, A. J., and Caldwell, J., *Clin. Pharmacokinet.*, **9**, 371 (1984).

5. Fried, J. H., and Harrison, I. T., *Br. Pat.*, 1 211 134, 1967.
6. *Scrip, Syntex–a Pharmaceutical Company Profile*, P. J. B. Publications, Richmond, 1988.
7. Lombardino, J. G., *Non-Steroidal Antiinflammatory Drugs*, Wiley, New York, 1985, Chap. 4, 253.
8. *The United States Pharmacopeia–The National Formulary*, USP XXII–NF XVII, US Pharmacopeial Convention, 1990, p. 917.
9. (a) Matson, S. L., *US Pat.*, 4 800 162, 1989; (b) Reynolds, S. D., Tung, H. H., and Waterson, S., *US Pat.*, 4 994 604, 1991; (c) Schuster, V. O., and Loew, D., *Dtsch. Apoth.-Ztg.*, **129**, 1390 (1989).
10. *Scrip*, No. 1570, November 28, 17 (1990); *Scrip*, No. 1348, September 30, 8 (1988).
11. *Scrip*, No. 1588, February 6, 5 (1991).
12. It is difficult to ascertain which chemical processes are practised industrially. In this chapter we have made suggestions based on scientific publications and patents.
13. *Scrip*, No. 1421, June 16, 15 (1989).
14. For reviews, see (a) Giordano, C., and Minisci, F., *Chim. Ind. (Milan)*, **64**, 340 (1982); (b) Giordano, C., Castaldi, G., and Uggeri, F., *Angew. Chem., Int. Ed. Engl.*, **23**, 413 (1984); (c) Rieu, J. P., Boucherle, A., Cousse, H., and Mouzin, G., *Tetrahedron*, **15**, 4095 (1986).
15. Syntex, *Fr. Pat.*, 78 18 143, 1978.
16. McEntee, T. E., *Kirk–Othmer Encyclopedia of Chemical Technology*, 3rd edn., Wiley, New York, Vol. 12, p. 30.
17. (a) Politechnika Lodzka, *Jpn. Pat. Appl.*, J5 3149–962, 1977; (b) Cannata, V., and Tamerlani, G., *Ger. Pat. Appl.*, 3 212 170, 1981; (c) Kogure, K., to Nisshin Flour Milling, *Jpn. Pat. Appl.*, 75 18 448, 1973.
18. (a) Giordano, C., Belli, A., Uggeri, F., and Villa, G., *Eur. Pat.*, 34 871, 1981; (b) Giordano, C., Belli, A., Uggeri, F., and Villa, G., *Eur. Pat.*, 35 305, 1981; (c) Castaldi, G., Belli, A., Uggeri, F., and Giordano, C., *J. Org. Chem.*, **48**, 4658 (1983).
19. (a) Giordano, C., and Castaldi, G., *Eur. Pat.*, 101 124, 1983; (b) Giordano, C., and Castaldi, G., *Eur. Pat.*, 151 817, 1985; (c) Castaldi, G., Giordano, C., Uggeri, F., *Synthesis*, 505 (1985).
20. (a) Tsuchihashi, G.-I., Kitajima, K., and Mitamura, S., *Tetrahedron Lett.*, **22**, 4305 (1981); (b) Tsuchihashi, G.-I., Mitamura, S., and Kitajima, K., *Eur. Pat.*, 48 136, 1980.
21. (a) Alvarez, F., *US Pat.*, 3 975 432, 1976; (b) Syntex, *Br. Pat.*, 1 422 015, 1972.
22. Grelan Pharmaceutical, *Jpn. Pat. Appl.*, J5 5055–135, 1978.
23. (a) Felder, E., Pitre, D., and Zutter, H., *Eur. Pat.*, 7116, 1979; (b) Dvorak, C. A., *US Pat.*, 4 395 571, 1983; (c) Alfa Chemicals Italiana, *Belg. Pat.*, 892 689, 1982; (d) Von Morzé, H., *Eur. Pat. Appl.*, 95 901, 1983.
24. (a) Cannata, V., and Tamerlani, G., *Eur. Pat.*, 143 371, 1984; (b) Cannata, V., and Tamerlani, G., *Eur. Pat.*, 182 279, 1985.
25. (a) Sih, C. J., *Eur. Pat. Appl.*, 227 078, 1986; (b) Gu, Q. M., Chen, C. S., and Sih, C. J., *Tetrahedron Lett.*, **27**, 1763 (1986); (c) Cesti, P., and Piccardi, P., *Eur. Pat. Appl.*, 195 717, 1986.
26. Honda, Y., Ori, A., and Tsuchihashi, G.-I., *Bull. Chem. Soc. Jpn.*, **60**, 1027 (1987).
27. Piccolo, O., Spreafico, F., Visentin, G., and Valoti, E., *J. Org. Chem.*, **52**, 10 (1987).
28. Bellucci, G.; Berti, G., Bianchini, R., and Vecchiani, S., *Gazz. Chim. Ital.*, **118**, 451 (1988).
29. (a) Larsen, R. D., Corley, E. G., Davis, P., Reider, P. J., and Grabowski, E. J. J., *J. Am. Chem. Soc.*, **111**, 7650 (1989); (b) Corley, E. G., Larsen, R. D., Grabowski, E. J. J., and Reider, P., *US Pat.*, 4 940 813, 1990.
30. Parrinello, G., and Stille, J. K., *J. Am. Chem. Soc.*, **109**, 7122 (1987).
31. (a) Ohta, T., Takaya, H., Kitamura, M., Nagai, K., and Noyori, R., *J. Org. Chem.*,

52, 3174 (1987); (b) Takaya, J., Ohta, T., Noyori, R., Sayo, N., Kumobayashi, H., and Akutagawa, S., *Eur. Pat. Appl.*, 272 787, 1987.
32. Monsanto, *Int. Pat. Appl.*, WO 90 15 790, 1990.
33. Phillips, G. T., Robertson, B. W., Watts, P. D., Matcham, G. W. J., Bertola, M. A., Marx, A. F., and Kogar, H. S., *Eur. Pat.*, 205 215, 1986, *US Patents* 4 697 036, 1987 and 4 739 507, 1989
34. Giordano, C., Castaldi, G., Uggeri, F., and Cavicchioli, S., *US Patents* 4 810 819, 1989, 4 855 464, 1989 and 4 888 433, 1989
35. (a) Castaldi, G., Cavicchioli, S., Giordano, C., and Uggeri, F., *Angew. Chem., Int. Ed. Engl.*, **25**, 259 (1986); (b) Castaldi, G., Cavicchioli, S., Giordano, C., and Uggeri, F., *J. Org. Chem.*, **52**, 3018, 5642 (1987); (c) Giordano, C., Coppi, L., and Restelli, A., *J. Org. Chem.*, **55**, 5400 (1990); (d) Giordano, C., *Actual. Chim.*, 203 (1990).
36. Giordano, C., Castaldi, G., Cavicchioli, S., and Villa, M., *Tetrahedron*, **45**, 4243 (1989).
37. (a) British Crop Protection Council, *The Pesticide Manual*, 8th edn, Lavenham Press, Lavenham, 1987, p. 1081; (b) *Adv. Pestic. Sci.*, Part II, 174 (1979).

16 A Practical Synthesis of (–)-Menthol with the Rh–BINAP Catalyst

S. AKUTAGAWA
Takasago Research Institute, Inc., Tokyo, Japan

16.1 THE PRODUCTION OF (–)-MENTHOL

There are many examples of food additives and fragrances where the desired organoleptic properties are related to a given absolute configuration. One of the most important of these is (–)-menthol, which is produced world-wide on a scale of 4500 tonnes per year, and finds outlets in many consumer products, of which cigarettes, chewing gum, toothpaste, pharmaceuticals and personal care products are examples. At present, (–)-menthol is obtained largely from the natural source *Mentha arvensis*, which is cultivated in China, although synthetic material is produced in Germany, the USA and Japan. Of the eight possible stereoisomers possible for the menthol structure, only the (1*R*, 3*R*, 4*S*)-isomer is important.

Formerly, Takasago used several routes to produce (–)-menthol. An example is the process shown in Scheme 1, in which optical activity was introduced in a late chemical resolution step. In April 1984, Takasago successfully commercialized a process for the manufacture of (–)-menthol by asymmetric synthesis. In this method, either of two readily available raw materials, isoprene or myrcene, may be utilized. The latter, which is also an important starting material

Chirality in Industry. Edited by A. N. Collins, G. N. Sheldrake and J. Crosby

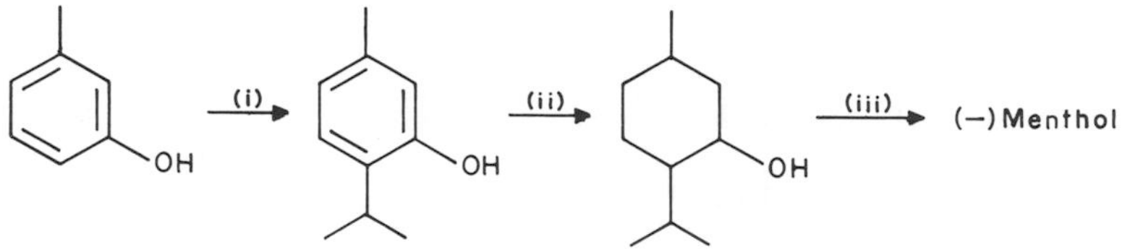

(i) propylene–SiO_2 Al_2O_3; (ii) H_2–copper chromate; (iii) optical resolution

Scheme 1

for the production of geraniol, is obtainable from turpentine. Key intermediates in this route to (−)-menthol are neryldiethylamine (**2**) and geranyldiethylamine (**1**) as shown in Scheme 2. These stereoisomeric amines are synthesized in high yield from isoprene and myrcene, respectively, by treatment with diethylamine and a catalytic quantity of lithium diethylamide. An important feature of the new process is the use of a chiral rhodium phosphine catalyst to isomerize these allylic amines to an optically active enamine (**3**). The catalyst for this reaction (Rh^I–BINAP) is recycled.

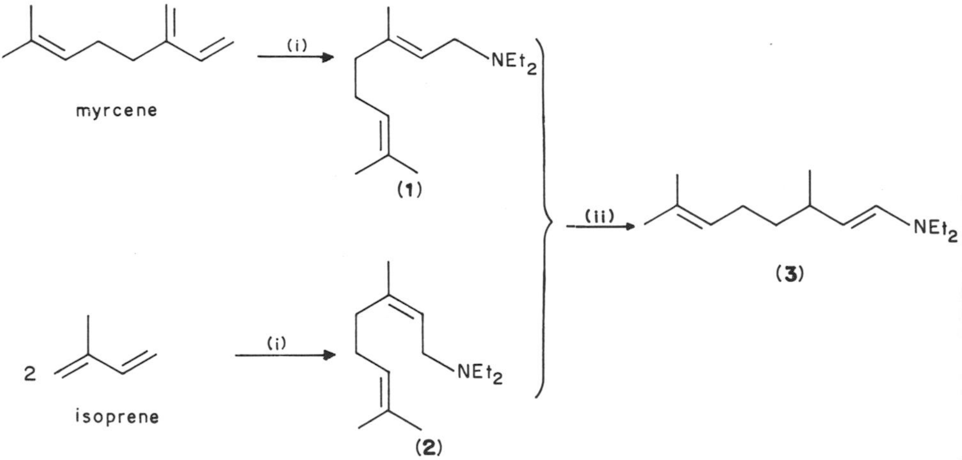

(i) $HNEt_2$–$LiNEt_2$; (ii) Rh^I–BINAP

Scheme 2

By hydrolysing the enamine so produced, the optically active aldehyde citronellal is produced in a very high yield, and with an enantiomeric purity greater than 98% (Scheme 3). An intramolecular ene reaction mediated by zinc dibromide then gives isopulegol. The latter reaction is selective for the

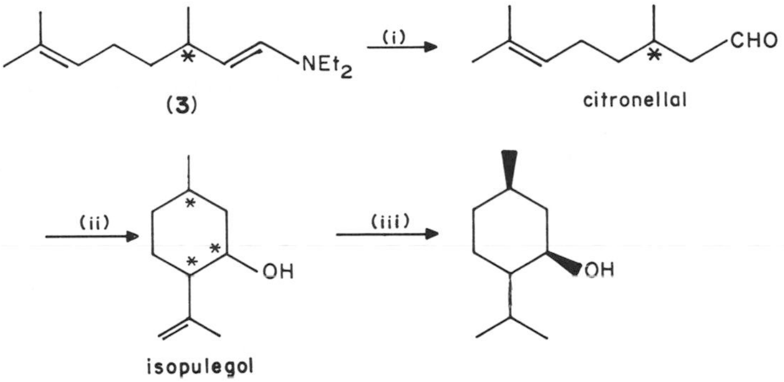

(3)

citronellal

isopulegol

(i) H_2SO_4; (ii) $ZnBr_2$; (iii) H_2–Nickel

Scheme 3

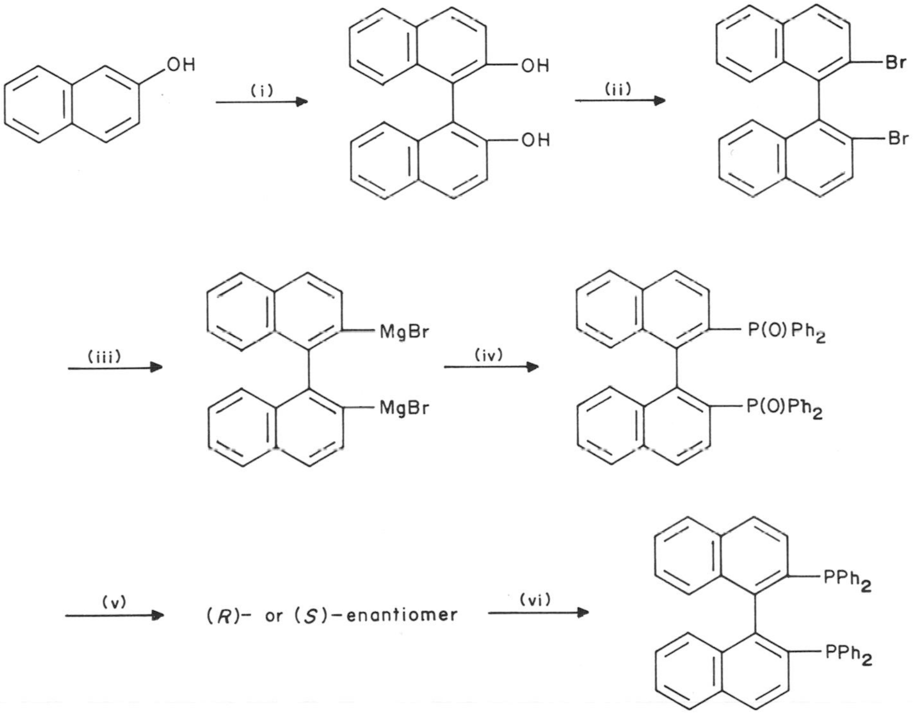

(R)- or (S)-enantiomer

(i) $FeCl_3$; (ii) Ph_3PBr_2; (iii) Mg–THF; (iv) $Ph_2P(O)Cl$; (v) optical resolution; (vi) $HSiCl_3$

Scheme 4

all-equatorial form of isopulegol; other Lewis acids produce no more than 65% of this material. The optical purity of the (−)-isopulegol may be improved to 100% by one recrystallization at −50 °C, and hydrogenation of the material so produced gives (−)-menthol which is essentially pure.

The process which is used to manufacture the chiral ligand (BINAP) from 2-naphthol employs a chemical method to resolve a bisphosphine oxide intermediate (Scheme 4).

16.2 SELECTION OF THE CATALYST SYSTEM FOR ASYMMETRIC ISOMERIZATION

Various metal ion complexes have been used in conjunction with bases to effect the migration of the double bonds of geranyl- or neryldiethylamine (**1** and **2**) to citronellal-(*E*)-enamine (**3**) and/or the dienamine (**4**).[1] The ratios of these products vary widely with different systems, as Table 1 illustrates. Several catalysts, such as $Co(acac)_2$-PPh_3-$Bu^i{}_2AlH$ (entry 4) or $[Rh(PPh_3)_2(COD)]^+$ (entry 6), exhibit an

Table 1. Metal-catalysed isomerizations

NEt_2 (1) , NEt_2 (2) → NEt_2 (3) + NEt_2 (4)

NHCy (5) → NCy (6)

Entry	Catalyst	Substrate	Product (%)
1	LDA–TMEDA	**1** or **2**	**4** (100)
2	Cp_2TiCl_2–Pr^iMgBr	**1**	**4** (100)
3	$CoH(N_2)(PPh_3)_3$	**1**	**3** (35), **4** (15)
4	$Co(acac)_2$–PPh_3–$Bu^i{}_2AlH$	**2**	**3** (81), **4** (19)
5	$Co(acac)_2$–PPh_3–$Bu^i{}_2AlH$	**5**	**6** (100)
6	$[Rh(PPh_3)_2(COD)]^+$	**1**	**3** (100)

Table 2. Asymmetric isomerization[a]

Catalyst	Substrate	Conversion (%)	Product (%)	*ee* (%)
Co(acac)$_2$–DIOP[b]–Bu$^i{}_2$AlH	**1**	45	**3**(87), **4**(13)	35(*R*)
	5	62	**6**(97), **3**(3)	57(*R*)
[Rh(DIOP)(COD)]$^+$	**1**	71	**3**(100)	22(*R*)
[Rh((*R*)-BINAP)(COD)]$^{+}$[c]	**1**	100	**3**(100)	97(*S*)
	2	97	**3**(100)	96(*R*)
[Rh((*S*)-BINAP)(COD)]$^+$	**1**	100	**3**(100)	97(*R*)
	5	100	**6**(100)	98(*R*)

[a] 40 °C, 23 h, [substrate]/[cat] = 100.

[b] DIOP = 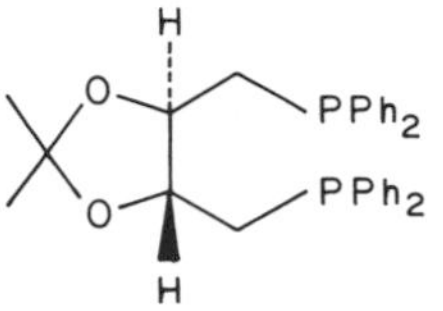 (2*R*,3*R*).

[c] BINAP = 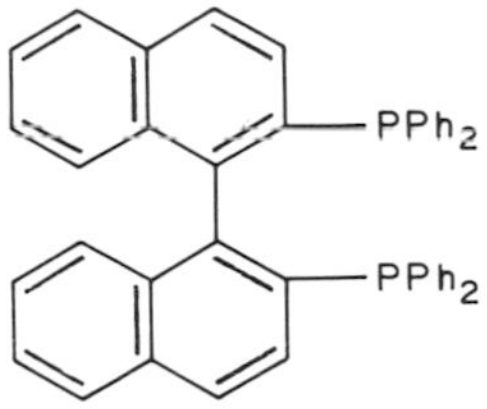 (*R*) or (*S*), $[\alpha]_D^{25}$ + or − 229°.

excellent product specificity for the desired enamine (**3**) in which the non-allylic double bond is not isomerized. Secondary amine substrates such as *N*-cyclohexylgeranylamine (**5**) give the corresponding imines (**6**), which are also potential precursors of citronellal.

The isomerization reaction may be rendered stereospecific by the presence of chiral ligands at the catalytic centre. Several combinations of chiral rhodium and cobalt complexes with substrates **1**, **2** and **5** were assessed, as shown in Table 2. Excellent conversions of both geranyl- and neryldiethylamine to the enamine **3** were observed with the [Rh((*R*)-BINAP)(COD)]$^+$ and [Rh((*S*)-BINAP)(COD)]$^+$ complexes. In addition, these species catalysed the stereospecific transformation of the secondary allylic amine **5** to the imine **6**. These reactions were complete within 23 h at a temperature of 40 °C and a substrate to catalyst ratio of 100. As can be seen from Table 2, enantiomeric excesses in the products ranged from 96% to 98%, and the relationship between stereochemical configuration of the product and that of the substrate–catalyst combination was very specific, as illustrated in Scheme 5. Thus, the combination of geranyldiethylamine (**1**) and the catalyst bearing the (*S*)-BINAP ligand produced the (*R*)-enamine (**3a**), whereas the (*R*)-BINAP ligand gave the (*S*)-enamine (**3b**). With neryldiethylamine (**2**), these specificities were reversed.

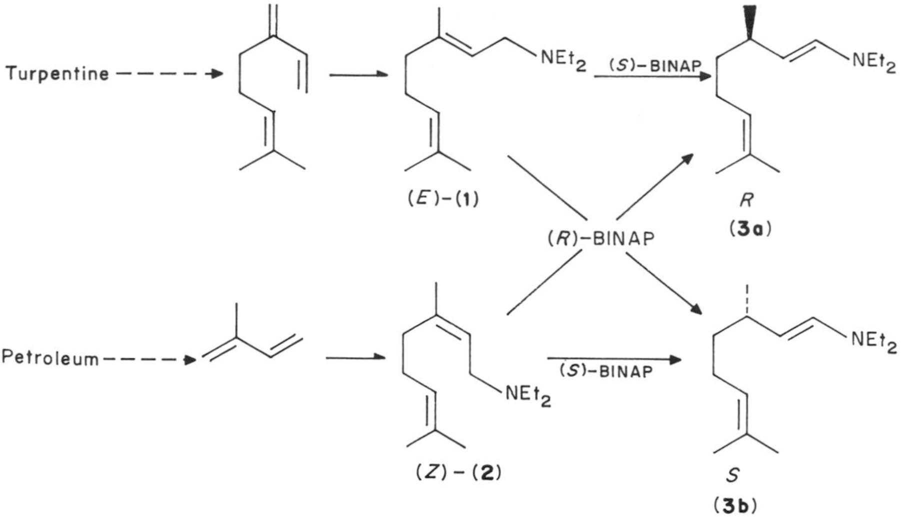

Scheme 5

As both enantiomers of BINAP may be obtained easily by an optical resolution method,[2] the stereochemical correlation described above gives rise to the economic benefits of easy access to both enantiomers of citronellal from a single intermediate and the option to utilize as a raw material either a renewable resource (turpentine) or petroleum.

16.3 PROCESS DEVELOPMENT

Although the Rh–BINAP catalysts display remarkably high selectivities in the above reactions, their commercial utility has, in the past, been compromised by their high price [Aldrich prices in 1991 were $RhCl_3$ $42.50 for 500 mg and (+)- or (−)-BINAP $25.40 for 100 mg].

To produce (+)-citronellal economically on a scale of 1000 tonnes per annum, the 'turnover number' (TON = moles of enamine produced by 1 mol of catalyst during 18 h) should be more than 50 000, while maintaining the selectivity of the catalyst. The following process developments enabled this criterion to be realized:

(a) The relationship between reaction temperature and the conversion/enantiomeric excess of the products was studied, and is summarized in Table 3. Although the optimum temperatures for conversion and enantiomeric excess were found to be different, a good compromise between these factors could be achieved at 100 °C.

Table 3. Reaction temperature *vs* conversion and enantiomeric excess

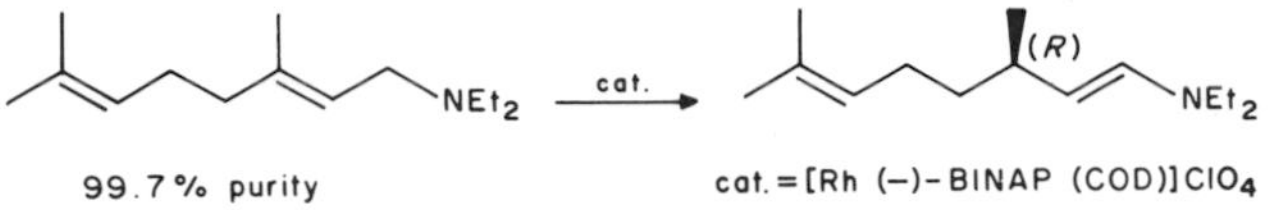

99.7% purity

cat. = [Rh (−)-BINAP (COD)]ClO_4

Temperature (°C)	Conversion[a] (%)	*ee* (%)
60	55	99.2
70	74	99.2
80	88	99.3
100	99	97.6
120	98[b]	96.7

[a] [Substrate] = 2.56 M, [cat] = 0.32×10^{-3} M, THF, 7 h.
[b] 2 h.

Table 4. Catalyst inhibitors

Substrate =

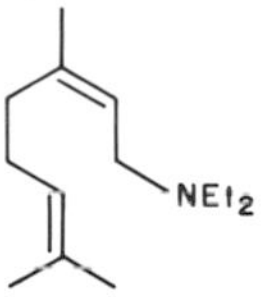

Inhibitor	Inhibitor/ catalyst	k_{obs} ($mol\,l^{-1}\,min^{-1}$)[a]
None		37
NEt_3	4	20
COD	2	8.5
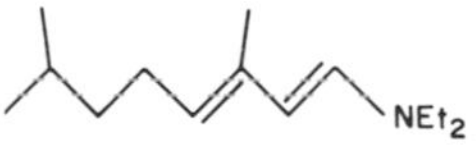 (7)	2	1.8

[a] THF, [Rh{(+)-BINAP}(COD)]$^+$, [Substrate] = 0.24 mol l^{-1} [Substrate]/[Catalyst] = 100.

Table 5. Progress towards an efficient isomerisation

Improvement	TON
[Rh(BINAP)(COD)]$^+$, initial	100
Vitride-treated substrate[a]	1000
Removal of amine isomer (**8**)[b]	8000
Re-use of **11** (10% loss)[c]	80 000
Re-use of **11** (2% loss)[c]	400 000

[a] Vitride[R] = sodium bis(2-methoxyethoxy)aluminium hydride.
[b] (**8**)
[c] **11** = $[Rh(I)(BINAP)_2]^+\ ClO_4^-$.

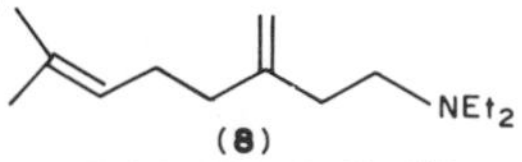

Table 6. Optically active terpenoids manufactured by Takasago

Compound	Formula	Chemical purity (%)	*e.e.* (%)	Volume (tonnes per annum)	Use
(+)-Citronellal	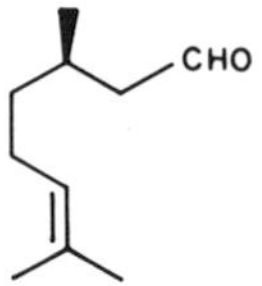 CHO	98.0	97 ± 1	(1500)	Intermediate
(−)-Isopulegol	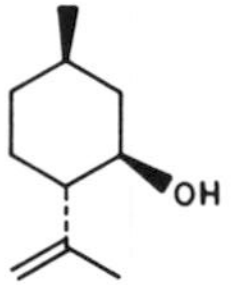 OH	100	100	(1100)	Intermediate
(−)-Menthol	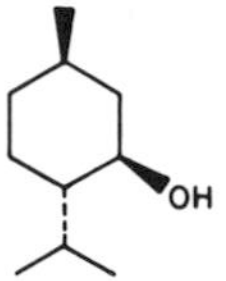 OH	100	100	1000	Pharmaceutical, tobacco, household
(−)-Citronellol	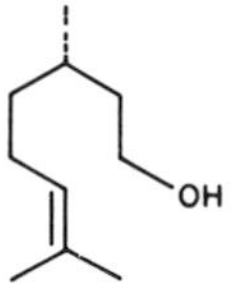 OH	99	98	20	Fragrance

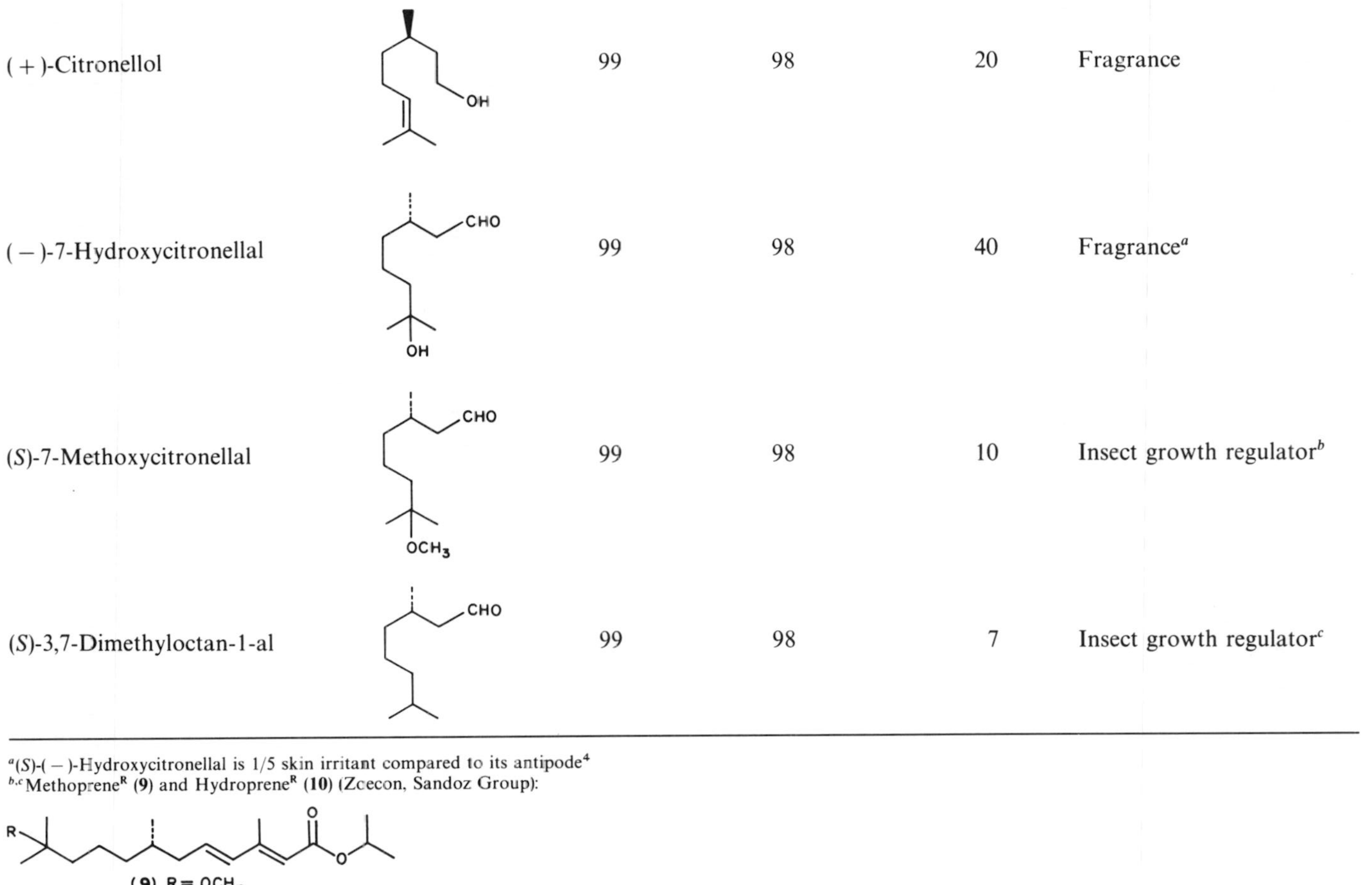

(+)-Citronellol	99	98	20	Fragrance
(−)-7-Hydroxycitronellal	99	98	40	Fragrance[a]
(*S*)-7-Methoxycitronellal	99	98	10	Insect growth regulator[b]
(*S*)-3,7-Dimethyloctan-1-al	99	98	7	Insect growth regulator[c]

[a] (*S*)-(−)-Hydroxycitronellal is 1/5 skin irritant compared to its antipode[4]
[b,c] Methoprene[R] (**9**) and Hydroprene[R] (**10**) (Zoecon, Sandoz Group):

(b) Several minor impurities in the neryldiethylamine were found to have a significant effect on the conversion. Triethylamine, COD and, particularly, *trans,trans*-dienamine (**7**) were all found to reduce markedly the rate of reaction (see Table 4). Control of the concentrations of these species is, therefore, essential for efficient conversion. The commercial production of geranyldiethylamine (**1**) is accompanied by 0.5–0.7% of another amine isomer (**8**) whose presence was also found to be detrimental to the isomerization, but which may be selectively removed.

(c) A novel Rh(I) cationic complex, $[Rh^{I}(BINAP)_2]^+ClO_4^-$ (**11**), in which the rhodium is coordinated by 2 mol of BINAP, was discovered.[3] The rhodium(I) and BINAP species may be recovered efficiently from the reaction mixture in this form, with losses of only 2%.

The combined effect of these developments has been to increase the turnover number from 100 for the original $[Rh^{I}(BINAP)(COD)]^+$ catalyst system to 400 000, as illustrated in Table 5. The efficiency of the isomerization now compares well with many enzyme-catalysed processes.[5]

Since 1984, Takasago has manufactured a series of optically active terpenoid compounds from allylamines with the isomerization technique described above. Table 6 gives a summary of these products and their purities, production volumes and uses.

16.4 CONCLUSION

During the past decade, chiral transition metal complexes have become an indispensable tool, both in industry and academia, for the production of a variety of optically active intermediates. The development of the Rh^{I}-BINAP system for asymmetric isomerization of the allylamines **1** and **2** is a prime example of the way in which relatively inefficient laboratory procedures may be refined to give economic industrial processes.

16.5 ACKNOWLEDGEMENTS

The author thanks the following colleagues, each of whom has played a role in the development of the industrial syntheses of terpenoids described above: Professors J. Takana (Shizuoka) and S. Watanabe (Chiba) (terpenoid amine syntheses), Professors R. Noyori (Nayoga) and H. Takaya (Kyoto) (BINAP synthesis), Professors S. Otsuka and K. Tani (Osaka) and Dr H. Kumobayashi (Takasago) (asymmetric isomerizations) and T. Sakaguchi, M. Yagi, H. Nagashima and N. Murakami (Takasago) (process development).

16.6 REFERENCES

1. Tani, K., Yamagata, T., Otsuka, S., Akutagawa, S., Kumobayashi, H., Taketomi, T., Takaya, H., Miyashita, A., and Noyori, R., *ACS Symp. Ser. 185*, **13**, 187 (1982).
2. Takaya, H., Mashima, K., Koyano, K., Yagi, M., Kumobayashi, H., Taketomi, T., Akutagawa, S., and Noyori, R., *J. Org. Chem.*, **51**, 629 (1986).
3. Tani, K., Yamagata, T., Tatsuno, Y., Yamagata, Y., Tomita, K., Akutagawa, S., Kumobayashi, H., and Otsuka, S., *Angew. Chem., Int. Ed. Engl.*, **24**, 217 (1985).
4. Watanabe, S., Kinosaki, A., Kawasaki, M., Yoshida, T., and Kaidbey, K., in *Proceedings of the 10th International Congress of Essential Oils, Fragrances and Flavors, Washington, DC, USA, 16–20 November 1986*, 1988, p. 1029.
5. Tani, K., Yamagata, T., Akutagawa, S., Kumobayashi, H., Taketomi, T., Takaya, H., Miyashita, A., Noyori, R., and Otsuka, S., *J. Am. Chem. Soc.* **106**, 5208 (1984).

17 Asymmetric Hydrogenation with Ru–BINAP Catalysts

S. AKUTAGAWA
Takasago Research Institute, Inc., Tokyo, Japan

17.1 INTRODUCTION

Transition metals are used extensively as catalysts for the reduction of double bonds in organic synthesis. In many cases the introduction of a chiral ligand into the metal complex results in asymmetric hydrogenation and occasionally these asymmetric hydrogenations are efficient enough to allow the synthesis of optically pure isomers with useful applications in industry (see Chapter 1, Section 1.2.3.1). An example is the hydrogenation of *N*-acylaminoacrylic acids to optically active α-amino acids utilizing rhodium–chiral phosphine complexes as catalysts (equation 1).[1] However, among the numerous ligands which have been applied in this process, only a few have demonstrated optical yields approaching 98–100% *ee*.

Chirality in Industry. Edited by A. N. Collins, G. N. Sheldrake and J. Crosby

$$\mathrm{RCH{=}C(NHAc)CO_2H} \xrightarrow[\text{Rh-chiral phosphine}]{H_2} \mathrm{RCH_2C^{*}H(NHAc)CO_2H} \quad (1)$$

After the successful commercialization of rhodium–BINAP complexes as asymmetric isomerization catalysts (described in Chapter 16), Takasago began to study potential industrial applications of ruthenium–BINAP complexes. The work reported here was carried out in collaboration with Professors R. Noyori (Nagoya University), H. Takaya (Kyoto University) and M. Saburi (Tokyo University) between 1984 and 1991.

17.2 THE SCOPE OF Ru–BINAP CATALYSTS

The potential of Ru–BINAP complexes as catalysts for asymmetric hydrogenations is apparent from the following characteristics:

(i) *A wide range of possible substrates.* A variety of prochiral alkenes are reduced to saturated chiral compounds, whilst carbonyl compounds give chiral secondary alcohols.
(ii) *Selectivity.* In many cases almost quantitative chemoselectivity and above 95% enantioselectivity are obtained.
(iii) *Catalyst activity.* The hydrogenation reaction can be carried out smoothly under moderate conditions (5–50 kg/cm^2 hydrogen pressure and reaction temperatures of 25–100 °C) using a very high substrate/catalyst molar ratio (approximately 10^5), which renders the processes extremely economical. In addition, the catalysts are convenient to prepare and isolation of the products is not difficult.

The versatility of Ru–BINAP catalysts was examined in the asymmetric hydrogenation of a series of unsaturated compounds, and is summarized in Table 1. Isolated ketones are unsuitable substrates for the reduction, although prochiral alkenes containing an activated double bond and prochiral substituted ketones are readily reduced to their saturated counterparts in high chemical yields (97–100%) and optical purities (92–99% *ee*). The asymmetric hydrogenation of several types of substrates with Ru–BINAP catalysts are described in more detail below.

17.2.1 ALLYLIC ALCOHOLS

The reduction of geraniol (**2**) or nerol (**3**) to optically active citronellol (**4**) is readily achieved by ruthenium-(II) dicarboxylate–BINAP complexes of type **1**.[2] As shown in Table 2, these reactions satisfy stringent economic requirements, including necessary chemo- and stereoselectivity. The isomeric form (*R* or *S*)

Table 1. Scope of asymmetric hydrogenation

Functional Group	Product	
1. Activated double bond		
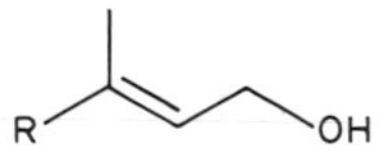	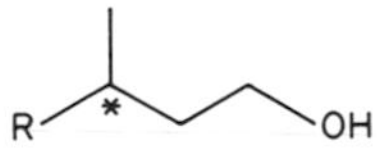	
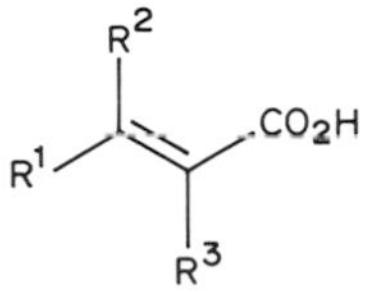	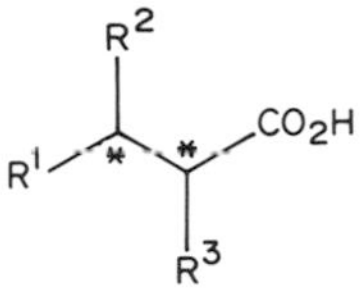	
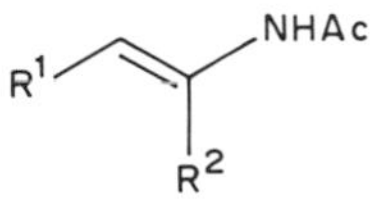	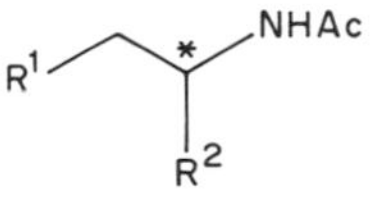	
2. Substituted ketones		
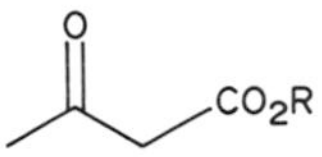	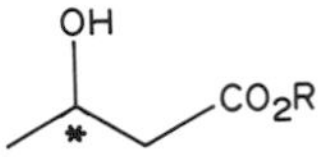	
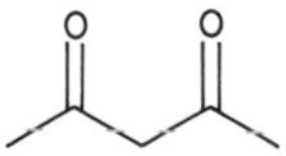	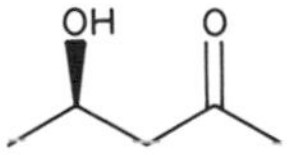 , 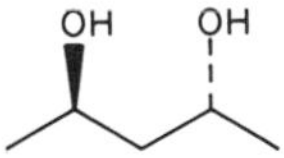	
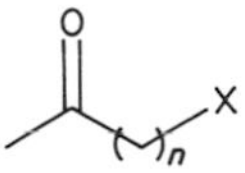 $n = 1, 2, 3$	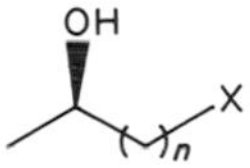	X = Cl, Br, OH, NR_2
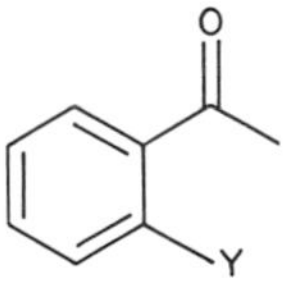	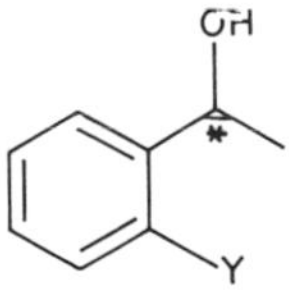	Y = Br, OH, CO_2H
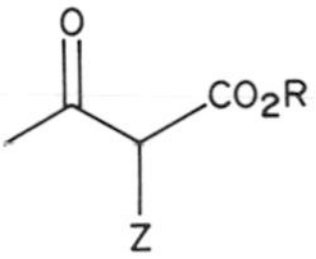	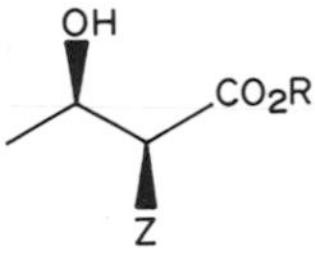	Z = NHAc, CH_2NHAc, halogen

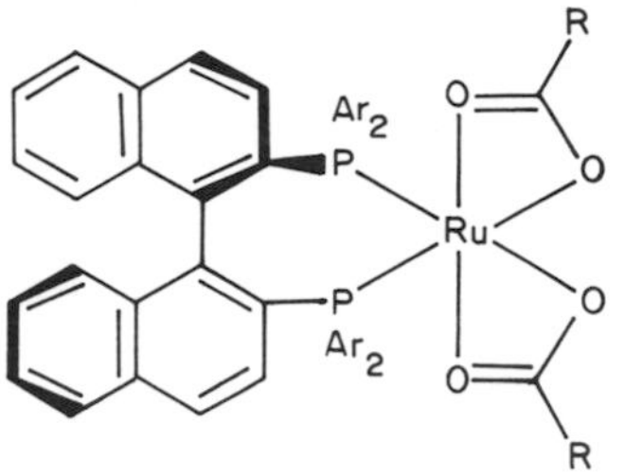

(R)-(1)

(S)-(1)

(a) Ar = C_6H_5 ; R = CH_3

(b) Ar = C_6H_5 ; R = $C(CH_3)_3$

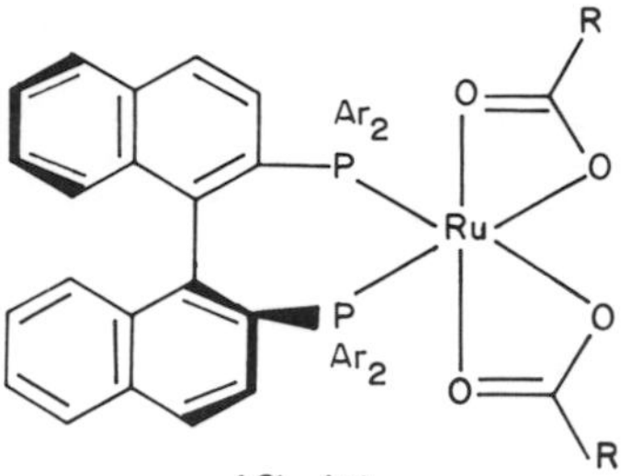

(c) Ar = p-$CH_3C_6H_4$; R = CH_3

Table 2. Asymmetric hydrogenation of geraniol (**2**) and nerol (**3**) catalysed by Ru^{II}–BINAP complexes[a]

Substrate	Catalyst	Substrate/catalyst molar ratio	Citronellol (**4**)	
			ee[b] (%)	Configuration
2	(*S*)-**1a**	530	96	*R*
2	(*S*)-**1b**	500	98	*R*
2[c]	(*S*)-**1c**	10000	96	*R*
2[c]	Ru[(*S*)-BINAP]$(OCOCF_3)_2$[d]	50000	96	*R*
2[c]	Ru[(*S*)-TOLBINAP]$(OCOCF_3)_2$[d]	50000	97	*R*
3	(*R*)-**1a**	540	98	*R*
3	(*S*)-**1c**	570	98	*S*

[a]The reaction was carried out with stirring in a stainless-steel autoclave at 18–20 °C using a 0.35–0.61 M solution of the substrate in methanol with exclusion of air. After removal of the Ru complex by precipitation by adding pentane followed by filtration through Celite 545, the whole mixture was concentrated and distilled to give **4** in 97–100% yields.

[b]Determined by HPLC analysis (chiral column).

[c]Reaction using a 5.8 M solution of **2** in methanol at an initial hydrogen pressure of 30 atm for 12–14 h.

[d]These complexes (tenative structures) were prepared from **1a** and **1c**, respectively, by ligand replacement by addition of excess of trifluoroacetic acid.

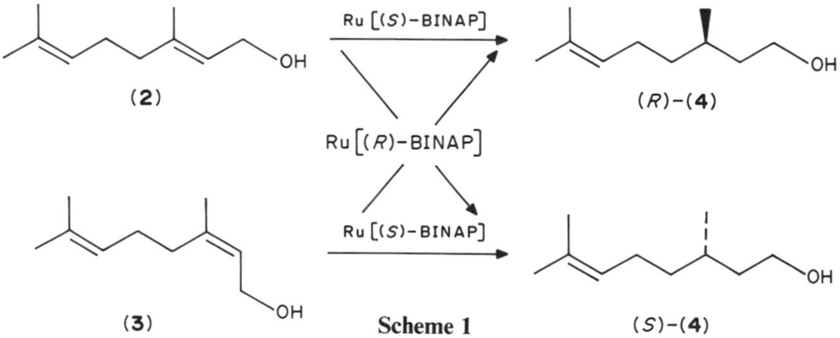

(2) Ru[(S)-BINAP] → (R)-(4)

Ru[(R)-BINAP]

(3) Ru[(S)-BINAP] → (S)-(4)

Scheme 1

of the product is determined by those of the ligand (*R* or *S*) and the substrate (*E* or *Z*). Scheme 1 illustrates this specific relationship observed between the substrate, the ligand and the product.

17.2.2 UNSATURATED CARBOXYLIC ACIDS

The Ru–BINAP complex **1** is also effective for the asymmetric hydrogenation of unsaturated carboxylic acids. (equation 2).[3] The results are summarized in

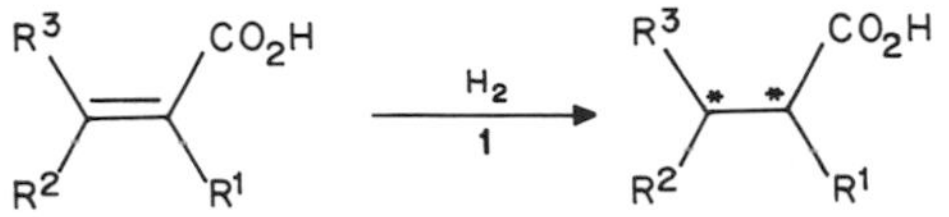

(2)

Table 3. These reactions are significant in that they can be used in the synthesis of a number of pharmaceutical products or their intermediates (**5**–**9**).

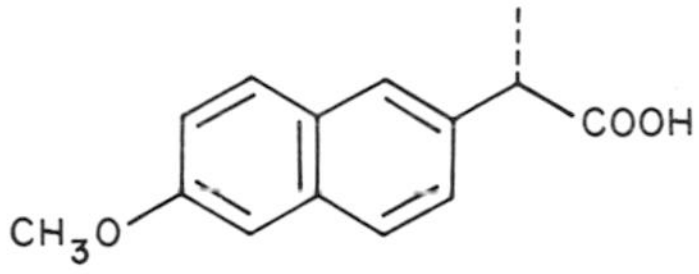

(*S*)–(**5**) 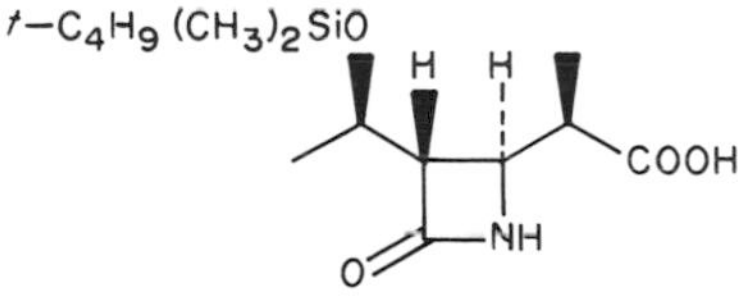 (**6**)

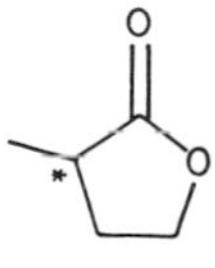

(**7**) 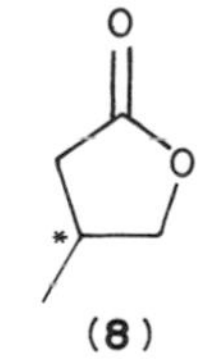 (**8**) 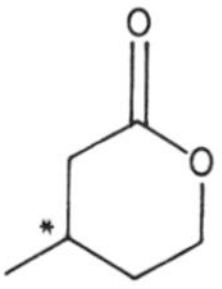 (**9**)

17.2.3 ENAMIDES

The carbon–carbon unsaturated system C=C—N—C=O in a cyclic structure is easily hydrogenated in the presence of complex **1** (equation 3).[4] This asymmetric reduction has a potential application for the synthesis of isoquinoline alkaloids (**10**–**12**), including the dextromethorphan intermediate **13**.

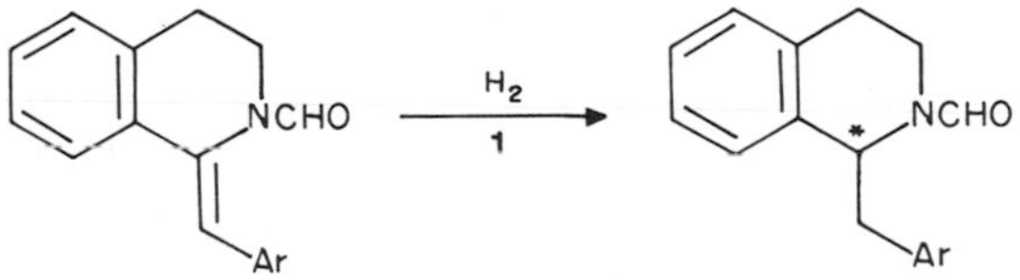

(3)

Table 3. Ru–BINAP-catalysed asymmetric hydrogenation of α,β-unsaturated carboxylic acids[a]

Substrate (equation 2)				Substrate/ catalyst	Conditions		Product	
R^1	R^2	R^3	Catalyst	molar ratio	H_2 (atm)	Time (h)	*ee* (%)[b]	Configuration[c]
CH_3	CH_3	H	(*R*)-**1a**	160	4	12	91	2*R*
H	$(CH_3)_2C{=}CH(CH_2)_2$	CH_3	(*R*)-**1a**	279	101	12	87	3*S*
H	CH_3	C_6H_5	(*R*)-**1a**	590	104	70	85	3*S*
C_6H_5	H	H	(*S*)-**1a**	397	112	24	92	2*S*
H	$HOCH_2$	CH_3	(*R*)-**1a**	145	86	16	93	3*R*
H	CH_3COOCH_2	CH_3	(*R*)-**1a**	106	98	12	95	3*R*
H	$HOCH_2CH_2$	CH_3	(*R*)-**1a**	129	100	12	93	3*S*
H	$CH_3COOCH_2CH_2$	CH_3	(*R*)-**1a**	110	100	12	88	3*S*
CH_3	CH_3COOCH_2	H	(*R*)-**1a**	106	4	12	83	2*R*
$HOCH_2$	CH_3	H	(*S*)-**1a**	145	4	12	95	[d]

[a]The reaction was carried out in a 0.05–0.3 M solution of the substrate (0.6–3.2 mmol) in degassed absolute methanol at 15–30 °C. The conversion was 100%.
[b]Determined by HPLC analysis of the corresponding amide derived from the product and (*R*)-1-(1-naphthyl)ethylamine.
[c]Determined by sign of rotation.
[d]Not determined. $[\alpha]_D^{17} - 3.3°$ (*c* 2.93, methanol).

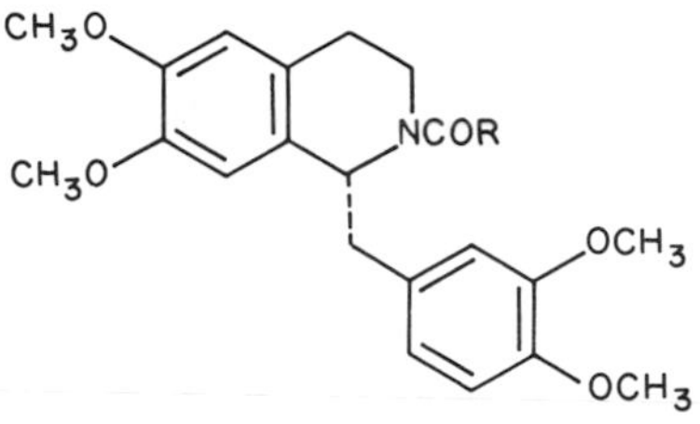

(10) R = CH_3

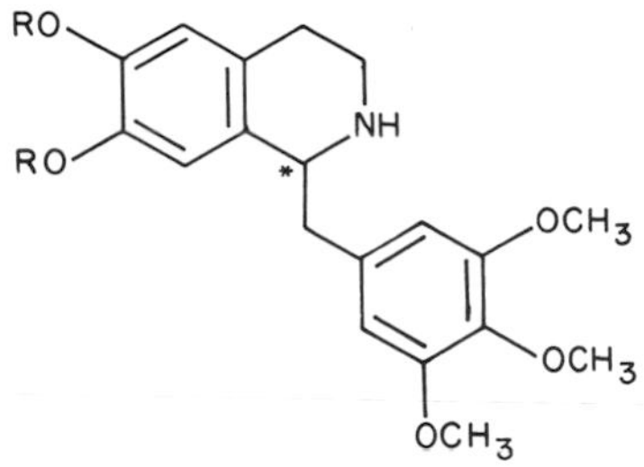

(11) R = H

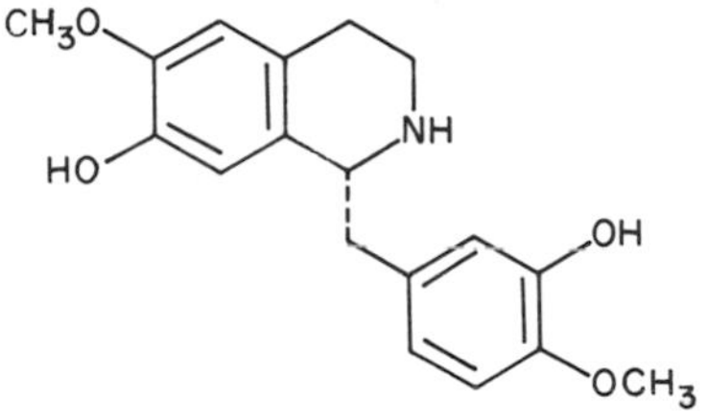

(12)

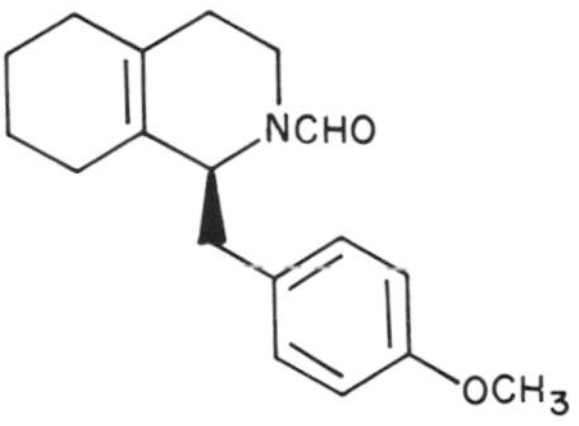

(13)

17.2.4 SUBSTITUTED KETONES

New cationic Ru^{II}–BINAP complexes bearing halogens as ligands (**14**–**16**) have been shown to have a high catalytic activity and enantioselectivity in the hydrogenation of substituted ketones to give optically active alcohols (Scheme 2).[5]

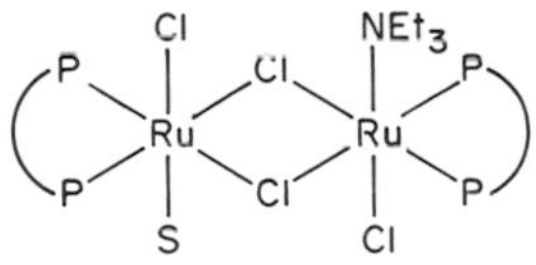

(14)

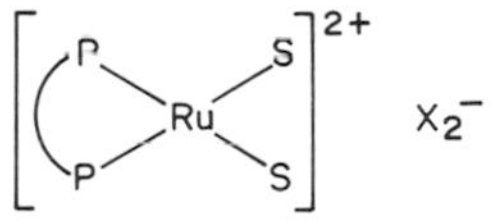

(15)

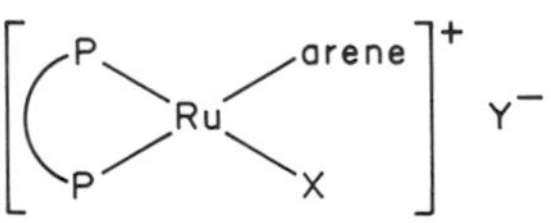

(16)

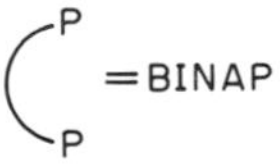

P⌒P = BINAP
S = solvent
X, Y = anion (halogen, etc.)

The asymmetric induction illustrated in Scheme 2 suggests that the key factor in the stereo differentiation is the simultaneous coordination of the carbonyl oxygen and the heteroatom X to the Ru atom to give a five- or six-membered

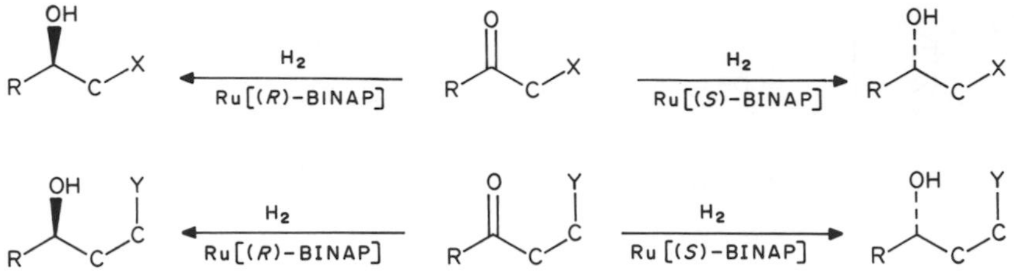

Scheme 2

chelate ring. The substituent X may be an oxygen- and/or nitrogen-containing functional group or a halogen. Typical examples of such substrates are listed in Table 4.

Similar chelating effects are also observed in the reduction of substituted aromatic ketones to give optically active benzyl alcohol derivatives (**17** and **18**) (equation 4).

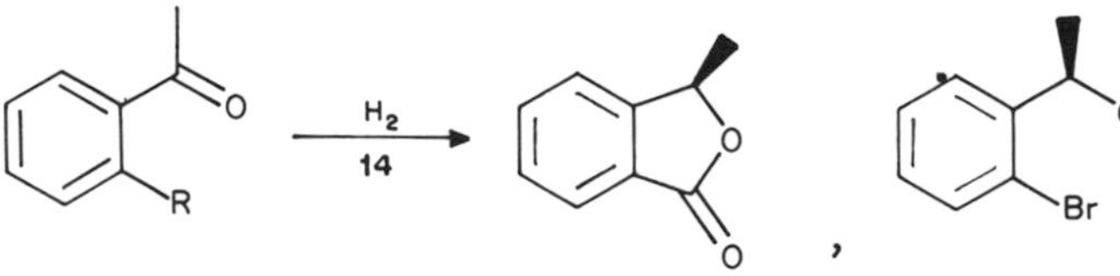

(4)

R = COOH

R = Br

(17) (18)

In the case of multi-substituted ketones, the competing coordination seems to result in a decrease in optical purity, as indicated by the examples listed in

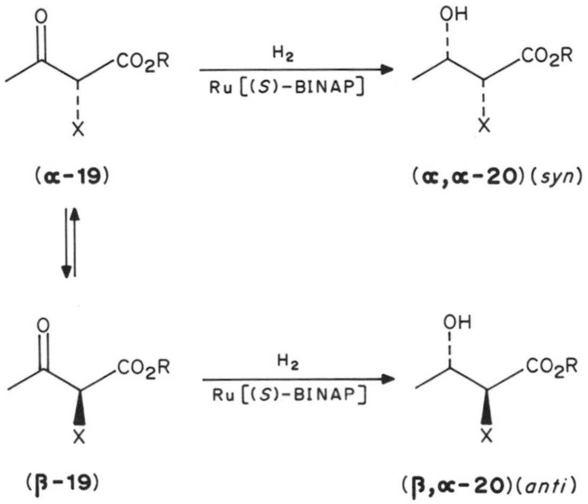

(α-19) (α,α-20) (*syn*)

(β-19) (β,α-20) (*anti*)

Scheme 3

Table 4. Catalytic asymmetric hydrogenation of substituted ketones[a]

Substrate	Catalyst	Substrate/ catalyst molar ratio	Conditions		Product		
			H_2 (atm)	Time (h)	Yield (%)	*ee* (%)[b]	Configuration[b]
CH_3COCH_2Cl	$RuBr_2$[(*S*)-BINAP]	780	50	12	72	81	*S*
CH_3COCH_2Br	$RuBr_2$[(*S*)-BINAP]	390	100	24	83	90	*S*
$C_6H_5COCH_2N(CH_3)_2$	$RuBr_2$[(*S*)-BINAP]	490	100	24	85	95	*S*
CH_3COCH_2OH	$RuCl_2$[(*R*)-BINAP]	230	93	32	100	92	*R*
$CH_3COCO_2CH_3$	$RuCl_2$[(*R*)-BINAP]	780	96	46	97	83	*R*
$CH_3COCH_2CH_2OH$	$RuCl_2$[(*R*)-BINAP]	900	70	42	100	98	*R*
$CH_3COCH_2CH_2C_2H_5$	$RuBr_2$[(*R*)-BINAP]	1260	86	51	100	>99	*R*
$CH_3COCH_2CON(CH_3)_2$	$RuBr_2$[(*S*)-BINAP]	680	63	86	100	96	*S*
$CH_3COCH_2COSC_2H_5$	$RuCl_2$[(*R*)-BINAP]	540	95	86	42	93	*R*
$CH_3COCOCH_3$	$RuBr_2$[(*S*)-BINAP]	680	80	61	100[c]	100	*S,S*
$CH_3COCH_2COCH_3$	$RuCl_2$[(*R*)-BINAP]	2000	72	89	100, 95[d]	100	*R,R*
$CH_3COCH(CH_3)COCH_3$	$RuCl_2$[(*S*)-BINAP]	2200	94	62	100[d]	99	*S,S*

[a]The reaction was carried out at 20–30 °C in a 1–4 M ethanol solution of the substrate (3–21 mmol).
[b]The enantiomeric excesses and absolute configurations of the products were determined by a combination of HPLC and ^1H NMR analysis of the appropriate MTPA esters and rotation measurement.
[c]*dl*:*meso* = 26:74.
[d]*dl*:*meso* = 99:1.

Table 5. Catalytic asymmetric hydrogenation of multi-substituted ketones[a]

		Substrate/ catalyst	Conditions		Product		
No.	Substrate	molar ratio	H_2 (atm)	Time (h)	Yield (%)	*ee* (%)[b]	Configuration[b]
1	$C_6H_5CH_2OCH_2COCH_2CO_2CH_3$	350	45	35	96	75	*R*
2	$C_6H_5CH_2OCH_2CH_2COCH_2CO_2CH_3$	370	50	85	94	99	*S*
3	$(^i\text{-}C_3H_7)_3SiOCH_2COCH_2CO_2C_2H_5$	290	100	86	100	95	*R*
4	$ClCH_2COCH_2CO_2CH_3$	320	100	25	95	70	*R*
5	$ClCH_2COCH_2CO_2CH_3$	450	100	0.1[c]	96	97	*R*

[a] The reactions were carried out at 20–30 °C in a 1–4 M ethanol solution of the substrate using complex **14**.
[b] The enantiomeric excesses and absolute configurations were determined by HPLC.
[c] The reaction temperature was 100 °C.

Table 6. Asymmetric hydrogenation of α-substituted β-keto esters

H_2, Ru[(*S*)-BINAP], CH_2Cl_2 ← (racemic α-X β-keto ester) → H_2, Ru[(*R*)-BINAP], CH_2Cl_2

X	*de* (%)	*ee* (%)
$NHCOCH_3$	99	98
NHCOPh	96	90
$CH_2NHCOPh$	94	90
Cl	80	95

Table 5. The above phenomena also support the hypothetical coordination pathway of Ru–BINAP-catalysed hydrogenations. For synthetic purposes, the low optical purities can be improved by increasing the steric effect (entry 3) or by changing the reaction temperature (entry 5).[6]

Derivatives of 3-oxo carboxylic esters with suitable substituents at position 2 give *syn* 3-hydroxy esters.[7] Typical examples are shown in Table 6. The reaction is considered to be a dynamic kinetic resolution as outlined in Scheme 3.

If the racemization of the enantiomers α-**19** and β-**19** is rapid enough with respect to the hydrogenation giving **20**, then when rates of the reaction of α-**19** and β-**19** are substantially different, the hydrogenation would form one isomer selectively among the four possible stereoisomeric hydroxy esters. With this procedure, a single chiral product possessing stereo defined vicinal asymmetric centres can be obtained from a racemic starting material in high yield.

17.3 PREPARATION OF Ru–BINAP CATALYSTS

A resolution method for the 1,1′-binapthalene-2,2′-diol (BINAP) ligand is discussed in Chapter 1 (Section 1.2.2.2). The ruthenium complexes may be prepared by simple ligand exchange, and the following preparation[8] of dicarboxylate complex (*S*)-**1** is representative of preparative methods for other ruthenium dicarboxylate complexes. A toluene solution of $[RuCl_2(COD)]_n$, (*S*)-BINAP and triethylamine is heated to prepare an intermediate BINAP complex. The crude product is filtered to remove insoluble byproducts, dissolved in *tert*-butanol and treated with sodium acetate. Recrystallization of the material thus obtained from a mixture of toluene and hexane gives complex (*S*)-**1**, which is pure enough for use as a catalyst in the processes described above.

Ru–BINAP halide complexes of the type **15** may be prepared[9] from (*S*)-BINAP and $[RuCl_2(C_6H_6)]_2$ simply by heating these materials in an ethanol–benzene mixture. Concentration of the solution affords complexes of type **15**, with ethanol as the coordinating solvent. Other counter ions may be obtained by exchange of the halide ions in this and similar complexes.

17.4 COMMERCIAL APPLICATIONS

Numerous possibilities exist for the adoption of Ru–BINAP catalysts in industrial asymmetric hydrogenations. Two typical examples of their commercial application by Takasago are described below.

17.4.1 α-TOCOPHEROL SIDE-CHAIN

The importance of the industrial manufacture of α-tocopherol in the naturally occurring stereochemical configuration is well known. The molecule contains two chiral moieties, the chroman part and the C-15 side-chain (Scheme 4). A precursor of the latter, (3*R*, 7*R*)-3,7,11-trimethyldodecan-1-ol, can be prepared economically utilizing BINAP catalyst (Scheme 5).[2] The synthetic strategy is based on the combination of asymmetric isomerization of an allylic amine by Rh–BINAP (see Chapter 16) for the chirality at the C-7 position, and asymmetric hydrogenation of an allylic alcohol by Ru–BINAP to give the (3*R*)-configuration.

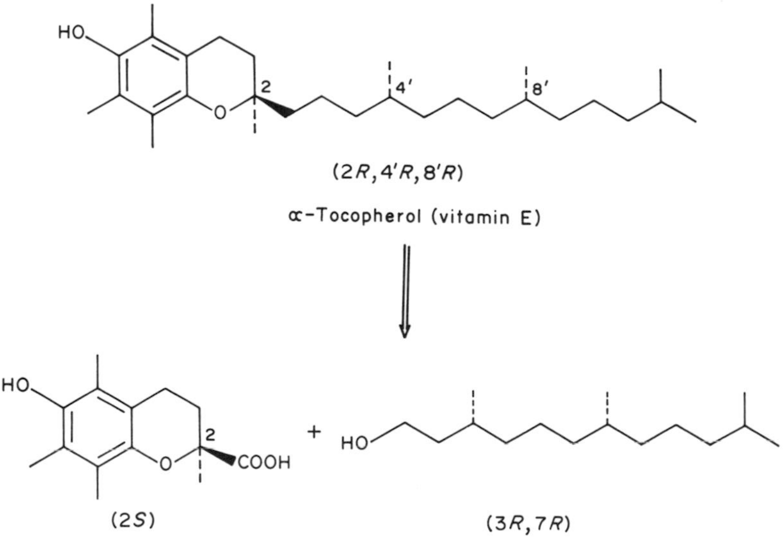

HO O 2 4′ 8′

(2*R*,4′*R*,8′*R*)

α-Tocopherol (vitamin E)

HO O 2 COOH + HO

(2*S*) (3*R*,7*R*)

Scheme 4

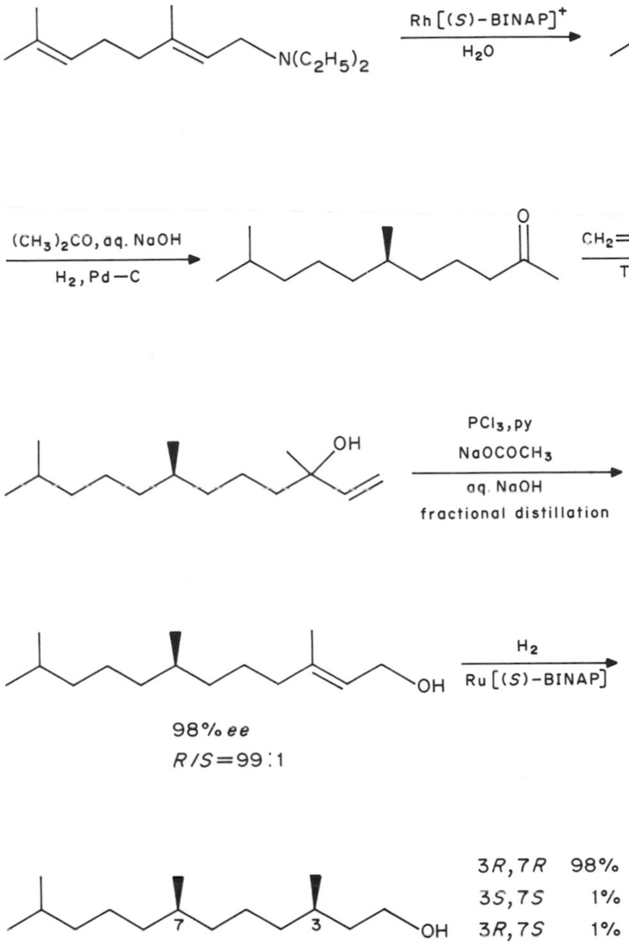

$N(C_2H_5)_2$ —Rh[(*S*)-BINAP]$^+$, H_2O→ CHO

98% *ee*

—$(CH_3)_2CO$, aq. NaOH; H_2, Pd–C→ O —CH_2=CHMgCl, THF→

OH —PCl_3, py; $NaOCOCH_3$; aq. NaOH; fractional distillation→

OH —H_2, Ru[(*S*)-BINAP]→

98% *ee*
R/*S*=99:1

7 3 OH

3*R*,7*R*	98%
3*S*,7*S*	1%
3*R*,7*S*	1%

Scheme 5. Synthesis of (3*R*, 7*R*)-3,7,11-trimethyldodecanol

17.4.2 β-LACTAM INTERMEDIATES

Many modified carbapenems and penems have been claimed as potential candidates for the next generation of antibiotics. For such development compounds, the demand for key intermediates is increasing (see also Chapter 13). A new, economically feasible synthesis of 4-acetoxyazetidin-2-one has been established and launched in Takasago (Scheme 6). The method utilizes a Ru–BINAP catalyst to produce two adjacent asymmetric centres in a desirable configuration.

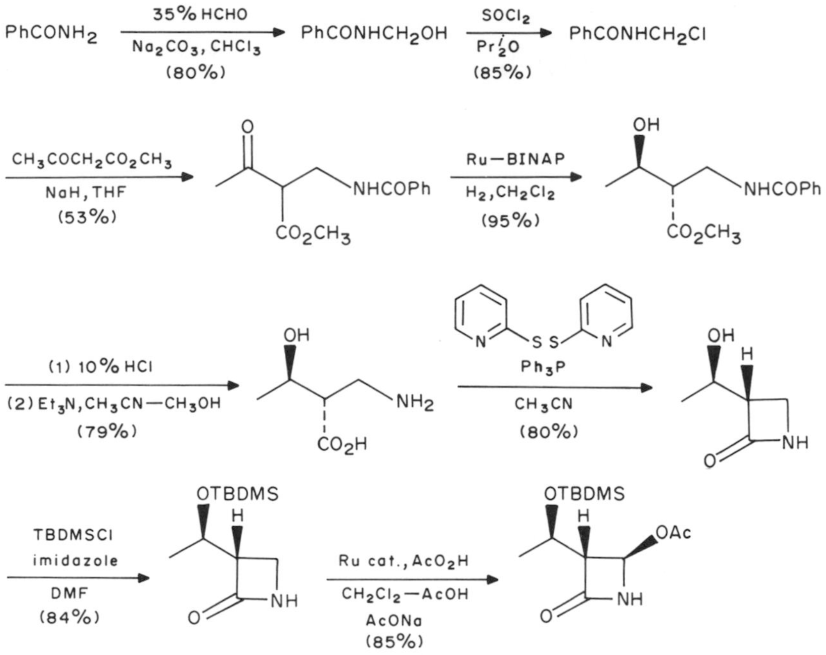

$PhCONH_2 \xrightarrow[\text{(80\%)}]{35\%\ HCHO,\ Na_2CO_3,\ CHCl_3} PhCONHCH_2OH \xrightarrow[\text{(85\%)}]{SOCl_2,\ Pr^i_2O} PhCONHCH_2Cl$

$\xrightarrow[\text{(53\%)}]{CH_3COCH_2CO_2CH_3,\ NaH,\ THF}$ NHCOPh, CO_2CH_3 $\xrightarrow[\text{(95\%)}]{Ru-BINAP,\ H_2,\ CH_2Cl_2}$ OH, NHCOPh, CO_2CH_3

$\xrightarrow[\text{(79\%)}]{(1)\ 10\%\ HCl;\ (2)\ Et_3N,\ CH_3CN-CH_3OH}$ OH, NH_2, CO_2H $\xrightarrow[\text{(80\%)}]{\text{2,2'-dipyridyl disulfide},\ Ph_3P,\ CH_3CN}$ OH, H, O, NH

$\xrightarrow[\text{(84\%)}]{TBDMSCl,\ imidazole,\ DMF}$ OTBDMS, H, O, NH $\xrightarrow[\text{(85\%)}]{Ru\ cat.,\ AcO_2H,\ CH_2Cl_2-AcOH,\ AcONa}$ OTBDMS, H, OAc, O, NH

Scheme 6

It is noteworthy that in this process the acetoxylation step is also carried out in the presence of a ruthenium catalyst.[10]

17.5 CONCLUSION

The utility of Ru–BINAP catalysts for the asymmetric hydrogenation of a range of prochiral compounds has been examined. These catalysts exhibit a high activity and excellent chemo- and enantioselectivities, thus leading to economic processes. The catalysts are currently employed in commercial production of a range of products in Takasago.

Asymmetric hydrogenation with Ru–BINAP catalysts is an extremely versatile method for the synthesis of a wide variety of optically pure compounds with extensive applications in industry and academia.

17.6 ACKNOWLEDGEMENTS

The author expresses his gratitude to Professors R. Noyori (Nagoyi University), H. Takaya (Kyoto University) and M. Saburi (Tokyo University). He also thanks his colleagues at Takasago, Drs H. Kumobayashi and N. Sayo and Mr T. Saito. Great appreciation is also extended to Dr A. Dodd (Research Fellow) and Miss T. Endo for their help with this chapter.

17.7 REFERENCES

1. Reviews: (a) Kagan, H. B., in *Asymmetric Synthesis* (ed. J. D. Morrison, Academic Press, New York, 1985, Vol. 5, Chap. 1; (b) Halpern, J., in *Asymmetric Synthesis*, Vol. 5, Chap. 2; (c) Koenig, K. E., in *Asymmetric Synthesis*, Vol. 5, Chap. 3; (d) Bosnich, B. (Ed.) *Asymmetric Catalysis*, Martinus Nijhoff, Dordrecht, 1986; (e) Brown, J. M., *Angew. Chem., Int. Ed. Engl.*, **26**, 190 (1987).
2. (a) Ohta, T., Takaya, H., and Noyori, R., *Inorg. Chem.*, **27**, 566 (1988); (b) Takaya, H., Ohta, T., Sayo, N., Kumobayashi, H., Akutagawa, S., Inoue, S., Kasahara, I., and Noyori, R., *J. Am. Chem. Soc.*, **109**, 1596 (1987); (c) Inoue, S., Osada, M., Koyano, K., Takaya, H., and Noyori, R., *Chem. Lett.*, 1007 (1985).
3. (a) Ohta, T., Takaya, H., Kitamura, M., Nagai, K., and Noyori, R., *J. Org. Chem.*, **52**, 3174 (1987); (b) Kawano, H., Ishii, Y., Ikariya, T., Saburi, M., Yoshikawa, S., Uchida, Y., and Kumobayashi, H., *Tetrahedron Lett.*, **28**, 1905 (1987).
4. (a) Noyori, R., Ohta, M., Hsiao, Y., Kitamura, M., Ohta, T., and Takaya, H., *J. Am. Chem. Soc.*, **108**, 7117 (1986); (b) Kitamura, M., Hsiao, Y., Noyori, R., and Takaya, H., *Tetrahedron Lett.*, **28**, 4289 (1987).
5. (a) Noyori, R., Ohkuma, T., Kitamura, M., Takaya, H., Sayo, N., Kumobayashi, H., and Akutagawa, S., *J. Am. Chem. Soc.*, **109**, 5856 (1987); (b) Kitamura, M., Ohkuma, T., Inoue, S., Sayo, N., Kumobayashi, H., Akutagawa, S., Ohta, T., Takaya, H., and Noyori, R., *J. Am. Chem. Soc.*, **110**, 629 (1988); (c) Kawano, H., Ishii, Y., Saburi, M., and Uchida, Y., *J. Chem. Soc., Chem. Commun.*, 87 (1988).
6. Kitamura, M., Ohkuma, T., Takaya, H., and Noyori, R., *Tetrahedron Lett.*, **29**, 1555 (1988).
7. (a) Noyori, R., Ikeda, T., Ohkuma, T., Widhalm, M., Kitamura, M., Takaya, H., Akutagawa, S., Sayo, N., Saito, T., Taketomi, T., and Kumobayashi, H., *J. Am. Chem. Soc.*, **111**, 9134 (1989); (b) Mashima, K., Matsumura, Y., Kumobayashi, H., Sayo, N., Hori, Y., Ishizaki, T., Akutagawa, S., and Takaya, H., *J. Chem. Soc., Chem. Commun.*, 609 (1991).
8. Ikariya, T., Ishii, Y., Kawano, H., Arai, T., Saburi, M., Yoshikawa, S., and Akutagawa, S., *J. Chem. Soc., Chem. Commun.*, 922 (1985).
9. Mashima, K., Kusano, K., Ohta, T., Noyori, R., and Takaya, H., *J. Chem. Soc., Chem. Commun.*, 1208 (1989).
10. Murahashi, S., Naota, T., Kuwabara, T., Saito, T., Kumobayashi, H., and Akutagawa, S., *J. Am. Chem. Soc.*, **112**, 7820 (1990).

18 The Preparation and Uses of Enantiomerically Pure β-Lactones

P. STUTTE
Lonza Ltd, Basle, Switzerland

A range of optically pure β-lactones and their derivatives malic acid and citramalic acid are marketed by the Swiss chemical company Lonza. The purpose of this chapter is to describe the methodology which has been adopted for the asymmetric synthesis of these versatile intermediates and to outline their potential as precursors to several important classes of compounds.

18.1 THE LARGE-SCALE SYNTHESIS OF OPTICALLY ACTIVE β-LACTONES

In 1982, Wynberg and Staring[1] reported the utility of the cinchona alkaloids quinine and quinidine for the catalysis of an asymmetric [2 + 2]cycloaddition of ketene to chloral and trichloroacetone. The process, outlined in Scheme 1, has been developed at Lonza to pilot-plant scale.

The reaction is carried out at a temperature of −50 °C in toluene. When quinidine or its esters are employed as catalysts, the (*R*)-enantiomers of the β-lactones are obtained, whereas the use of quinine gives access to the corresponding (*S*)-enantiomers. Thus, the reaction of 1,1,1-trichloroacetone (**2**) with ketene in the presence of quinidine gives the β-lactone (*R*)-4-methyl-4-(trichloromethyl) oxetan-2-one [(*R*)-TCMMO] (**4**), from which (*S*)-citramalic

Chirality in Industry. Edited by A. N. Collins, G. N. Sheldrake and J. Crosby

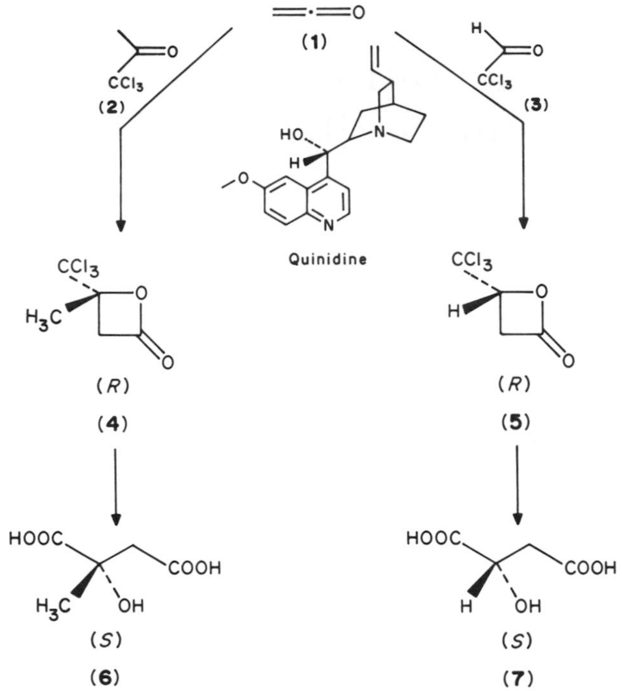

Scheme 1. Synthesis of (S)-malic acid (7) and citramalic acid (6) via the (R)-oxetanones 5 and 4 by asymmetric cycloaddition[1]

acid (**6**) may be derived by alkaline hydrolysis. Substitution of chloral (**3**) for trichloroacetone in the cycloaddition reaction leads to the β-lactone (**5**), an intermediate in the production of (S)-malic acid (**7**).

The alkaloid catalysts are used at concentrations of 0.5–2.0%, and may be recovered with an efficiency of up to 95% by extraction from the reaction mixture. Nearly quantitative chemical yields of the β-lactone products have been obtained routinely by this procedure, the enantiomeric excess being at least 96% for each of the examples in Scheme 1. This optical purity can be improved further by a final recrystallization of the β-lactones from methylcyclohexane.

18.2 THE PRODUCTION OF (S)-MALIC AND (S)-CITRAMALIC ACIDS AND THEIR DERIVATIVES

In addition to their usefulness as precursors of (S)-citramalic acid, β-lactones (**4** and **5**) are versatile chiral building blocks in their own right. Some examples of useful derivatives of (R)-TCMMO (**4**) are given in Scheme 2.

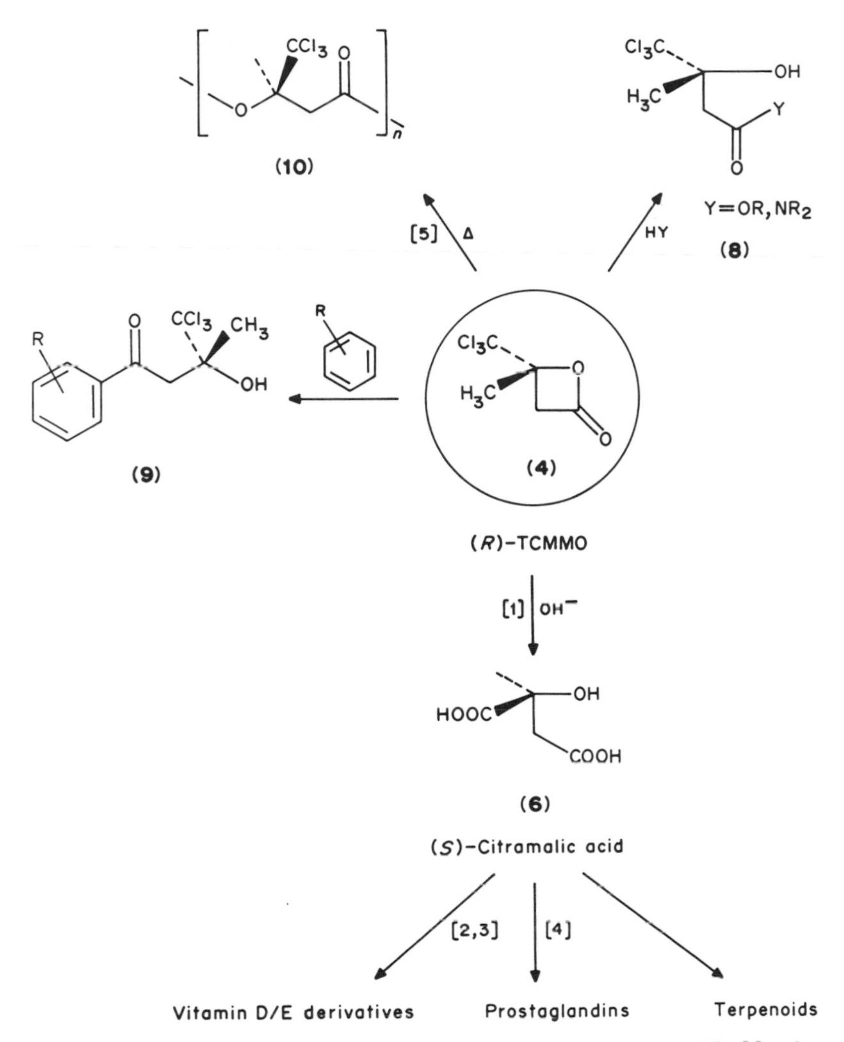

Scheme 2. Reactions of (*R*)-4-methyl-4-(trichloromethyl)oxetan-2-one(**4**). Numbers in square brackets are references

Ring opening of the β-lactones without hydrolysis of the trichloromethyl group is achieved by acidification with aqueous inorganic acids, and yields trichlorohydroxybutyric acids of type **8**. Alkaline hydrolysis with aqueous sodium hydroxide opens the β-lactone ring of **4**, and also hydrolyses the trichloromethyl group to give citramalic acid (**6**) as its (*S*)-enantiomer. Careful control of the reaction conditions ensures that racemization is avoided.

The β-lactone ring may be opened with nucleophiles such as water, alcohols or amines to yield the acids, esters and amides, respectively, represented by

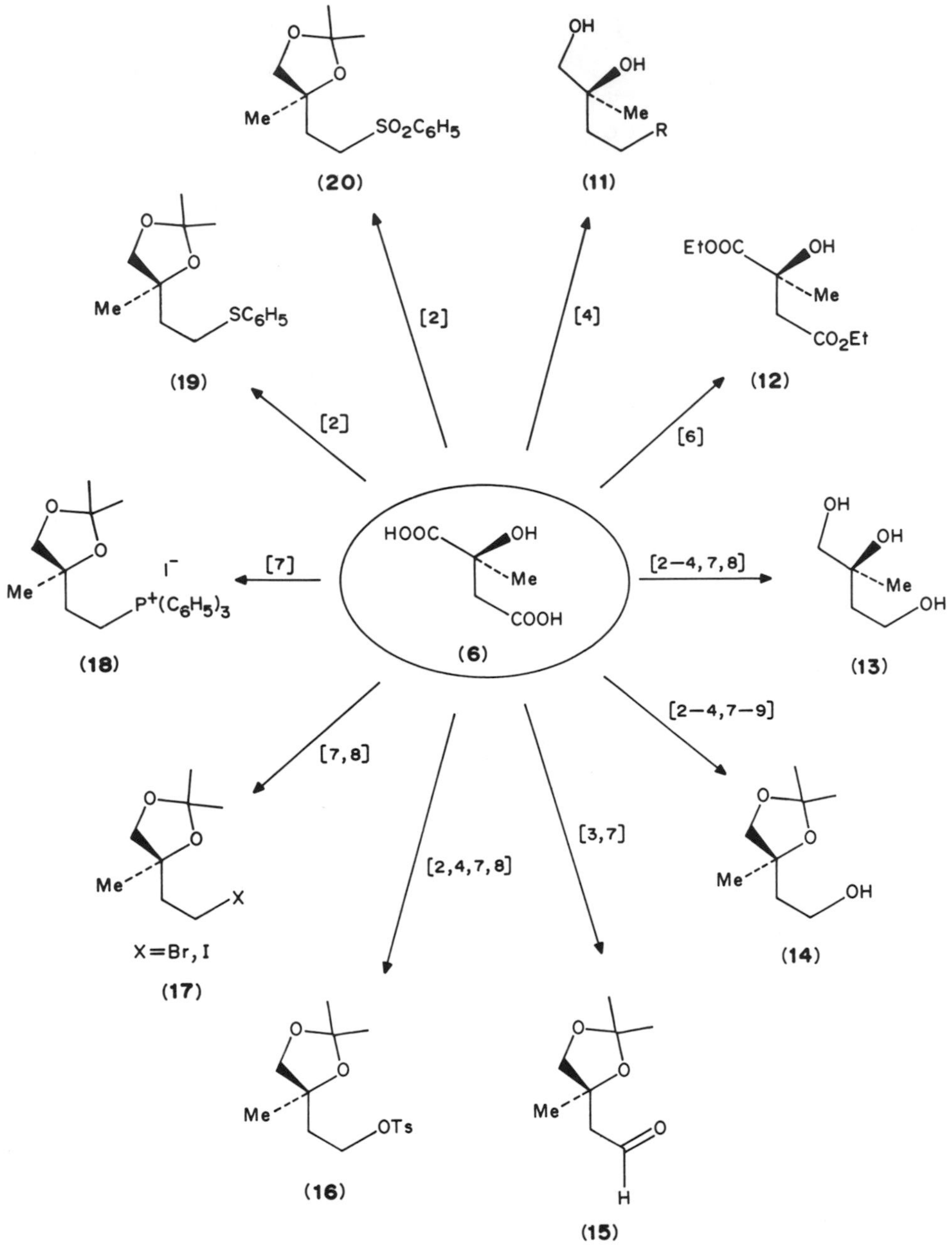

(20) (11) (12) (13) (14) (15) (16) (17) (18) (19) (6)

[2] [4] [6] [2–4, 7, 8] [2–4, 7–9] [3, 7] [2, 4, 7, 8] [7, 8] [7] [2]

HOOC, OH, Me, COOH

X = Br, I

Scheme 3. Reactions of citramalic acid (**6**). Numbers in square brackets are references

structure **8**. (*R*)-TCMMO will participate directly in a Friedel–Crafts acylation of electron-rich benzene derivatives, giving access to β-hydroxy alkyl aryl ketones (**9**). Polymeric derivatives (**10**) of (*R*)-TCMMO have also been reported.[5] No appreciable loss of enantiomeric purity is observed in any of the above reaction types.

The strategies which have been most frequently adopted for the futher manipulation of citramalic acid are illustrated for the (*S*)-isomer in Scheme 3. An initial reduction of the carboxylic acid groups to give a triol (**13**) and protection of the *vicinal* hydroxy groups by reaction with acetone gives the ketal (**14**). A variety of C-5 intermediates may then be obtained by further transformation of the remaining hydroxy group of this versatile compound. Thus, aldehyde (**15**), tosylate (**16**), halogen (**17**), phosphonium salt (**18**) thioether (**19**) and sulphone (**20**) compounds are all easily synthesized. Various carbon—carbon bond-forming reactions (e.g. Wittig methodology in the case of aldehyde **15**) may then be utilized to incorporate these C-5 synthons into industrial targets.[2–10] Schemes 4–7 outline the important steps in some of the many syntheses which have been accomplished with chiral intermediates of the above type.

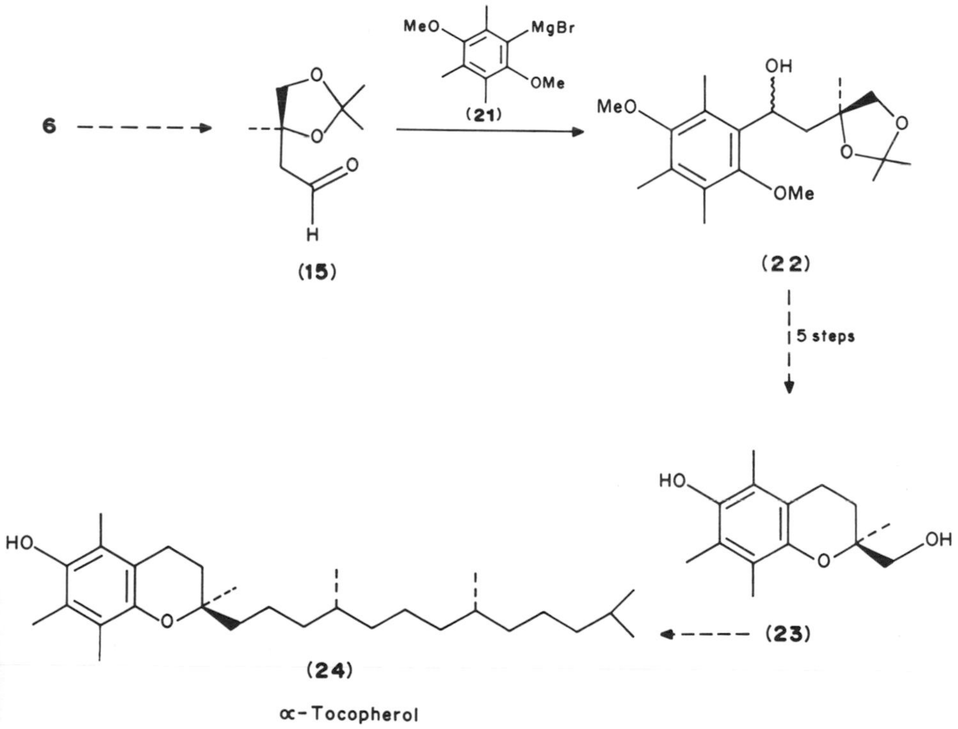

MeO MgBr OMe (21)

6 (15) H O

OH MeO OMe O O (22)

5 steps

HO OH O (23)

HO O (24)

α-Tocopherol

Scheme 4. Synthesis of α-tocopherol

Scheme 4 illustrates the use of Grignard methodology in the synthesis of α-tocopherol (**24**) from the aldehyde **15**. This aldehyde may be derived easily from (*S*)-citramalic acid via the corresponding alcohol (**14**) and contains, in latent form, all of the functionality required to construct the saturated ring of the optically active chroman unit. Thus, when aldehyde-ketal **15** was treated with the Grignard reagent **21**, the alcohol **22** was produced as a mixture of diastereoisomers.[3] Subsequent elaboration of **22** to α-tocopherol (**24**) via the chroman intermediate **23** was accomplished with standard methodology in several high-yielding steps.

The stereochemistry of the vitamin D_3 metabolite (25*S*)-25-hydroxyvitamin D_3 26,23-lactone (**28**) was confirmed by an elegant synthesis in which the side-chain of the sterol was derived from (*S*)-citramalic acid[2] (Scheme 5). In this case, a condensation of the sulphone **20** with an aldehyde serves to attach the chiral C-5 unit under mild conditions. The sulphone **20** was prepared by

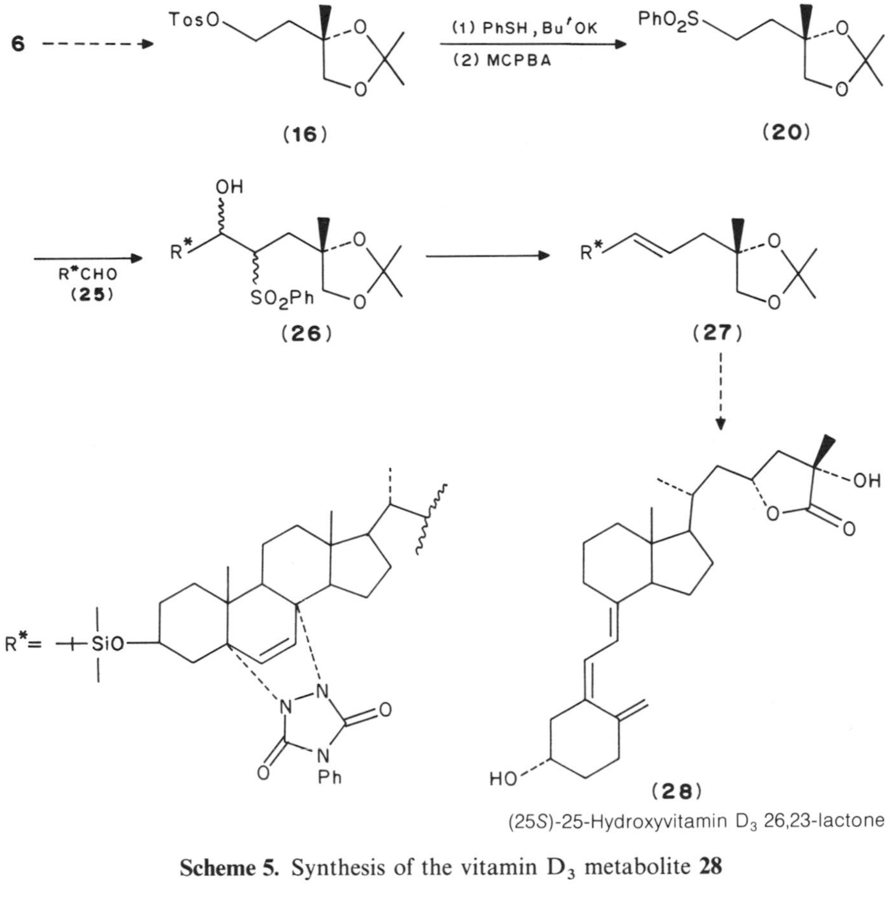

(25*S*)-25-Hydroxyvitamin D_3 26,23-lactone

Scheme 5. Synthesis of the vitamin D_3 metabolite **28**

treatment of the toluenesulphonate **16** with thiophenol and base, followed by oxidation of the resultant thioether. The sulphone **20** was treated with the sterol derivative **25** under basic conditions to give the condensation product **26**, from which the phenylsulphonyl and hydroxyl groups were removed by reduction with sodium amalgam. A further sequence of steps involving deprotection, oxidation and iodolactonization of the ketal **27** and a photochemical rearrangement of the sterol backbone furnished the desired vitamin D_3 metabolite in good yield.

A simple synthesis of the insect pheromone frontalin (**32**) has been developed in which the stereochemistry of both chiral centres of the product are defined by the (*S*)-configuration of the chiral centre in the citramalic acid starting material (Scheme 6). The aldehyde **15** was treated with the phosphorus ylid **29** to yield the α,β-unsaturated ketone **30**. Reduction of the double bond gave a ketone-ketal (**31**) which, on deprotection, cyclized to form (1*S*, 5*R*)-(−)-frontalin (**32**).

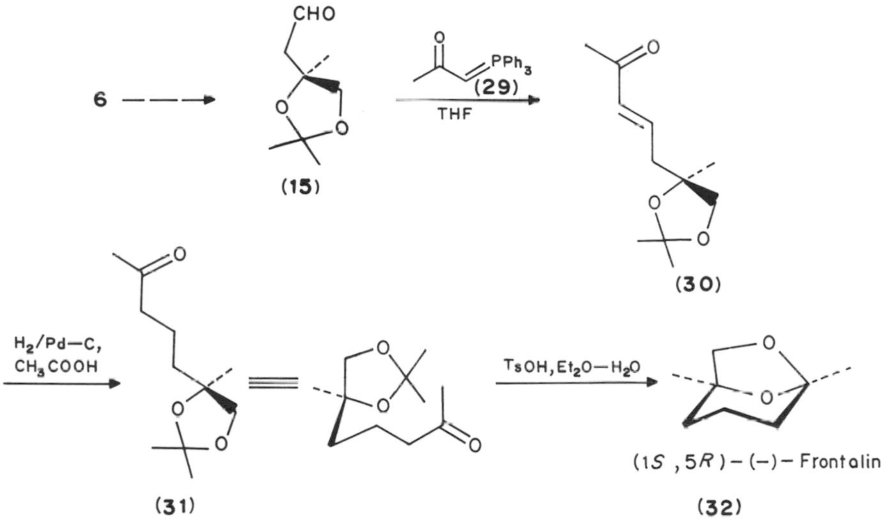

Scheme 6. Synthesis of Frontalin

A final example is taken from the active and important industrial field of prostaglandin synthesis. 15-Deoxy-16-methyl-16-α,β-hydroxyprostaglandin E_1 methyl ester (PGE_1) (**35**) possesses gastric anti-secretory activity in man, and compounds of this type are therefore of great interest as potential anti-ulcer drugs. The activity of **35** has been shown to be associated solely with the (16*S*)-isomer shown in Scheme 7. Alkylation of the toluenesulphonate ester **16** with copper (II) catalysis, and deprotection of the ketal product, furnished the key intermediate diol **33**. The nine-carbon side-chain of PGE_1 was derived from this precursor in straightforward steps via the alkyne alcohol **34**.

(16) (33) (34)

15-Deoxy-16(*S*)-hydroxy-16-methylprostaglandin

(35)

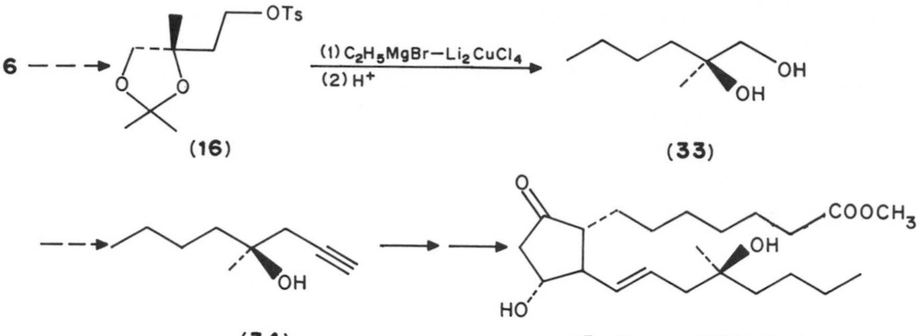

Scheme 7. Synthesis of the prostaglandin **35**

18.3 CONCLUSION

It is often cost-effective, in the manufacture of complex targets on a large scale, to introduce chirality at an early stage using a simple, chiral raw material, rather than to resolve an advanced intermediate by chemical methods. The β-lactones **4** and **5** and their derivatives, which are now available on a commercial scale by asymmetric synthesis, may thus be utilized to make various sophisticated chiral products of industrial interest. The use of the cycloaddition technology to produce other optically active β-lactones is under active development, and this programme should eventually widen the range of readily accessible chiral synthons of this type.

18.4 REFERENCES

1. Wynberg, H., and Staring, E. G. J., *J. Am. Chem. Soc.*, **104**, 166 (1982); *J. Org. Chem.*, **50**, 1977 (1985).
2. Nakayama, K., Yamada, S., and Takayama, H., *Tetrahedron Lett.*, **22**, 2591 (1981).
3. Barner, R., and Schmid, M., *Helv. Chim. Acta*, **62**, 2384 (1979).
4. Fujimoto, Y., Yadava, J. S., and Sih, C. J., *Tetrahedron Lett.*, **21**, 2481 (1980).
5. Grenier, D., and Prud'Homme, R. E., *J. Polym. Sci., Polym. Chem. Ed.*, **19**, 1781 (1981).
6. For general derivatives, see *Beilstein*, H **3**, 443; E I, **3**, 157; E II, **3**, 294; E III, **3**, 930; E IV, **3**, 1149.
7. Barner, R., Hübscher, J., Daly, J. J., and Schönholzer P., *Helv. Chim. Acta*, **64**, 915 (1981).
8. Synthesis: Staring, E. J. C., and Wynberg, H., *Int. Pat. Appl.*, PCT/NL 83/000 40, 1983.
9. Morrison, J. D., and Scott (Eds), *Asymmetric Synthesis, Vol. 4, The Chiral Carbon Pool*, Academic Press, New York, 1984.
10. Barner, R., and Hübscher, J., *Helv. Chim. Acta*, **66**, 880 (1983).

IMMOBILIZATION TECHNIQUES AND MEMBRANE BIOREACTORS

19 The Industrial Production of Optically Active Compounds by Immobilized Biocatalysts

I. CHIBATA, T. TOSA and T. SHIBATANI
Tanabe Seiyaku Co., Ltd, Osaka, Japan

Chirality in Industry. Edited by A. N. Collins, G. N. Sheldrake and J. Crosby

19.1 INTRODUCTION

The extraordinary capabilities of enzymes as catalysts for asymmetric reactions have been discussed in Chapter 5, and should be apparent from the wealth of examples in Chapter 1. However, enzymes are not always ideal catalysts for practical applications. Often, they are either inherently unstable, or incompatible with organic solvents or elevated temperatures. As one of the methods employed to make enzymes more suitable for the production of useful compounds, the immobilization of enzymes has been extensively studied since the late 1960s, and several new techniques have emerged. Immobilized biocatalysts are defined as enzymes, microbial cells, animal cells or plant cells which retain their catalytic activities when physically confined or localized in a defined region of space, and which can be used repeatedly and continuously.

The purpose of this chapter is to review briefly the development and scope of immobilizaion techniques, and to describe their use in the manufacture of several important optically active products at Tanabe Seiyaku.

19.2 IMMOBILIZED BIOCATALYSTS AND REACTORS

19.2.1 IMMOBILIZED BIOCATALYSTS

Methods for the immobilization of enzymes and microbial cells can be classified into three categories: carrier-binding, crosslinking and entrapping methods.[1]

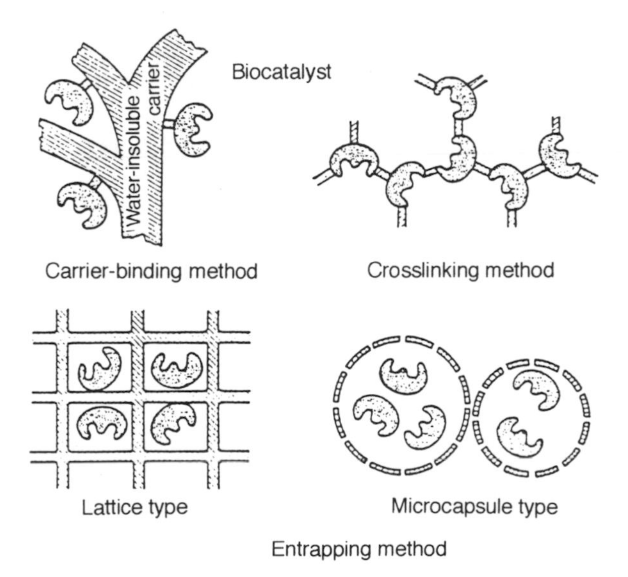

Figure 1. Schematic diagrams of immobilized biocatalysts[1]

These are shown schematically in Figure 1. Among them, the carrier-binding method is applied mainly to the immobilization of enzymes, and the entrapping method has been most extensively investigated for the immobilization of microbial cells.

19.2.1.1 The Carrier-binding Method

This technique involves the linkage of enzymes or cells directly to water-insoluble matrices, and can be further divided into three categories according to the binding mode of the enzymes or cells, that is, physical adsorption, ionic binding or covalent binding. As matrices, water-insoluble polysaccharides (cellulose, dextran and agarose derivatives), proteins (gelatin and albumin), synthetic polymers (ion-exchange resins and polyacrylamide gel), inorganic materials (porous glass, silica and alkaline earth metal ions), etc., may be used.

The immobilization of enzymes by physical adsorption and ionic binding methods can be simply achieved under mild conditions, and preparations with relatively high activity are often obtained. However, the binding forces between enzyme and matrix are weak in comparison with those in the covalent binding method, so leakage of the enzyme from the matrix may occur as a result of changes in the ionic strength or pH of the substrate or product solution. The activity of these types of immobilized enzymes can be restored when the enzyme activity decreases after prolonged operation. Thus, physical adsorption and ionic binding methods are advantageous in comparison with covalent binding methods, particularly when expensive matrices or enzymes are used.

The immobilization of enzymes by covalent binding is carried out under relatively severe conditions in comparison with physical adsorption or ionic binding. Accordingly, unless immobilization of an enzyme by covalent binding is carried out under well controlled conditions, an immobilized enzyme with high activity will not be obtained. The binding forces between the enzyme and matrix are strong, and the enzyme cannot easily leak out from matrices, even in the presence of substrates or salts at high concentration. However, when the activity of enzymes immobilized by covalent binding decreases during long-term operation, regeneration is impossible.

19.2.1.2 The Crosslinking Method

As in the covalent binding technique, the crosslinking method is based on the formation of chemical bonds, but in this case water-insoluble matrices are not used. The immobilization of enzymes or cells occurs by the formation of intermolecular crosslinkages between the enzyme molecules or the cells by means of bi- or multi-functional reagents, such as glutaraldehyde, diisocyanate derivatives and bisdiazobenzidine.

19.2.1.3 The Entrapping Method

The entrapping method entails the confinement of enzymes or cells in the lattice of a polymer matrix or enclosing them in semipermeable membranes. This technique differs from covalent binding and crosslinking methods in that the enzyme or cell itself does not bind to the matrix or membrane. The method may therefore have wide applicability. The matrices employed include collagen, gelatin, agar, alginate, carrageenan, cellulose triacetate, polyacrylamide, photo-crosslinkable resins and polyurethanes. Alginate, carrageenan and polyacrylamide, in particular, have been extensively used, because the immobilization method is easy, these matrices are not toxic to the cells and immobilized cells with high enzyme activity can be obtained.

19.2.2 REACTORS

Among the applications of immobilized biocatalysts, their use for chemical processes is the most active and important field. The design of enzyme reactors and the choice of the reactor type are perhaps the key elements in any process using immobilized biocatalysts. However, there are no simple rules for choosing the reactor type for a specific process. In practice, it is necessary to analyse in detail the advantages and disadvantages of a particular reactor and to design it according to the intended application.

A number of different types of reactors have been used with immobilized biocatalysts on the laboratory and industrial scales. These reactors are the stirred tank reactor, the packed bed reactor, the fluidized bed reactor, the ultrafiltration membrane reactor, the enzyme film and the enzyme tube, according to the mode of the operation, the flow pattern and the type of reactor.

The characteristics of these reactors are briefly described.

19.2.2.1 Stirred-tank Reactors

A batch stirred tank reactor is the simplest type, and consists of a container and a stirrer, which may be of turbine wing or propeller design. Baffles are fitted to the sides of the tank to improve the efficiency of stirring.

This type of reactor is useful for substrate solutions of high viscosity and for immobilized biocatalysts of relatively low activity, although immobilized biocatalysts tend to degrade on stirring.

A batch mode of operation is generally suitable for the production of small amounts of chemicals. Although a continuous stirred tank reactor is more efficient than a batch stirred tank reactor, the reactor equipment is slightly more complicated.

19.2.2.2 Packed Bed Reactors

In the design of packed bed reactors, which are the most widely used type with immobilized biocatalysts, important considerations are the pressure drop across the packed bed and the effect of the bed dimensions on the reaction rate. For industrial applications, the flow direction of the substrate solution is also important; in some cases downward flow causes compression of the bed, so upward flow is generally the method of choice. Upward flow is also preferred when gas is produced in the course of an enzyme-catalysed reaction.

A continuous mode of operation has the following advantages over batch processes: (i) it is more amenable to automation and, consequently, is associated with lower labour costs, (ii) a stabilization of the operating conditions, and (iii) easier quality control procedures for the products.

19.2.2.3 Fluidized Bed Reactors

When a substrate solution of high viscosity or a gaseous substrate or product are used in a continuous reaction system, a continuous fluidized bed reactor is often the most suitable. The particle size of the immobilized biocatalyst is an important factor in the formation of a smooth fluidized bed, and care must be taken to avoid decomposition of the biocatalyst.

19.2.2.4 Ultrafiltration Membrane Reactors

Ultrafiltration membranes may be used to retain enzymes in continuous reactors by virtue of the size difference between the enzymes and the reaction products. In these systems, the enzymes are not necessarily immobilized, but rather confined by an ultra-filtration membrane with a suitable molecular weight cut-off, with low molecular weight products being removed through the membrane. In order to optimize the yield of a given enzyme-catalysed reaction the membranes may be arranged either as flat sheets or hollow fibres. The characteristics of membrane reactors will be considered in more detail in the next chapter.

19.2.2.5 Enzyme Tubes and Films

When enzymes are immobilized on the inner or outer surface of tubular supports or flat membranes, the dynamics of substrate conversion depend on the mass transport conditions as well as the enzyme kinetics. Although a smooth, laminar flow of substrate solution along the surface of the membrane may be achieved relatively easily with flat membranes, a tubular configuration offers a much greater ratio of surface area to volume. Several designs in which the membrane is arranged, for example, in a spiral configuration, have been used to improve the surface area of enzyme films.

19.3 OPTICAL RESOLUTION OF RACEMIC AMINO ACIDS BY IMMOBILIZED AMINOACYLASE

The industrial production of L-amino acids has traditionally been dominated by fermentation processes, although an increasing number of processes for the resolution of racemic materials made by chemical synthesis are now operated on an industrial scale (Chapters 8, 9 and 20). Among many resolution methods, an enzymic method using a mould aminoacylase developed by Tanabe Seiyaku has proved to be one of the best procedures. A chemically synthesized racemic acylamino acid is hydrolysed stereoselectively by an aminoacylase to give L-amino acid and unhydrolysed D-acylamino acid as shown in equation 1. After concentration, the compounds are easily separated owing to the difference in their solubilities. The D-acylamino acid is racemized and re-used in the resolution procedure.

$$\underset{\text{N-Acyl-DL-amino acid}}{RCH(NHCOR')COOH} + H_2O \xrightarrow{\text{aminoacylase}} \underset{\substack{\text{L-Amino acid} \\ +\,R'COOH}}{RCH(NH_2)COOH} + \underset{\text{N-Acyl-D-amino acid}}{RCH(NHCOR')COOH} \qquad (1)$$

(N-Acyl-D-amino acid → racemization → N-Acyl-DL-amino acid)

About 30 years ago, extensive studies made at Tanabe Seiyaku on aminoacylases from various microorganisms and other sources showed that the aminoacylase of *Aspergillus oryzae* was easily isolated, and had a particularly broad substrate specificity for various acylamino acids. From 1954 to 1969, this enzymic resolution method was employed by Tanabe Seiyaku for the production of several L-amino acids.

In order to achieve the goal of continuous optical resolution, a variety of immobilization methods were tested, and aminoacylase immobilized by ionic binding to DEAE-Sephadex was chosen.[2] An enzyme reactor which was designed for continuous production is shown in Figure 2. Since 1969, Tanabe Seiyaku have operated a series of enzyme reactors for the production of, *inter alia*, L-methionine, L-valine and L-phenylalanine. With this immobilized enzyme system, L-amino acids can be produced economically; the overall production cost being more than 40% lower than that of conventional batch systems using native enzyme. This constituted the first industrial application of immobilized

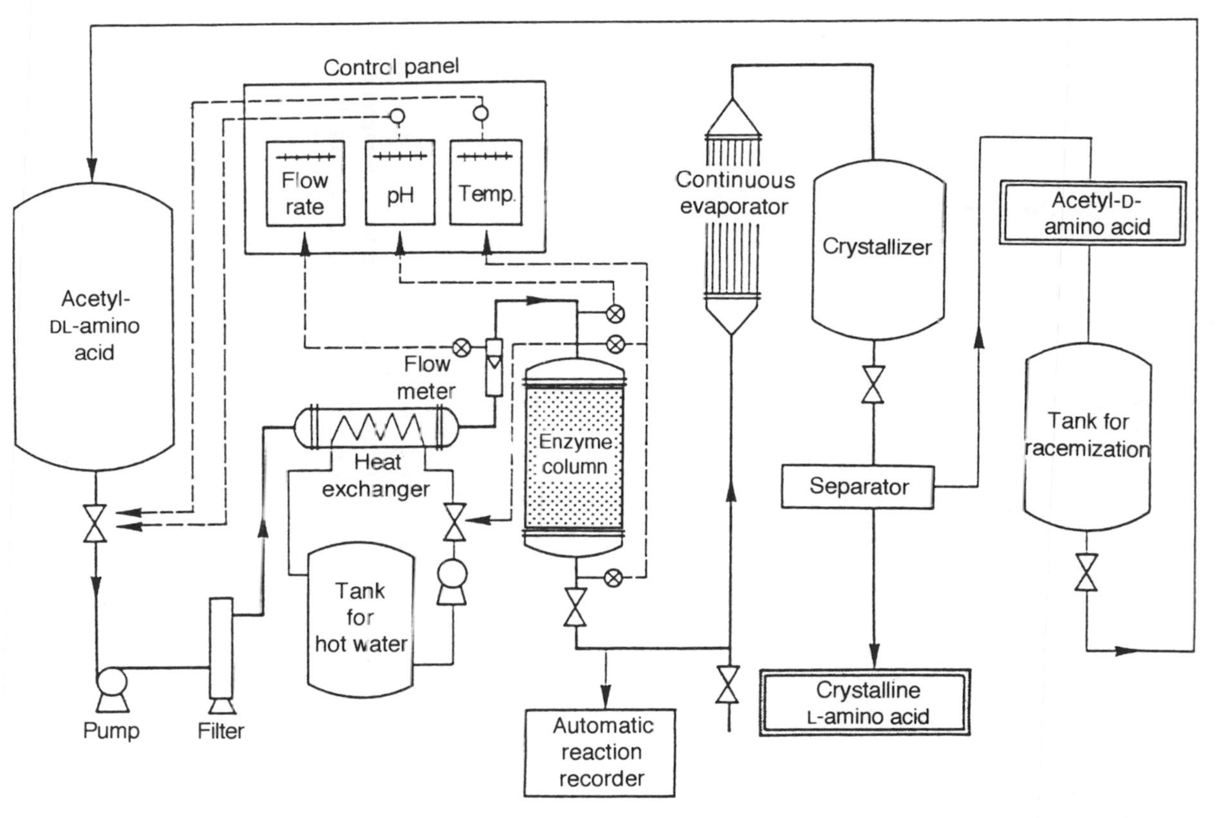

Figure 2. Flow diagram for the continuous production of L-amino acids by immobilized aminoacylase[1]

enzymes worldwide. Another continuous acylase process which employs a membrane reactor is discussed in the next chapter (Section 20.3.1).

19.4 PRODUCTION OF L-ASPARTIC ACID BY IMMOBILIZED *E. COLI* CELLS

L-Aspartic acid is widely used in medicines and as a food additive. Recently, aspartame, a dipeptide comprising L-aspartic acid and L-phenylalanine methyl ester, was commercialized as a low-calorie, synthetic sweetener (see Chapter 11). Demand for L-aspartic acid as a raw material for the synthesis of the dipeptide is, therefore, increasing.

Aspartic acid has been manufactured by fermentative or enzymatic batch processes from fumaric acid and ammonia, using an aspartase enzyme, according to the following equation:

$$\text{HOOC–CH=CH–COOH} \underset{\text{aspartase}}{\overset{NH_3}{\rightleftharpoons}} \text{HOOC–CH}_2\text{–CH(NH}_2\text{)–COOH} \quad (2)$$

Fumaric acid   L-Aspartic acid

Since 1960, L-aspartic acid has been produced by Tanabe Seiyaku by a batchwise reaction, using intact *Escherichia coli* having high aspartase activity. However, this procedure has some disadvantages in industrial use because the preparation of each batch involves the incubation of a mixture containing substrate and enzyme (or microbial cells). In 1973, the continuous production of L-aspartic acid by an immobilized microbial cell system was commercialized by Tanabe Seiyaku,[3] and this section summarizes the development of the process.

At the outset of the study, various immobilization methods for aspartase extracted from *E. coli* were tested. Relatively active immobilized aspartase was obtained by the entrapping method using polyacrylamide gel, and the stability of the immobilized enzyme column was investigated by operating it continuously for a long period. The activity of the column decreased by about 50% when using 1 M ammonium fumarate as a substrate solution after operation for about 30 days at 30 °C. Furthermore, extraction of the enzyme from microbial cells before immobilization added an extra stage, and the activity and stability of the immobilized enzyme were not considered satisfactory for industrial purposes.

It was surmised that if microbial cells having aspartase activity were directly immobilized, the disadvantages of immobilized aspartase might be overcome. Various methods for the immobilization of *E. coli* cells were tried as follows: (1) entrapping in the polyacrylamide gel matrix; (2) encapsulating in semipermeable polyurea membrane produced from toluene-2,4-diisocyanate and hexamethylenediamine; and (3) crosslinking with a bifunctional reagent, such as glutaraldehyde or toluene-2,4-diisocyanate. Among these methods, the most active immobilized *E. coli* cells were obtained by entrapping them in a polyacrylamide gel lattice.

Using a column packed with immobilized *E. coli* cells, the operational stability was investigated. It was found that the immobilized cell column was very stable; its half-life being 120 days at 37 °C. As the reaction is exothermic, the column reactor used for industrial production of L-aspartic acid was designed as a multistage system with cooling, as shown in Figure 3.

The overall production costs of this system were reduced to about 60% of those of the conventional batchwise reaction using intact cells owing to the

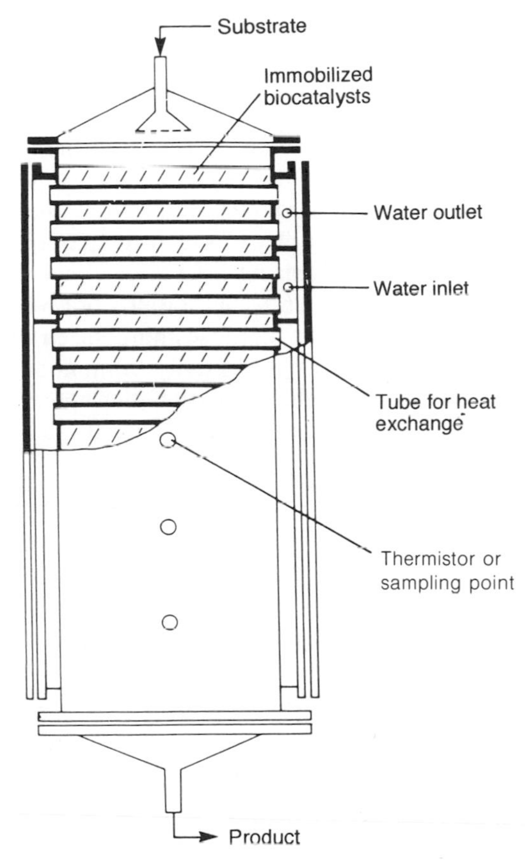

Figure 3. Schematic diagram of heat-exchange type column for production of L-aspartic acid[4,5]

marked increase in productivity of L-aspartic acid per unit of cells, the reduction in labour costs resulting from automation and an increase in the yield of L-aspartic acid. Further, a procedure employing immobilized cells is advantageous from the standpoint of waste treatment. This was the first industrial application of immobilized cells as a solid catalyst.

As stated above, the polyacrylamide gel method is advantageous for the immobilization of microbial cells for industrial applications. However, there are some limitations to this method. Some enzymes are inactivated during the immobilization procedure by the action of the acrylamide monomer, β-dimethylaminopropionitrile or potassium peroxodisulphate.

In an attempt to find a more general immobilization technique and to improve the productivity of the immobilized microbial cell system, various synthetic and natural polymers were screened. It was discovered that several polymers formed a gel lattice suitable for entrapping microbial cells. Among the polymers tested, κ-carrageenan was the most suitable for immobilization of microbial cells.

κ-Carrageenan is a naturally occurring hydrocolloid, consisting of a high molecular weight, linear-sulphated polysaccharide which is mainly composed of D-galactose and 3,6-anhydro-D-galactose and their sulphate ester derivatives. It is produced by extraction from red algae seaweeds, and is widely used in the food and cosmetic industries as a gelling, thickening and stabilizing agent. κ-Carrageenan becomes a gel on cooling, and also on contact with a solution containing potassium ions. The rigidity of κ-carrageenan gel in the presence of potassium ions increases with increasing potassium concentration.

The aspartase activity and the operational stability of *E. coli* cells immobilized with κ-carrageenan were compared with those immobilized with polyacrylamide (Table 1). It was found that the enzyme activity of immobilized cells prepared with κ-carrageenan was much higher, and that the operational stability was increased by a hardening treatment with glutaraldehyde and hexamethylenedia-

Table 1. Comparison of productivities of *E. coli* immobilized with polyacrylamide and κ-carrageenan for production of L-aspartic acid

Immobilization method	Aspartase activity (Ug^{-1} cells)	Stability at 37 °C (half-life, days)	Relative productivity[a] (%)
Polyacrylamide	18 850	120	100
Carrageenan	56 340	70	174
Carrageenan (GA)[b]	37 460	240	397
Carrageenan (GA + HMDA)[b]	49 400	680	1498

[a] Productivity $= \int_0^t E_0 \exp(-k_d t)dt$; E_0 = initial activity; k_d = decay constant; t = operation period.
[b] GA = glutaraldehyde; HMDA = hexamethylenediamine.

mine, which extended the half-life to almost 2 years. When the productivity of the immobilized preparation with polyacrylamide was taken as 100, that of cells immobilized with κ-carrageenan and hardened with glutaraldehyde and hexamethylenediamine was 1500, i.e. 15 times higher. Tanabe Seiyaku changed from the conventional polyacryalmide method to the new carrageenan method for industrial production of L-aspartic acid in 1978. When a 1000 l column is used, the theoretical yield of L-aspartic acid is 3.4 tonnes per day and 100 tonnes per month.

Several strains having higher aspartase activity were bred by mutation and recombinant DNA techniques. The results are summarized in Table 2. The highest aspartase activity was obtained with strains AT202 and AT202 (pNK101), but these strains obtained by transduction were less stable when used over long periods, and therefore strain EAPc7 was adopted in the subsequent investigations. This strain was obtained by screening for high aspartase activity in the course of continuous cultivation, and the genetic stability of the strain was sufficient for use over a long period.

The aspartase activity of the strain EAPc7 was about seven times higher than that of the parent *E. coli* ATCC 11303. However, this strain also had fumarase activity, which results in the conversion of fumaric acid to L-malic acid. Obviously, this would be disadvantageous in the production of L-aspartic acid according to equation 2, and therefore in order to employ this strain in an industrial process it was necessary to eliminate the fumarase activity. Several treatments for specifically eliminating fumarase activity from the strain were tested, and it

Table 2. Breeding of aspartase-high producing strain from *E. coli*

Strain	Mutation and plasmid	Aspartase activity (μmol min^{-1} mg^{-1} protein)
I. Mutation and transduction		
ATCC 11303	Wild type	14
EAPc7	*asp-1*	98 (7-fold)
EAPc244	*cya-1*	93 (7-fold)
AT202	*asp-1*, *cya-1*	240 (18-fold)
II. Recombinant plasmid from E.coli *k-12* aspA		
1. Stabilization of recombinant plasmid of aspA		
MM 294	Wild Type	
MM 294 (pNK101)	pBR 322-*aspA-par*	
2. Runaway vector		
MM 294 (pYT 125)	pSY 343-*aspA*	
III. Combination of mutation, transduction and recombinant DNA		
ATCC 11303	Wild type	14
AT 202 (pNK 101)	pBR 322-*aspA*	1016 (80-fold)

was found that when the strain was treated in a culture broth (pH 4.9) containing 50 mM L-aspartic acid at 45 °C for 1 h the fumarase activity was almost completely eliminated without inactivating aspartase.

The aspartase activity of treated EAPc7 cells immobilized with κ-carrageenan was about four times higher than that of the parent *E. coli* cells immobilized with κ-carrageenan. The immobilized preparation was very stable in comparison with the preparations of untreated EAPc7 cells and the parent *E. coli* cells, its half-life being 126 days. The reason for its increased stability is thought to be the inactivation of some proteases in the intact EAPc7 cells because, when the intact cells were treated to remove fumarase activity, their caseinolytic activity was also reduced to about 10% of its initial level.

The concentrations of L-aspartic acid, fumaric acid and L-malic acid in the effluent from columns packed with untreated and treated EAPc7 cells immobilized with κ-carrageenan were measured, and the results are shown in Table 3. In the case of treated cells, the concentration of L-malic acid was below 2 mM and the concentraion of L-aspartic acid was about 30 mM higher than in the case of the untreated cells.

Table 3. Concentration of acid in the effluent from immobilized cell columns[a]

	Concentration of acids in effluent (mM)					
	Untreated cell column			Treated cell column		
Operation period (days)	L-Aspartic acid	Fumaric acid	L-Malic acid	L-Aspartic acid	Fumaric acid	L-Malic acid
1	956.3	12.1	31.6	987.4	10.7	1.9
10	954.1	17.4	28.5	987.7	10.4	1.9
20	953.2	12.1	34.7	988.1	9.9	2.0

[a] A solution of 1 M ammonium fumarate (pH 8.5) containing 1 mM Mg^{2+} was applied to the column packed with immobilized cell preparations at the flow rate of space velocity = 1 h^{-1} at 37 °C.

Table 4. Comparison of productivities of various immobilized *E. coli* cells for production of L-aspartic acid

Immobilized preparation	Aspartase activity (μmol h^{-1} g^{-1} cells)	Half-life at 37 °C (days)	Relative productivity[a] (%)
Parent cells	56 300	70	100
Untreated EAPc7 cells	203 060	70	361
Treated EAPc7 cells	192 140	126	614

[a] Productivity of immobilized parent cells was taken as 100%. Productivity = $\int_0^t E_0 \exp(-k_d t) dt$; E_0 = initial activity; k_d = decay constant; t = operation period.

The efficiency of parent *E. coli* cells and untreated and treated EAPc7 cells immobilized with κ-carrageenan for production of L-aspartic acid was compared. The results show that when the productivity of the parent *E. coli* cell preparation was taken as 100, the treated EAPc7 cell preparation exhibited the highest productivity (Table 4), i.e. almost six times that of the parent cell preparation.

In 1982, Tanabe Seiyaku changed from the conventional method for the large-scale production of L-aspartic acid to the improved method using treated EAPc7 cells having higher aspartase activity, with very encouraging results.

19.5 PRODUCTION OF L-MALIC ACID BY IMMOBILIZED *BREVIBACTERIUM FLAVUM*

L-Malic acid is employed in the food industry as an acidulant in fruit and vegetable juice, carbonated soft drinks, jams, sweets, etc. In the pharmaceutical industry, L-malate salts of basic amino acids are used in amino acid infusions. Worldwide, production of L-malic acid is now approximately 500 tonnes per year (see also Chapter 18).

In 1974, Tanabe Seiyaku succeeded in the industrial production of L-malic acid from fumaric acid by *Brevibacterium ammoniagenes* cells immobilized by the polyacrylamide gel method. The asymmetric reaction catalysed by the fumarase of the cells is as follows:

$$\text{HOOC-CH=CH-COOH} \underset{\text{fumarase}}{\overset{H_2O}{\rightleftharpoons}} \text{HOOC-CH}_2\text{-C(OH)(H)-COOH} \quad (3)$$

Fumaric acid L-Malic acid

As in the case of L-aspartic acid production, Tanabe Seiyaku investigated the κ-carrageenan method to improve the productivity for L-malic acid. After screening various microorganisms for higher fumarase activity, *Brevibacterium flavum* was found to show a higher enzyme activity after immobilization with κ-carrageenan than the formerly used *B. ammoniagenes*, as shown in Table 5.[6,7] Therefore, this polyacrylamide method was also modified to the carrageenan method in 1977. Further, it was found that the stability of fumarase activity of *B. flavum* increased when the cells were immobilized with carrageenan in the presence of polyethyleneimine or Chinese gallotannin (Table 5). Since the heat stability and operational stability of the immobilized preparation were increased, the column could be operated at a higher temperature of 50–55 °C for long periods.

The stabilization of the fumarase activity of *B. flavum* immobilized by this improved method is considered to be due to three-way interactions between

Table 5. Comparison of productivity of L-malic acid in various immobilized preparations

Microbial cells and immobilization method	Fumarase activity (unit ml^{-1} gel)	Operational stability at 37 °C (half-life, days)	Relative productivity[a] (%)
B. ammoniagenes			
Polyacrylamide	8.8	53	100
B. flavum			
Polyacrylamide	10.2	94	273
κ-Carrageenan	15.0	160	897
κ-Carrageenan + polyethyleneimine	16.3	243	1,587
κ-Carrageenan + Chinese gallotannin	18.5	310	2,460

[a] Productivity $\int_0^t E_0 \exp(-k_d t)dt$; E_0 = initial fumarase activity, k_d = decay constant, t = operation period. Each immobilized cell column was operated with the same fumarase activity (4.4 units ml^{-1} gel) as that at the half-life of *B. ammoniagenes* immobilized with polyacrylamide, and the productivity of *B. ammoniagenes* immobilized with polyacrylamide, 485 mmol ml^{-1} gel, was taken as 100%.

κ-carrageenan, polyethyleneimine and *B. flavum*. The addition of polyethyleneimine enhanced the productivity of the system by 80%, and the resultant process is nearly 21 times more efficient than that which had used *B. ammoniagenes* immobilized with polyacrylamide. The new method is a satisfactory procedure for the industrial production of L-malic acid.

19.6 PRODUCTION OF L-ALANINE AND D-ASPARTIC ACID BY IMMOBILIZED *PSEUDOMONAS DACUNHAE*

Since 1965, Tanabe Seiyaku had manufactured L-alanine by a batch process using L-aspartate β-decarboxylase of *Pseudomonas dacunhae*. The reaction proceeds as shown in equation 4.

$$\text{L-aspartic acid (HOOC–CH}_2\text{–CH(NH}_2\text{)–COOH)} \xrightarrow{\text{L-aspartate } \beta\text{-decarboxylase}} \text{L-alanine (CH}_3\text{–CH(NH}_2\text{)–COOH)} + CO_2 \quad (4)$$

In the continuous system[8] using immobilized *P. dacunhae* cells, it was difficult to maintain the plug-flow of the substrate solution and to keep a constant pH in the reactor because of the evolution of CO_2. Tanabe Seiyaku therefore designed a closed column reactor (Figure 4) which performs the enzyme reaction

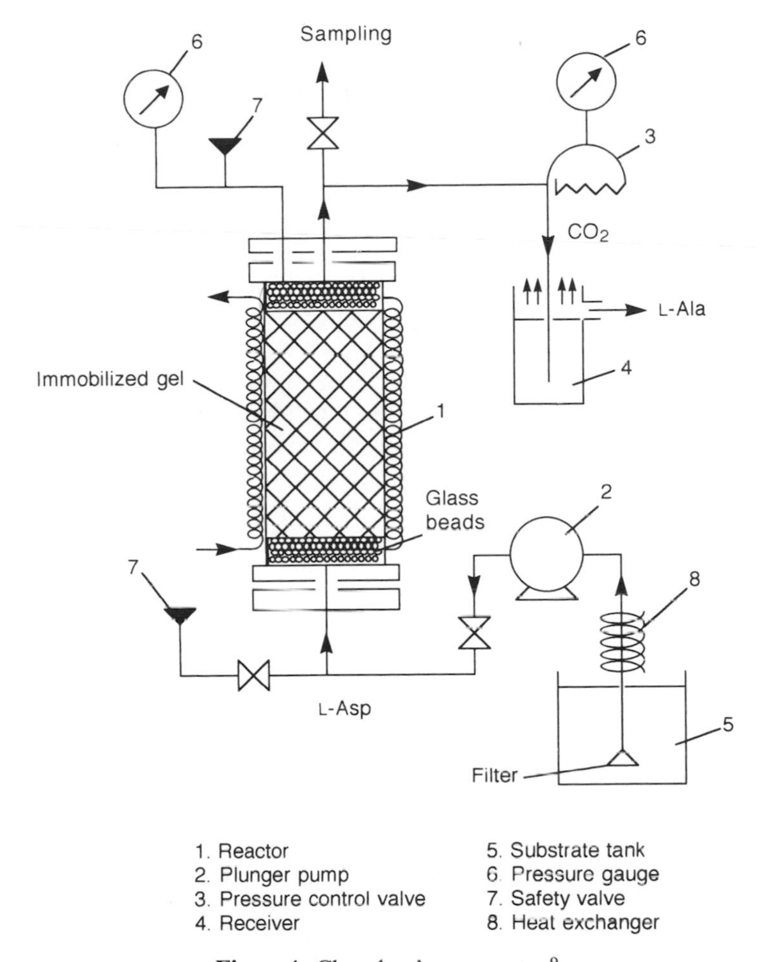

Figure 4. Closed column reactor[9]

at an elevated pressure such as 10 kg cm^{-2}. In this system, the liberated CO_2 gas is dissolved into reaction mixture, and therefore complete plug-flow of the substrate solution is maintained and the pH of reaction mixture is stabilized. The efficiency of immobilized cells for production of L-alanine in the closed column system is much higher than that in the conventional column system at normal pressure.

The continuous production of the raw material for this process, L-aspartic acid, from ammonium fumarate by immobilized cells of *E. coli* has already been discussed in Section 19.4. To investigate the possibility of an efficient and direct process for the production of L-alanine from ammonium fumarate, the combination of two types of immobilized microbial cells was studied. Three permutations were investigated: the use of immobilized *E. coli* cells and immobilized *P. dacunhae* cells, the use of co-immobilized *E. coli* and *P. dacunhae* cells and the sequential use of immobilized *E. coli* and *P. dacunhae* cells.

From the standpoint of industrial application, L-alanine was produced most efficiently by the sequential use of an immobilized *E. coli* column followed by a *P. dacunhae* column. The main reason for this is the difference in the optimum pH for these enzymes (the optimum pH for immobilized *E. coli* aspartase is 8.5, whereas that of *P. dacunhae* aspartate β-decarboxylase is 6.0). The process, as illustrated in Figure 5, has been operated on an industrial scale since 1982.[9]

The decarboxylase enzyme is highly selective for the L-enantiomer of aspartic acid. Thus, when racemic aspartic acid is used as the substrate, L-alanine and D-aspartic acid may be produced simultaneously (D-aspartic acid is an important intermediate for a synthetic penicillin developed by Tanabe Seiyaku[10]). Continuous processes for L-alanine and D-aspartic acid using *P. dacunhae* have been operated commercially by Tanabe Seiyaku since 1989.

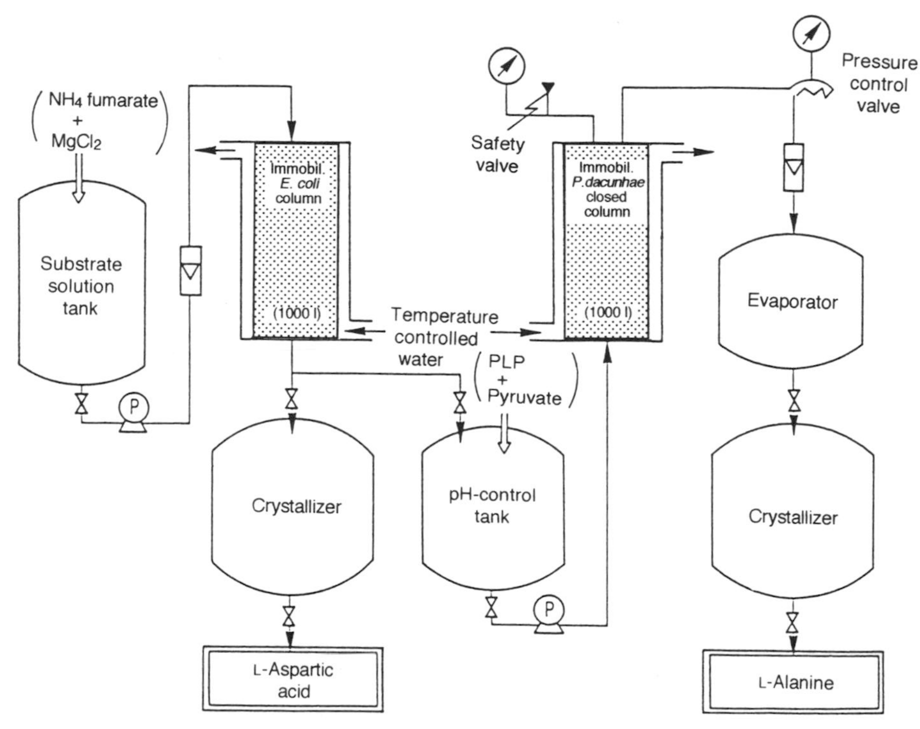

Figure 5. Flow diagram for the continuous production of L-aspartic acid and L-alanine by immobilized cells[10]

Table 6. Studies on enzymatic production of L-phenylalanine at Tanabe Seiyaku

Raw material	Enzyme
N-Acetyl-DL-phenylalanine	Aminoacylase
DL-Phenylalanine	D-Amino acid oxidase and aminotransferase
DL-Phenylserine	Phenylserine dehydratase and aminotransferase
trans-Cinnamic acid	Phenylalanine ammonia-lyase
Phenylpyruvic acid	Aminotransferase
Acetamidocinnamic acid	Acetamidocinnamate amidohydrolase and aminotransferase
DL-5-Benzylhydantoin	Hydantoinase and carbamoylase

19.7 PRODUCTION OF L-PHENYLALANINE BY IMMOBILIZED *CORYNEBACTERIUM* SP. AND *PSEUDOMONAS DENITRIFICANS*

L-Phenylalanine is important in human nutrition and as a component of amino acid infusions for medical purposes. Owing in part to its use as a raw material for aspartame (Chapter 11), the market for L-phenylalanine is increasing rapidly in line with that of L-aspartic acid. The worldwide annual demand for L-phenylalanine was only 50 tonnes in 1981, but grew to 3000 tonnes in 1985 owing to the demand for aspartame. By the end of the 1990s it is expected to increase to about 2.5 times this level.

Many companies have investigated methods for the economical production

$$PhCH{=}C(NHCOCH_3)COOH \text{ (ACA)} \xrightarrow{\text{ACA aminohydrolase}} PhCH_2COCOOH \text{ (Phenylpyruvic acid)} \xrightarrow[\text{R—CH(NH}_2\text{)—COOH}]{\text{Aminotransferase}} PhCH_2CH(NH_2)COOH \text{ (L-Phenylalanine)} \quad (5)$$

of L-phenylalanine. Table 6 shows the possibilities for the enzymatic production of L-phenylalanine which have been studied by Tanabe Seiyaku.[11] Among these methods, the route from acetamidocinnamic acid (ACA) through phenylpyruvic acid to L-phenylalanine was found to be the best for the industrial production of L-phenylalanine.[12] The reaction is mediated by two types of enzyme as given in previous page.

First an acetyl group of ACA is hydrolysed by acetamidocinnamate aminohydrolase, a kind of peptidase or aminoacylase, then the amino group is cleaved non-enzymatically to form phenylpyruvic acid. Following this, the phenylpyruvic acid is converted into L-phenylalanine by an aminotransferase, using a second amino acid as the amino donor.

The first attempts to establish the production of L-phenylalanine from ACA by one species of microorganism failed owing to the difficulty in increasing both enzyme activities at the same time. However, a successful process employing two microorganisms was developed, and is described below.

Microorganisms were selected from two screens, the first for acetamidocinnamate aminohydrolase activity and the second for aminotransferase activity. As a result of the first screen for microorganisms with acetamidocinnamate aminohydrolase, a strain belonging to the genus *Corynebacterium* from soil was found and named *C.* sp. S-5.[13] Strain S-5 could grow abundantly in a simple chemically defined medium consisting of a source of carbon, inorganic nitrogen and minerals. However, acetamidocinnamate aminohydrolase activity was not observed in such a medium. Induction of acetamidocinnamate aminohydrolase activity was observed after addition of ACA to the medium, although the production of this enzyme was inhibited by substrates such as glucose (catabolite repression). When the strain was cultured in a medium containing glucose and ACA as the sole nitrogen source, it exhibited low enzyme activity and an elongated doubling time. To increase the productivity of L-phenylalanine, a *C.* sp. strain S-5 mutant with high enzyme activity was isolated by selecting cells that grew faster in a medium containing ACA as a nitrogen source in the presence of glucose. The mutant strain C-23 with the highest activity was thus obtained.

An organism from culture collections at Tanabe Seiyaku, *Paracoccus denitrificans*, performed well in the second screen for aminotransferase activity. In addition, an aminotransferase-hyperproducing mutant, pFPr-1, was produced by treatment with chemical mutagens.

The combination of *C.* sp. strain C-23 as an acetamidocinnamate aminohydrolase source and *P. denitrificans* strain pFPr-1 as an aminotransferase source converted ACA smoothly into L-phenylalanine. After 48 h, the reaction was complete, and 760 μmol per ml of L-phenylalanine had been produced in a molar yield of 95% from 0.8 M ACA. The difference between the amount of ACA consumed and that of L-phenylalanine produced in this process was accounted for by phenylpyruvic acid which accumulated in the reaction mixture. Subsequently, the L-phenylalanine transaminase gene was cloned by a 'shotgun' method using *E. coli* K-12 to produce *E. coli* strain HB101 (pPAP142) harbouring the plasmid which produced 30 times more transaminase activity than wild-type *P. denitrificans* cells.[14]

To improve on this batchwise conversion of ACA by two microorganisms,

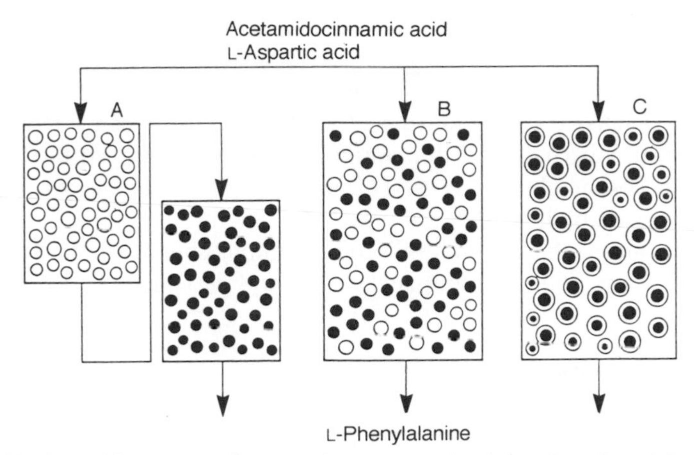

Figure 6. Various bioreactors for continuous production of L-phenylalanine from acetamidocinnamic acid by a two-step enzyme reaction.[13] ○, Immobilized *C.* sp. cells; ●, immobilized *P. denitrificans* cell; ◉, co-immobilized *C.* sp.–*P. denitrificans* cells

the continuous conversion of ACA by immobilized cells was attempted. Immobilized *C* sp. cells, immobilized *P. denitrificans* cells and co-immobilized cells of these species were prepared by the κ-carrageenan method.

Figure 6 shows various types of bioreactors considered for L-phenylalanine production from ACA using these immobilized cells. In the sequential system (A), the acetamidocinnamate aminohydrolase reaction proceeds in the first reactor and the aminotransferase reaction occurs in the second. Methods B and C are one-bed systems packed with a mixture of two immobilized cell types and co-immobilized cells, repectively. The productivity of co-immobilized cells was found to be higher than that of the mixture of the two immobilized cells as shown in Table 7.

It is reasonable to assume that phenylpyruvic acid produced by the acetamidocinnamate aminohydrolase reaction is immediately converted into

Table 7. Comparison of L-phenylalanine productivity in three bioreactors

Bioreactor: Immobilized cells[a]	Bioreactor: Column	L-Phenylalanine productivity[b] (μmol min^{-1} g gel^{-})
Two immobilized cells	Two beds	0.70
Mixture of two immobilized cells	One bed	0.74
Co-immobilized cells	One bed	1.28

[a] Activity yield of κ-carrageenan method was about 55% in each case.
[b] A solution containing 0.15 M ACA, 0.3 M L-aspartic acid and 0.1 mM PLP (pH 8.0) was passed through the columns each packed with immobilized cells at a flow of space velocity = 2.0 h^{-1} at 30 °C, and the amount of L-phenylalanine in the effluent was measured.

L-phenylalanine by the amino transferase reaction without diffusion resistance of phenylpyruvic acid within the gel.

The operational stability of the co-immobilized cells was studied, and the half-life of the activity was found to be 3 weeks at 30 °C. When 150 mM of ACA and 300 mM of L-aspartic acid (the amino donor) were charged into a column packed with co-immobilized cells, 147 mM of L-phenylalanine was produced with a conversion of 95%.

In the near future, this co-immobilized continuous column method is expected to replace the immobilized aminoacylase method at Tanabe Seiyaku.

19.8 CONCLUSION

In this chapter, some current examples of the industrial production of optically active compounds by immobilized biocatalysts have been described. Immobilized living cell systems are now being expanded to include plant and animal cells, which are expected to find applications in biotechnology. As more novel microorganisms with useful characteristics are produced by recombinant DNA techniques, so the development of methods which facilitate their adoption in manufacturing processes will continue to be of vital importance to the fine chemicals industry.

19.9 REFERENCES

1. Chibata, I. (Ed.) *Immobilized Enzymes. Research and Development* Kodansha, Tokyo and Halsted Press, New York, 1978.
2. Tosa, T., Mori, T., Fuse, N., and Chibata, I., *Enzymologia*, **31**, 214 (1966).
3. Chibata, I., Tosa, T., and Sato, T., *Appl. Biochem. Biotechnol.*, **13**, 231 (1986).
4. Nishimura, N., and Kisumi, M., *J. Biotechnol.*, **7**, 11 (1988).
5. Nishimura, N., Taniguchi, T., and Komatsubara, S., *J. Ferment. Bioeng.*, **67**, 107 (1989).
6. Chibata, I., Tosa, T., and Takata, I., *Trends Biotechnol.*, **1**, 9 (1983).
7. Takata, I., Tosa, T., and Chibata, I., *Appl. Biochem. Biotechnol.*, **8**, 31, 39 (1983).
8. Yamamoto, K., Tosa, T., and Chibata, I., *Biotechnol. Bioeng.*, **22**, 2045 (1980).
9. Furui, M., and Yamashita, K., *J. Ferment. Technol.*, **61**, 587 (1983).
10. Senuma, M., Otsuki, O., Sakata, N., Furui, M., and Tosa, T., *J. Ferment. Bioeng.*, **67**, 233 (1989).
11. Tosa, T., in *Bioproducts and Bioprocesses* (ed. A. Fiechter, H. Okada and R. D. Tanner), Springer, Berlin, Heidelberg, 1989, pp. 155–167.
12. Nakamichi, K., Nishida, Y., Nabe, K., and Tosa, T., *Appl. Biochem. Biotechnol.*, **11**, 367 (1985).
13. Nakamichi, K., Nabe, K., and Tosa, T., *J. Biotechnol.*, **4**, 293 (1986).
14. Takagi, T., Taniguchi, T., Yamamoto, Y., and Shibatani, T., *Biotechnol. Appl. Biochem.*, **13**, 112 (1991).

20 Membrane Bioreactors for the Production of Enantiomerically Pure α-Amino Acids

A. S. BOMMARIUS, K. DRAUZ and U. GROEGER
Degussa AG, Hanau, Germany

and

C. WANDREY
Research Centre Jülich, Jülich, Germany

Chirality in Industry. Edited by A. N. Collins, G. N. Sheldrake and J. Crosby

20.1 INTRODUCTION

This chapter describes some of the methods for large-scale production of enantiomerically pure α-amino acids and the role of membrane reactors within these processes. Following the discussion of the acylase, amidase and hydantoinase processes some novel developments with enzymes to be used in membrane bioreactors are presented: utilization of racemases, production of L-ornithine with arginase and L-proline with proline acylase, the *C*-terminal deamidation of peptides with peptide amidase and novel processes for enzymatic peptide synthesis.

20.1.1 ENANTIOMERICALLY PURE PRODUCTS WITH BIOCATALYSTS

Despite the advances of asymmetric synthesis with organic and inorganic reagents or auxiliaries in recent years, the ever more stringent requirements for optical purity of pharmaceuticals, plant and animal protection agents or biochemical intermediates are still difficult to achieve. Many enzymes, however, possess enantiospecificities unmatched elsewhere in the synthetic chemists' repertoire. Based mostly on work at the Research Centre in Jülich, enzymes have been used at Degussa to generate enantiomerically pure α-amino acids on a large scale, of both L- and D-configuration. Novel developments also include racemases and hydrolases for enzymatic peptide synthesis. All such enzymatic reactions necessitate specific process technology: enzymes have to be conditioned to be optimally active and stable under reaction conditions and, in most cases, the biocatalyst has to be recovered for better overall process economy. Both the biocatalytic reactions and the concomitant reactors are the subject of this Chapter.

Enantiomerically pure amino acids are interesting compounds for the pharmaceutical and food industry. L-Amino acids are used as components in parenteral nutrition, as food additives for high-stress diets (for athletes), as flavour enhancers (e.g. L-glutamate) or as animal feed supplements to deficient low-protein feedstuff (e.g. L-lysine supplementation of corn). Alternatively, they are used, just as are D-amino acids, as intermediates for pharmaceuticals, cosmetics and pesticides, and as chiral synthons for organic synthesis.

There are several routes to enantiomerically pure α-amino acids: both enantiomers of each can be obtained from enantioselective crystallization of racemates via diastereomeric salt pairs and from enzymatic synthesis. In addition, L-amino acids can be produced from extraction of protein hydrolyzates or from fermentation. Production by enzymic synthesis has a number of attractions as discussed in earlier chapters. A particular advantage over many fermentation processes is the wide range of possible substrates.

20.1.2 PROCESS TECHNOLOGIES FOR BIOCATALYTIC PROCESSES

Owing to the high cost of biocatalysts, their productivity has to be optimized. One of the most important measures towards that end is separation of the enzyme from product and/or reactant stream with subsequent reuse. For both whole cell and isolated enzyme biocatalysts, the two main methods are: immobilization of the biocatalyst onto reactor walls or macroscopic beads and membrane ultrafiltration of the medium with retention of the biocatalyst (Figure 1). Both methods can be used in batch or continuous mode. Immobilized biocatalysts are almost exclusively used in continuous mode in fixed or fluidized bed reactors (see Chapter 19). The authors have utilized soluble native enzymes or resting cells in a membrane reactor, either in batch or repeated-batch mode with subsequent ultrafiltration (batch-UF reactor) or in continuous mode as an enzyme-membrane recycle reactor (EMR reactor). The options are described in more detail and compared in the next section.

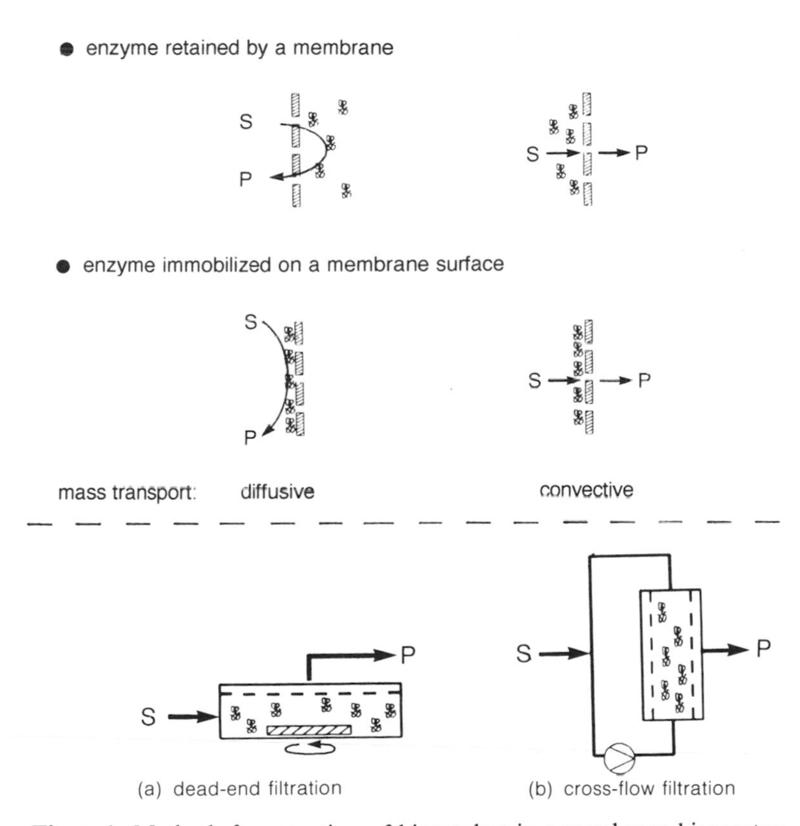

Figure 1. Methods for retention of biocatalyst in a membrane bioreactor

20.2 REACTION ENGINEERING

For enzyme-catalysed reactions to reach a commercial scale, engineering may become decisive. The main goal of enzyme reaction engineering is to achieve high space–time yield at high substrate conversion with high chemical selectivity and/or enantioselectivity. Membrane bioreactors are well suited for this goal because there are hardly any heat and mass transfer limitations, even at high catalyst concentrations. A membrane reactor can be operated under sterile conditions, which is very difficult with carrier-fixed enzymes. Since the product stream passes an ultrafiltration (UF) membrane it is mostly pyrogen-free, a major advantage in downstream processing for pharmaceutical products.

In starting to determine the reaction conditions, the pH value and temperature are set according to both reactant and product as well as enzyme properties (Figure 2). Initial substrate concentrations or, in a bimolecular reaction, the ratio of the two substrate concentrations have to be fixed. Enzyme activity and stability have to be checked as functions of reaction conditions prior to kinetic measurements. Biocatalyst concentration is important for achieving a desired space–time yield and also for influencing selectivity if there is competition between the enzymatically catalysed reaction and an undesired parallel or consecutive side reaction. In this regard, a membrane bioreactor offers the option of operating with extreme concentrations of soluble catalyst. In multienzyme systems, the optimal activity ratio of all enzymes has to be found.

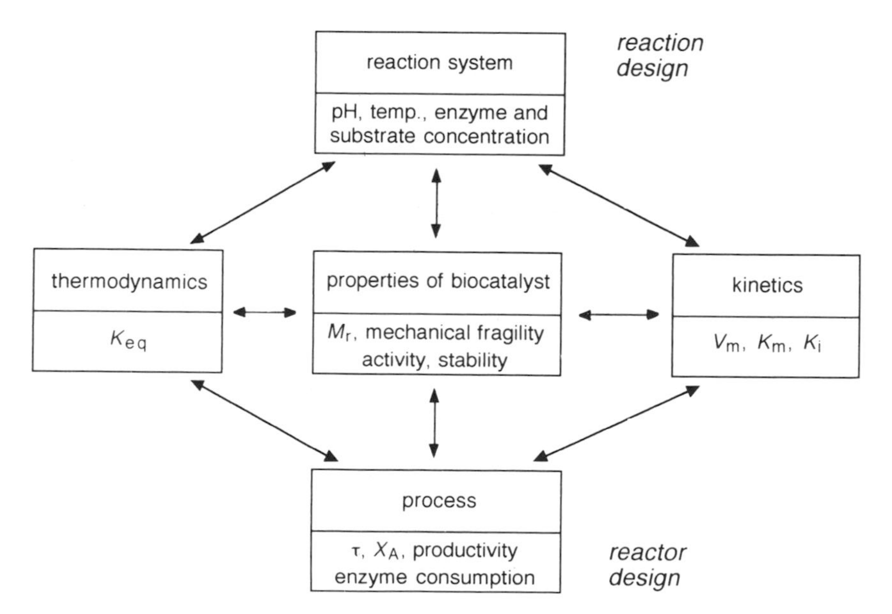

Figure 2. Relationship between different aspects of enzyme reaction engineering

Kinetic measurements have to be carried out under conditions relevant to processes, whereas most kinetic measurements found in the literature refer to conditions of optimum enzyme pH and temperature and to very dilute solutions. Rather than measuring initial rate kinetics only, the influence of all products has to be determined as function of conversion. In cases of a distinct state of equilibrium below 100% conversion, the reverse reaction has to be integrated into the kinetic model because the same active site of an enzyme catalyses both the forward and reverse reactions, so the forward reaction cannot be treated separately from the back reaction. Parameter estimation for kinetic models should be obtained by non-linear regression and not by linearization, as in the Lineweaver–Burk method (Figure 3). The best test of a kinetic model is a proper fit of concentration versus time along the entire range of conversion in a batch experiment (integral method). Sometimes it may be difficult to integrate the equations of the kinetic models, in which cases numerical methods such as the well known Runge–Kutta method are required. Kinetic modelling has been extended to multi-enzyme systems such as redox reactions with cofactor regeneration[1] (see below).

Enzyme stability is an important aspect in biocatalytic processes, and is expressed as biocatalyst consumption per unit weight of product. In multi-enzyme systems, the stability of all enzymes has to be balanced for optimum overall rate and space–time yield. All numbers are meaningful at specified reaction conditions only. A membrane reactor offers the option to run the enzyme reaction at constant conversion with constant residence time, despite

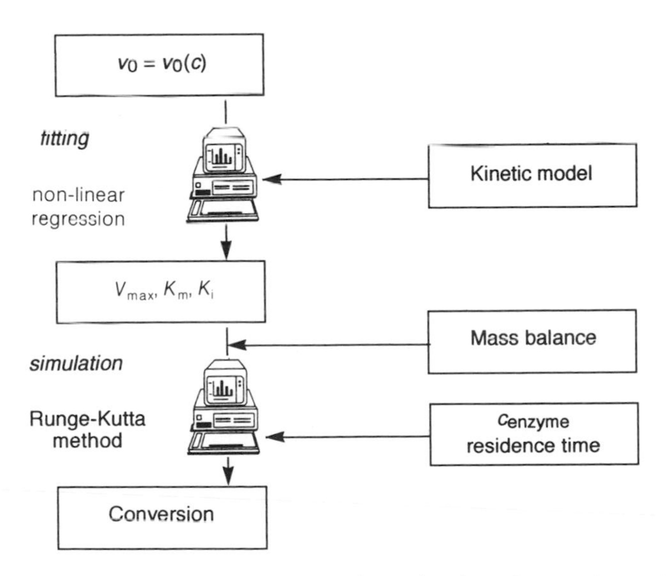

Figure 3. Parameter estimation and determination of operating points

biocatalyst deactivation, by supplementing catalyst into the reactor because a membrane bioreactor normally operates as a continuous stirred tank reactor (CSTR). While being advantageous for reactions with substrate inhibition, as in the acylase process, this is a disadvantage in cases of severe product inhibition. A system of two membrane reactors in series has been proposed as a solution to the product inhibition problem but the reduced catalyst cost does not justify the additional investment in most cases.

The scale-up of membrane bioreactors to the cubic metre range has been accomplished. Potential concentration polarization problems can be minimized by a stirrer on the laboratory scale or a loop reactor with resulting tangential flow on the full production scale; any design with enforced flow across the membrane is superior to an approach with diffusive mass transfer (Figure 1).

While reactor operation and downstream processing cost are not negligible, substrate and enzyme costs per unit weight of product are the decisive influences on cost optimization. If the necessary cofactors can be retained in the membrane reactor, as described in Section 20.3.2, their cost can be minimized. Sterile operation and removal of pyrogens by the UF membrane reduce fouling of catalyst and minimize the procedures required to validate the product. The cost of membranes is not overriding because modern membranes possess excellent process stability. Thus, the initial goal of minimizing membrane area at given flux was replaced by the strategy of providing surplus membrane area to enhance the total operation time of a membrane bioreactor. The main challenge to membrane stability is fouling by high total protein concentrations stemming from repeated additions of the enzyme to counter deactivation. Sufficient membrane area and occasional cleaning prevent any limitations from concentration polarization by protein or other debris.

20.3 ENZYMATIC PROCESSES

20.3.1 HYDROLASES

20.3.1.1 Aminoacylases

The most widespread method for enzymatic L-amino acid synthesis is the resolution of racemates of *N*-acetyl-DL-amino acids by the enzyme acylase I (aminoacylase; E.C. 3.5.1.14). In this process, the *N*-acetyl-L-amino acid is cleaved to yield L-amino acid whereas the *N*-acetyl-D-amino acid does not react. The first industrial process was operated by Tanabe[7,8] and is described in the preceding chapter (Section 19.3).

After separation of the L-amino acid by ion exchange or crystallization, the remaining *N*-acetyl-D-amino acid can be racemized by acetic anhydride in alkaline solution or by adding a racemase[2] to achieve very high overall conversions to the L-amino acid. The *N*-acetyl-DL-amino acid substrates are conveniently accessible on a laboratory as well as an industrial scale through acetylation of DL-amino acids with acetyl chloride or acetic anhydride in a Schotten–Baumann reaction.[3]

Acylase I is available from two sources: porcine kidney and *Aspergillus oryzae* fungus. The properties of both enzymes have been compared:[4] both are dimeric zinc-containing enzymes. Although the price per unit activity is comparable (however, the specific activity of renal enzyme is usually much higher than that of fungal acylase), the *Aspergillus* enzyme is the only one used in large-scale processing for two reasons: the process has grown to a scale where supply of acylase from kidney would pose an insurmountable logistical problem, and the stability of fungal acylase is vastly superior to that of the renal enzyme.

Most tests of acylase stability have been conducted in repeated-batch mode;[4b,c,5] it was found that resistance to deactivation and oxidation is superior for the fungal enzyme. Stability results from repeated-batch experiments are obtained at varying conversion, however, and therefore do not yield proper deactivation data. A superior method is the investigation of enzyme activity over time in a recycle reactor at constant conversion[5] with reaction conditions as close as possible to intended large-scale conditions. Results on the operational stability of both acylases in a recycle reactor demonstrated the much better stability of the *Aspergillus* enzyme, while renal enzyme is not stable enough for long-term operation.[5,6]

The substrate specificity of aminoacylase I is unusually wide:[4c] enzymes from both sources prefer long straight-chain, hydrophobic substrates but fungal acylase also readily accepts aromatic, and many substituted substrates. Figure 4 lists both proteinogenic and non-proteinogenic amino acids prepared at Degussa in bulk quantities by the resolution of the respective N-acetyl amino acids.

As mentioned previously, the first industrial process was started by Tanabe Seiyaku with an immobilized acylase in a fixed-bed reactor[8] (see Chapter 19). In a different approach, based on the work by Wandrey and Flaschel,[5b] Degussa introduced a continuous acylase process employing an enzyme–membrane–reactor (EMR) configuration in 1981.[9] The configuration corresponds to scheme (b) in Figure 1: the reactor is operated as a CSTR with flow rates in the recycle loop up to 200 times higher than in the substrate feedstream through the sterile filters. The soluble enzyme is retained in the reactor by a UF membrane while substrate and product molecules pass through the membrane. For pilot- and large-scale operation, the necessary membrane area is configured into a hollow-fibre module with the fibres made of polysulphone or regenerated cellulose; a molecular weight cut-off (MWCO) of 10 000 is sufficient to obtain a rejection rate of much higher than 99% for the fungal acylase (molecular weight 73 000).

At Degussa, several enzyme membrane reactor set-ups are in operation covering six orders of magnitude from laboratory via pilot stage to full produc-

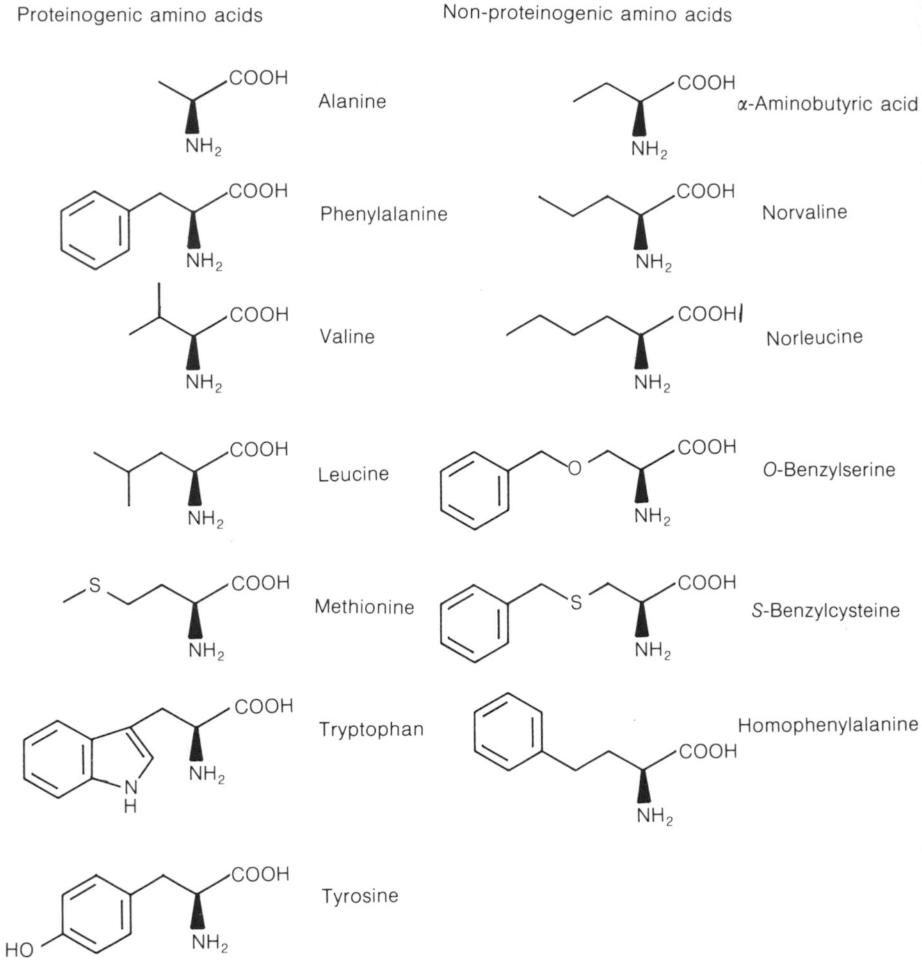

Proteinogenic amino acids

Alanine

Phenylalanine

Valine

Leucine

Methionine

Tryptophan

Tyrosine

Non-proteinogenic amino acids

α-Aminobutyric acid

Norvaline

Norleucine

O-Benzylserine

S-Benzylcysteine

Homophenylalanine

Figure 4. L-Amino acids prepared in bulk amounts by acylase I-catalysed resolution of *N*-acetyl-DL-amino acids

tion scale; the process has been scaled up to an annual production level of several 100 tonnes of enantiomerically pure α-amino acids.

20.3.1.2 Amidases

Another industrial process for amino acids involves the resolution of DL-amino acid amides by whole cells of *P. putida*.[10] The development of this process is described in Chapter 8 (Section 8.3).

20.3.1.3 Hydantoinases

Both the aminoacylase process resolving *N*-acetyl-DL-amino acids and the amidase process resolving DL-amino acid amides give only 50% conversion after the first pass. The remaining D-isomer has to be racemized in an external step to obtain high overall chemical yields. Since external racemization adds complexity and cost to any process, reaction systems are sought which either convert a prochiral substrate to one enantiomer (Section 20.3.2) or which utilize an internal racemization procedure. A useful class of compounds for the second option are the 5-monosubstituted hydantoins which can be regarded as α-amino acids cyclically protected at both the carboxyl- and the α-amino groups. Enantiospecific D- or L-hydantoinases hydrolyse the respective hydantoin isomer to yield D- or L-carbamoylic acid (hydantoic acid). In presence of a second enzyme, carbamoylase, the D- or L-carbamoylic acid is then cleaved to the respective α-amino acid. While L-hydantoinases are rare, D-hydantoinases are quite common (see below). Since 5-monosubstituted D- or L-hydantoins racemize via keto–enol tautomerism[11] under basic conditions, an *in situ* racemization of the unreacted hydantoin isomer leads to conversions of hydantoins of up to 100%. Figure 5 illustrates the hydantoinase–carbamoylase system.

Many hydantoins, notably 5-phenyl-substituted compounds, racemize so rapidly (in the order of minutes) that no catalyst for the racemization step is required in order to utilize all of the substrate. Other hydantoins with bulky substituents separated from the hydantoin ring by alkyl groups racemize more slowly; in this case the reaction can be accelerated by a racemase (Section 20.4.1.3).

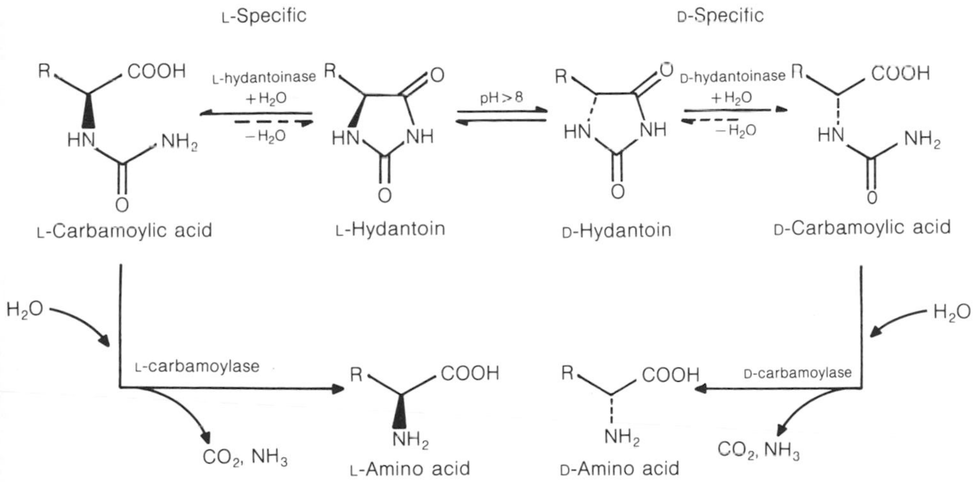

Figure 5. Schematic diagram of the hydantoinase–carbamoylase system

D-Hydantoinases and D-Carbamoylases

Many organisms possess D-specific hydantoinase activity. The cause is the identity of hydantoinase with the enzyme dihydropyrimidinase (E.C. 3.5.2.2), an important enzyme of the ubiquitous pyrimidine catabolism acting on dihydropyrimidines which are six-membered ring analogues of hydantoins. While hydantoinase activity had already been observed in milk in 1926[12] and in *Aspergillus* sp. in 1932,[13] D-selective hydantoinase activity was first observed by Yamada *et al.*[14] in 1978. Microorganisms showing both hydantoinase and carbamoylase activity were first characterized by Olivieri *et al.*[15] (1981), Yokozeki *et al.*[16] (1987) and Möller *et al.*[17] (1988).

Processes leading to D-amino acids via hydantoins have been developed to commercial scale by DEBI Recordati[18] and Kanegafuchi Industries.[19] The products are the highest volume D-amino acids, D-phenylglycine (from DL-phenylhydantoin) and D-*p*-hydroxyphenylglycine (from DL-*p*-hydroxyphenylhydantoin), with more than 1000 tonnes of each being made annually worldwide. They are intermediates for the semi-synthetic antibiotics ampicillin and amoxicillin, respectively (see Chapter 1). Whereas the Kanegafuchi biocatalyst

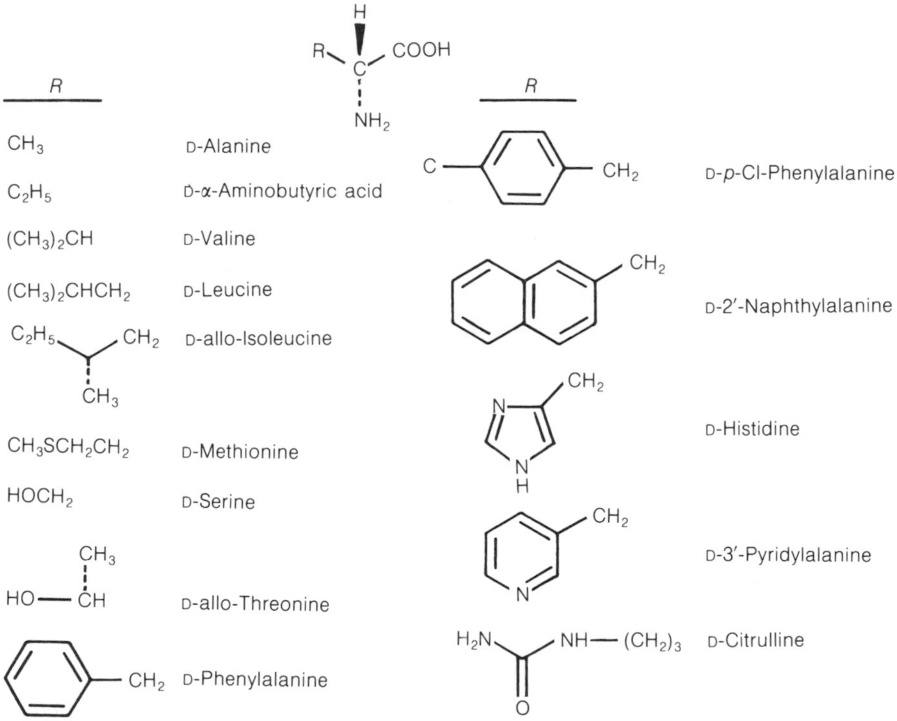

H, R, COOH, C, NH_2

R		*R*	
CH_3	D-Alanine	C (phenyl) CH_2	D-*p*-Cl-Phenylalanine
C_2H_5	D-α-Aminobutyric acid	(naphthyl) CH_2	D-2′-Naphthylalanine
$(CH_3)_2CH$	D-Valine	(imidazolyl, N, N H) CH_2	D-Histidine
$(CH_3)_2CHCH_2$	D-Leucine	(pyridyl, N) CH_2	D-3′-Pyridylalanine
C_2H_5 CH_2 CH_3	D-allo-Isoleucine	H_2N NH— $(CH_2)_3$ O	D-Citrulline
$CH_3SCH_2CH_2$	D-Methionine		
$HOCH_2$	D-Serine		
CH_3 HO— CH	D-allo-Threonine		
(phenyl) CH_2	D-Phenylalanine		

Figure 6. D-Amino acids produced in bulk amounts by Degussa

using *Bacillus brevis* only has D-hydantoinase activity, which necessitates chemical decarbamoylation of the D-carbamoylic acid intermediate,[19b] the process developed by DEBI-Recordati employing *Agrobacterium radiobacter* resting cells with hydantoinase and carbamoylase activity yields D-amino acids directly. The system has very broad substrate specificity:[20] Figure 6 lists some of the D-amino acids produced with the DEBI-Recordati biomass in pilot and bulk quantities jointly with Degussa.

In addition to application as intermediates for antibiotics, D-amino acids have found uses as enantioselective catalysts, such as cyclo[D-Phe-D-His] for the asymmetric (*R*)-hydrocyanation of *m*-phenoxybenzaldehyde to the corresponding (*R*)-cyanohydrin,[21] or as intermediates for herbicides, such as D-valine in fluvalinate (see Chapter 8), which also contains the *m*-phenoxybenzaldehydecyanohydrin, or as part of peptide hormone agonists and antagonists. In the latter, D-amino acids impart higher stability, especially in orally administered drugs, to enzymatic proteolytic cleavage in the liver, kidney or bloodstream, and thus to inactivation. An example of such a pharmaceutical is Cetrorelix, an LHRH (luteinizing hormone-releasing hormone) antagonist from Asta Medica (Figure 7).

Ac-D-Nal-D-(*p*-Cl)-Phe-D-Pal-Ser-Tyr-D-Cit-Leu-Arg-Pro-D-Ala-NH_2

Figure 7. LHRH antagonist Cetrorelix

Hydantoinases and Carbamoylases for L-Amino Acids

The hydantoinase–carbamoylase enzyme system is also a potentially interesting route to L-amino acids, but L-hydantoinase–L-carbamoylase systems in nature are confined to specific organisms and their physiological significance is uncertain. After the initial detection of activity in *Bacillus brevis* towards L-carboxyethylhydantoin,[22] organisms using L-hydantoins with aryl-bearing substituents in the 5-position were found,[23] and finally also *Bacillus* and *Nocardia* species utilizing 5-alkylhydantoins.[24] The successful demonstration of a commercial process has been recently reported for an *Arthrobacter* sp. which cleaves indolylmethylhydantoin to L-tryptophan.[25] The specific activity of the system was 1.08 mmol (g^{-1} h^{-1}),[26] so racemization as the rate-determining step had to be accelerated more than 1000-fold by a racemase (Section 20.4.1.3).[25b,27] The substrate range of the L-hydantoinase–L-carbamoylase system found by Wagner *et al.* is confined to aromatic species and listed in Table 1.[26]

20.3.2 DEHYDROGENASES

Racemization of the remaining enantiomer of the substrate is always a problem when using simple hydrolase systems for the production of enantiomerically

Table 1. Substrate specificity of *Arthrobacter* sp. DSM 3747 towards L-amino acids[26b]

Residue R	L-Amino acid	Productivity $[mM(g\,DCW)^{-1}\,h^{-1}]^{a}$
Benzyl	Phenylalanine	1.1
p-Cl-benzyl	*p*-Cl-phenylalanine	3.9
3,4-OCH_3-benzyl	3,4-Dimethoxyphenylalanine	0.1
Phenylethyl	Homophenylalanine	0.2
Indolylmethyl	Tryptophan	1.1[26a]
Benzyloxymethyl	*O*-Benzylserine	1.0
S-Benzylmercaptomethyl	*S*-Benzylcysteine	1.0
Phenyl	Phenylglycine	n.a.[b]
p-OH-phenyl	*p*-OH-phenylglycine	n.a.

[a]DCW = dry cell weight.
[b]n.a. = not accepted.

pure α-amino acids. In contrast, dehydrogenases offer the chance to start from prochiral compounds: reductive amination of α-keto acids by means of L-amino acid dehydrogenases (E.C. 1.4.1) yields the corresponding α-amino acids. The reductive amination reaction depends on NADH, thus cofactor regeneration has to be considered. An elegant solution for the regeneration of NADH from NAD^+ is the use of the system formate–formate dehydrogenase (FDH; E.C. 1.2.1.2).[28] The enzyme is produced very efficiently by the yeast *Candida boidinii* in a chemostat with methanol as a carbon source, and production rates of up to 7500 units $l^{-1}\,day^{-1}$ have been achieved.[29] Today, FDH is available commercially, and the price per unit will probably decrease with rising demand. The oxidation of formate to carbon dioxide catalysed by FDH proceeds quantitatively owing to the removal of carbon dioxide from the reaction mixture. This feature, and the coupled quantitative reductive amination of α-keto acid substrates catalysed by an L-amino acid dehydrogenase, are very attractive features of this system. With a keto acid concentration of 100 mM and a coenzyme concentration of 0.1 mM (sufficient for saturation), a total cycle number of 1000 can be achieved in a batch experiment; for substrates of higher solubility, appropriately higher values are obtained.

After completion of the reaction, the enzymes are retained by an ultrafiltration (UF) membrane while native cofactor permeates through the membrane. UF membranes cannot be built size-specifically enough to retain cofactor while letting amino acid molecules permeate through the membrane. This fact led to the idea of enlarging the size of the cofactor with the goal of retention by the UF membrane: NAD can be covalently bound to a water-soluble polymer such as polyethylene glycol (PEG) with a molecular weight of about 20 000.[30] As a water-soluble cofactor, the PEG-bound NAD still functions as a transport

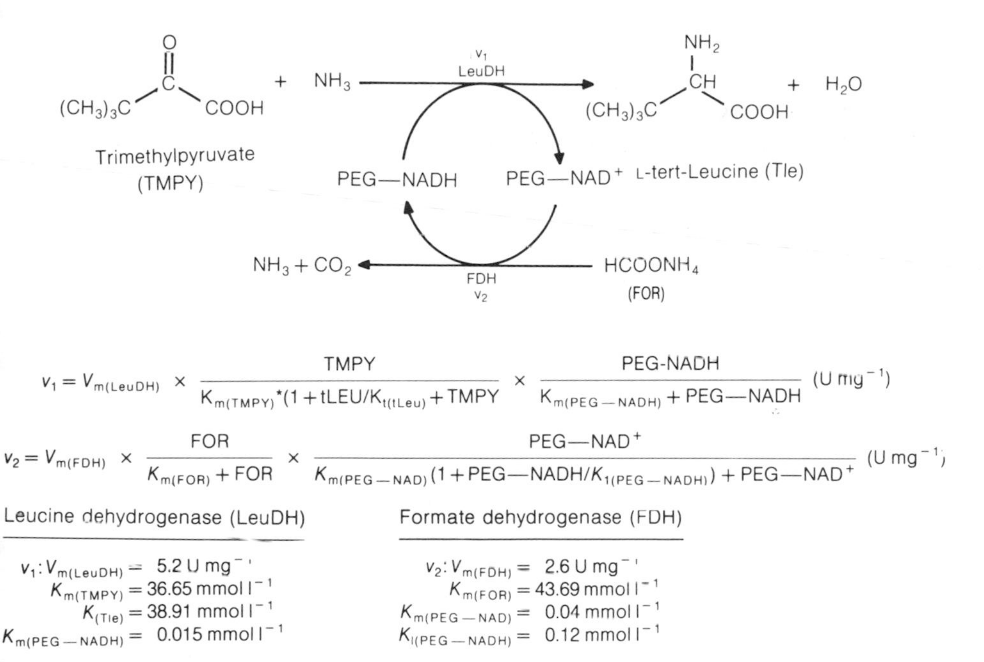

Figure 8. Enzymatic synthesis of L-*tert*-leucine by reductive amination of trimethyl pyruvic acid with cofactor regeneration: rate expressions and kinetic parameters

metabolite between the production and the regeneration enzymes; its degree of retention is higher than 99.9%.

The synthesis of L-*tert*-leucine from trimethylpyruvic acid has been studied in detail[31] (Figure 8). L-*tert*-Leucine as an unnatural amino acid cannot be made by a fermentative route but there is a strong demand for this compound owing to its use as chiral auxiliary in a number of syntheses.[32] Substrate and product are very soluble, so space–time yields are high. As can be seen from the kinetic parameters given in Figure 8, the coenzyme derivatives are readily accepted by both LeuDH (E.C. 1.4.1.9) and FDH. The appropriate K_M values are higher in comparison with those for native cofactor, however, as for native cofactor 0.1 mM is sufficient for saturation.

In Figure 9, the reaction rate of the coupled system LeuDH–FDH is displayed as a function of both initial substrate concentration and cofactor concentration. All numbers are for 90% conversion of substrate. The rate increases smoothly with cofactor concentration but shows a maximum with increasing substrate concentration owing to severe product inhibition by L-*tert*-leucine.

With a substrate concentration of 0.5 M, L-*tert*-leucine was produced continuously in a membrane reactor with an average conversion of 85% and a space–time yield of 638 $g\,l^{-1}\,day^{-1}$. Total turnover number over a period of

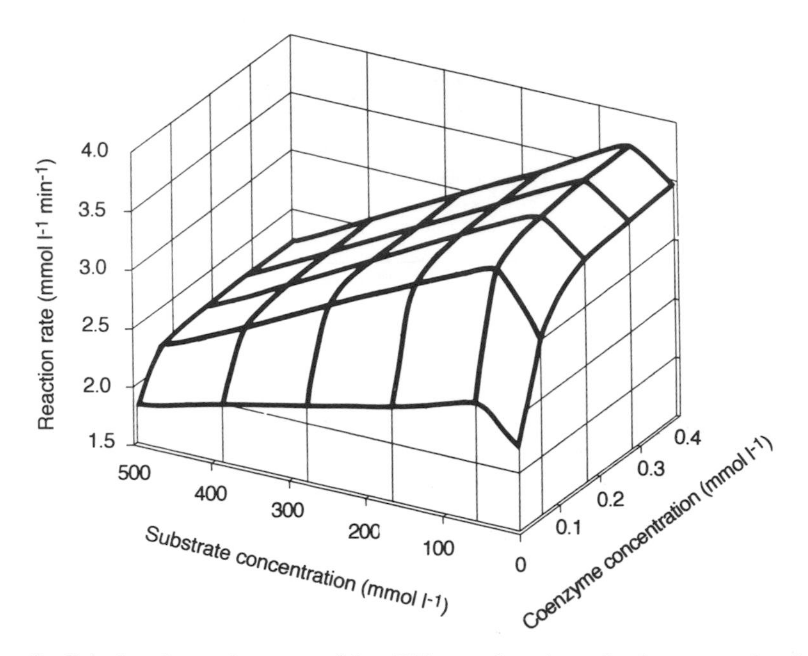

Figure 9. Calculated reaction rate of LeuDH as a function of substrate and cofactor (PEG–NADH) concentration in a continuous process. Conditions: LeuDH, $30\,\mathrm{U\,ml^{-1}}$; FDH, $30\,\mathrm{U\,ml^{-1}}$; conversion, 90%, controlled by variation of residence time; $[\mathrm{HCOO^-\,NH_4^+}] = 1\,\mathrm{mol\,l^{-1}}$

2 months was 125 000. This result should help to overcome the notion that cofactor cost is still a limiting factor for enzymatic redox reactions.

In an analogous reaction, phenylpyruvate can be reductively aminated to L-phenylalanine. However, despite an impressive total turnover number of 600 000,[33] such a system cannot compete with the fermentative route to L-phenylalanine starting from glucose and ammonia. An economic analysis reveals that substrate cost (phenylpyruvate versus glucose) is dominating, and not the cost of the cofactor. For this reason, commercialization of continuous cofactor regeneration was achieved with a non-proteinogenic amino acid such as L-*tert*-leucine. Figure 10 shows a pilot plant for continuous cofactor regeneration at Degussa, where multi-kilogram amounts of enantiomerically pure non-proteinogenic α-amino acids have been produced by the dehydrogenase route.

20.4 NEW DEVELOPMENTS

20.4.1 RACEMASES

Processes which involve the enzyme-catalysed resolution of racemic substrates to the enantiomerically pure amino acids usually lead to one preferred enantiomer.

Figure 10. Pilot plant for the production of L-amino acids by reductive amination of keto acids with cofactor regeneration (Degussa)

Although both enantiomers could be produced from such processes, the market rarely needs both enantiomers in equal volume, so the resolution should be complemented by a racemization to allow a high chemical yield and thus favourable economics.[34]

20.4.1.1 *N*-Acylamino Acid Racemase

In the case of the acylase-catalysed resolution of *N*-acetyl-DL-amino acids, the racemization of the remaining *N*-acetyl-D-amino acid after separation of the L-amino acid[9b] has to be undertaken but adds complexity and cost, and therefore a procedure without the need for external recycling of the unwanted enantiomer is desirable. Such an '*in situ* racemization' is possible with a racemase acting

Table 2. Substrate specificity of *N*-acylamino acid racemase from *Streptomyces* sp. Y-53[2]

Substrate	Relative activity	Substrate	Relative activity
N-Acetyl-D-methionine	100	*N*-Acetyl-L-methionine	100
N-Formyl-D-methionine	40	*N*-Formyl-L-methionine	63
N-Acetyl-D-alanine	33	*N*-Acetyl-L-alanine	21
N-Benzoyl-D-alanine	14	*N*-Benzoyl-L-alanine	ND[a]
N-Acetyl-D-leucine	37	*N*-Acetyl-L-leucine	74
N-Acetyl-D-phenylalanine	64	*N*-Acetyl-L-phenylalanine	84
N-Chloroacetyl-D-phenylalanine	90	*N*-Chloroacetyl-L-phenylalanine	112
N-Acetyl-D-tryptophan	10	*N*-Acetyl-L-tryptophan	8
N-Acetyl-D-valine	35	*N*-Acetyl-L-valine	19
N-Chloroacetyl-D-valine	80	*N*-Chloroacetyl-L-valine	105
N-Acetyl-D-alloisoleucine	33	*N*-Acetyl-L-alloisoleucine	ND
D-Methionine	0	L-Methionine	0
D-Alanine	0	L-Alanine	0
D-Leucine	0	L-Leucine	0
D-Phenylalanine	0	L-Phenylalanine	0
D-Tryptophan	0	L-Tryptophan	0
D-Valine	0	L-Valine	0

[a]ND = not determined.

specifically on *N*-acylamino acids without affecting the stereochemistry of the product L-amino acids. Such an *N*-acylamino racemase was recently found by Takeda: microbiological screening yielded a bacterium from soil from the actinomycetes order.[2] The pentameric enzyme from *Streptomyces* sp. Y-53 (molecular weight 2×10^5) displays optimum activity in a similar pH range (6–8.5) but at a slightly lower temperature (40 °C) than acylase. The racemase is even more strongly activated than acylase by zinc, cobalt and manganese ions. As shown in Table 2, the enzyme specifically catalyses the racemization of *N*-acylamino acids without acting on amino acids.

The data prove the high potential of this racemase to allow nearly quantitative conversion of racemic *N*-acylamino acids into enantiomerically pure L-amino acids. If a D-acylase was employed, use of the racemase would lead to D-amino acids only. It is also possible that the acylase and the racemase could be used as a two-enzyme system in the continuous enzyme–membrane reactor set-up currently used for the production of L-amino acids.

20.4.1.2 Amino Acid Amide Racemase

As has been pointed out, the limitation to 50% yield in one pass holds not just for the acylase process but also for the amidase process developed by DSM. However, organisms with amino acid amide racemase activity in addition to L-amidase activity have been isolated by this company[34,35] (see Chapter 8).

20.4.1.3 Hydantoin Racemase

The '*in situ* racemization' desired for racemic acylase and amidase substrates is already known for 5-aryl-monosubstituted hydantoins, i.e. for hydantoinase substrates (Section 20.3.1.3). However, with 5-monoalkylsubstituted hydantoins, the rate of racemization might become rate-limiting for the overall conversion of L- or racemic substrates into D-carbamoylic or D-amino acids.[26] First hints of the involvement of a racemase came from Guivarch *et al.* in 1980,[24b] and later from other groups.[24c,36] Wagner *et al.* in collaboration with Rütgerswerke[27a] discovered that *Arthrobacter* sp. DSM 3747 converts racemic 5-arylalkyl-hydantoins and also the corresponding *N*-carbamoyl-D-amino acids exclusively into the corresponding L-amino acids. However, the hydantoinase reaction is reversible, and on further investigation it was found that, instead of the originally postulated *N*-carbamoylamino acid racemase, a coupled action of three enzymes (an unspecific hydantoinase, a hydantoin racemase and an L-specific carbamoylase), allows *N*-carbamoylamino acids to be converted into L-amino acids (Figure 11).[25b,27b]

Remarkably, this tetrameric hydantoin racemase (molecular weight 84 000, p*I* 4.5) does not need pyridoxal phosphate.[27c] It acts preferentially on 5-arylalkyl-hydantoins (Table 3) but not at all on amino acids, *N*-acetylamino acids, dipeptides or D-α-amino-ε-caprolactam.

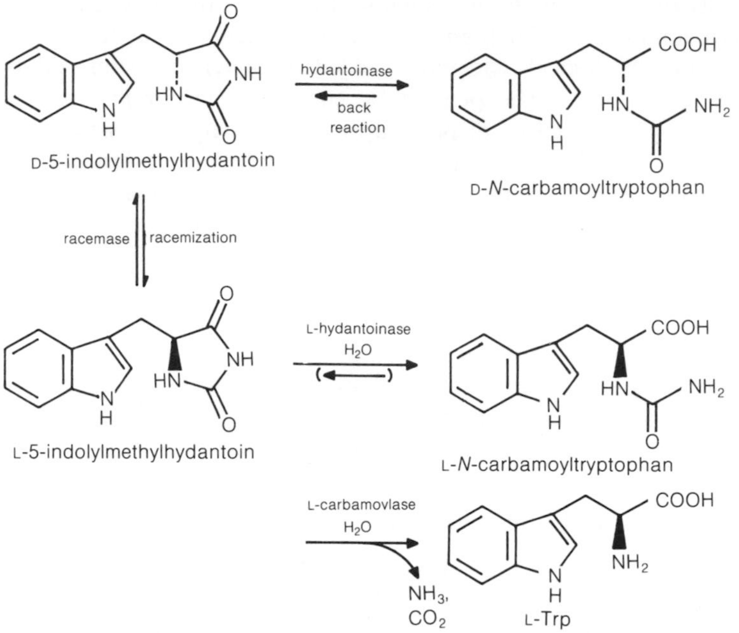

Figure 11. Quantitative conversion of DL-indolylmethylhydantoin or D-*N*-carbamoyltryptophan to L-tryptophan by coupled action of an unspecific hydantoinase, hydantoin racemase, and L-specific carbamoylase[25b]

20.4.2 ARGINASE

L-Ornithine and its salts are of growing importance for the pharmaceutical industry; in parenteral nutrition (L-Orn·HCl and L-Orn·acetate), in the treatment of hepatic diseases (L-Orn·L-Asp and L-Orn·α-ketoglutarate) or as starting material for the chemoenzymatic synthesis of citrulline.[21b] In large-scale synthesis, L-Orn can be produced by fermentation processes[37] or by enzymatic hydrolysis of L-arginine using L-arginase (L-arginine amidinohydrolase, E. C. 3.5.3.1) applied by Degussa AG[38] (equation 1).

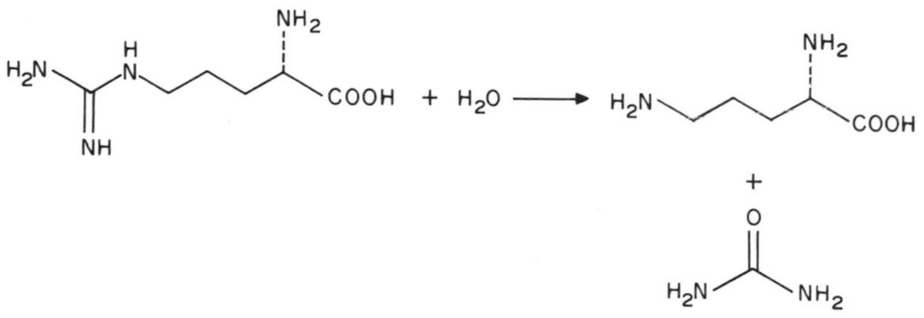

$$H_2N\text{-}C(=NH)\text{-}NH\text{-}(CH_2)_3\text{-}CH(NH_2)\text{-}COOH + H_2O \longrightarrow H_2N\text{-}(CH_2)_3\text{-}CH(NH_2)\text{-}COOH + H_2N\text{-}C(=O)\text{-}NH_2 \quad (1)$$

Table 3. Substrate specificity of hydantoin racemase from *Arthrobacter sp.* DSM 3747[27c]

Substrate: HN, N, R^1, R^2, O, O

R^1	R^2	Relative activity (%)	R^1	R^2	Relative activity (%)
CH_2 N H	H	100.0	CH_2	H	9.8
			S CH_2	H	20.4
CH_2 N H	CH_3	20.2	S CH_2	CH_3	0
			$(CH_3)_2CH$	H	0
HO CH_2	H	76.7	H_3C	H	0
			HO CH_2	H	0
CH_2	H	62.7	O HO CH_2	H	0

To attain a sufficiently high activity level, native arginase from calf or bovine liver is used instead of immobilized preparations. The soluble native enzyme and the high solubility of reactants and products point to an enzyme membrane reactor as the process design of choice. However, since arginase from calf or bovine liver is sensitive to mechanical agitation, the usual recycle reactor with a pumped cycle leads to rapid inactivation of the enzyme during the production process. To increase the stability of the enzyme, a new reactor concept was developed which uses a quiescent medium with hydraulic transfer of substrate and enzyme solution[39] (Figure 12).

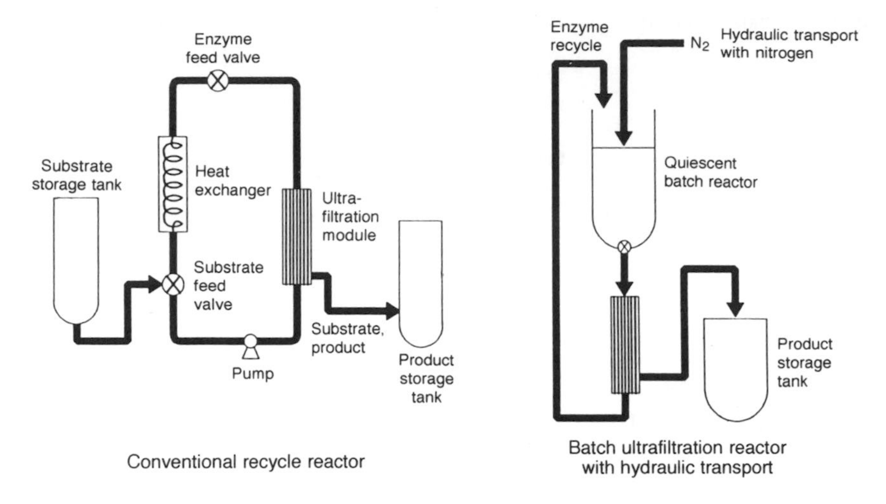

Figure 12. Alternative process design of the enzyme–membrane reactor without mechanical agitation[39]

By using a batch ultrafiltration reactor with hydraulic transport and by the addition of activating and stabilizing agents such as manganese ions and ascorbic acid, deactivation of the enzyme could be decreased by a factor of about 20. On a pilot scale, the process was conducted at enzyme consumption levels of less than 400 units per kelogram of L-ornithine at 83–88% conversion.[38,39]

Another enzymatic method for preparing 2-substituted L-ornithine was developed by Merrell Dow. The method is based on the enantioselective hydrolysis of racemic α-amino-δ-valerolactams, such as DL-α-amino-α-difluoromethylornithine, catalysed by L-α-amino-ε-caprolactam hydrolase[40] (equation 2).

$$F_2HC(H_2N)\text{-lactam (NH, C=O)} + H_2O \longrightarrow H_2N\text{-}(CH_2)_3\text{-}C(CHF_2)(NH_2)COOH \qquad (2)$$

20.4.3 PROLINE ACYLASE

The acylase-catalysed resolution of *N*-acyl-DL-amino acids (Section 20.3.1.1) has some restrictions. Although acylase I from both porcine kidney and *Aspergillus oryzae* fungus has broad substrate specificity and high enantioselectivity, the enzyme does not accept *N*-acylated substrates where the hydrogen atom at the amide nitrogen is replaced by an alkyl group. Therefore, *N*-acylated secondary

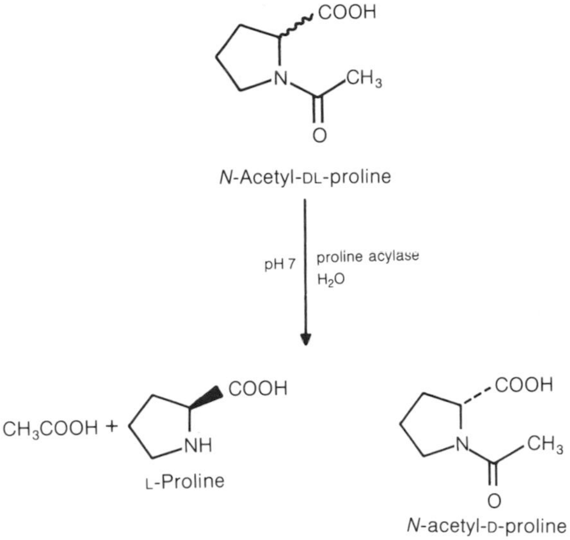

COOH
N
CH_3
O
N-Acetyl-DL-proline

pH 7 proline acylase H_2O

CH_3COOH + COOH NH
L-Proline

COOH N CH_3 O
N-acetyl-D-proline

Figure 13. Enantioselective hydrolysis of *N*-acetyl-DL-proline to L-proline catalysed by proline acylase[43]

amines such as *N*-acetylproline and *N*-acetyl-*N*-alkylamino acids are not hydrolysed by this enzyme[4c,41] or by aminoacylases from other sources.[42] This gap in the substrate specificity of aminoacylases was successfully closed by the isolation of aminoacylases which act specifically on *N*-acetyl-L-proline and its derivatives[43] (Figure 13).

The proline acylase from *Comamonas testosteroni* has a high thermostability,[43d,e] which is an advantage for industrial application. Besides a broad substrate specificity towards *N*-acylated cyclic amino acids for the enantioselective synthesis of L-azetidine-2-carboxylic acid, L-proline, L-thiazolidine-4-carboxylic acid or L-pipecolinic acid,[46] this enzyme affords for the first time the enzyme-catalysed resolution of racemic *N*-alkylamino acids[45] (Table 4).

20.4.4 PEPTIDE AMIDASE

During their work on the isolation and application of carboxypeptidase C from orange flavedo for peptide synthesis[46] Steinke and Kula[47] detected a new amidase which selectively hydrolyzes peptide amides to the corresponding peptides without cleaving the peptide bond (Figure 14).

The hexameric peptide amidase from orange flavedo (molecular weight 162 000, p*I* 9.5) is stable and active up to 30 °C in the pH range 6–9. It hydrolyses selectively the amide bond of peptide amides and *N*-acylated amino acid amides without cleaving the peptide bond or acyl amino acid itself[47] (Table 5). Although

Table 4. Substrate specificity of proline acylase from *Comamonas testosteroni* DSM 5416 towards *N*-chloroacetyl-*N*-alkylamino acids[45]

Substrate	Relative Activity (%)	Conversion (%)
N-Clac-L-proline	100	100
N-Clac-*N*-methyl-L-alanine	175	100
N-Clac-*N*-methyl-DL-alanine	115	49
N-Clac-*N*-ethyl-DL-alanine	82	nd[b]
N-Clac-*N*-propyl-DL-alanine	27	nd
N-Clac-*N*-methyl-DL-2-aminobutyric acid	10	50
N-Clac-*N*-ethyl-DL-2-aminobutyric acid	2	nd
N-Clac-L-azetidine-4-carboxylic acid	141	100
N-Clac-L-thiazolidine-2-carboxylic acid	115	100
N-Clac-DL-pipecolinic acid	164	54

[a]Clac = chloroacetyl.
[b]nd = not determined.

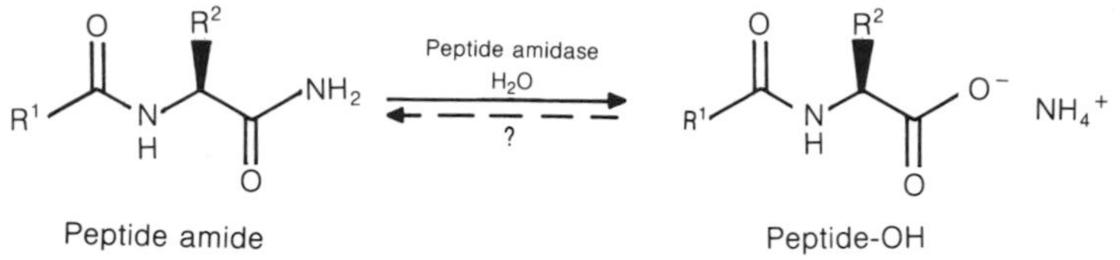

Figure 14. Amide cleavage catalysed by peptide amidase from orange flavedo[47b] (R^1 = amino acid or peptide residue or *N*-terminal protecting group of an amino acid amide, R^2 = amino acid side-chain)

the substrate specificity towards the *C*-terminal amino acid is broad, the enantioselectivity is high, so D-amino acid amides in this position are not accepted. Furthermore, non-*N*-acylated amino acid amides are not hydrolysed. Hence this new and highly selective amidase can be used for the resolution of diastereomeric peptides and the selective *C*-terminal deprotection of peptide amides.

20.4.5 PEPTIDE SYNTHESIS

Since peptide amidase from orange flavedo can be used in water-miscible organic solvents[48] and does not cleave amino acid amides, it is a useful tool for enzymatic peptide syntheses and enables process-integrated deprotection of the resulting peptide at the *C*-terminus[47a] (Figure 15).

Further, in thermodynamically controlled peptide syntheses, the peptide amide as the primary reaction product can be continuously removed from the

Table 5. Substrate specificity of peptide amidase of orange flavedo[47b]. Reproduced by permission of VCH

Substrate[a]	Conversion after 6 h (%)
Ac-L-Trp-NH_2	100
Bz-L-Arg-NH_2	100
Bz-L-Tyr-NH_2	100
H-L-Val-L-Phe-NH_2	100
H-L-Asp-L-Phe-NH_2	100
H-L-Ala-L-Phe-NH_2	100
H-L-Arg-L-Met-NH_2	100
H-L-Phe-L-Leu-NH_2	100
Z- Gly-L-Tyr-NH_2	100
Bz-L-Tyr-L-Thr-NH_2	100
Bz-L-Tyr-L-Ser-NH_2	100
Bz-L-Tyr-L-Ala-NH_2	100
Boc-L-Leu-L-Val-NH_2	20
Trt-Gly-L-Leu-L-Val-NH_2	80
Z-L-Pro-L-Leu-Gly-NH_2	100
Z-Gly-Gly-L-Leu-NH_2	100
H-Gly-D-Phe-L-Tyr-NH_2	100
H-Gly-L-Phe-D-Phe-NH_2	0[b]
H-L-Arg-L-Pro-D-Ala-NH_2	0[b]
Z-L-Arg-L-Arg-pNa	0[b]

[a]Ac = acetyl, Bz = benzoyl, Boc = *tert*-butyloxycarbonyl, Trt = trityl, Z = benzyloxycarbonyl, pNa = *p*-nitroanilide; amino acid amides are not hydrolysed.
[b]No conversion in 12 h.

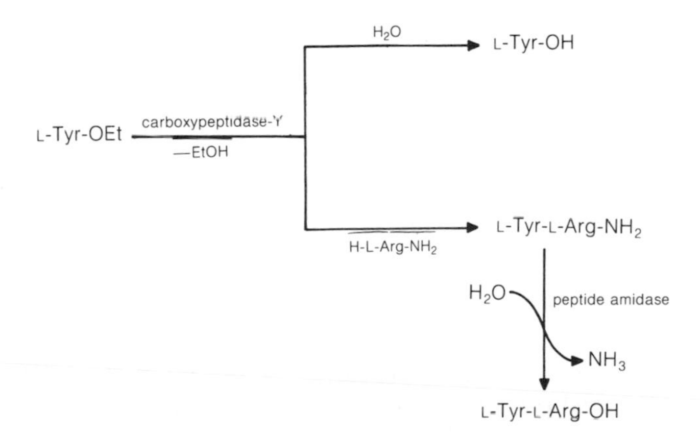

Figure 15. Kinetically controlled synthesis of *C*-terminally unprotected peptides catalysed by combined action of carboxypeptidase Y (CPD-Y) and peptide amidase (PA)[47a]

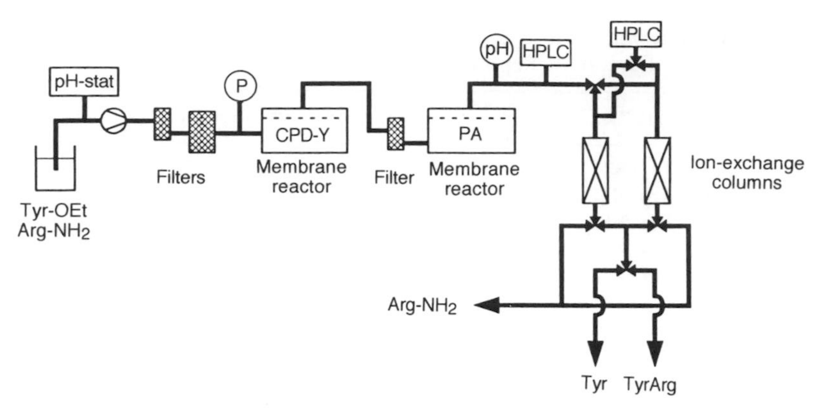

Figure 16. Two-stage enzyme membrane reactor for continuously operated enzymatic peptide synthesis with integrated product recovery[49]

reaction equilibrium by the action of peptide amidase, thus permitting high yields of the peptides. Based on these findings, a continuous process for enzymatic peptide synthesis catalysed by subsequent action of carboxypeptidase Y (CPD-Y) and peptide amidase (PA) has been developed using a two-stage enzyme membrane reactor.[49] This process includes an integrated recovery of the peptide and by-products and separation and recycling of the nucleophilic amino acid amide by ion-exchange chromatography (Figure 16).

An EMR-based process has also been developed and partially scaled up for the continuous α-chymotrypsin (α-CT)-catalysed, kinetically controlled synthesis of the dipeptide kyotorphin (L-Tyr-L-Arg)[50] (equation 3).

$$\text{Mal-L-Tyr-OEt} + \text{L-Arg-OEt} \xrightarrow{\alpha\text{-CT}} \text{Mal-L-Tyr-L-Arg-OEt} + \text{EtOH} \quad (3)$$

The reaction in an EMR system was improved over conventional enzymatic peptide synthesis by the introduction of new, highly water-soluble protecting groups (Mal = maleyl), resulting in one of the highest space–time yields ever seen in continuously operated enzymatic peptide syntheses with esters, 13.36 kg l^{-1} day^{-1}.[51] With L-Arg-NH_2 as the nucleophile, a space–time yield on the laboratory scale of 25.04 kg l^{-1} day^{-1} has been achieved.

20.5 CONCLUSION

In this chapter, some important applications of membrane bioreactors for the production of L- and D-α-amino acids have been outlined. At Degussa, several L-amino acids for parenteral nutrition are produced in multi-100-tonne quantities in such reactors.

In a membrane reactor system, native enzymes, pure or of technical grade, are used in homogeneous solution; multi-enzyme systems in addition to resting cell biocatalysts can also be applied. A membrane reactor is particularly well suited for cofactor-dependent enzyme reactions, especially if the cofactor is regenerated by another enzyme reaction and retained by the membrane in a modified form.

Biocatalysis in membrane reactors has advantages over heterogeneous enzymatic catalysis: there are no mass transfer limitations, enzyme deactivation can be compensated for by adding soluble enzyme and the reactors can be kept sterile more easily than immobilized enzyme systems. The product is mostly pyrogen free (a major advantage for producing pharmaceuticals), because the product stream has to pass an ultrafiltration membrane.

There are two major reactor configurations which incorporate a membrane ultrafiltration step: (i) the continuous reactor with convective flow through flat-sheet or hollow-fibre membrane modules, or (ii) a batch ultrafiltration reactor with a stirred vessel or a quiescent reactor with hydraulic transport. Scale-up of membrane reactors is simple because large units with increased surface area can be created by combining several modules.

In the main, hydrolases and oxidoreductases have been used in membrane bioreactor systems, although other classes of enzymes are, in principle, suited to this approach.

20.6 REFERENCES

1. Vasic-Racki, D., Jonas, M., Wandrey, C., Hummel, W., and Kula M. -R., *Appl. Microbiol. Biotechnol.*, **31**, 215 (1989).
2. Takeda Chemical Industries, *Eur. Pat. Appl.*, 0 304 021, 1989.
3. For a review, see Sonntag, N.O.V., *Chem. Rev.*, **52**; 237 (1953) (see pp. 312–324).
4. (a) Gentzen, I., Löffler, H. -G., and Schneider, F., *Z. Naturforsch.*, **35c**, 544 (1980); (b) Gentzen, I., Löffler, H. -G., and Schneider, F., in *Metalloproteins* (ed. U. Eser), Thieme, Stuttgart, 1979, pp. 270–274; (c) Chenault, H. K., Dahmer, J., and Whitesides, G. M., *J. Am. Chem. Soc.*, **111**, 6354 (1989).
5. (a) Wandrey, C., *Habilitationsschrift*, TU Hannover, 1977; (b) Wandrey, C., and Flaschel, E., *Adv. Biochem. Eng.*, **12**, 147 (1979).
6. Bommarius, A. S., Drauz, K., Klenk, H., and Wandrey, C., Enzyme Engineering 11, *Ann. N. Y. Acad. Sci.*, **929** (1992).
7. Tosa, T., Mori, T., Fuse, N., and Chibata, I., *Enzymologia*, **31**, 214 (1966).
8. (a) Tosa, T., Mori, T., Fuse, N., and Chibata, I., *Agric. Biol. Chem.*, **33**, 1047 (1969)., (b) Chibata, I., Tosa, T., Sato, T., and Mori, T., *Methods Enzymol.*, **44**, 746 (1976).
9. (a) Degussa/GBF, *US Pat.*, 4 304 858, 1981; (b) Leuchtenberger, W., Karrenbauer, M., and Plöcker, U., *Enzyme Engineering 7, Ann. N. Y. Acad. Sci.*, **434**, 78 (1984).
10. (a) DSM/Stamicarbon, *US Pat.*, 4 172 846, 1975; (b) DSM/Stamicarbon, *US Pat.*, 3 971 000, 1976; (c) DSM/Stamicarbon, *Eur. Pat.* 0 199 407, 1990.
11. Bovarnick M., and Clark, H. T., *J. Am. Chem. Soc.*, **60**, 2426 (1938).
12. Gaebler, O. H., and Keltch, A. K., *J. Biol. Chem.*, **70**, 763 (1926).
13. Sobotka, H., *US Pat.*, 1 861 458, 1932.

14. Yamada, H., Takahashi, S., Kii, Y., and Kumagai, H., *J. Ferment. Technol.*, **56**, 484 (1978).
15. Olivieri, R., Fascetti, E., Angelini, L., and Degen, L., *Biotechnol. Bioeng.*, **23**, 2173 (1981).
16. Yokozeki, K., Nakamori, S., Eguchi, C., Yamada, K., and Mitsugi, K., *Agric. Biol. Chem.*, **51**, 355 (1987).
17. Möller, A., Syldatk, C., Schulze. M., and Wagner, F., *Enzyme Microb. Technol.*, **10**, 618 (1988).
18. DEBI/Recordati, personal communication.
19. (a) Yamada, H., Takahashi, S., Kii, Y., and Kumagai, H., *J. Ferment. Technol.*, **56**, 484 (1978); (b) Kanegafuchi Chemical Industries, *Jpn. Pat.*, 62 25 990, 1987; (c) Yamada, H., Takahashi, S., Yoneda K., and Amagasaki, H., *Ger. Pat.*, 2 757 980, 1991.
20. (a) Degussa, *Ger. Pat. Appl.*, DE 3 917 057, 1992; (b) Drauz, K., Kottenhahn, M., Makryaleas, K., Klenk, H., and Bernd, M., *Angew. Chem.*, **103**, 704 (1991); *Angew. Chem. Int., Ed. Engl.*, **30**, 712 (1991); (c) Bommarius, A. S., Kottenhahn, M., and Drauz, K., *NATO ASI Ser. C*, **381**, 161 (1992).
21. Tanaka, K., Mori, A., and Inoue, S., *J. Org. Chem.*, **55**, 181 (1990).
22. Tsugawa, R., Okumura, S., Ito, T., and Katsuga, N., *Agric. Biol. Chem.*, **30**, 27 (1966).
23. (a) Klages, U., Weber, A., and Wilschowitz. L., *Ger. Pat.*, 3 702 384, 1988; (b) Yamashiro, A., Yokozeki, K., Kano, H., and Kubota, K., *Agric. Biol. Chem.*, **52**, 2851 (1988); (c) Yamashiro, A., Kubota, K., and Yokozeki, K., *Agric. Biol. Chem.*, **52**, 2857 (1988).
24. (a) Sano, K., Yokozeki, K., Eguchi, C., Kagawa, T., Noda, I., and Mitsugi, K., *Agric. Biol. Chem.*, **41**, 819 (1977); (b) Guivarch, M., Gillonier, C., and Brunie, J. -C., *Bull. Soc. Chim. Fr.*, 91 (1980); (c) Nishida, Y., Nakamichi, K., Nabe, K., and Tosa. T., *Enzyme Microb. Technol.*, **9**, 721 (1987); (d) Yokozeki, K., Sano, K., Eguchi, C., Yamada, H., and Mitsugi, K., *Agric. Biol. Chem.*, **51**, 819 (1987).
25. (a) Syldatk, C., Mackowiak, V., Höke, H., Gross, C., Dombach, G., and Wagner, F., *J. Biotechnol.*, **14**, 345 (1990); (b) Gross, C., Syldatk, C., Mackowiak, V., and Wagner, F., *J. Biotechnol.*, **14**, 363 (1990); (c) Wagner, T., Syldatk, C., Lehmensiek, V., Krohn, K., Höke, H., and Läufer, A., *Eur. Pat. Appl.*, 0 377 083, 1990.
26. (a) Syldatk, C., Läufer, A., Müller, R., and Höke, H., *Adv. Biochem. Eng.*, **41**, 29 (1990); (b) Syldatk, C., Lehmensiek, V., Ulrichs, G., Bilitewski, U., Krohn, K., Höke, H., and Wagner, F., *Biotechnol. Lett.*, **14**, 99 (1992).
27. (a) Rütgerswerke; *Ger. Pat. Appl.*, 3 712 539, 1988; (b) Syldatk, C., and Wagner, F., *Food Biotechnol.*, **4**, 87 (1990); (c) Wagner, F., Pietzsch, M., and Syldatk, C., paper presented at the *XIth Enzyme Engineering Conference*, Kailua-Kona, USA, 1991.
28. Schütte, H., Flossdorf, J., Sahm, H., and Kula, M. -R., *Eur. J. Biochem.*, **62**, 151 (1976).
29. Wedy, M., *Dissertation*, RWTH, Aachen, 1992.
30. Bückmann, A. F., Kula, M.-R., Wichmann, R., and Wandrey, C., *J. Appl. Biochem.*, **3**, 301 (1981).
31. Kragl, U., Vasic-Racki, D., and Wandrey, C., *Chem. Ing. Tech.*, **64**, 499 (1992).
32. (a) Schöllkopf, U., and Neubauer, H. J., *Synthesis*, 861 (1982); (b) Schöllkopf, U., *Tetrahedron*, **39**, 2085 (1983), (c) Schöllkopf, U., and Schrever, R., *Liebigs Ann. Chem.*, 939 (1984).
33. Wandrey, C., in *Proceedings of the 4th European Congress on Biotechnology.*, (ed. O. M. Neijssel, R. R. van der Meer and K. Ch. A. M. Luyben), Elsevier, Amsterdam, 1987, vol. 4, pp. 171–188.
34. Kamphuis, J., Boesten, W. H. J., Broxterman, Q. B., Hermes, H. F. M., van Balken, J. A. M., Meijer, E. M., and Schomaker, H. E., *Adv. Biochem. Eng.*, **42**, 133 (1990).

35. (a) DSM/Stamicarbon/Novo-Nordisk, *Eur. Pat. Appl.*, 0 307 023, 1989; (b) Stamicarbon/Novo-Nordisk, *Eur. Pat. Appl.*, 0 383 403, 1990.
36. Battilotti, M., and Barberini, U., *J. Mol. Cat.*, **43**, 343 (1988).
37. Ajinomoto, *Eur. Pat. Appl.*, 0 393 708, 1990.
38. (a) Degussa, *Ger. Pat. Appl.*, 4 020 980, 1992; (b) Degussa, *Ger. Pat. Appl.*, 4 119 029, 1992.
39. Bommarius, A. S., Makryaleas, K., and Drauz, K., *Biomed. Biochim. Acta*, **50**, 249 (1991).
40. (a) Merrell Dow Pharmaceuticals, *US Pat.*, 4 902 719, 1990; (b) Merrell Dow Pharmaceuticals, *Eur. Pat. Appl.* 0 357 029, 1990.
41. (a) Birnbaum, S. M., Levintow, L., Kingsley, R. B., and Greenstein, J. P., *J. Biol. Chem.*, **194**, 455 (1952); (b) Fu, S.-C. J., and Birnbaum, S. M., *J. Am. Chem. Soc.*, **75**, 918 (1953); (c) Kang, S., Minematsu, Y., Shimohigashi, Y., Waki, M., and Izumiya, N., *Mem. Fac. Sci. Kyushi Univ., Ser. C*, **16**, 61 (1987).
42. Sugie, M., and Suzuki, H., *Agric. Biol. Chem.*, **44**, 1089 (1980).
43. (a) Noda Sangyo Kagaku Kenkyusho, *Jpn. Pat. Appl.*, 80 007 015, 1981; (b) Kikuchi, M., Koshiyama, I., and Fukushima, D., *Biochim. Biophys. Acta*, **744**, 180 (1983); (c) Daicel Chemical Industries, *Jpn. Pat. Appl.*, 87 232 381, 1989; (d) Degussa, *Eur. Pat. Appl.* 0 416 282, 1991; (e) Groeger, U., Drauz, K., and Klenk, H., *Angew. Chem.*, **102**, 428 (1990); *Angew. Chem., Int. Ed. Engl.*, **29**, 417 (1990).
44. Drauz, K., Groeger, U., Schäfer, M., and Klenk, H., Chem.-*Ztg.*, **115**, 97 (1991).
45. Groeger, U., Drauz, K., and Klenk, H., *Angew. Chem.*, **104**, 222 (1992); *Angew. Chem., Int. Ed. Engl.*, **31**, 195 (1992).
46. Steinke, D., and Kula, M.-R., *Enzyme Microb. Technol.*, **12**, 836 (1990).
47. (a) Forschungszentrum Jülich/Degussa, *Ger. Pat.*, 4 014 564, 1991; (b) Steinke, D., and Kula, M, -R., *Angew. Chem.*, **102**, 1204 (1990); *Angew. Chem., Int. Ed. Engl.*, **29**, 1139 (1990).
48. Steinke, D., and Kula, M.-R., *Biomed. Biochim. Acta.*, **50**, 143 (1991).
49. Schwarz, A., *Dissertation*, University of Bonn, 1991.
50. (a) Flörsheimer, A., Kula, M.-R., Schütz, H.-J., and Wandrey, C., *Biotechnol. Bioeng.*, **33**, 1400 (1989) (b) Herrmann, G., Schwarz, A., Wandrey, C., Kula, M.-R., Knaup, G., Drauz, K., and Berndt, H., *Biotechnol. Appl. Biochem.*, **13**, 346 (1991).
51. (a) Forschungszentrum Jülich/Degussa, *Ger. Pat. Appl.*, 4 101 895, 1992; (b) Fischer, A., Schwarz, A., Wandrey, C., Bommarius, A. S., Knaup, G., and Drauz, K., *Biomed. Biochim. Acta.*, **50**, 169 (1991).

21 Epilogue

There is a tendency for the industrial process chemist to resort to well proven, conventional procedures for making optically active molecules. This arises because of the obvious commercial need to get new products into the market as rapidly as possible. So that registration procedures are not delayed, it is desirable to define the ultimate quality of the manufactured product, and therefore the route, as soon as possible. This mitigates against *de novo* development of, for example, methods of catalytic asymmetric synthesis for reactions or substrates for which no close parallel exists, as this is unlikely to be rewarded on the right time scale.

Catalytic asymmetric syntheses still comprise only a small proportion of the total portfolio of industrial processes for optically active materials. Whilst they have received deserved publicity, this may have led to an unrealistic perception of their current utility. Significant achievements have, not atypically, required between 5 and 10 years of development in the laboratory (many decades of man years) before scale-up could be contemplated.

In the biological arena, examples of the application of enzyme-catalysed reactions, particularly those which employ hydrolases, to the synthesis of molecules of commercial interest are appearing at an increasing rate, reflecting the relative ease of implementation. Their successful use in purely organic media is already widespread at the research level and we can expect soon to see many large scale applications. Enzymes which require cofactors have an even more attractive range of capabilities; the technical problems associated with their use have been largely overcome and the present handful of large-scale examples is expected to increase significantly.

Advances in biological methodology are taking place on several fronts: protein engineering, catalytic antibodies and protein construction from first principles. The last of these, which relies on predictions of protein structure from a knowledge of the interactions and characteristics imparted by individual amino acids, seems furthest from fruition and practical impact. Antibodies have already been produced which catalyse stereospecific reactions[1] and the level of expression of first generation antibodies has been increased by mutagenesis.[2] It is difficult to predict whether antibodies or the protein engineer will be first to come to the aid of the large-scale operator; probably the latter through provision of enzymes with increased robustness to operating conditions.

Crystallization phenomena will continue to provide the keystones in many efficient processes.

Whilst the occurrence of desirable crystal behaviour and solubilities are in large measure unpredictable, a systematic search for exploitable properties at all relevant points in a synthesis will repay the effort and should be part of the *modus operandi* of the process development chemist. For example, if the substance is readily racemized and a crystallization-induced asymmetric transformation, 'deracemization,' is possible it offers an extremely attractive industrial option (cf. Section 1.2.2).

Commercial targets will continue to provide the major driving force for method development across the whole field and pharmaceuticals will continue to dominate. There will be added impetus from the recent trend to consider more favourably 'racemic switches' where drugs formerly marketed as racemates are brought to the market as single enantiomers.[3] Applications in agrochemicals have hitherto been less numerous, largely owing to much tighter cost constraints than apply to pharmaceuticals and more rapid development schedules necessary for success in this highly competitive and increasingly mature market. The challenge to develop appropriate cost-effective methods and so extend the range of single enantiomer agrochemicals will undoubtably be met.

In areas where chirality is not an absolute prerequisite for activity, and may well be viewed as a development nuisance, the search for 'flat' molecules will be made. However, the need for single enantiomers will, most certainly, not go away. Desirable activity is too closely intertwined with the chiral nature of life processes.

The future will see an interesting race between biology, conventional chemistry and separation technology. The field is in an exciting and dynamic state; faster progress is guaranteed because of the beneficial rivalry between the different camps.[4] There will be synergies; for example, one procedure may allow the production of a chiral auxiliary, not necessarily cheaply, which can then become part of a chiral catalyst.

Many of the technologies described in this volume are still in their infancy. One thing is certain: the future will bring advances of which we cannot even dream.

NOTES AND REFERENCES

1. Nakayama, G. R., and Schultz, P. G., *J. Am. Chem. Soc.*, **114**, 780 (1992). Kitazume, T., Lin, J. T., Takeda, M., and Yamazaki, T., *J. Am. Chem. Soc.*, **113**, 2123 (1991).
2. Hilvert, D., Tang, Y., and Hicks, J. B., *Proc. Natl. Acad. Sci. USA*, **88**, 8784 (1991).
3. Trend endorsed by several speakers at *Chiral 92 Symposium*, Spring Innovations, Manchester, UK, 24–25 March, 1992.
4. cf. Correspondence between Pryce, R. J., Roberts, S. M., and Brown, J. M., Davies, S. G., *Nature* (*London*), **345**, 582 (1990).

Index

(Entries in bold type are references to tables or figures)